Intermodulation Distortion in Microwave and Wireless Circuits

For a listing of recent titles in the *Artech House Microwave Library*, turn to the back of this book.

Intermodulation Distortion in Microwave and Wireless Circuits

José Carlos Pedro
Nuno Borges Carvalho

Artech House
Boston • London
www.artechhouse.com

Library of Congress Cataloging-in-Publication Data

Pedro, José Carlos.

Intermodulation distortion in microwave and wireless circuits / José Carlos Pedro, Nuno Borges Carvalho.

p. cm. — (Artech House microwave library)

Includes bibliographical references and index.

ISBN 1-58053-356-6 (alk. paper)

1. Microwave circuits. 2. Radio circuits. 3. Electric distortion—Mathematical models. 4. Electric circuits, Nonlinear. 5. Signal theory (Telecommunication) I. Carvalho, Nuno Borges. II. Title. III. Series.

TK7876.P43 2003

621.381'32—dc21 2003052295

British Library Cataloguing in Publication Data

Pedro, José Carlos

Intermodulation distortion in microwave and wireless circuits. — (Artech House microwave library)

1. Microwave circuits—Design 2. Wireless communication systems

3. Modulation (Electronics) 4. Electric interference I. Title

II. Carvalho, Nuno Borges

621.3'81326

ISBN 1-58053-356-6

Cover design by Igor Valdman

685 Canton Street

Norwood, MA 02062

International Standard Book Number: 1-58053-356-6

Library of Congress Catalog Card Number: 2003052295

10 9 8 7 6 5 4 3 2 1

To our wives
Maria João
and
Raquel

Contents

Foreword

The effects of nonlinearity on microwave communications became a serious concern in the late 1950s and early 1960s. At that time, most research focused on Volterra methods as the primary tool for nonlinear circuit analysis, and considerable progress was made in developing those techniques. As often happens, however, improvements in practical hardware moved faster than advances in theory. Low-distortion transistors (both FET and bipolar) and, especially, the Schottky-barrier diode made much of that theory unnecessary: through the 1970s, distortion in microwave circuits was a relatively minor problem, and most research was devoted to reducing noise. It would be an overstatement to say that the 1960s' research on nonlinearity was forgotten; it is accurate, however, to note that it was little used.

By the late 1980s, the development of digital mobile telephones introduced complex communication systems into consumer electronics. Such systems were notoriously sensitive to distortion. At the same time, advances in solid-state devices resulted in transistors having such low noise that it no longer limited the performance of communication systems. Unfortunately, these same low-noise devices generated high levels of distortion. Distortion of complex signals again became a serious problem, and nonlinearity became an important research subject. It is ironic to see how we have come full circle.

Research in nonlinear high-frequency circuits has a dual focus. The first is on the design of nonlinear circuits, in which nonlinearity is exploited for some particular function. Among these circuits are frequency multipliers and mixers; one could also include such circuits as class AB "linear" power amplifiers, in which nonlinearity is exploited to improve efficiency. In those circuits, nonlinearity is a desirable characteristic. The second focal point is on the deleterious effects of undesired nonlinearity on otherwise linear systems, which we properly call *pseudolinear* systems. The analysis and optimization of such systems is complicated by the complex nature of the signals that they must accommodate; typically, carriers that are digitally modulated in sophisticated formats. The signals are stochastic, not deterministic. Viewed in the frequency domain, the signals have multiple frequency components, or continuous spectra. Most circuit-analysis methods are not well suited for such excitations; clearly, new knowledge is needed.

Perhaps because of the subject's complexity, nonlinear circuit analysis and optimization have been addressed by only a few books. Most have been concerned with simple, sinusoidal excitations of nonlinear circuits, and occasionally with relatively simple distortion phenomena. One or two have been quite academic, sadly detached from the needs of practicing engineers. Few have dealt with multitone excitation of pseudolinear circuits, which, at present, is a pressing problem; with complex interconnections of circuit blocks to form systems; or with the design and optimization of such circuits. This book attacks those problems head-on, and as such, is an important contribution to the professional literature.

The book follows a logical development from fundamental concepts, through multitone characterization and analysis, to modeling and design. Readers will find parts of Chapter 2 familiar, but the more familiar two-tone concepts are quickly extended to multitone problems. Chapter 3, which is almost a third of the book, includes the most comprehensive treatment of the application of Volterra methods in the technical literature. The remaining chapters address modeling and system design from a very broad view, again with an eye on the response to multitone excitations.

I am enthusiastic about this book, and I am confident that it will be valuable to anyone dealing with the frustrations of making modern communication systems work as well in reality as they should in theory.

Stephen Maas
Applied Wave Research, Inc.
July 2003

Preface

The explosive deployment of new digital wireless services has turned bandwidth into an invaluable telecommunications commodity. Therefore, RF circuit design engineers are continuously being confronted with tougher and tougher linearity specifications, so that systems can show smaller nonlinear signal perturbation and adjacent-channel spurious responses. Unfortunately, and despite the amount of scientific material available on this matter, there is still an enormous gap between the restricted club of experts on nonlinear analysis, and the much wider group of practitioners.

Even if the rapid growth of wireless markets could be thought as momentary—and we do not think it is—the difficulty of incorporating scientific knowledge in real circuit design is determined by a pervasive problem: the lack of preparation most engineers have on nonlinear phenomena. Actually, it is widely recognized by engineers and scholars that the vast majority of electronics and telecommunications engineering programs almost exclusively address linear circuits and systems, leaving uncovered the effects of nonlinearity. So, nowadays, engineers feel a significant difficulty in dealing with those aspects, as they are tied down by an insuperable incapability when struggling to overcome their basic knowledge deficiencies.

Although *Intermodulation Distortion in Microwave and Wireless Circuits* was primarily written for those engineers working in RF and microwave circuits design, it is also appropriate to researchers, academics, or graduate students. In fact, its tutorial coverage of the basic aspects of nonlinearity, nonlinear analysis tools, and circuit design methods was intended to turn it into a valuable tool for a broad range of technical readers. Hence, the only prerequisites assumed are the equivalent of a bachelor's degree in electrical engineering. Nevertheless, the main purpose of the book is to present a broad and in-depth view of nonlinear distortion phenomena seen in microwave and wireless systems.

Chapter 1 starts by addressing the intermodulation distortion problem, in the most general terms, and from a system's perspective.

Chapter 2 deals with nonlinear distortion characterization from a practical point of view. It presents the most commonly used distortion figures of merit as defined from one-tone, two-tone, and multitone tests, and their correspondent laboratory measurement setups.

Chapter 3 is the chapter dedicated to nonlinear analysis mathematical tools. Although its emphasis is mainly theoretical, it also provides an overview of the methods now available for nonlinear analysis of practical circuits and systems, showing some of their more important comparative advantages and pitfalls.

As nonlinear distortion analysis requires the use of extensive computer aided design tools, models of the electronic elements, circuits, and systems play a determinant role on the success of any analysis or design procedure. So, Chapter 4 is dedicated to the mathematical representation of those electronic devices.

Finally, Chapter 5 addresses circuit design methods for distortion minimization. It starts by a systemic view of the signal-to-noise ratio problem, to recall the traditional discussion on dynamic-range optimization and highly linear low-noise amplifier design. After that, nonlinear distortion generated in high-power amplifiers is addressed. Because of the importance of RF and microwave mixers as nonlinear distortion sources, Chapter 5 also addresses the analysis of these circuits. It concludes with an analysis of distortion arising in balanced circuits providing the design engineer with the basic information to direct most practical designs.

We could not end this brief note without expressing our most sincere gratitude to many people that directly, or indirectly, helped us carry on this task.

First of all, we would like to thank our family for their patience and emotional support provided along these 3 years of short weekends and long, sleepless nights.

In addition we are especially in debt to a group of our students, or simply collaborators, who were determinant in disclosing some of the results described in the text, or who contributed with experimental data. For their special influence on the final result we include the names of Jose Angel Garcia, Christian Fager, Pedro Cabral, Pedro Lavrador, Paulo Gonçalves, Ricardo Matos Abreu, Emigdio Malaver, and João Paulo Martins. We should also mention the colleagues at other universities with whom we have had scientific research collaborations, which helped greatly in our own studies, namely the Group of Microwaves and Radar of Polytechnic University of Madrid and the Communications Engineering Dept. of University of Cantábria.

We would like to also acknowledge the financial and institutional support provided by both the Portuguese national science foundation (FCT) and the Telecommunications Institute-Aveiro University.

Finally, the authors would like to specially thank Dr. Steve Maas for his encouragement in writing the book and his suggestions while reviewing it.

CHAPTER 1

Introduction

1.1 Signal Perturbation—General Concepts

This book deals with the nonlinear distortion phenomena seen in microwave and wireless systems. As its name indicates, nonlinear distortion is a form of signal perturbation originated in the system's nonlinearities.

To understand this concept, let's suppose we want to send some amount of data from a transmitter to a receiver through a wireless medium, as shown in Figure 1.1. Under this scenery, we would naturally define *signal perturbation* as being any component, other than the sought data, the receiver detects, since it poses difficulties in the correct decoding of the information received. Signal perturbation can thus be either due to the addition of new components, or to the modification of the original signal characteristics.

In the first set, we find all *additive random noise* components—either internal or external to the system—but also any other *additive deterministic interferences* uncorrelated with the desired data. These can be originated by another system, or even by any other communications channel of the same system, sharing the same transmission medium. This group of additive perturbation components is represented in our wireless system of Figure 1.1 by the interferer transmitter block.

The second set of perturbations includes any form of signal distortion. Contrary to noise and interference, which are independent perturbation sources of additive nature, distortion cannot be dissociated from the signal. That is, *distortion is a modification of the signal,* and thus, cannot be detected when the signal source is shutdown.

In this sense, we can conceive as many forms of distortion as the number of different ways the signal can be modified. For reasons that will become clear later, it is useful to classify those into linear and nonlinear distortions, whether they result from a linear or nonlinear signal transformation.

Linear distortion can be manifested as a simple change of scale, or as a much more obvious change of signal form. The first case only implies a variation of the gain factor, which can be important in electronic measurement instruments, but is almost irrelevant in telecommunication systems. A change of signal form arises in dynamic circuits, as filters, and can result in severe signal spectrum shaping in

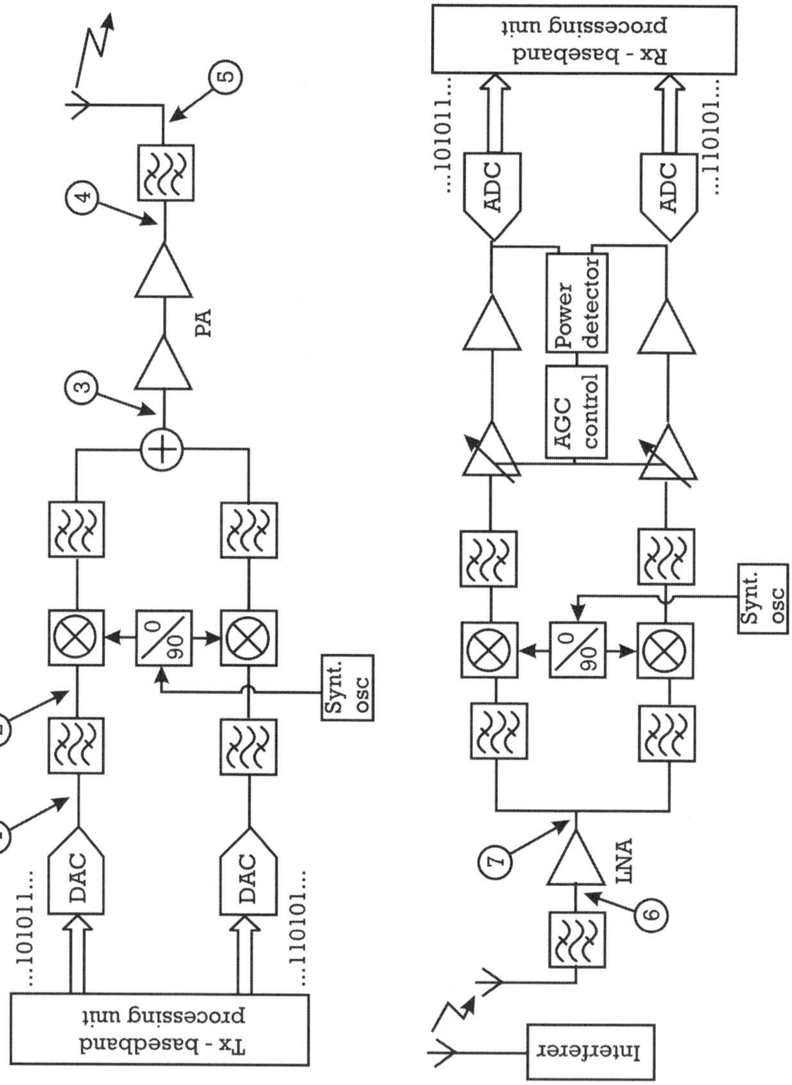

Figure 1.1 Block diagram of a typical wireless communications transmitter-receiver link.

analog chains, or in intersymbol interference in digital transmission chains. Two known examples of these are the modification of voice tone imposed by the traditional fixed telephone network, and the presence of tails on the output of a lowpass filter driven by a stream of rectangular pulses.

This form of linear distortion is patent, for instance, on the output of the pulse-shaping filter present in the system of Figure 1.1 [whose signals are identified as (1) and (2)], but also on the ports of the bandpass filter located at the transmitter power amplifier (PA) output [(4) and (5)]. A sample of these signals is depicted in Figures 1.2 and 1.3, respectively.

Nonlinear distortion can produce modifications of gain, signal shape, and much more.

Indeed, a nonlinear device, like the transmitter PA of our wireless system, can even generate components that are totally uncorrelated with the original signal (i.e., behaving as random noise to the desired information). An illustration of this property is clear if the spectrum of the PA input [signal (3)] (Figure 1.4) is compared

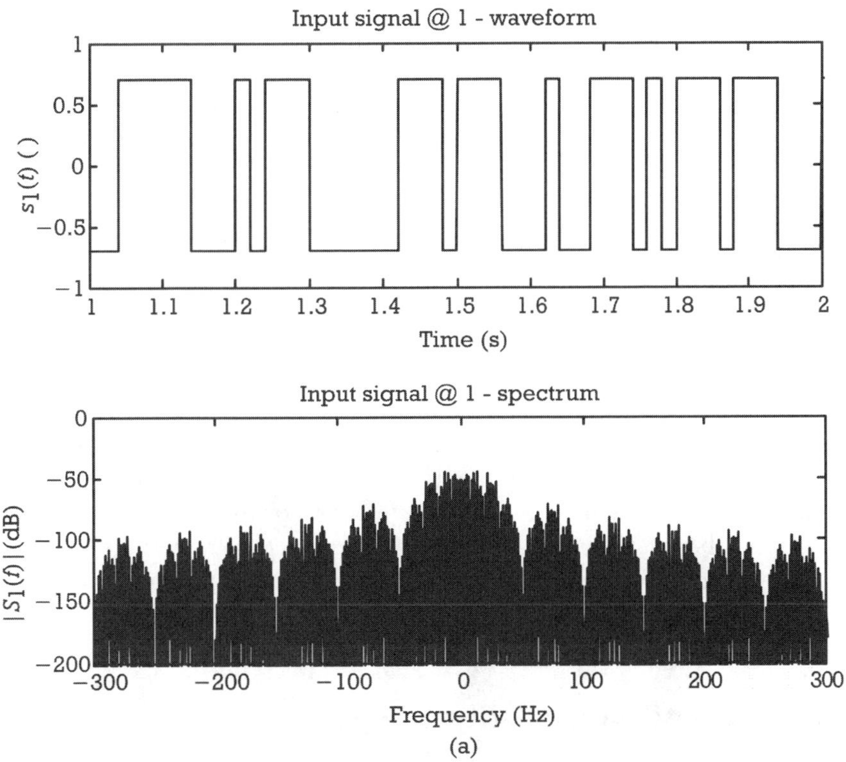

Figure 1.2 Example of linear distortion caused by the pulse-shaping filter of the wireless system described in Figure 1.1. (a) Time-domain waveform and spectrum of the input signal. (b) Time-domain waveform and spectrum of the output signal.

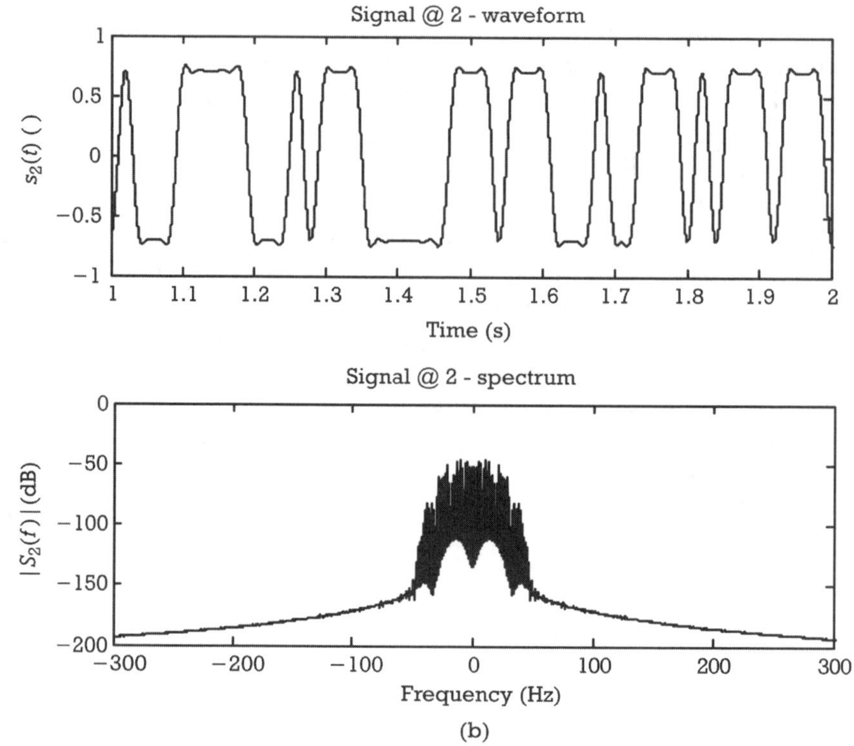

Figure 1.2 (continued).

to the frequency-domain representation of its output [signal (4)] (Figure 1.5). The generation of harmonics, but also of other spectral lines located around the original signal spectrum, is an obvious indication that there are certain output components that carry no useful information at all.

But, as depicted in Figure 1.6, a nonlinear system can also generate cross-talk between communication channels by pressing information carried on one channel onto another one. It can also transfer data from a certain spectral position to a different band, as in the mixers of Figure 1.1, or completely eliminate the data signal, and simply extract its average power, as in the power detector of the automatic gain control loop.

1.2 Linearity and Nonlinearity

Before we start detailing nonlinear distortion effects, it is important to briefly introduce the fundamental properties of systems from which we expect this form of distortion generation (i.e., nonlinear systems). As their name indicates, nonlinear

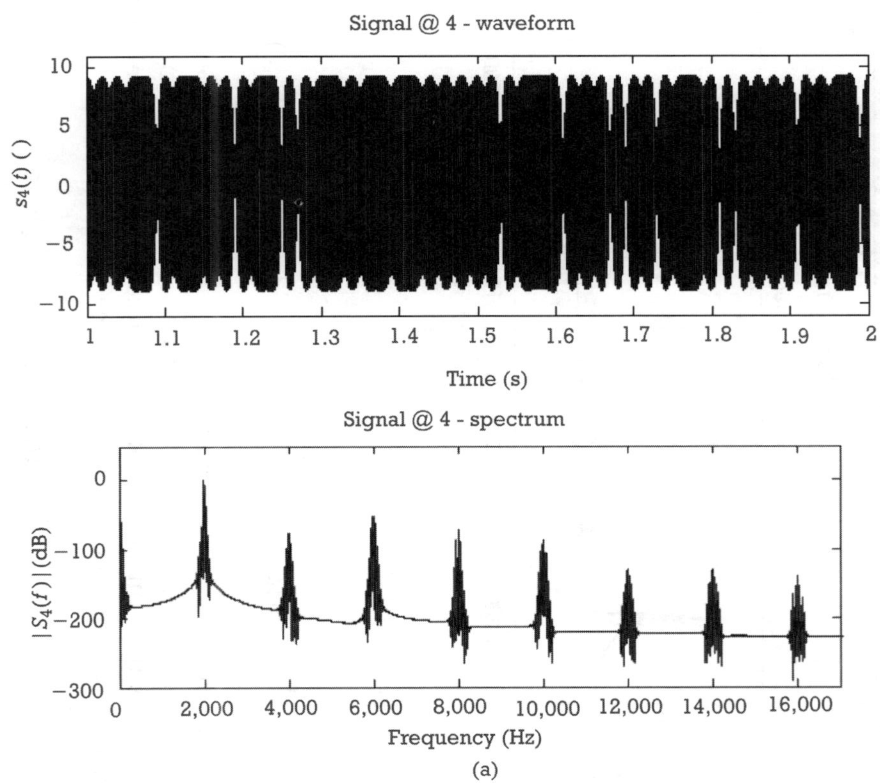

Figure 1.3 Example of linear distortion caused by the bandpass filter located at the PA output of the wireless system described in Figure 1.1. (a) Time-domain waveform and spectrum of the input signal. (b) Time-domain waveform and spectrum of the output signal.

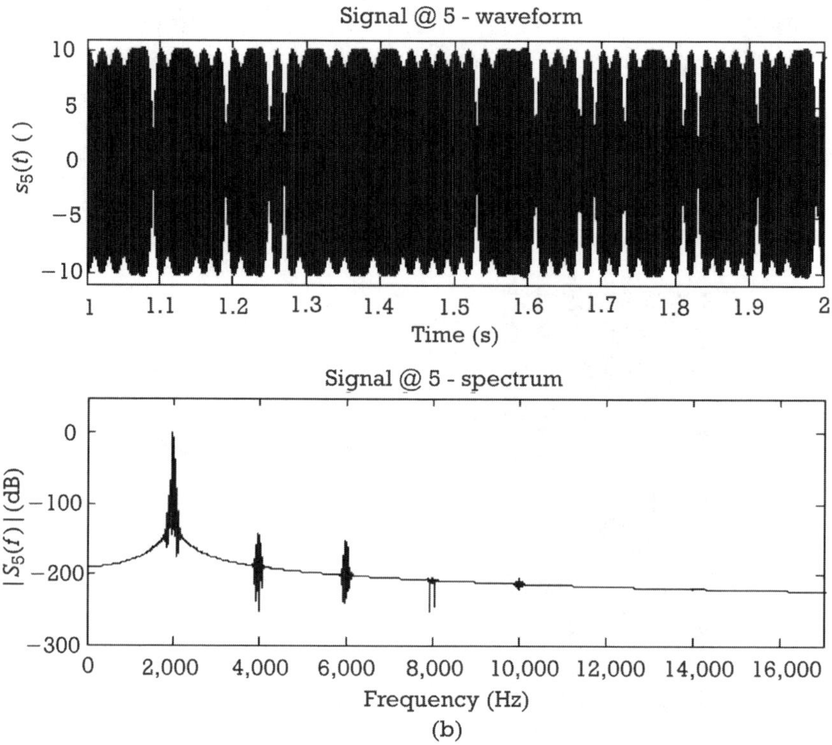

Figure 1.3 (continued).

systems are systems that are not linear. So, it is better to start by defining linear systems.

Linear systems are signal operators, $S_L[.]$, that obey superposition—that is, whose output to a signal composed by the sum of other more elementary signals can be given as the sum of the outputs to these elementary signals when taken individually. In mathematical terms, this can be stated as

$$y(t) = S_L[x(t)] = k_1 y_1(t) + k_2 y_2(t) \tag{1.1}$$

if

$$x(t) = k_1 x_1(t) + k_2 x_2(t) \quad \text{and} \quad y_1(t) = S_L[x_1(t)],\ y_2(t) = S_L[x_2(t)] \tag{1.2}$$

Any system that does not obey superposition is said to be a nonlinear system.

Stated in this way, it seems that nonlinear systems are the exception, whereas they are really the general rule. For example, while we have always been told in

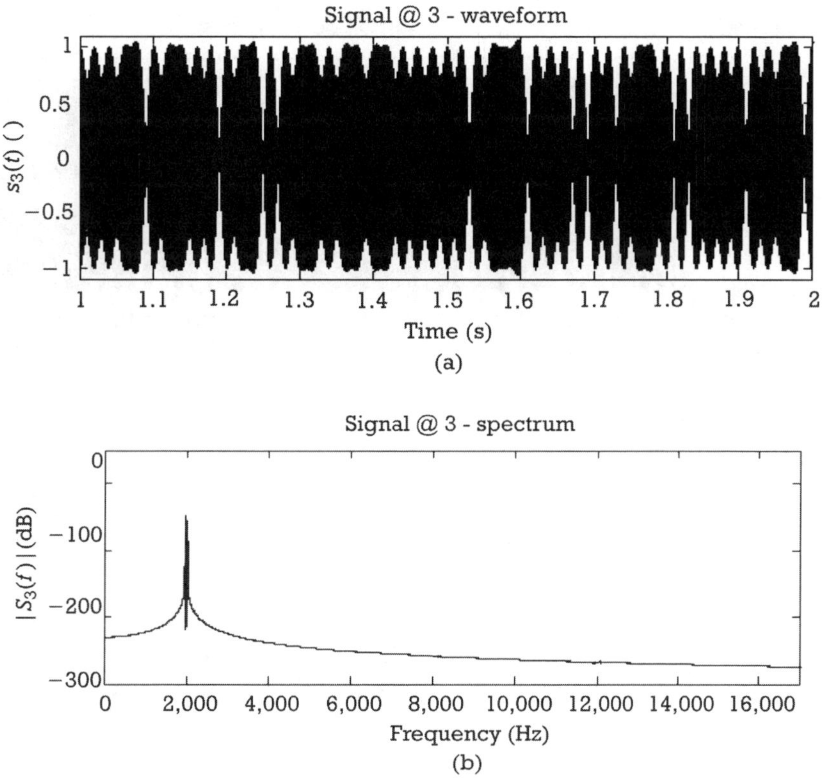

Figure 1.4 Time-domain (a) waveform and (b) spectrum of the signal driving the PA of the wireless system described in Figure 1.1.

our undergraduate studies that the low noise electronic amplifier located at the receiver input of our wireless link of Figure 1.1 is a linear system, it can easily be shown that even this simple active device may be quite far from being linear.

To see that, consider, for instance, the general active system of Figure 1.7, where P_{in} and P_{out} are the signal powers flowing from the source to the amplifier, and from this to the load, respectively; P_{dc} is the dc power delivered to the amplifier by the power supply; and P_{diss} is the total lost power, either dissipated in the form of heat or in any other signal form that has not been considered as signal (e.g., harmonic components).

Defining the amplifier power gain as the ratio between the signal power delivered to the load to the signal power delivered to the amplifier:

$$G_P = \frac{P_{out}}{P_{in}} \tag{1.3}$$

and noting that the fundamental energy conservation principle requires that

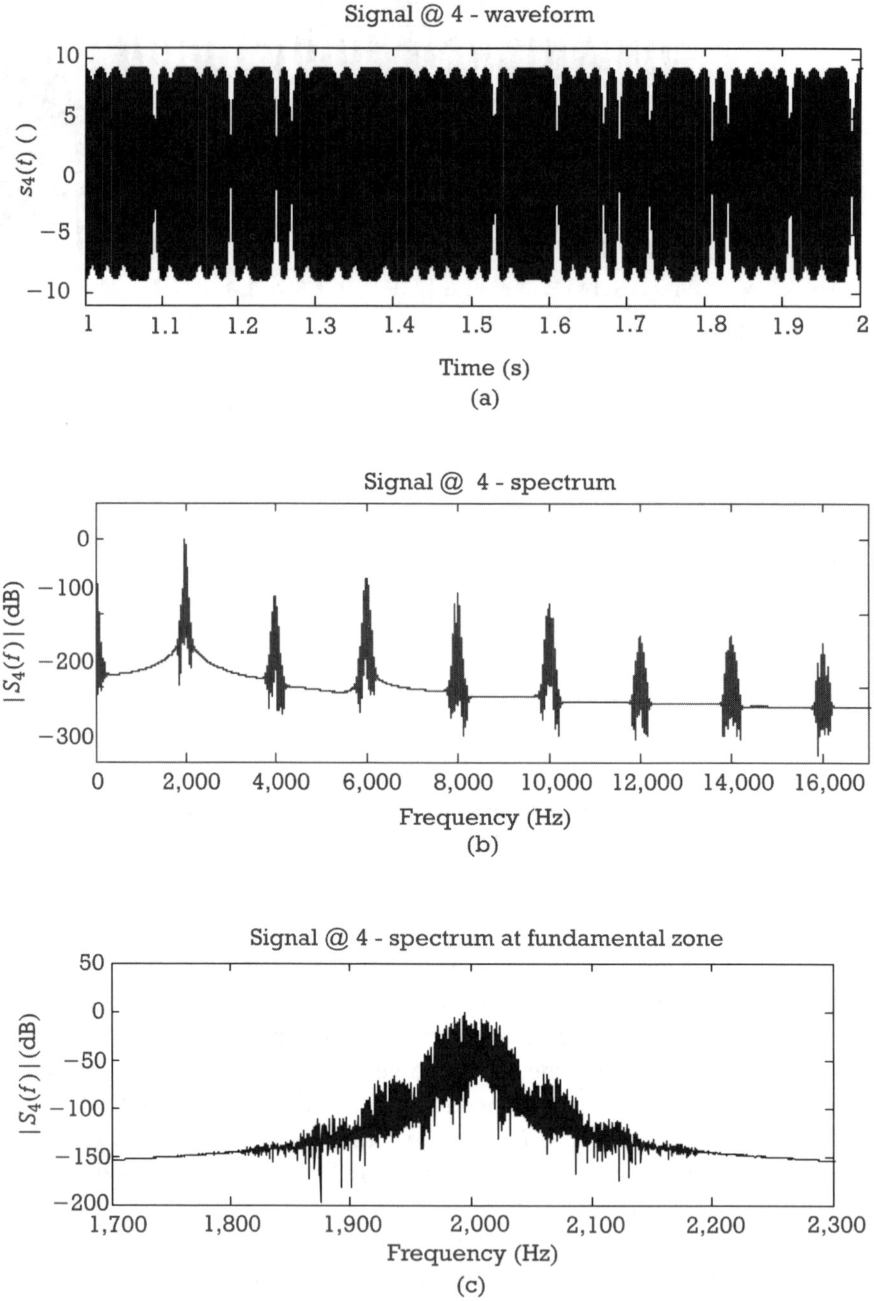

Figure 1.5 Time-domain waveform and spectrum of the signal located at the PA output of the wireless system described in Figure 1.1. (a) Time-domain waveform. (b) Complete spectrum up to the eighth harmonic. (c) Close view of the spectrum fundamental zone.

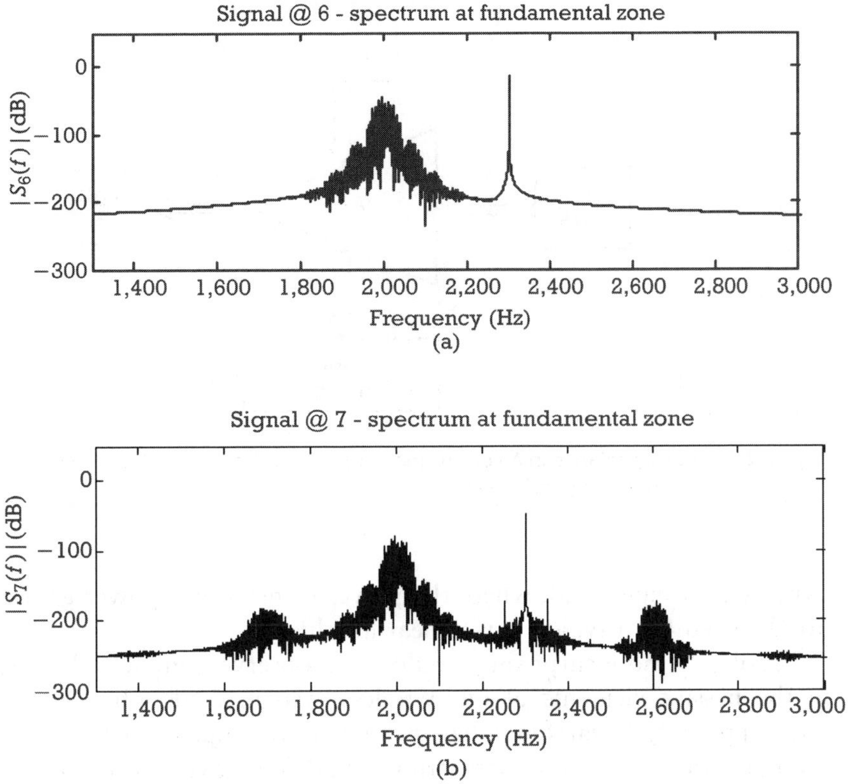

Figure 1.6 Example of cross-talk generated in the nonlinearities of the small-signal low noise amplifier (LNA) of the wireless receiver of Figure 1.1. (a) LNA input spectrum showing the desired information signal in presence of an unmodulated interferer. (b) LNA output spectrum in which the presence of spectral lines around the interferer is a clear indication of nonlinear cross-talk.

$$P_{out} + P_{diss} = P_{in} + P_{dc} \tag{1.4}$$

and so,

$$G_P = 1 + \frac{P_{dc} - P_{diss}}{P_{in}} \tag{1.5}$$

we immediately conclude that, since P_{diss} has a theoretical minimum of zero and P_{dc} is limited by the finite available power from the supply, it is impossible for the amplifier to keep a constant gain for any increasingly high input power. And that means there is a minimum level of input power beyond which the amplifier will manifest an increasingly noticeable nonlinear behavior. This is exactly what is

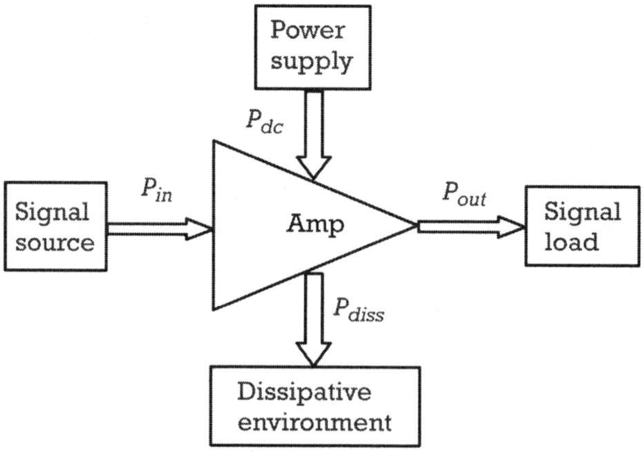

Figure 1.7 Energy balance in an electronic amplifier used to prove that all active electronic devices are inherently nonlinear.

expressed in Figure 1.8, where the power transfer and power gain characteristics are depicted for a typical quasilinear amplifier.

Probably more surprising would be to observe nonlinear distortion generated in the passive elements of our wireless system. And it happens! For example, any supposedly linear filter that includes ferromagnetic cored coils will generate nonlinear distortion in the saturating magnetic flux versus current core curve. And even more exotic is the nonlinear characteristics associated with stainless steel RF connectors (again because of their magnetic flux saturation) or with contacts of different conductor materials as bolts and turning screws in antennas, almost all types of connectors, and rusty contact surfaces [1].

In fact, nature is continuously showing us evidence that the above classification of linear and nonlinear systems should be read more in the sense that from all the nonlinear systems, only the ones that can be forced, or approximated, to obey superposition are classified as pertaining to the subset of linear systems. All the others must be treated as nonlinear. The necessity of forcing a nonlinear system to obey superposition, and thus to become linear, is simply due to the abundance of mathematical tools developed for those systems, and the lack of similar theoretical instruments for treating nonlinearity. Actually, nonlinearity is significantly more difficult because it also produces much richer responses.

1.3 Overview of Nonlinear Distortion Phenomena

To get a first glance into the richness of nonlinearity, let us compare the responses of simple linear and nonlinear systems to typical inputs encountered in our wireless

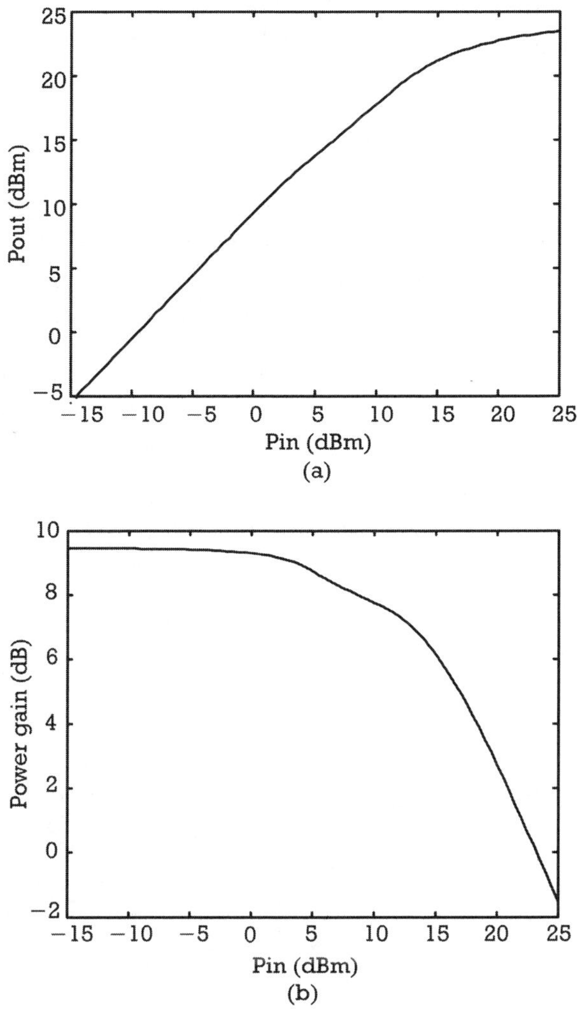

Figure 1.8 (a) Power transfer, and (b) gain characteristics of a typical RF quasilinear amplifier.

telecommunications environment example. Those stimulus inputs are usually sinusoids, amplitude and phase modulated by some baseband information signals, which take the form of

$$x(t) = A(t) \cos[\omega_c t + \theta(t)] \tag{1.6}$$

For that, we will restrict the systems to be represented by a low-degree polynomial, $y_{NL}(t) = S_{NL}[x(t)]$, such as

$$y_{NL}(t) = a_1 x(t - \tau_1) + a_2 x(t - \tau_2)^2 + a_3 x(t - \tau_3)^3 + \ldots \tag{1.7}$$

which we will assume is truncated to third degree.

Although this polynomial of the delayed stimulus is only a short example of all the nonlinear operators we could possibly imagine, modifying its coefficients and delays allows us to approximate many different continuous functions. Furthermore, if the input signal level is decreased enough, so that $x(t) >> x(t)^2, x(t)^3$, the polynomial smoothly tends to a linear system of $y_L(t) = S_L[x(t)] = a_1 x(t - \tau_1)$.

So, while the response of this linear system to (1.6) is

$$y_L(t) = a_1 A(t - \tau_1) \cos[\omega_c t + \theta(t - \tau_1) - \phi_1] \tag{1.8}$$

the response of the nonlinear system would be

$$\begin{aligned} y_{NL}(t) = {} & a_1 A(t - \tau_1) \cos[\omega_c t + \theta(t - \tau_1) - \phi_1] \\ & + a_2 A(t - \tau_2)^2 \cos[\omega_c t + \theta(t - \tau_2) - \phi_2]^2 \\ & + a_3 A(t - \tau_3)^3 \cos[\omega_c t + \theta(t - \tau_3) - \phi_3]^3 \end{aligned} \tag{1.9}$$

which, using the following trigonometric relations,

$$\cos(\alpha) \cos(\beta) = \frac{1}{2} \cos(\alpha - \beta) + \frac{1}{2} \cos(\alpha + \beta) \tag{1.10}$$

$$\Rightarrow \begin{cases} \cos(\alpha)^2 = \frac{1}{2} + \frac{1}{2} \cos(2\alpha) \\ \cos(\alpha)^3 = \frac{3}{4} \cos(\alpha) + \frac{1}{4} \cos(3\alpha) \end{cases}$$

can be rewritten as

$$\begin{aligned} y_{NL}(t) = {} & a_1 A(t - \tau_1) \cos[\omega_c t + \theta(t - \tau_1) - \phi_1] \\ & + \frac{1}{2} a_2 A(t - \tau_2)^2 + \frac{1}{2} a_2 A(t - \tau_2)^2 \cos[2\omega_c t + 2\theta(t - \tau_2) - 2\phi_2] \\ & + \frac{3}{4} a_3 A(t - \tau_3)^3 \cos[\omega_c t + \theta(t - \tau_3) - \phi_3] \\ & + \frac{1}{4} a_3 A(t - \tau_3)^3 \cos[3\omega_c t + 3\theta(t - \tau_3) - 3\phi_3] \end{aligned} \tag{1.11}$$

where $\phi_1 = \omega_c \tau_1$, $\phi_2 = \omega_c \tau_2$, and $\phi_3 = \omega_c \tau_3$.

The case of most practical interest to microwave and wireless systems is the one in which the amplitude and phase modulating signals, $A(t)$ and $\theta(t)$, are slowly varying signals, as compared to the RF carrier $\cos(\omega_c t)$. If the system's time delays are comparable to the carrier period (a simple case where the system does not exhibit memory to the modulating signals), they are thus negligible when compared to the envelope amplitude and phase evolution with time. Hence, (1.8) and (1.11) can be rewritten as

$$y_L(t) = a_1 A(t) \cos[\omega_c t + \theta(t) - \phi_1] \tag{1.12}$$

and

$$\begin{aligned} y_{NL}(t) = {} & a_1 A(t) \cos[\omega_c t + \theta(t) - \phi_1] \\ & + \frac{1}{2} a_2 A(t)^2 + \frac{1}{2} a_2 A(t)^2 \cos[2\omega_c t + 2\theta(t) - 2\phi_2] \\ & + \frac{3}{4} a_3 A(t)^3 \cos[\omega_c t + \theta(t) - \phi_3] \\ & + \frac{1}{4} a_3 A(t)^3 \cos[3\omega_c t + 3\theta(t) - 3\phi_3] \end{aligned} \tag{1.13}$$

The first notorious difference between the linear and the nonlinear responses is the number of terms present in (1.12) and (1.13). While the linear response to a modulated sinusoid is a similar modulated sinusoid, the nonlinear response includes many other terms, usually named as *spectral regrowth,* beyond that linear component. Actually, this is a consequence of one of the most important and distinguishing properties between linear and nonlinear systems:

> Contrary to a linear system, which can only operate quantitative changes to the signal spectra (i.e., modifying the amplitude and phase of each spectral component present at the input), nonlinear systems can qualitatively modify spectra, as they eliminate certain spectral components, and generate new ones.

Two of the best examples for illustrating this rule are the rectifier (or ac/dc converter) response to a pure sinusoid, and the corresponding output of a linear filter. While the latter can, at most, modify the amplitude and phase of the input sinusoid (but can neither destroy it completely nor generate any other frequency component), the ac/dc converter eliminates the ac frequency component and transfers its energy to a new component at dc.

In our wireless nonlinear PA example, the nonlinear output components presented energy near dc, or $0\omega_c$, the second and third harmonics, $2\omega_c$ and $3\omega_c$, etc., but also over the linear response, ω_c, as was shown in Figure 1.5.

The component at dc shares the same origin as the dc output in the mentioned rectifier. In practical systems, it manifests itself as a *shift in bias from the quiescent point* (defined as the bias point measured without any excitation) to the actual bias point measured when the system is driven at its rated input excitation power. This bias point shifting effect has been for long time recognized in class B or C power amplifiers, which draw a significant amount of dc power when operated at full signal power, but remain shut down when the input is quiet.

Looking from the spectral generation view point, that dc component comes from all possible *mixing, beat or nonlinear distortion products* of the form $\cos(\omega_i t)\cos(\omega_j t)$, whose outputs are located at $\omega_x = \omega_i - \omega_j$, and where $\omega_i = \omega_j$.

The other components located around dc constitute a distorted version of the amplitude modulating information, $A(t)$, as if the composite signal of (1.6) had suffered an amplitude demodulation process. They are, therefore, called the *baseband components* of the output. Their frequency lines are also generated from mixing products at $\omega_x = \omega_i - \omega_j$, but now where $\omega_i \neq \omega_j$.

The components located around $2\omega_c$ and $3\omega_c$ are, for obvious reasons, known as the second and third-order *nonlinear harmonic distortion,* or simply the harmonic distortion. Note that they are, again, high-frequency sinusoids amplitude modulated by distorted versions of $A(t)$.

The cluster of spectral lines located around $2\omega_c$ is generated from all possible mixing products of the form $\cos(\omega_i t)\cos(\omega_j t)$, whose outputs are located at $\omega_x = \omega_i + \omega_j$, and where $\omega_i = \omega_j$ ($\omega_x = 2\omega_i = 2\omega_j$) or $\omega_i \neq \omega_j$. The third harmonic cluster has its roots on all possible mixing products of the form $\cos(\omega_i t)\cos(\omega_j t)\cos(\omega_k t)$, whose outputs are located at $\omega_x = \omega_i + \omega_j + \omega_k$, and where $\omega_i = \omega_j = \omega_k$ ($\omega_x = 3\omega_i = 3\omega_j = 3\omega_k$), $\omega_i = \omega_j \neq \omega_k$ ($\omega_x = 2\omega_i + \omega_k = 2\omega_j + \omega_k$), or even $\omega_i \neq \omega_j \neq \omega_k$.

Finally, the components located near ω_c are distorted versions of the input. They include newly generated lines that fall around the original spectrum, but also lines that share exactly the same position as the linear response, and thus are indistinguishable from it. Contrary to the baseband or harmonic distortion, which are forms of *out-of-band distortion,* and thus could be simply discarded by bandpass filtering, some of these new *inband distortion* components are unaffected by any linear operator that, naturally, must preserve the fundamental components. Thus, they constitute the most important form of distortion in bandpass microwave and wireless subsystems.[1] Actually, the impairment of nonlinear distortion in telecommunication systems is so high, when compared to linear distortion, that it is common use to reserve the name "distortion" for nonlinear distortion. Accordingly, in the

1. Strictly speaking, the distinction between inband and out-of-band distortion components only makes sense when the excitation has already a distinct bandpass nature, as in the RF parts of microwave and wireless systems. In baseband subsystems, the various clusters of mixing products overlap, and they all perturb the expected linear output.

remainder of this text, we will use the terms "nonlinear distortion" or simply "distortion" as synonyms, unless otherwise expressly stated.

Referring again to the wireless system example of Figure 1.1, Figure 1.9 shows exactly that inband distortion effect, by comparing the bandpass filtered version of our PA nonlinear response to a scaled (or linearly processed) replica of its input. Although the bandpass filter has recovered the sinusoidal shape of the carrier—a clear indication that the harmonics have effectively been filtered out [Figure 1.9(b)]—the amplitude envelope is still notoriously distorted, which is a manifestation that inband distortion was unaffected by filtering.

For studying these inband distortion components, we have to first distinguish between the spectral lines that fall exactly over the original ones, and the lines that constitute distortion sidebands. In wireless systems, the former are known as

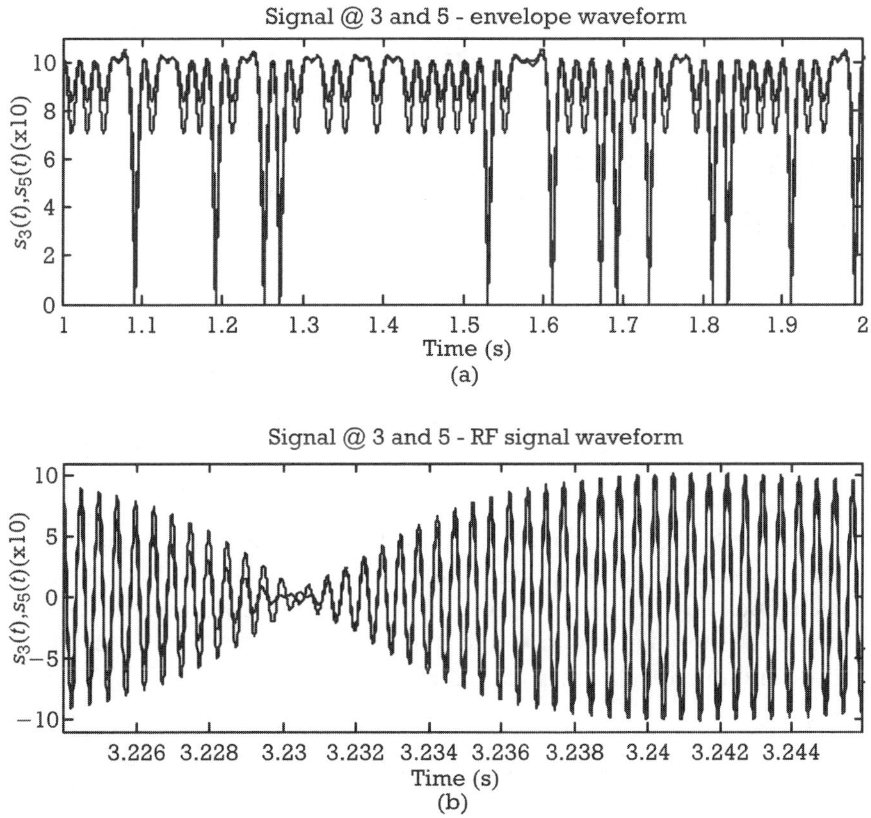

Figure 1.9 The effect of bandpass filtering on the inband and out-of-band distortion. (a) Time-domain waveforms of the wireless system's PA input and filtered output signal amplitude envelopes. (b) Close view of the actual modulated signals showing the detailed RF waveforms.

cochannel distortion and the latter as *adjacent-channel distortion,* since they perturb the wanted and the adjacent-channels, respectively.

In our third-degree polynomial system, all inband distortion products share the form of $\cos(\omega_i t)\cos(\omega_j t)\cos(\omega_k t)$, whose outputs are located at $\omega_x = \omega_i + \omega_j - \omega_k$. And, while both cochannel and adjacent-channel distortion can be generated by mixing products obeying $\omega_i = \omega_j \neq \omega_k$ ($\omega_x = 2\omega_i - \omega_k = 2\omega_j - \omega_k$) or $\omega_i \neq \omega_j \neq \omega_k$, only cochannel distortion arises from products observing $\omega_i = \omega_j = \omega_k$ ($\omega_x = \omega_i$) or $\omega_i \neq \omega_j = \omega_k$ ($\omega_x = \omega_i$).

To get a better insight into these inband distortion products, let us imagine we have a stimulus that is a combination of the modulated signal of (1.6) plus another unmodulated carrier, as was conceived in the system of Figure 1.1:

$$x(t) = A_1(t)\cos[\omega_1 t + \theta(t)] + A_2\cos(\omega_2 t) \tag{1.14}$$

Although this excitation can be viewed as our modulated signal plus an interfering carrier, it could be also understood as two of the spectral lines of (1.6), or even as the addition of two similar modulated signals, with the exception that now we are explicitly showing the amplitude and phase variation of one of the carriers and omitting that for the other.

Since the input is now composed of two different carriers, many more mixing products will be generated. Therefore, it is convenient to count all of them in a systematic manner. For that, we first substitute the temporal input of (1.14) by a phasorial representation using the Euler expression for the cosine:

$$\begin{aligned} x(t) &= A_1(t)\cos[\omega_1 t + \theta(t)] + A_2\cos(\omega_2 t) \\ &= A_1(t)\,\frac{e^{j[\omega_1 t + \theta(t)]} + e^{-j[\omega_1 t + \theta(t)]}}{2} + A_2\,\frac{e^{j\omega_2 t} + e^{-j\omega_2 t}}{2} \end{aligned} \tag{1.15}$$

which leads us to the conclusion that the input can now be viewed as the sum of four terms, each one involving a different frequency. That is, we are assuming that each sinusoidal function involves a positive and a negative frequency component (the correspondent positive and negative sides of the Fourier spectrum), so that any combination of tones can be represented as

$$x(t) = \sum_{q=1}^{Q} A_q\cos(\omega_q t) = \frac{1}{2}\sum_{q=-Q}^{Q} A_q e^{j\omega_q t} \tag{1.16}$$

where $q \neq 0$, and $A_q = A_{-q}^*$ for real signals.

Having $x(t)$ in this form, the desired output is determined as the sum of various polynomial contributions of the form

$$y_{NL_n}(t) = \frac{1}{2^n} a_n \left[\sum_{q=-Q}^{Q} A_q e^{j\omega_q t} \right]^n \tag{1.17}$$
$$= \frac{1}{2^n} a_n \sum_{q_1=-Q}^{Q} \dots \sum_{q_n=-Q}^{Q} A_{q_1} \dots A_{q_n} e^{j(\omega_{q_1} + \dots + \omega_{q_n})t}$$

whose frequency components are all possible combinations of the input ω_q:

$$\omega_{n,v} = \omega_{q_1} + \dots + \omega_{q_n} \tag{1.18}$$
$$= m_{-Q}\,\omega_{-Q} + \dots + m_{-1}\,\omega_{-1} + m_1\,\omega_1 + \dots + m_Q\,\omega_Q$$

where $v = [m_{-Q} \dots m_{-1} m_1 \dots m_Q]$ is the nth order mixing vector, which must verify

$$\sum_{q=-Q}^{Q} m_q = m_{-Q} + \dots + m_{-1} + m_1 + \dots + m_Q = n \tag{1.19}$$

For example, a two-tone input like the one of (1.14) will produce the following mixing products of order 1, $\omega_{1,v}$:

$$\omega_{1,v} = -\omega_2, -\omega_1, \omega_1, \omega_2 \tag{1.20}$$

the following of order 2, $\omega_{2,v}$:

$$\omega_{2,v} = -2\omega_2, -\omega_2 - \omega_1, -2\omega_1, \omega_1 - \omega_2, dc, \omega_2 - \omega_1, 2\omega_1, \omega_1 + \omega_2, 2\omega_2 \tag{1.21}$$

and the following ones of order 3, $\omega_{3,v}$:

$$\omega_{3,v} = -3\omega_2, -2\omega_2 - \omega_1, -\omega_2 - 2\omega_1, -3\omega_1, -2\omega_2 + \omega_1, -\omega_2, -\omega_1, -2\omega_1 + \omega_2,$$
$$2\omega_1 - \omega_2, \omega_1, \omega_2, 2\omega_2 - \omega_1, 3\omega_1, 2\omega_1 + \omega_2, \omega_1 + 2\omega_2, 3\omega_2 \tag{1.22}$$

Obviously, each of these mixing products can be generated by different arrangements of the same input tones. For instance, $2\omega_1 - \omega_2$ can be generated from three different manners as: $\omega_1 + \omega_1 - \omega_2$, $\omega_1 - \omega_2 + \omega_1$ and $-\omega_2 + \omega_1 + \omega_1$, whereas ω_1 can be generated from the following different combinations: $\omega_1 + \omega_1 - \omega_1$, $\omega_1 - \omega_1 + \omega_1$, $-\omega_1 + \omega_1 + \omega_1$, involving only $\pm\omega_1$; and $\omega_1 + \omega_2 - \omega_2$, $\omega_1 - \omega_2 + \omega_2$, $\omega_2 + \omega_1 - \omega_2$, $\omega_2 - \omega_2 + \omega_1$, $-\omega_2 + \omega_2 + \omega_1$, $-\omega_2 + \omega_1 + \omega_2$, involving ω_1 and $\pm\omega_2$.

Actually, the number of these possible combinations can be directly calculated from the multinomial coefficient:

$$t_{n,v} = \frac{n!}{m_{-Q}! \ldots m_{-1}! m_1! \ldots m_Q!} \tag{1.23}$$

In fact, since the spectral line at $2\omega_1 - \omega_2$ is characterized by the mixing vector $v = [1\ 0\ 2\ 0]$, it will lead to a multinomial coefficient of

$$t_{n,v} = \frac{n!}{m_{-Q}! \ldots m_{-1}! m_1! \ldots m_Q!} = \frac{3!}{1!0!2!0!} = 3 \tag{1.24}$$

while the spectral line at ω_1 can be given by a mixing vector of $v_1 = [0\ 1\ 2\ 0]$ and another one of $v_2 = [1\ 0\ 1\ 1]$ leading to the following multinomial coefficients:

$$t_{n,v_1} = \frac{3!}{0!1!2!0!} = 3 \quad \text{and} \quad t_{n,v_2} = \frac{3!}{1!0!1!1!} = 6 \tag{1.25}$$

So, according to these derivations, the output of (1.7) to (1.14) can be calculated from the polynomial response to (1.15) and then converted again to cosines using the Euler relation. Alternatively, noting that the output spectrum must be symmetrical, this result may also be determined by calculating all the possible mixing vectors generating only positive frequencies, and their corresponding multinomial coefficients, and then recovering the cosine representation simply multiplying these coefficients by 2. That is, the amplitude of each mixing product will be $t_{n,v}/2^{n-1}$ except, naturally, if it falls at dc where it will be $t_{n,v}/2^n$. Using this procedure [and again the assumption of slowly varying $A(t)$ and $\theta(t)$], the desired output of (1.7) to (1.14) was found to be

$$\begin{aligned}
y_{NL}(t) = {} & a_1 A_1(t) \cos[\omega_1 t + \theta(t) - \phi_{110}] + a_1 A_2 \cos(\omega_2 t - \phi_{101}) \\
& + \frac{1}{2} a_2 [A_1(t)^2 + A_2^2] + a_2 A_1(t) A_2 \cos[(\omega_2 - \omega_1)t - \theta(t) - \phi_{2-11}] \\
& + a_2 A_1(t) A_2 \cos[(\omega_1 + \omega_2)t + \theta(t) - \phi_{211}] \\
& + \frac{1}{2} a_2 A_1(t)^2 \cos[2\omega_1 t + 2\theta(t) - \phi_{220}] + \frac{1}{2} a_2 A_2^2 \cos(2\omega_2 t - \phi_{202}) \\
& + \frac{3}{4} a_3 A_1(t)^2 A_2 \cos[(2\omega_1 - \omega_2)t + 2\theta(t) - \phi_{32-1}] \\
& + \left[\frac{3}{4} a_3 A_1(t)^3 + \frac{6}{4} a_3 A_1(t) A_2^2\right] \cos[\omega_1 t + \theta(t) - \phi_{310}]
\end{aligned}$$

$$
\begin{aligned}
&+\left[\frac{6}{4}a_3A_1(t)^2A_2+\frac{3}{4}a_3A_2^3\right]\cos(\omega_2t-\phi_{301})\\
&+\frac{3}{4}a_3A_1(t)A_2^2\cos[(2\omega_2-\omega_1)t-\theta(t)-\phi_{3-12}]\\
&+\frac{1}{4}a_3A_1(t)^3\cos[3\omega_1t+3\theta(t)-\phi_{330}]\\
&+\frac{3}{4}a_3A_1(t)^2A_2\cos[(2\omega_1+\omega_2)t+2\theta(t)-\phi_{321}]\\
&+\frac{3}{4}a_3A_1(t)A_2^2\cos[(\omega_1+2\omega_2)t+\theta(t)-\phi_{312}]\\
&+\frac{1}{4}a_3A_2^3\cos(3\omega_2t-\phi_{303})
\end{aligned}
\tag{1.26}
$$

where $\phi_{110}=\omega_1\tau_1$, $\phi_{101}=\omega_2\tau_1$, $\phi_{2-11}=\omega_2\tau_2-\omega_1\tau_2$, $\phi_{220}=2\omega_1\tau_2$, $\phi_{211}=\omega_1\tau_2+\omega_2\tau_2$, $\phi_{202}=2\omega_2\tau_2$, $\phi_{32-1}=2\omega_1\tau_3-\omega_2\tau_3$, $\phi_{310}=\omega_1\tau_3$, $\phi_{301}=\omega_2\tau_3$, $\phi_{3-12}=2\omega_2\tau_3-\omega_1\tau_3$, $\phi_{330}=3\omega_1\tau_3$, $\phi_{321}=2\omega_1\tau_3+\omega_2\tau_3$, $\phi_{312}=\omega_1\tau_3+2\omega_2\tau_3$, and $\phi_{303}=3\omega_2\tau_3$, and whose inband components are only

$$
\begin{aligned}
&a_1A_1(t)\cos[\omega_1t+\theta(t)-\phi_{110}]+a_1A_2\cos(\omega_2t-\phi_{101})\\
&+\frac{3}{4}a_3A_1(t)^2A_2\cos[(2\omega_1-\omega_2)t+2\theta(t)-\phi_{32-1}]\\
&+\left[\frac{3}{4}a_3A_1(t)^3+\frac{6}{4}a_3A_1(t)A_2^2\right]\cos[\omega_1t+\theta(t)-\phi_{310}]\\
&+\left[\frac{6}{4}a_3A_1(t)^2A_2+\frac{3}{4}a_3A_2^3\right]\cos(\omega_2t-\phi_{301})\\
&+\frac{3}{4}a_3A_1(t)A_2^2\cos[(2\omega_2-\omega_1)t-\theta(t)-\phi_{3-12}]
\end{aligned}
\tag{1.27}
$$

As expected, (1.27) includes two linear outputs proportional to the first-degree coefficient a_1, and six more nonlinear components arranged in four different frequencies. From these, the sideband components at $2\omega_1-\omega_2$ and $2\omega_2-\omega_1$ are usually known as the *intermodulation distortion* (*IMD*). Strictly speaking, every mixing product can be denominated an *intermodulation component* since it results from intermodulating two or more different tones. But, although it cannot also be said to be of uniform practice, the term IMD is usually reserved for those particular sideband components. Similarly to what we have already discussed for the amplitude modulated one-tone excitation, they constitute a form of adjacent-channel distortion.

Beyond these IMD products, (1.27) also shows four cochannel distortion components located around ω_1 and ω_2. Two of those are given as

$$\frac{3}{4} a_3 A_1(t)^3 \cos[\omega_1 t + \theta(t) - \phi_{310}] \quad (1.28)$$

and

$$\frac{3}{4} a_3 A_2^3 \cos(\omega_2 t - \phi_{301}) \quad (1.29)$$

which are similar in the form. They are both the cochannel distortion outcomes that would appear if the tones at ω_1 and ω_2 were used, one by one, as independent excitations. Noting that (1.28) can be rewritten as

$$\left[\frac{3}{4} a_3 A_1(t)^2\right] A_1(t) \cos[\omega_1 t + \theta(t) - \phi_{310}] \quad (1.30)$$

and that $A_1(t)^2$ must include a dc term plus baseband and second harmonics of $A(t)$ own frequency components, we must conclude that (1.28) actually includes many distortion components that are inherently distinct from the input, but also some other ones that constitute an exact replica of the input. In mathematical terms, this means that the cochannel distortion has components that are uncorrelated with the input and the linear output, and others that are correlated with these [2, 3].[2]

Since part of the output is uncorrelated with the input signal, it does not contain the desired information and thus behaves towards it as random noise. Its presence is a major source of perturbation to the processed data—a reason why it is sometimes called *intermodulation noise.*

On the other hand, the correlated components carry exactly the same information as the linear output. The only difference they have to the true first-order components is that they are not a linear replica of the input as their proportionality constant, or gain, varies with the signal amplitude squared. That is, from a certain viewpoint, they should be considered nonlinear distortion since they are, actually, a nonlinear deviation of the ideal linear behavior. But, from another perspective, they can be also considered as useful signal since, added with the first-order linear components and the term proportional to $A_1(t) A_2{}^2$, they are simply making the overall system gain dependent on the average excitation power.

2. Rigorously speaking, two signals, $x(t)$ and $y(t)$, are said to be uncorrelated when the cross-correlation between them is zero: $R_{xy}(\tau) = \int_{-\infty}^{\infty} x(t)\, y(t+\tau)\, dt = 0$. If $R_{xy}(\tau) \neq 0$, the signals are correlated.

This duality of roles can be perfectly accepted if we think of what we expect from an electronic measurement system and from a wireless system. In the first case, since we want the system's output to be a scaled replica of the measured quantity, any deviation from linearity is a direct source of measurement error. Therefore, in this scenery, we would be pushed to consider those third-order signal correlated components as distortion. In the second case, since we are not too worried about the overall system gain, whose variations are, after all, generally corrected by an *automatic gain control* (AGC) loop, we would be pushed to consider those components as desired signal and not distortion.

Because, in general, ϕ_{110} is different from ϕ_{310}, and ϕ_{101} is different from ϕ_{301}, the addition of the signal correlated third-order components to the linear components constitutes a vector addition, which means that variations in input amplitude will produce changes in output amplitude, but also in output phase. These two effects, whose graphical illustration is depicted in Figure 1.10, are two of the most significant properties of nonlinear telecommunication systems. They are traditionally characterized with sinusoidal excitations by the so-called *AM-AM conversion*—meaning that input amplitude modulation induces output amplitude modulation—and *AM-PM conversion,* which describes the way input amplitude modulation can also produce output phase modulation.

In general, since AM-AM and AM-PM conversions are driven by amplitude envelope variations, they could be induced by ω_1 onto ω_1 and ω_2 onto ω_2, but also from ω_2 onto ω_1 and ω_1 onto ω_2. This is, for instance, the case of the term

$$\frac{6}{4} a_3 A_1(t)^2 A_2 \cos(\omega_2 t - \phi_{301}) \tag{1.31}$$

where the amplitude variation of one of the signals (in the present example at ω_1) induces amplitude and phase variations on the other (at ω_2). In telecommunication systems this is known as *cross-modulation,* which is responsible for undesired channel cross-talk, as was already seen in Figure 1.6.

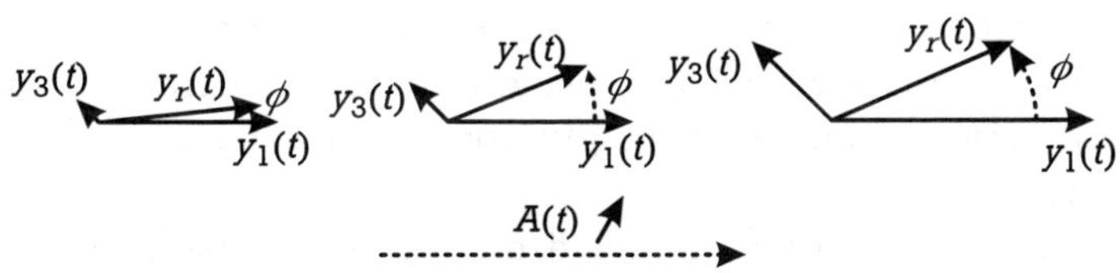

Figure 1.10 Illustration of AM-AM and AM-PM conversions in a nonlinear system driven by a signal of increasing amplitude envelope. $y_1(t)$: linear component; $y_3(t)$: third-order signal correlated distortion component; $y_r(t)$: resultant output component; and ϕ: resultant output phase.

Finally, the term

$$\frac{6}{4} a_3 A_1(t) A_2^2 \cos[\omega_1 t + \theta(t) - \phi_{310}] \tag{1.32}$$

is used to model *desensitization*—that is, the compression of gain (supposing a_3 and ϕ_{310} result in an opposing phase to a_1 and ϕ_{110}), and thus system's sensitivity degradation to one signal (in this case ω_1), caused by another one stronger in amplitude (at ω_2). When the difference in amplitudes between the desired signal and the *interferer* is so high that a dramatic desensitization is noticed, the small-signal is said to be *blocked* and the interferer is named as a *blocker* or *jammer*. Probably, the most obvious reflection of this desensitization or blocking effects is the dazzle we have all already experienced when a strong source of light is pointed at us at night.

Table 1.1 summarizes the above definitions by identifying all the distortion components present in the output of our third-degree polynomial subject to a two-tone excitation signal as (1.26).

1.4 Scope of the Book

After having addressed the intermodulation problem of microwave and wireless systems in general terms, the following chapters will detail most of these concepts.

Chapter 2 addresses the characterization of nonlinear distortion from a practical perspective, focusing on the most widely used figures of merit identified by one-tone, two-tone, and multitone tests. So, for instance, it addresses the above-referred AM-AM and AM-PM characteristics, the intercept point concept, and the cochannel and adjacent-channel power distortion ratios. And, for each of these figures, it discusses the existing laboratory setups normally used to measure them.

Chapter 3 deals with the mathematical techniques for nonlinear circuits and systems' analysis. Despite its theoretical emphasis, it also provides a compendium of the techniques currently on hand for the analysis of nonlinear microwave and wireless circuits, discussing some of their more important advantages and pitfalls. This will help the reader choose, for each particular problem, one from the available commercial software packages using time-step integration, harmonic-balance, or Volterra series. It can also be helpful for someone deciding to write his own analysis software.

Chapter 4 is a brief chapter dedicated to the mathematical representation of electronic systems. Because the analysis of nonlinear distortion demands the extensive use of computer-aided design tools, accurate models of the electronic elements, circuits, and systems are of paramount importance for the success of any analysis or design task. Unfortunately, modeling nonlinear electron devices constitutes, by

Table 1.1 Summary of the Various Forms of Nonlinear Distortion Arising from a Third-Degree Polynomial Subject to a Two-Tone Excitation of Amplitudes A_1 and A_2. (a) First-Order Response. (b) Second-Order Response. (c) Third-Order Response

(a)	*Mixing Vector*			*Frequency*		
m_{-2}	m_{-1}	m_1	m_2	*Component—*ω_x	*Output Amplitude*	*Type of Response*
1	0	0	0	$-\omega_2$	$1/2\ a_1 A_2$	Linear
0	1	0	0	$-\omega_1$	$1/2\ a_1 A_1$	Linear
0	0	1	0	ω_1	$1/2\ a_1 A_1$	Linear
0	0	0	1	ω_2	$1/2\ a_1 A_2$	Linear

(b)	*Mixing Vector*			*Frequency*		
m_{-2}	m_{-1}	m_1	m_2	*Component—*ω_x	*Output Amplitude*	*Type of Response*
2	0	0	0	$-2\omega_2$	$1/4\ a_2 A_2^2$	Second-order harmonic distortion
0	2	0	0	$-2\omega_1$	$1/4\ a_2 A_1^2$	
0	0	2	0	$2\omega_1$	$1/4\ a_2 A_1^2$	
0	0	0	2	$2\omega_2$	$1/4\ a_2 A_2^2$	
1	1	0	0	$-\omega_1 - \omega_2$	$1/2\ a_2 A_1 A_2$	Second-order intermodulation distortion
1	0	1	0	$\omega_1 - \omega_2$	$1/2\ a_2 A_1 A_2$	
0	1	0	1	$\omega_2 - \omega_1$	$1/2\ a_2 A_1 A_2$	
0	0	1	1	$\omega_1 + \omega_2$	$1/2\ a_2 A_1 A_2$	
0	1	1	0	$\omega_1 - \omega_1$	$1/2\ a_2 A_1^2$	Shift of bias point
1	0	0	1	$\omega_2 - \omega_2$	$1/2\ a_2 A_2^2$	

(c)	*Mixing Vector*			*Frequency*		
m_{-2}	m_{-1}	m_1	m_2	*Component—*ω_x	*Output Amplitude*	*Type of Response*
3	0	0	0	$-3\omega_2$	$1/8\ a_3 A_2^3$	Third-order harmonic distortion
0	3	0	0	$-3\omega_1$	$1/8\ a_3 A_1^3$	
0	0	3	0	$3\omega_1$	$1/8\ a_3 A_1^3$	
0	0	0	3	$3\omega_2$	$1/8\ a_3 A_2^3$	
2	1	0	0	$-2\omega_2 - \omega_1$	$3/8\ a_3 A_1 A_2^2$	
1	2	0	0	$-2\omega_1 - \omega_2$	$3/8\ a_3 A_1^2 A_2$	
2	0	1	0	$-2\omega_2 + \omega_1$	$3/8\ a_3 A_1 A_2^2$	Third-order intermodulation distortion
0	2	0	1	$-2\omega_1 + \omega_2$	$3/8\ a_3 A_1^2 A_2$	
1	0	2	0	$2\omega_1 - \omega_2$	$3/8\ a_3 A_1^2 A_2$	
0	1	0	2	$2\omega_2 - \omega_1$	$3/8\ a_3 A_1 A_2^2$	
0	0	2	1	$2\omega_1 + \omega_2$	$3/8\ a_3 A_1^2 A_2$	
0	0	1	2	$2\omega_2 + \omega_1$	$3/8\ a_3 A_1 A_2^2$	
2	0	0	1	$-2\omega_2 + \omega_2$	$3/8\ a_3 A_2^3$	AM/AM conversion (gain compression or expansion AM/PM conversion)
0	2	1	0	$-2\omega_1 + \omega_1$	$3/8\ a_3 A_1^3$	
0	1	2	0	$2\omega_1 - \omega_1$	$3/8\ a_3 A_1^3$	
1	0	0	2	$2\omega_2 - \omega_2$	$3/8\ a_3 A_2^3$	
1	1	1	0	$-\omega_2 + \omega_1 - \omega_1$	$3/4\ a_3 A_1^2 A_2$	Cross-modulation and desensitization
1	1	0	1	$-\omega_1 + \omega_2 - \omega_2$	$3/4\ a_3 A_1 A_2^2$	
1	0	1	1	$\omega_1 + \omega_2 - \omega_2$	$3/4\ a_3 A_1 A_2^2$	
0	1	1	1	$\omega_2 + \omega_1 - \omega_1$	$3/4\ a_3 A_1^2 A_2$	

itself, enough material to fill up many books. So, the adopted strategy was not to present a (necessarily sketchy) view of all possible element nonlinear models, but to discuss a set of criteria to help the reader distinguish their ability to accurately predict nonlinear distortion. Therefore, issues like local versus global representation capabilities, physical versus empirical models, and their associated parameter extraction procedures are first discussed, in the distortion simulation context. Then, the most important models of some nonlinear elements common in microwave and wireless circuits are briefly discussed. Furthermore, due to the rapidly increasing importance of system-driven nonlinear simulation, a section dedicated to behavioral, or black box, modeling of telecommunication subsystems is also included.

Finally, Chapter 5 is devoted to circuit design techniques appropriate for distortion mitigation. Beginning with a system level view, it brings in basic concepts of signal-to-noise ratio protection, dynamic-range optimization, and low-noise amplifier design. This introduces the analysis of the most important sources of nonlinear distortion in small-signal amplifiers based on either field effect or bipolar transistors. After that, nonlinear distortion generated in high-power amplifiers is addressed. Here, also the basic concepts of power amplifier design are first presented to then explore the compromises between maximum output power, power-added efficiency, and nonlinear distortion. By doing that, a set of general rules for highly linear power amplifier design are proposed. Because of the importance of RF and microwave mixers as nonlinear distortion sources, Chapter 5 concludes with the analysis of these circuits. However, the increased problem complexity, as compared to amplifiers, determined that only some simple general rules could be presented. Anyway, the analysis of distortion arising in balanced or unbalanced mixers using passive Schottky diodes and active FETs is believed to give the designer the basic information to direct most practical designs.

References

[1] Liu, P., "Passive Intermodulation Interference in Communication Systems," *Electronics & Communication Engineering Journal*, Vol. 2, No. 3, 1990, pp. 109–118.

[2] Minkoff, J., "The Role of AM-to-PM Conversion in Memoryless Nonlinear Systems," *IEEE Transactions on Communications*, Vol. 33, No. 2, 1985, pp. 139–144.

[3] Schetzen, M., *The Volterra and Wiener Theories of Nonlinear Systems*, New York: John Wiley & Sons, Inc., 1980.

CHAPTER 2

IMD Characterization Techniques

2.1 Introduction

Electronic devices are specified by their figures of merit. These are determined by characterization procedures that are thus of primary importance to the industry manufacturers. Take the case, for instance, of a power amplifier, where its gain, power-added efficiency, or nonlinear distortion are significant figures of merit, representing the observable properties of the device. Evaluating these quantities, then, plays a fundamental role on the correct specification of the power amplifier.

While figures of merit for linear behavior have been extensively studied and are already well established, their nonlinear counterparts still continue to be developed and debated.

The main objective of this chapter is to present an overview of the basic characterization techniques, and associated measurement setups, that enable the correct definition of most significant nonlinear distortion figures of merit.

Nonlinear devices do not comply with superposition. This fundamental truth obviates the use of any set of basis functions as a convenient means for describing their outputs to a general stimulus. So, the system's response to a certain input is as much useful as the input tested is closer to the excitation expected in real operation. But, since it is supposed that the system must handle information signals—which, by definition, are unpredictable—the input representation is a very difficult task. Indeed, although electrical engineers are used to test their linear systems with sinusoids (a methodology determined by Fourier analysis), now their probing signals should typically approximate band-limited power spectral density functions, PSD.

The first and simpler approximation we will consider for this PSD is to concentrate all the power distributed in the channel's bandwidth, *Bw,* into a single spectral line, and then to excite the system with that sinusoid. This corresponds to the single-tone tests, in which fundamental output power and phase versus input power are measured, along with the output at a few of the first harmonics.

Because well-behaved nonlinear systems subject to a sinusoid can only produce output spectral components that are harmonically related to the input frequency, the one-tone test is very poor as a characterization tool of those systems. For example, no spectral regrowth can be observed in normal narrowband wireless telecommunication systems, and so, no interference can be measured either inside the tested spectral channel—cochannel interference—or in any other closely located channel—adjacent-channel interference.

To overcome that difficulty, the one-tone characterization was replaced by the two-tone test. In that case, the input PSD is represented by two tones of equal amplitude and located at the *Bw* extremes, or somewhere in between. Now, although all even-order nonlinear components still constitute out-of-band distortion, there are a large number of odd-order combinations that produce inband spectral regrowth. As we will explain later, this led to the definition of some of the most widely used nonlinear distortion standards as the *intermodulation distortion ratio* (IMR), or the *third-order intercept point* (IP_3).

The main drawback associated with two-tone tests is their difficulty in evaluating cochannel distortion. Actually, since some of the odd-order mixing terms fall exactly at the same frequencies as the fundamentals, and the first-order, or linear, output components have much stronger amplitude than the distortion, there is no possibility of independently measuring cochannel distortion. Again, the way found to circumvent that weakness was to increase the resolution with which the input PSD is sampled. Although a multichannel stimulus approximation with a restricted number of tones is sometimes adopted (as in cable TV systems [1]), nonlinear distortion tends to be specified from multitone or band-limited noise tests. So, the last part of the text will be devoted to these more involved multitone characterization procedures.

Finally, to illustrate and compare the various presented procedures, a real microwave wideband medium power amplifier will be characterized using the various defined figures of merit.

2.2 One-Tone Characterization Tests

A linear device is identified by its frequency-domain transfer function, $H(j\omega)$. To measure it, an excitation signal consisting of a sinusoid is inserted at the input of the device under test (DUT),

$$x(t) = A_i \cos(\omega t) \tag{2.1}$$

and the output is measured at the same input frequency, called the fundamental frequency. Due to the device's linearity, a frequency sweep of that stimulus can only produce output changes in amplitude and phase, and the output must be expressed as

$$y(t) = A_o(\omega) \cos[\omega t + \phi_o(\omega)] \tag{2.2}$$

This is shown in Figure 2.1.

Although this sinusoidal test procedure can be directly extended to a nonlinear device under test (DUT), the test becomes substantially more involved. In fact, beyond the output dependence on frequency, common to the previous linear situation, now the output amplitude, A_o, will no longer be a scaled replica of the input level, A_i, nor the relative phase, ϕ_o, will only be determined by the frequency of the sinusoid: both A_o and ϕ_o will also nonlinearly vary with the stimulus level. Furthermore, that DUT will also generate new frequency components precisely located at the harmonics of the input. So, a more convenient way to represent its output would be

$$y(t) = \sum_{r=0}^{\infty} A_{o_r}(\omega, A_i) \cos [r\omega t + \phi_{o_r}(\omega, A_i)] \tag{2.3}$$

Figure 2.2 illustrates typical output amplitude and phase response characteristics of a nonlinear DUT versus input drive (for constant frequency), while Figure 2.3 shows an illustration of the output spectrum.

The observed output amplitude and phase variation versus drive manifest themselves as if the nonlinear device could convert input amplitude variations into output amplitude and phase changes—or, in other words, as if it could transform possible amplitude modulation (AM) associated to its input, into output amplitude modulation (AM-AM conversion) or phase modulation (AM-PM conversion). AM-AM conversion is particularly important in systems based on amplitude modulation; while AM-PM has its major impact in modern telecommunication and wireless systems that rely on phase modulation formats.

As will be referred to later in Section 4.4, the main application of this type of characterization is the extraction of behavioral models suitable to describe the nonlinear system performance at the excitation envelope [2]. Nevertheless, since this is a static step-by-step characterization, the extracted behavioral models cannot present any memory to those envelopes [3].

Finally, the DUT's capability for generating new harmonic components is characterized by the ratio of the integrated power of all the harmonics to the measured power at the fundamental, a figure of merit named *total harmonic distortion* (THD).

These three figures of merit will be detailed in the following sections. For that, we will assume that our nonlinear system can again be represented by the power series with memory of (1.7), herein rewritten for convenience:

$$y_{NL}(t) = a_1 x(t-\tau_1) + a_2 x(t-\tau_2)^2 + a_3 x(t-\tau_3)^3 + \ldots \tag{2.4}$$

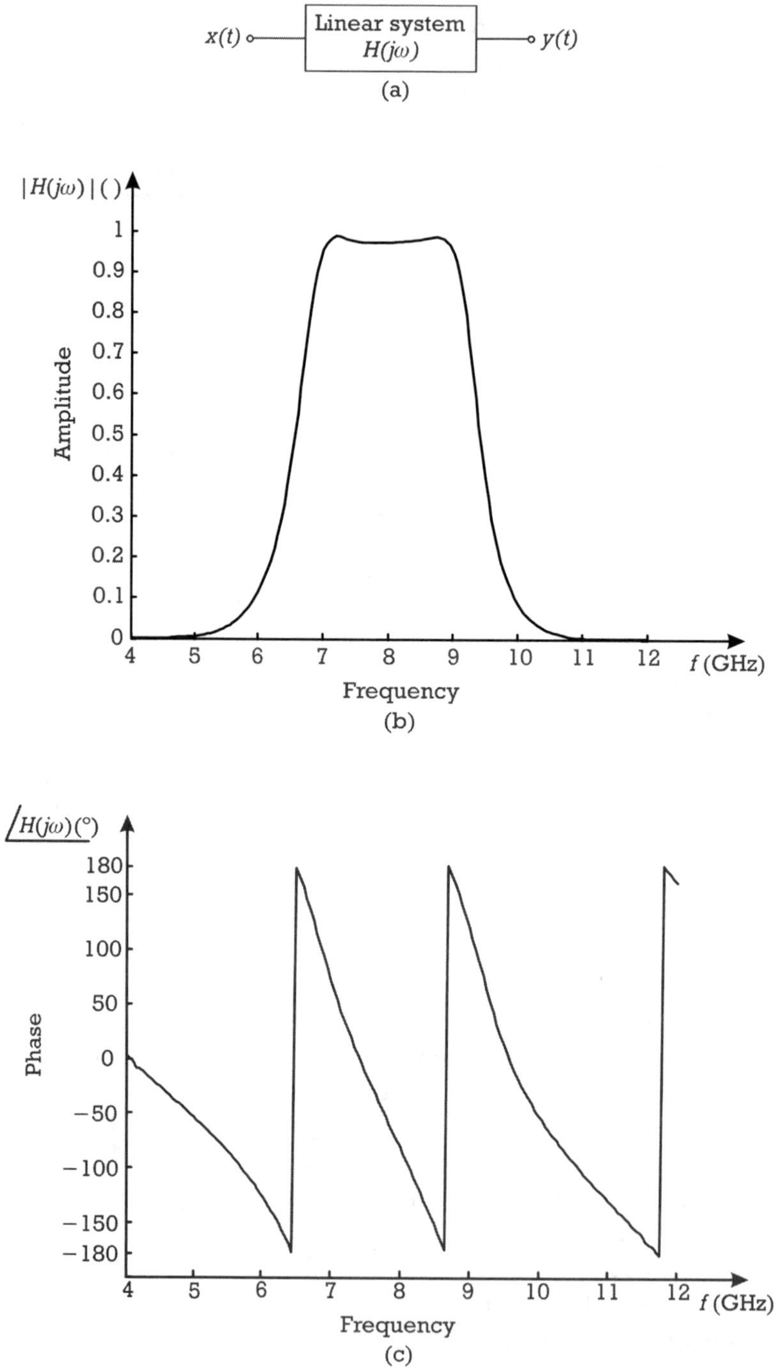

Figure 2.1 (a) Linear system representation, (b) output amplitude, and (c) phase changes during an input frequency sweep.

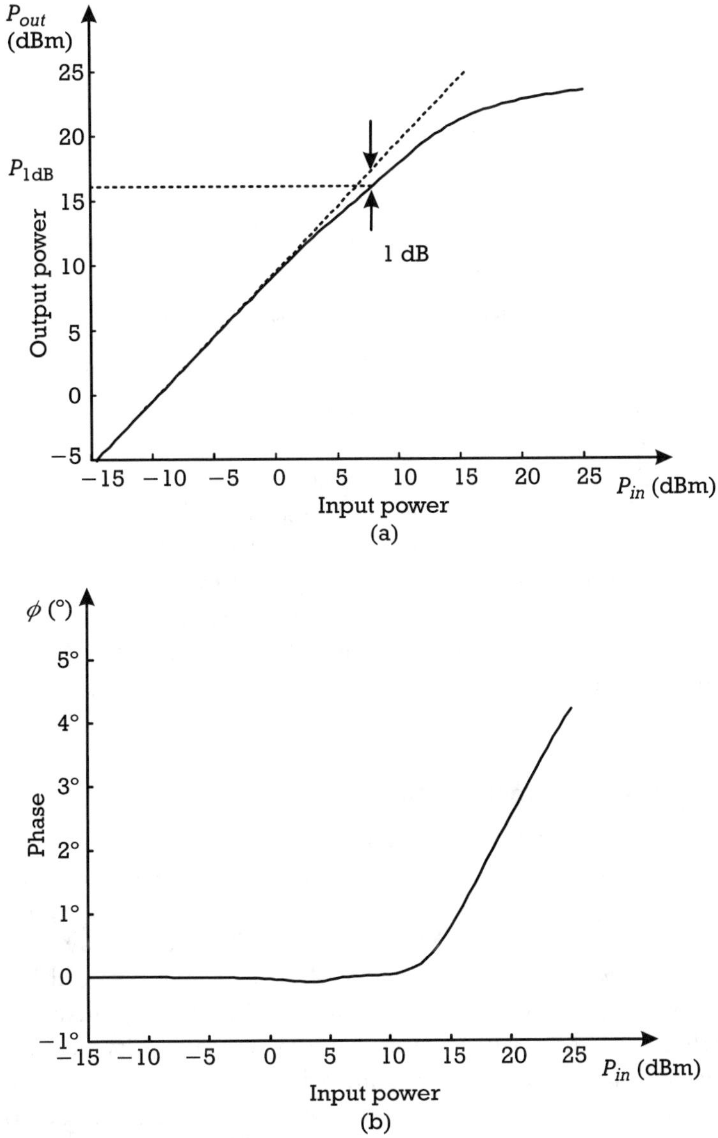

Figure 2.2 Nonlinear DUT output (a) amplitude and (b) phase characteristics versus input drive level.

2.2.1 AM-AM Characterization

AM-AM characterization describes the relation between the output amplitude of the fundamental frequency, $r = 1$ in (2.3), with the input amplitude of a fixed input frequency [4].

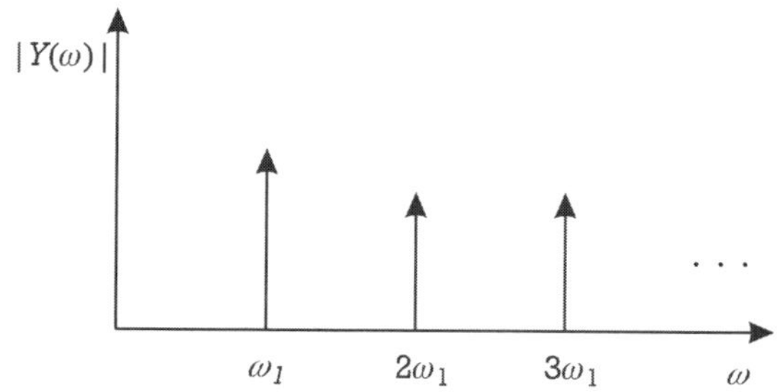

Figure 2.3 Typical nonlinear DUT's harmonic generation characteristics.

Thus, it characterizes gain compression or expansion of a nonlinear device versus input drive level.

AM-AM characterization enables the evaluation of an important figure of merit called the *1-dB compression point,* $P_{1\text{dB}}$. It is defined as the output power level at which the signal output is already compressed by 1 dB, as compared to the output that would be obtained by simply extrapolating the linear system's small-signal characteristic. Thus, the $P_{1\text{dB}}$ figure also corresponds to a 1-dB gain deviation from its small-signal value, as depicted in Figure 2.2. AM-AM characterization is sometimes expressed as a certain dB/dB deviation at a predetermined input power [5].

2.2.2 AM-PM Characterization

Another interesting property of nonlinear systems is that vector addition of the output fundamental with distortion components also determines a phase variation of the resultant output, when the input level varies (see Figure 1.10). This is the outcome of the expected AM-PM characteristics of our system.

Note, however, that although AM-AM behavior would be visible whether or not the system presented memory effects, AM-PM is exclusive of dynamic systems. Actually, as is shown in Section 4.4, not only memory is essential, as it must be intrinsically mixed with the nonlinearity. For example, a system whose memory would only be the effect of a linear delay just in front of a memoryless nonlinearity [case of equal τ_1, τ_2, and τ_3 in (2.4)] would not show any AM-PM conversion.

AM-PM characterization consists of studying the variation of the output signal phase, $\phi_{o1}(\omega, A_i)$, with input signal amplitude changes for a constant frequency, and may be expressed as a certain phase deviation, in degrees/dB, at a predetermined input power [5].

2.2.3 Total Harmonic Distortion Characterization

The third characterization technique, especially used in multioctave systems (as audio amplifiers), measures THD [6]. This figure of merit is defined as the ratio between the square roots of total harmonic output power and output power at the fundamental signal. Therefore, and according to (2.3), *THD* can be expressed by

$$THD = \frac{\sqrt{\frac{1}{T}\int_0^T \left[\sum_{r=2}^{\infty} A_{o_r}(\omega, A_i)\cos\left[r\omega t + \phi_{o_r}(\omega, A_i)\right]\right]^2 dt}}{\sqrt{\frac{1}{T}\int_0^T \left[A_{o_1}(\omega, A_i)\cos\left(\omega t + \phi_{o_1}(\omega, A_i)\right)\right]^2 dt}} \quad (2.5)$$

In the simple polynomial nonlinearity model of (2.4), *THD* would be given by

$$THD = \frac{\sqrt{\frac{1}{8}a_2^2A_i^4 + \frac{1}{32}a_3^2A_i^6 + \ldots}}{\sqrt{\frac{a_1^2A_i^2}{2}}} = \frac{1}{2}\frac{A_i}{a_1}\sqrt{a_2^2 + \frac{1}{4}a_3^2A_i^2 + \ldots} \quad (2.6)$$

2.2.4 One-Tone Characterization Setups

AM-AM and AM-PM characterizations are performed reading output signal components whose frequency is equal to the input excitation. Therefore, a usual amplitude controlled sinusoidal—or continuous-wave (CW)—generator connected to a vector network analyzer are sufficient for these tasks. The corresponding setup is depicted in Figure 2.4.

Since the network analyzer simultaneously measures DUT's gain and phase, it is possible to characterize both AM-AM and AM-PM with a single amplitude power sweep. For that, relative gain is first converted into absolute output power, and then, that value, along with measured phase difference, is plotted against input drive level.

However, if a gain plot is directly used, it provides an immediate way for evaluating the DUT's 1-dB compression point. Since this P_{1dB} is nothing more than the output power at which the gain is already compressed 1 dB from its small-signal value, a gain plot inspection directly gives the corresponding input power level, which can be readily converted to output power, adding the actual measured gain. This procedure is exemplified in Figure 2.5.

Alternative, and less expensive, AM-AM characterization setups use a scalar network analyzer, or even a spectrum analyzer. Unfortunately, since neither of these pieces of equipment is able to measure phase, AM-PM characterization would no longer be possible.

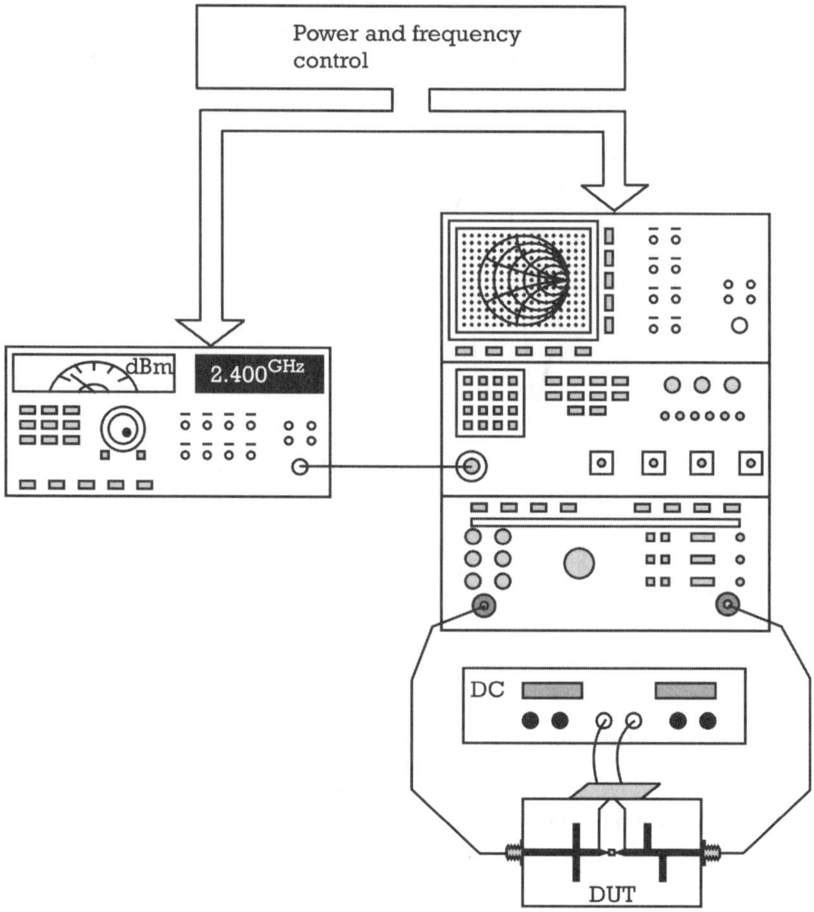

Figure 2.4 AM-AM and AM-PM characterization setup based on a vector network analyzer.

THD characterization can only be performed with a spectrum analyzer, as the measured output includes frequency components that are different from the input excitation.

Figure 2.6 presents one such setup that can perform both AM-AM and *THD* characterization. For that, the input generator is swept in amplitude and the output is measured at the fundamental frequency, for AM-AM, or at the harmonic components, for *THD*. Obviously, this *THD* evaluation method relies on individual output power measurements at each harmonic, thus requiring a subsequent calculation according to (2.5).

Another simple AM-AM characterization setup relies on a power meter for measuring the DUT's input and output powers. However, some care must be taken when using this setup because the power meter integrates all the power generated

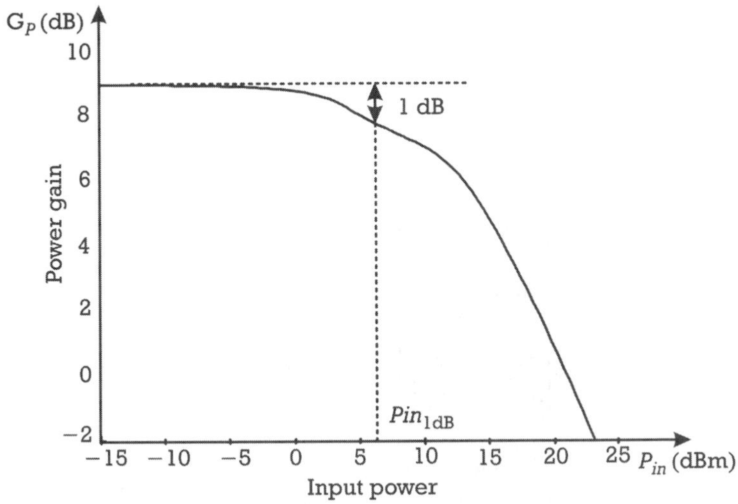

Figure 2.5 DUT's gain versus input drive level, showing P_{1dB} evaluation.

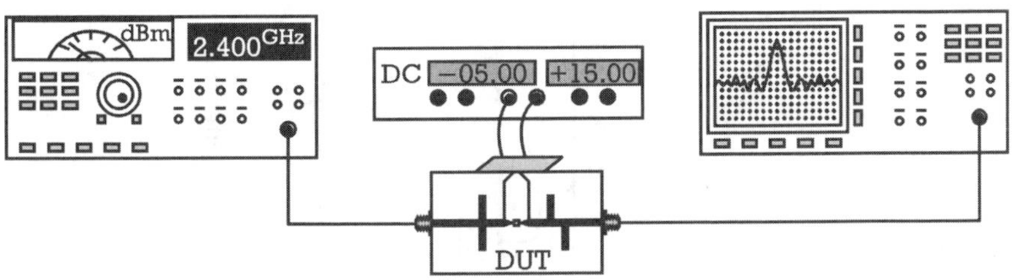

Figure 2.6 AM-AM and/or *THD* characterization setup using a spectrum analyzer.

by the DUT. Therefore, accurate AM-AM characterization requires that the DUT's harmonic content be negligible in comparison to the fundamental output power, in the whole input power sweep span.

For the AM-PM characterization, the simple setup represented in Figure 2.7, which is based on a calibrated phase shifter and a spectrum analyzer, could also be used.

In this case, the output of the DUT—signal plus distortion—will be added to a sample of the input signal shifted by α degrees. The objective of the bridge network is to cancel the output fundamental signal. The α degrees introduced by the phase shifter, for perfect bridge adjustment, correspond to $\phi_{o1} = \alpha + (2k + 1)\pi$, where k is any integer number and ϕ_{o1} is the DUT's output phase. If this procedure is repeated for an input power sweep, the resulting $\Delta\phi_{o1}$ versus P_{in} function constitutes the sought AM-PM characterization. The main problems associated with this setup are due to the phase shifter's finite resolution, and to

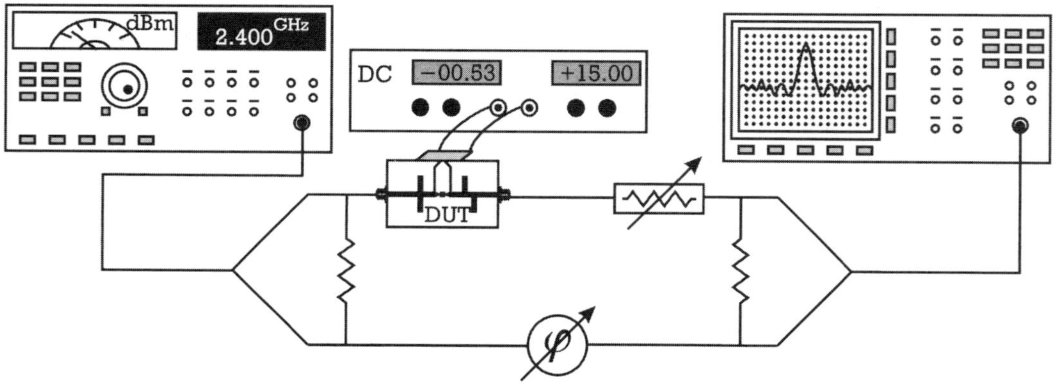

Figure 2.7 AM-PM characterization setup using a calibrated phase shifter and a spectrum analyzer.

the possible variable phase shift introduced by the attenuator present in the DUT's branch. Furthermore, since the auxiliary branch is supposed to provide a signal that is an exact replica of the input, it must be guaranteed that its phase shifter does not generate any distortion components.

Other one-tone characterization setups, relying on dedicated or special laboratory equipment, are possible. From these, the use of the microwave transition analyzer deserves to be mentioned. This modern piece of equipment not only combines the vector network analyzer operation with a spectrum analyzer, as some of its software options directly allow AM-AM, AM-PM, and *THD* automated measurements. Indeed, its two-port high-speed sampling oscilloscope, with built-in Fourier transform software, turns it into a revolutionary spectrum analyzer with phase measurement capabilities.

A final remark on these setups should assert that excessive signal generator phase noise or long-term frequency instability, as well as reduced signal analyzer dynamic range, can create severe impairments on the quality of the results. These difficulties, which are shared by almost all distortion measurement methods, and will be discussed later in greater detail, are especially notorious when large signal to distortion components ratios are involved.

2.3 Two-Tone Characterization Tests

As said in the introduction of this chapter, a better representation of true telecommunication signal excitations than the pure sinusoid considered above is the two-tone stimulus. Similarly to the one-tone tests, this type of signal allows the characterization of generated harmonics—which, in bandpass systems, are usually attenuated by the output matching networks—but it also enables the identification of new mixing components close to the fundamentals. These inband components play a dominant role in bandpass systems, as they constitute the main sources of nonlinear distortion impairments.

As seen in Section 1.3, if our nonlinear polynomial model of (2.4) were excited by a two-tone excitation like

$$x(t) = A_{i1} \cos(\omega_1 t) + A_{i2} \cos(\omega_2 t) \tag{2.7}$$

the output would be given by

$$y_{NL}(t) = \sum_{r=1}^{\infty} A_{o_r} \cos(\omega_r t + \phi_{o_r}) \quad \text{where } \omega_r = m\omega_1 + n\omega_2 \text{ and } m, n \in Z \tag{2.8}$$

which shows that the output would be composed of a very large number of mixing terms involving all possible combinations of $\pm\omega_1$ and $\pm\omega_2$.

Referring to a usual narrowband RF subsystem, as the ones found in wireless transmission channels, two types of information can be extracted from a two-tone test: the so-called inband distortion measurements, in which $m + n = 1$, and the out-of-band components' evaluation, where $m + n \neq 1$. The next sections will be devoted to detailing these two different sets of characterizations.

2.3.1 Inband Distortion Characterization

Inband distortion products are the mixing components falling exactly over, or very close to, the output fundamental frequencies. Therefore, and according to (2.8), the inband distortion frequencies will be those satisfying

$$m + n = 1 \tag{2.9}$$

For example, if a system represented by (2.4) is considered, inband measurements would have to be performed at the fundamental frequencies: ω_1, ω_2; third-order components ($|m| + |n| = 3$) at: $2\omega_1 - \omega_2$, $2\omega_2 - \omega_1$; fifth-order components ($|m| + |n| = 5$) at: $3\omega_1 - 2\omega_2$, $3\omega_2 - 2\omega_1$; seventh-order components ($|m| + |n| = 7$) at: $4\omega_1 - 3\omega_2$, $4\omega_2 - 3\omega_1$; and so forth.

These distortion products constitute a group of lower and upper sidebands, separated from the signals and from each other by the tones' frequency difference $\omega_2 - \omega_1$. As said in Section 1.3, they are known as the IMD.

Accordingly, *the signal-to-intermodulation distortion ratio,* or simply *the intermodulation ratio* (*IMR*), is defined as the ratio between the fundamental and IMD output power[1]:

$$IMR \equiv \frac{P_{fund}}{P_{IMD}} = \frac{P(\omega_1)}{P(2\omega_1 - \omega_2)} = \frac{P(\omega_2)}{P(2\omega_2 - \omega_1)} \tag{2.10}$$

Figure 2.8 presents a logarithmic plot of the fundamental output power at one of the fundamental signals and the IMD power measured in one of the distortion sidebands, versus input power, as would be observed from an equal amplitude two-tone test. At sufficiently small-signal levels, the fundamental output power increases 1 dB for each decibel rise of input power, while a 3 dB per decibel rate is noticed for the IMD power. This is dictated by the dominance of our system model's first and third-degree terms. However, at very large-signal levels, where the contribution of the higher order terms is no longer negligible, both curves tend to compress towards constant fundamental and IMD output power values.

1. Even though (2.10) presents an equality between lower and upper *IMR,* there are some cases where these values are different. In those situations, often called IMD asymmetries [7], *IMR* must be defined as upper or lower.

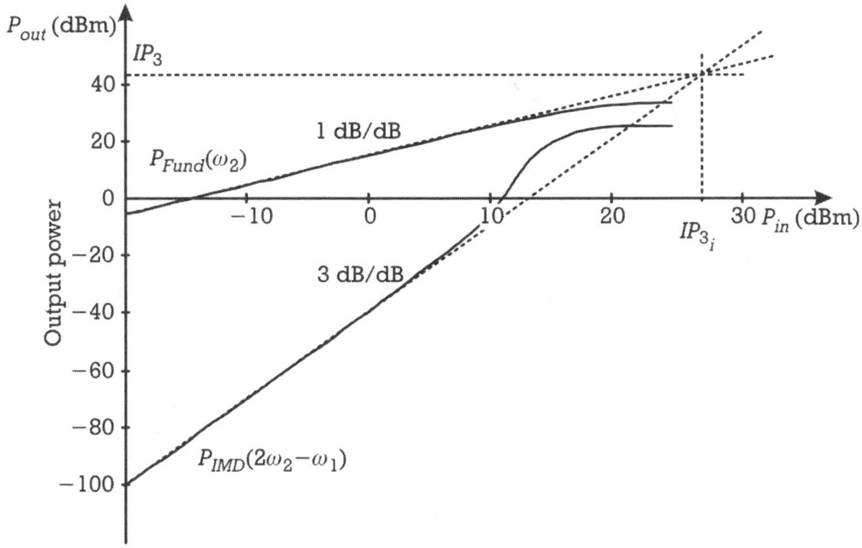

Figure 2.8 Output fundamental power per tone and distortion power in one IMD sideband for an equal amplitude two-tone excitation.

This behavior, common to the large majority of microwave and wireless systems, enables the definition of a very important figure of merit for characterizing the IMD in nonlinear devices: the third-order intercept point IP_3. IP_3 is a fictitious point that is obtained when the extrapolated 1-dB/dB slope line of the output fundamental power intersects the extrapolated 3-dB/dB slope line of the IMD power. Since IP_3 is determined by the system's third-order distortion behavior, it cannot be used for IMD characterization unless it is guaranteed that no large-signal effects are involved. In other words, and contrary to a loose practice seen in various product specifications and sometimes even in scientific publications, IP_3 can only be extrapolated from the small-signal zone where IMD presents a distinct and constant 3-dB/dB slope.

Mathematically, IP_3 can be directly calculated from the results of (1.26), assuming an excitation of unmodulated tones of equal amplitude $A_{i1} = A_{i2} = A_i$, and keeping only terms up to third order. Thus, third-order IMD output power at one of the sidebands (e.g., at $2\omega_1 - \omega_2$) will be given by

$$P_{IMD}(2\omega_1 - \omega_2) = \frac{1}{T_{2\omega_1-\omega_2}} \int_0^{T_{2\omega_1-\omega_2}} \left\{\frac{3}{4} a_3 A_i^3 \cos\left[(2\omega_1 - \omega_2)t - \phi_{32-1}\right]\right\}^2 dt$$

$$= \frac{9}{32} a_3^2 A_i^6 \tag{2.11}$$

while the linear output power at ω_1 will be

$$P_{linear}(\omega_1) = \frac{1}{T_{\omega_1}} \int_0^{T_{\omega_1}} [a_1 A_i \cos(\omega_1 t - \phi_{110})]^2 \, dt = \frac{1}{2} a_1^2 A_i^2 \tag{2.12}$$

Now, applying the definition of IP_3, which is the extrapolated linear output power of one of the fundamentals that equals the extrapolated power of the considered third-order IMD sideband, we would get

$$\frac{1}{2} a_1^2 A_i^2 = \frac{9}{32} a_3^2 A_i^6 \Rightarrow A_i^2 = \frac{4}{3} \frac{a_1}{a_3} \tag{2.13}$$

and thus, substituting this A_i^2 into $P(\omega_1)$,

$$IP_3 = P(\omega_1) = \frac{2}{3} \frac{a_1^3}{a_3} \tag{2.14}$$

Sometimes—as in nonlinear devices presenting power loss instead of gain, like in passive mixers—the input IP_3, or IP_{3i}, is preferred. Its definition follows exactly the one given for the output IP_3 except that now the referred extrapolated power is measured at the input. Therefore, IP_{3i} and IP_3 only differ by the system's linear gain. Another variant definition of this figure of merit, although more rarely used, relates extrapolated total output linear power with total third-order IMD power. It is simply the double (3 dB higher in logarithmic power units) of the herein defined IP_3.

A final remark on this IMD specification standard should state that, despite rarely being seen, some other intercept figures of merit could be defined for fifth-order (IP_5) or seventh-order (IP_7) distortion.

As is shown in Figure 2.8, small-signal IMR decreases 2 dB per decibel of input power rise, up to an extrapolated point, the IP_3, where it would collapse to 0 dB. So, restricted to the small-signal region, one of these figures can always be obtained from the other by

$$IP_{3\text{dB}} = P_{0\text{dB}} + \frac{1}{2} IMR_{\text{dB}} \tag{2.15}$$

or

$$IMR_{\text{dB}} = 2(IP_{3\text{dB}} - P_{0\text{dB}}) \tag{2.16}$$

where P_{0dB} is the fundamental output power per tone (in logarithmic units) at which IMR_{dB} was measured.

2.3.2 Out-of-Band Distortion Characterization

As already said, out-of-band components are the mixing products of (2.8), obeying

$$m + n \neq 1 \tag{2.17}$$

These include harmonics of each of the fundamentals, like in the one-tone case, but also new mixing products at $m\omega_1 + n\omega_2$ that fall, either near dc ($n + m = 0$), or close to the various harmonics ($n + m = 2, 3, 4, \ldots$). Table 2.1 illustrates such out-of-band products generated by a third-order nonlinearity subject to a two-tone excitation.

The product located at dc describes the bias point shift from the quiescent point, when input driving level increases. Then, the one at $\omega_2 - \omega_1$ is usually called the baseband. The reason for this designation comes from the fact that if the two-tone excitation were considered as a carrier at $(\omega_1 + \omega_2)/2$, amplitude modulated in double-sideband format (suppressed carrier) by a baseband modulating signal, then $\omega_2 - \omega_1$ would be the double of that baseband frequency signal.

According to what was defined for the inband distortion components, these out-of-band components can be also described by corresponding intercept points as is depicted in Figure 2.9 for IP_2.

As their name indicates, out-of-band distortion components appear at zones of the output spectrum quite far from the fundamental signals. So, rigorously speaking, they are only out-of-band in narrowband systems, but not in multioctave ones. Furthermore, because they become relatively simple to be filtered out in narrowband systems, their importance, as a transmission quality impairment, is only evident on ultrawideband applications.

2.3.3 Two-Tone Characterization Setups

Because mixing products involving a combination of both ω_1 and ω_2 have frequencies that are different from either ω_1 or ω_2 inputs, two-tone test measurements use a spectrum analyzer. The most commonly used arrangement for such a setup is shown in Figure 2.10.

Table 2.1 Out-of-Band Distortion Components Generated from a Two-Tone Excitation

Mixing Product	*dc*	$\omega_2 - \omega_1$	$2\omega_1$	$\omega_2 + \omega_1$	$2\omega_2$	$3\omega_1$	$2\omega_1 + \omega_2$	$2\omega_2 + \omega_1$	$3\omega_2$
Order	2nd	2nd	2nd	2nd	2nd	3rd	3rd	3rd	3rd

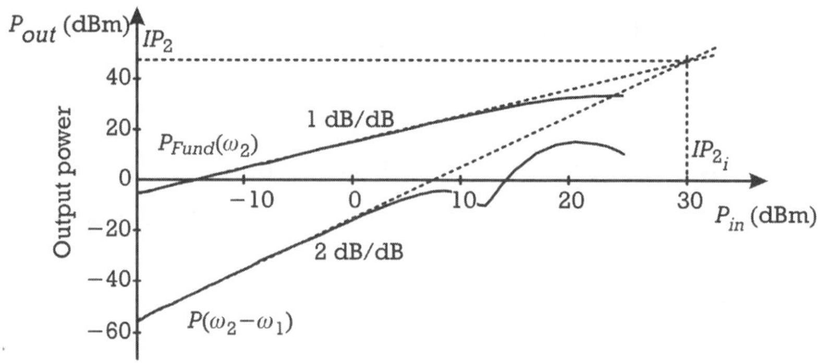

Figure 2.9 Output fundamental power per tone and distortion power at the baseband for an equal amplitude two-tone excitation.

Although, in its simplest form, this setup would only involve two signal generators of variable level, a power combiner and the spectrum analyzer, the implementation depicted in Figure 2.10 requires many more laboratory components. They are intended to guarantee accurate measurements of very high signal-to-distortion ratios.

First of all, it is assumed that the combination of the two input signals is not made with a simple T-junction, but by using a true power combiner. This not only guarantees port matching, as it profits from adjacent port isolation. And that is of paramount importance as it prevents each signal to mix with the other in the nonlinear output stages of the generators. The measurement error that may be induced by this parasitic IMD is so dramatic that it may be found useful to artificially boost combiner isolation with the two isolators shown in Figure 2.10. Moreover, harmonics of the generated signals can also mix with the other fundamental to produce further residual distortion either in the DUT, the signal generators' output stages, or both. Since the amplitude and phase of this residual IMD is unknown, but it may be amplified by the DUT's linear gain, it will add to the wanted DUT's IMD producing an unpredictable error. That is why two lowpass filters were included at the signal generator outputs to improve their signal spectral purity.

Beyond those sources of IMD measurement error, there is also the spectrum analyzer nonlinear input stage. Remember that if you are characterizing a very linear device, everything in the setup (including the spectrum analyzer) should be even more linear. Obviously, one way to get rid of that additional distortion caused by the DUT's fundamentals in the spectrum analyzer front-end stage would be to use a large attenuator at its input. However, that attenuator adds a certain amount of noise, as it masks, to the same amount, the very small DUT's IMD components. The solution is to take advantage, as much as possible, of the available spectrum analyzer's dynamic range.

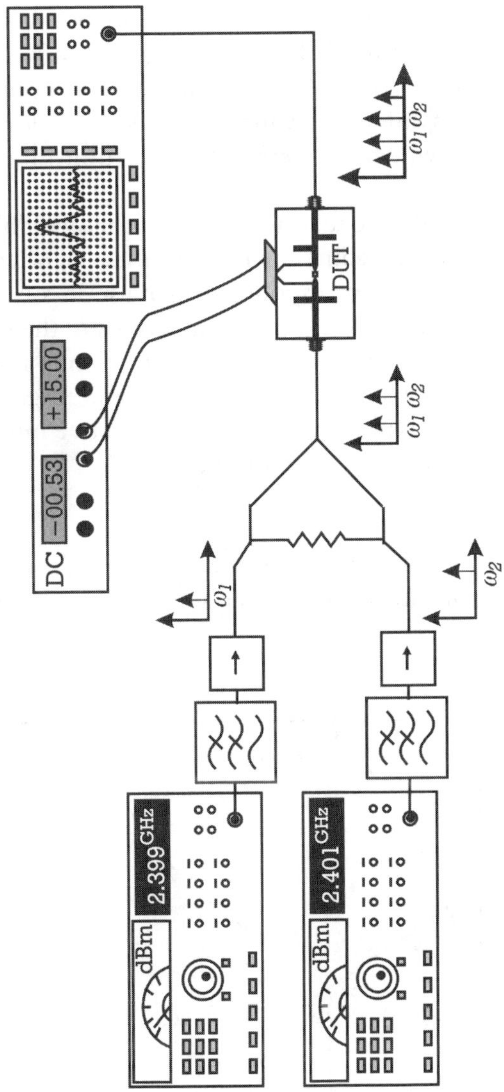

Figure 2.10 Most commonly used two-tone test measurement setup.

This can be done by studying the dynamic range characteristics of that equipment, and then choosing the optimum input power level (adjusting the spectrum analyzer's input attenuator value) and convenient sweep time (determined by the selected resolution bandwidth, and thus, correspondent noise floor) [8], as illustrated in Figure 2.11.

An alternative way to accomplish the same objective consists of tuning the spectrum analyzer's settings until the best results are met. For that, the first step should be to empirically choose the minimum required input attenuation. Starting with the highest end of available attenuation values, the desired attenuation value is the minimum one in which the carrier-to-distortion ratio actually read is still unchanged.

The second step should be to take advantage of the available spectrum analyzer's sensitivity. This demands a reduction in resolution bandwidth, which must be inevitably accompanied by a correspondent increase in sweep time and probable reduction in frequency span. This, in turn, may obviate the simultaneous observation of the output fundamentals and the distortion components, as it may call for low phase noise and highly stable frequency-synthesized generators.

In the next section we will show how an imaginative modification of the setup may circumvent the majority of these difficulties.

2.3.3.1 Bridge Setup

The above discussion, on the distortion measurement difficulties, has shown that the main source of error introduced by the spectrum analyzer is due to the DUT's

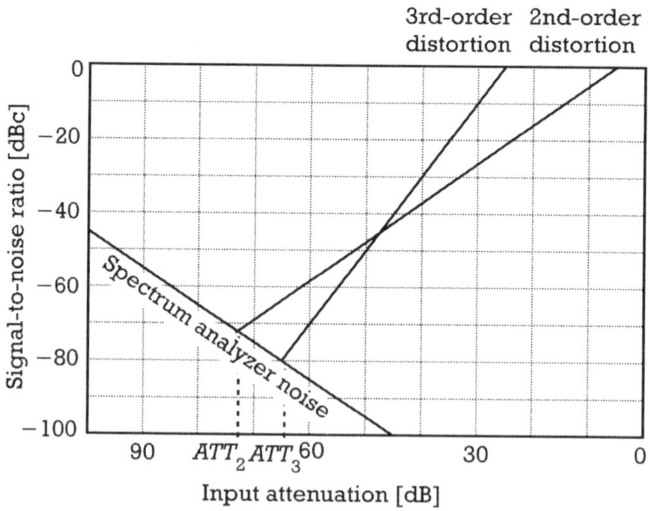

Figure 2.11 Different spectrum analyzer error entities when measuring nonlinear distortion. ATT_2 (ATT_3) represents the optimum spectrum analyzer input attenuation value for maximized signal to noise plus second (third)-order distortion ratio.

fundamentals. Therefore, eliminating those components, without perturbing the desired distortion, would constitute a foremost benefit. Nevertheless, except for very special cases where out-of-band distortion is sought, in which tone separation is so high that the fundamentals can be rejected without significantly perturbing the closely located distortion components, or, eventually, where the measurement band is previously fixed, filtering is out of the question. Therefore, the elimination of the DUT's output fundamentals demands more ingenious solutions.

The bridge setup presented in Figure 2.12 is one of such possible methods [9]. It relies on the fact that, since the signal present at the DUT's input is an exact replica of the output fundamentals, but includes no distortion components, it can be used to cancel out those fundamentals, while preserving the desired distortion. In doing that, the amplitude and phase of a sample of the excitation are tuned in the auxiliary bridge arm and then combined with the DUT's output to produce the sought cancellation. Naturally, it is herein also assumed that both the attenuator and phase shifter are linear components, and so they cannot introduce any distortion of their own.

Note that, except for the bridge network composed by a 3-dB power splitter, linear gain and phase control cells—a variable attenuator and phase shifter—and the final 3-dB power combiner, the bridge setup is essentially equal to the one of Figure 2.10.

2.4 Multitone or Continuous Spectra Characterization Tests

As was already explained in the beginning of this chapter, the nonlinear nature of our problem determines a close relation between the usefulness of a certain characterization technique and the similarity of the test signal with the real equipment's excitation. Therefore, and although one-tone and two-tone techniques still represent the industry standards in intermodulation distortion characterization, nowadays, engineers seek for alternative test procedures closer to the system's final operation regime. Actually, telecommunications signals are usually composed of one or more carriers modulated by information signals (i.e., signals that are necessarily aperiodic in time), and presenting band-limited continuous spectra [10].

However, there are general-purpose devices whose final application is not known a priori, as situations are found where a sample of the real signal is not available at test time. In these, and many other practical cases, the DUT is simply tested against signals that mimic those band-limited continuous spectra.

Examples of this type of signal include digitally modulated carriers with pseudorandom baseband signals, multitones, and band-limited noise [11]. These can be grouped in two distinct sets: test signals of discrete spectrum, like the ones generated by pseudorandom modulation, or equally separated multitones; and test signals of

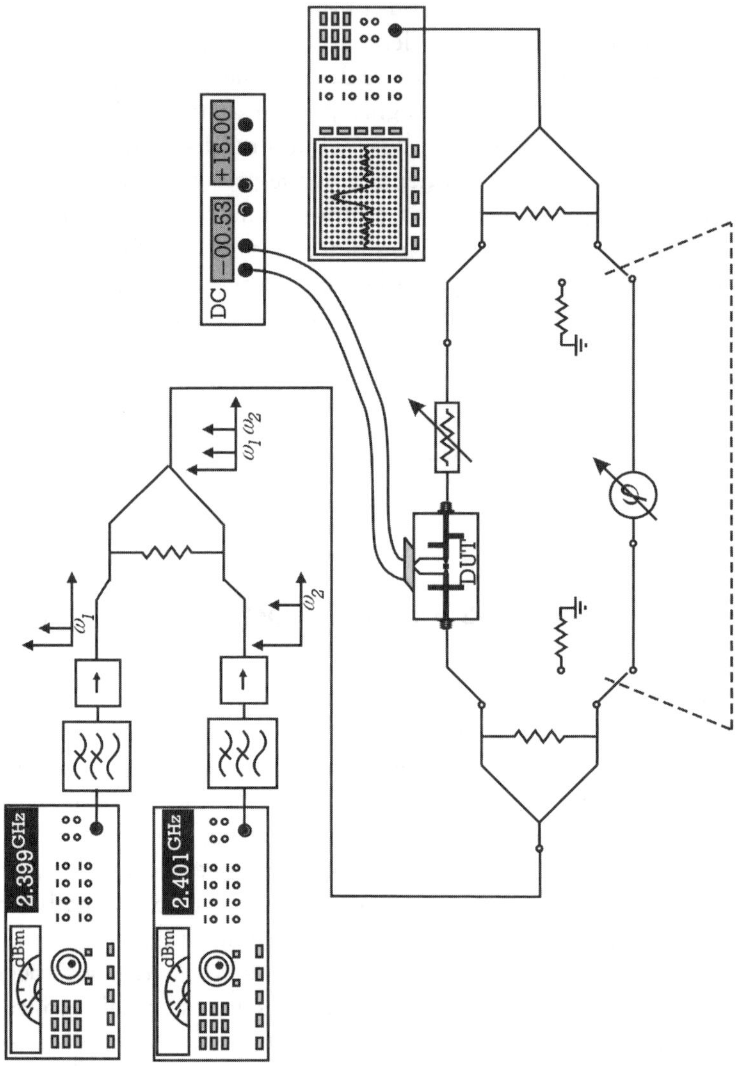

Figure 2.12 Two-tone test bridge setup for improved IMD measurement quality.

true continuous spectrum, as band-limited noise or any carrier modulated with real aperiodic data.

That classification is particularly important from the time waveform point of view. While continuous spectra describe aperiodic data—and so are random in a certain sense—uniformly discretized spectra represent periodic signals whose time-domain characteristics are completely different.

To exemplify, imagine a class of signals composed of 10 evenly spaced tones of equal amplitude, described by

$$x(t) = \sum_{q=1}^{Q} A_q \cos(\omega_q t + \theta_q) \tag{2.18}$$

where $Q = 10$, and $\omega_q = \omega_0 + (q - 1)\Delta\omega$, with ω_0 the position of the first tone and $\Delta\omega$ the constant frequency separation between them. Despite all of these 10-tone sets having exactly the same power spectrum, their time characteristics may be extremely different.

For example, if one instance of that class were generated by adding 10 uncorrelated (in phase) tones, then the inherent and independent phase noises of each of the 10 carriers would determine a random time waveform, similar to band-limited white noise, except that now the spectral components would no longer be equally distributed within the bandwidth, but arranged in PSD clusters around the Q carriers. This noisy waveform can really be viewed as a particular discretization of a band-limited white noise excitation, where the power density function was integrated to become Q times the power of each spectral line. Indeed, the discretized version would exactly tend to the true noise excitation if Q were made infinitely large. A possible way to implement this excitation using the representation of (2.18) would be to consider a random phase for each tone and then repeat this procedure for several phase arrangements, finishing with an average of all the resulting output time waveforms [12]. The resulting waveform of one of such randomized phase arrangements can be seen in Figure 2.13(a).

Now, suppose another 10-tone signal were generated in a way that all the carriers are related in phase, as is the case of multitones produced by phase-locked synthesized sources or by digital arbitrary waveform generators. Because all the tones are now correlated, there may be time spots where they all attain their highest maximum, and other time spots where they may cancel each other, as shown in Figure 2.13(b). Such a waveform has a peak power to average power ratio of 13.01 dB, whereas it is only 7.83 dB for the one of Figure 2.13(a). These differences in peak-to-average ratio would be even bigger if the number of tones considered were increased.

So, despite these two multitone signals sharing same power spectrum and average power, it is obvious that their extremely different signal excursions will induce quite distinct distortion behavior. This is a very important issue to bear in

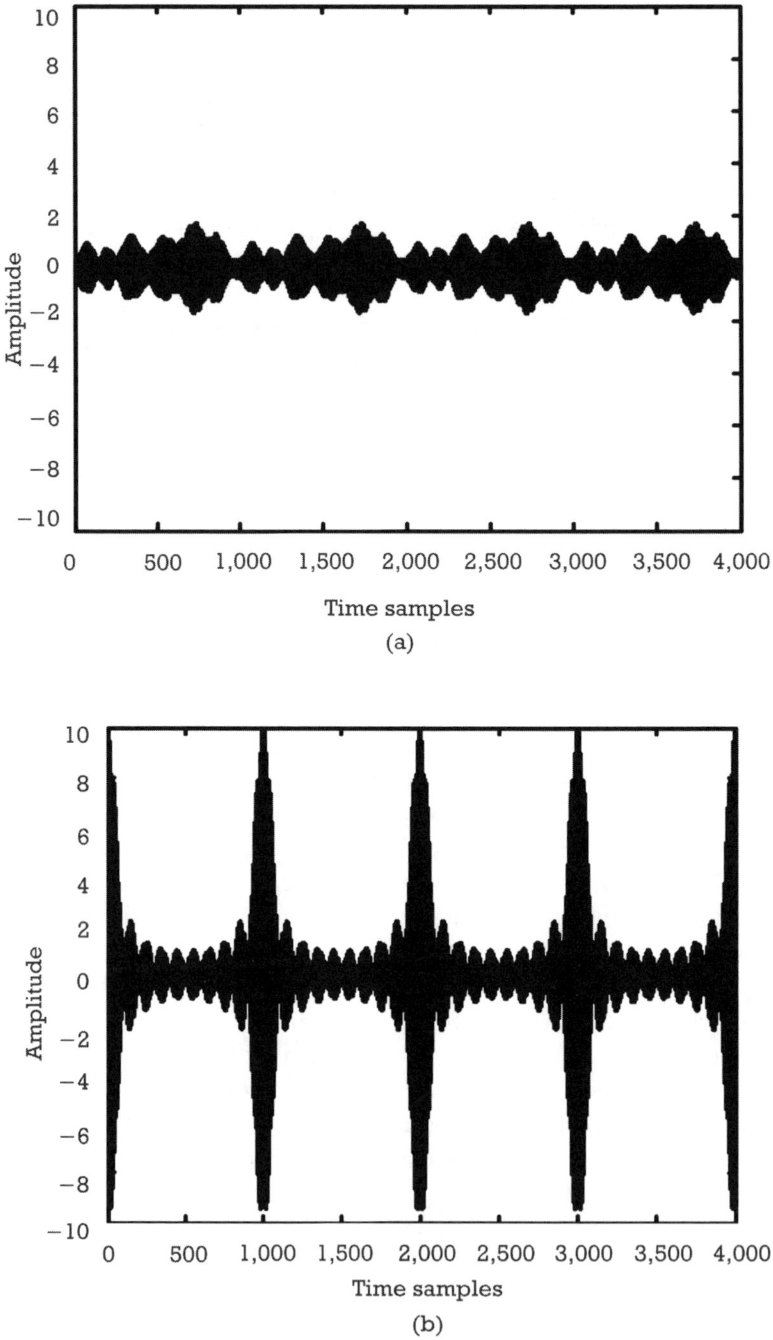

Figure 2.13 Time-domain waveforms of two signals composed of 10 evenly spaced tones of equal amplitude: (a) independent tones with a randomized phase arrangement; and (b) all 10 tones phase-locked to a common reference.

mind any time a multitone signal is used as a nonlinear distortion characterization test signal.

If that multitone signal were used to stimulate our nonlinear system, we would expect an output given by a large number of line-clusters placed near dc (the generalized multitone baseband), colocated with the output fundamentals and close to it (the generalized multitone IMD), and near all the harmonics. So, the output power spectrum would be given by [13]

$$y_{NL}(t) = \sum_{r=1}^{R} A_r \cos(\omega_r t + \phi_r) \tag{2.19}$$

where $\omega_r = m_0 \omega_0 + \ldots + m_q \omega_q + \ldots + m_{Q-1} \omega_{Q-1}$, $|m_0| + \ldots + |m_q| + \ldots + |m_{Q-1}| \le O$, $\omega_0, \ldots, \omega_q, \ldots, \omega_{Q-1}$, are the input frequencies and O is the maximum order of the mixing product under consideration. A power spectrum like this resembles the one represented in Figure 2.14.

According to what was already stated for the two-tone excitation, in a narrowband system the inband distortion is the distortion cluster falling exactly over, or close to, the fundamentals. So, let us focus on that zone, viewed in more detail in Figure 2.15.

Comparing the input and output spectra shown in Figure 2.15, it is clear that the output contains many more frequency components. These are usually named spectral regrowth [11], because they are a consequence of the property of nonlinear systems in generating, or "growing," new frequency lines. In general, (i.e., whenever the input tones are not evenly spaced) this spectral regrowth includes not only the components adjacent to the signal, as seen in Figure 2.15, but also new mixing products located among the fundamentals, but not coincident with them. The first type of spectral regrowth components is the adjacent-channel distortion [or alternate-channel distortion, in case the mixing product is located at a distance greater than $(Q-1)\Delta\omega$ and lesser than $2(Q-1)\Delta\omega$, from the input tone of highest or lowest frequency], whereas the second type constitutes cochannel distortion.

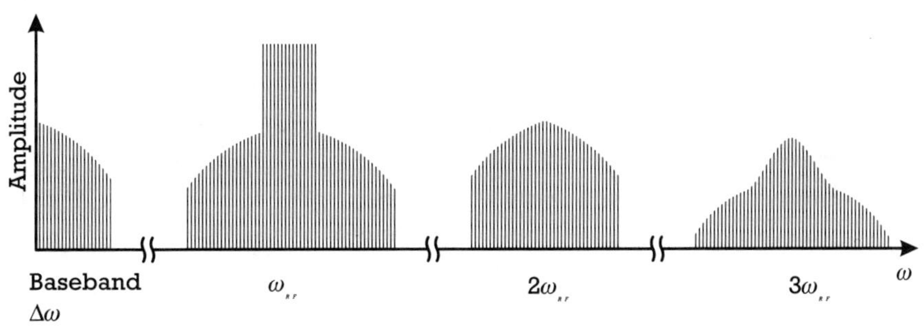

Figure 2.14 Spectrum response of a third-order system to a narrowband multitone excitation.

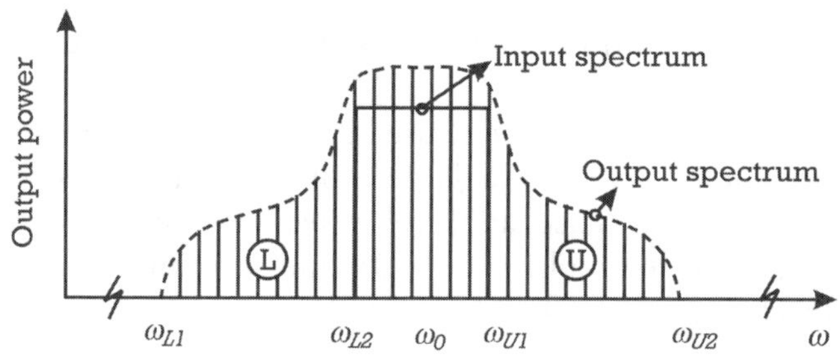

Figure 2.15 Input and inband output spectra as observed in a system excited by a narrowband multitone stimulus.

Beyond spectral regrowth, Figure 2.15 also includes cochannel distortion components that are exactly coincident with the fundamentals. For example, looking at the ω_5 fundamental, and taking only third-order products, these can be generated from mixing terms as $\omega_5 = \omega_5 + \omega_q - \omega_q$ for all q, and, in case the tones are equally spaced, even by mixing terms of the form $\omega_5 = \omega_{q_1} + \omega_{q_2} - \omega_{q_3}$, with $q_1 \neq q_2$ and $q_1 \neq q_3$. Note that if the input tones are all uncorrelated in phase (the most appropriate situation for emulating a true random telecommunications signal), then, despite all these cochannel products will contribute to distortion at ω_5 fundamental, only the first group is correlated with it. Therefore, their components will add in amplitude and phase, while all the other mixing products will only add in power. Nevertheless, either phase correlated or uncorrelated, these distortion components are always very difficult to be measured because they are coincident with the fundamentals, which, usually, have much higher amplitude.

Having already classified the distortion components arising from a multitone test, we will now describe its most used characterization standards. This naturally includes multitone intermodulation ratio, cochannel and adjacent-channel power ratios.

2.4.1 Multitone Intermodulation Ratio

Beginning with a generalization of the *IMR* concept introduced with two-tone tests, we now bring in the *multitone intermodulation ratio* (M-IMR). As shown in Figure 2.16, this figure of merit is defined, for multitone tests, as the ratio of the common fundamental power per tone, $P_{o/T}$, to the power of the ω_r distortion component present in the lower or upper adjacent bands, $P_{L/U}(\omega_r)$:

$$M - IMR(r) = \frac{P_{o/T}}{P_{L/U}(\omega_r)} \tag{2.20}$$

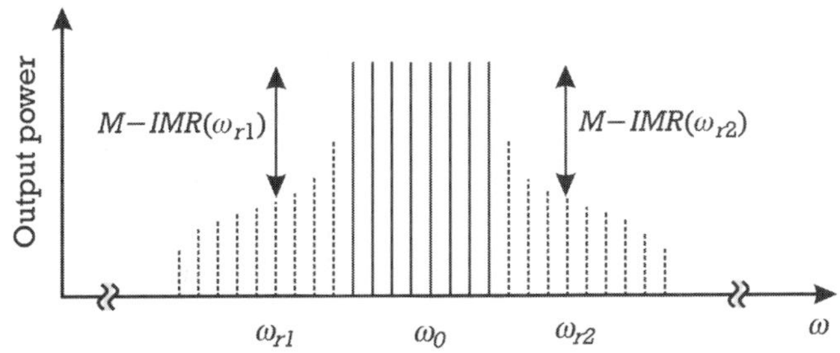

Figure 2.16 Illustration of multitone intermodulation ratio definition.

2.4.2 Adjacent-Channel Power Ratio

According to the explanation given above, adjacent-channel distortion is composed of all distortion components falling on the adjacent-channel location. It behaves, therefore, as interference to a possible adjacent-channel. Because of the youth of this subject, various proposed figures of merit are still accepted, and investigated, to characterize this form of distortion [11, 14].

One of these, *total adjacent-channel power ratio* ($ACPR_T$), is the ratio of total output power measured in the fundamental zone, P_o, to the total power integrated in the lower, P_{LA}, and upper, P_{UA}, adjacent-channel bands, shown in Figure 2.15 as "L" and "U." Thus, if $S_o(\omega)$ is taken as the power spectral density function of the inband system's output, total $ACPR$ is expressed as

$$ACPR_T \equiv \frac{P_o}{P_{LA} + P_{UA}} = \frac{\int_{\omega_{L_2}}^{\omega_{U_1}} S_o(\omega)\, d\omega}{\int_{\omega_{L_1}}^{\omega_{L_2}} S_o(\omega)\, d\omega + \int_{\omega_{U_1}}^{\omega_{U_2}} S_o(\omega)\, d\omega} \tag{2.21}$$

If the excitation were a multitone, then the output spectrum would be discrete, and the integrals of (2.21) would become summations of spectral regrowth line powers.

If only the lower or upper adjacent-channels are of concern, then we can use the *adjacent-channel power ratio* (*lower or upper*), defined as the ratio between total output power measured in the fundamental zone, P_o, and the lower or upper adjacent-channel power, $P_{AL/U}$:

$$ACPR_{L/U} \equiv \frac{P_o}{P_{AL/U}} = \begin{cases} \dfrac{\displaystyle\int_{\omega_{L2}}^{\omega_{U1}} S_o(\omega)\,d\omega}{\displaystyle\int_{\omega_{L1}}^{\omega_{L2}} S_o(\omega)\,d\omega} & \text{Lower} \\[2ex] \dfrac{\displaystyle\int_{\omega_{L2}}^{\omega_{U1}} S_o(\omega)\,d\omega}{\displaystyle\int_{\omega_{U1}}^{\omega_{U2}} S_o(\omega)\,d\omega} & \text{Upper} \end{cases} \tag{2.22}$$

Finally, an alternative definition, particularly used in the wireless equipment industry, is herein called *spot adjacent-channel power* (ACP_{SP}), to distinguish it from the previously referred $ACPR$. According to the illustration presented in Figure 2.17, it assumes the DUT is excited by a real information signal. So, ACP_{SP} is given by the ratio of total output power measured in the fundamental zone, P_o, to the power integrated in a band of predefined bandwidth and distance from the center frequency of operation $P_{SPL/U}$.

$$ACP_{SP} \equiv \frac{P_o}{P_{SPL/U}} = \begin{cases} \dfrac{\displaystyle\int_{\omega_{L2}}^{\omega_{U1}} S_o(\omega)\,d\omega}{\displaystyle\int_{\omega_{NBL1}}^{\omega_{NBL2}} S_o(\omega)\,d\omega} & \text{Lower} \\[2ex] \dfrac{\displaystyle\int_{\omega_{L2}}^{\omega_{U1}} S_o(\omega)\,d\omega}{\displaystyle\int_{\omega_{NBU1}}^{\omega_{NBU2}} S_o(\omega)\,d\omega} & \text{Upper} \end{cases} \tag{2.23}$$

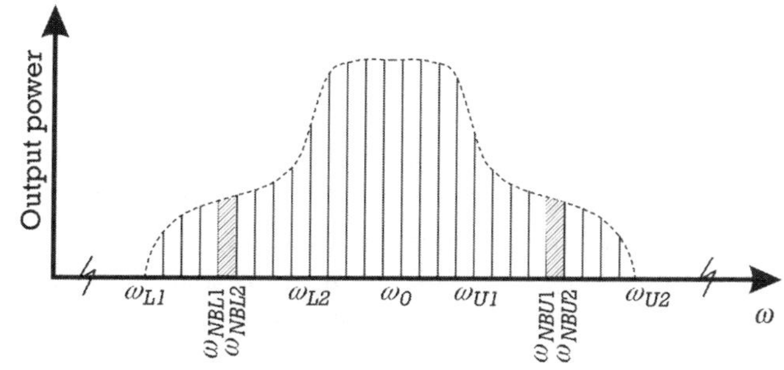

Figure 2.17 Illustration of spot adjacent channel power ratio definition.

2.4.3 Noise Power Ratio

Noise power ratio (NPR) was proposed as an indirect means of characterizing cochannel distortion. Because this form of distortion is intricately mixed with the fundamentals of much higher amplitude, the noise power ratio test eliminates the fundamental components from the zone where the test is to be made. For that, the DUT is no longer excited by a full bandwidth noise spectrum, but with one in which a slice was previously deleted. That is usually done by passing the excitation through a very narrow notch filter, before it is fed to the DUT. If the notch bandwidth is sufficiently narrow, it is believed that the required measurement window is created without significantly perturbing the test conditions. In this way, any frequency component, or power spectral density function, observed at the output within the notch position, constitutes spectral regrowth, and is thus the desired cochannel distortion.

A sample output obtained with a typical noise power ratio test is shown in Figure 2.18.

Noise power ratio is, therefore, defined as the ratio of the output power spectral density function measured in the vicinity of the test window position, ω_T, $S_o(\omega_T)$, to the power spectral density observed within that window, $S_{wd}(\omega_T)$:

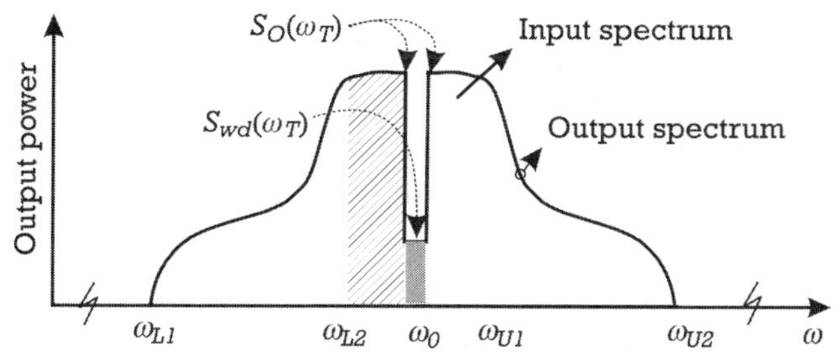

Figure 2.18 Illustration of a noise power ratio test, where the corresponding input and output power spectral densities are shown.

$$NPR(\omega_T) \equiv \frac{S_o(\omega_T)}{S_{wd}(\omega_T)} \tag{2.24}$$

As a remark on the NPR test, note that, despite having assumed a continuous spectrum, the noise power ratio concept can also be implemented with multitone signals. The only requirement to be imposed in this case is that there has to be a very large number of uncorrelated tones in order to guarantee that the elimination of one still does not affect significantly the power and statistical properties of the excitation.

2.4.4 Cochannel Power Ratio

In Chapter 1, we have already studied the various types of third-order cochannel mixing products appearing in a certain frequency position, ω_r. So, taking into account that in an NPR test the input tone at $\omega_q = \omega_r$ is deleted, these tests will preserve all products of the form $\omega_r = \omega_{q_1} + \omega_{q_1} - \omega_{q_2}$ (whose phase is $\phi_r = 2\phi_{q_1} - \phi_{q_2}$) or $\omega_r = \omega_{q_1} + \omega_{q_2} - \omega_{q_3}$, (whose phase is $\phi_r = \phi_{q_1} + \phi_{q_2} - \phi_{q_3}$ and where $\omega_{q_1}, \omega_{q_2}, \omega_{q_3} \neq \omega_r$), but will inevitably destroy any mixing product involving ω_r, $\omega_r = \omega_{q_1} + \omega_{q_2} - \omega_{q_2}$ [whose phase is simply $\phi_r = \phi_{q_1}$ (i.e., exactly the one of the input) and where $\omega_{q_1} = \omega_r$ and ω_{q_2} is any tone]. As a consequence, a first conclusion that may be anticipated is that NPR is blind to any distortion mechanism whose products are correlated in phase with the input signal. Moreover, since the number of the eliminated mixing products ($\omega_r = \omega_r + \omega_{q_2} - \omega_{q_2}$) is of the order of the total number of tones, and they all add in amplitude, not in power, we may also foresee that the amplitude of these eliminated components will be far from being negligible.

Quantifying the amount of perturbation imposed by the measurement window in the signal-correlated distortion, demands for calculating the response of a nonlinear system to a band-limited noise spectrum. In order to simplify the algebraic manipulations, we will restrict the excitation, $x(t)$, to be a band-limited white (constant power spectral density function) Gaussian noise and the system to be memoryless and of third degree.

Since $x(t)$ is a noise signal, it must be represented in the time and frequency-domains by the Fourier pair of its autocorrelation function $R_{xx}(\tau) = E\{x(t)x(t+\tau)\}$ and power spectral density function, $S_{xx}(\omega)$, respectively. It can be shown [11] that the response of our memoryless third-degree system, $y(t)$, can be represented by a correspondent autocorrelation function, $R_{yy}(\tau)$, such that

$$\begin{aligned} R_{yy}(\tau) &= a_2^2 R_{xx}(0)^2 + [a_1^2 + 6a_1a_3R_{xx}(0) + 9a_3^2R_{xx}(0)^2]R_{xx}(\tau) \\ &\quad + 2a_2^2R_{xx}(\tau)^2 + 6a_3^2R_{xx}(\tau)^3 \end{aligned} \tag{2.25}$$

whose power spectral density function is thus

$$S_{yy}(\omega) = a_2^2 R_{xx}(0)^2 \delta(\omega) + [a_1^2 + 6a_1 a_3 R_{xx}(0) + 9a_3^2 R_{xx}(0)^2] S_{xx}(\omega)$$
$$+ 2a_2^2 S_{xx}(\omega) * S_{xx}(\omega) + 6a_3^2 S_{xx}(\omega) * S_{xx}(\omega) * S_{xx}(\omega) \quad (2.26)$$

in which $\delta(\omega)$ is a Dirac delta function at $\omega = 0$, and "*" stands for spectral convolution.

Assuming the power spectral density function of the input, $S_{xx}(\omega)$, is centered at ω_0, has a constant amplitude of $N_0/2$, spanning from $-\omega_h = -\omega_0 - Bw/2$ to $-\omega_l = -\omega_0 + Bw/2$ and $\omega_l = \omega_0 - Bw/2$ to $\omega_h = \omega_0 + Bw/2$, the power spectral density of the fundamental linear components is thus

$$S_{yy1}(\omega) = a_1^2 S_{xx}(\omega) \quad (2.27)$$

which is, as expected, simply a scaled replica of the input. Components of third order can be identified as the ones depending on the third-degree coefficient a_3 as

$$S_{yy3}(\omega) = 9a_3^2 R_{xx}(0)^2 S_{xx}(\omega) + 6a_3^2 S_{xx}(\omega) * S_{xx}(\omega) * S_{xx}(\omega) \quad (2.28)$$

and include two different parts. Although the first one has the same shape of the input, it is by no means a linear component as it rises cubically, not linearly, with the input signal level. It actually represents third-order signal-correlated cochannel perturbation. (The remaining term of (2.26), $6a_1 a_3 R_{xx}(0) S_{xx}(\omega)$, is, indeed, a symptom of that correlation.) The second term describing third-order distortion comes from a three-fold convolution and is thus composed of two packs of three parabolic shaped spectral regrowth bands of $3Bw$ bandwidth. One of these packs is centered at $\pm 3\omega_0$ and represents system's third-harmonic distortion. The other one appears at $\pm\omega_0$ and constitutes the desired inband distortion. So, total signal-correlated and uncorrelated inband perturbation power spectral density is given by

$$S_{yy3}(\omega) = 18a_3^2 \left(\frac{N_0}{2}\right)^3 \left[\frac{\omega^2}{2} - (\omega_l - Bw)\omega + \frac{1}{2}(\omega_l - Bw)^2\right]$$
$$\text{for } \omega_l - Bw < \omega < \omega_l \quad (2.29a)$$

$$= \frac{9}{2} a_3^2 N_0^3 Bw^2 + 18a_3^2 \left(\frac{N_0}{2}\right)^3 \left[-\omega^2 + (\omega_l + \omega_h)\omega + \frac{1}{2} Bw^2 - \omega_l \omega_h\right]$$
$$\text{for } \omega_l < \omega < \omega_h \quad (2.29b)$$

$$= 18a_3^2 \left(\frac{N_0}{2}\right)^3 \left[\frac{\omega^2}{2} - (\omega_h + Bw)\omega + \frac{1}{2}(\omega_h + Bw)^2\right]$$
$$\text{for } \omega_h < \omega < \omega_h + Bw \quad (2.29c)$$

For clarity, Figure 2.19 shows a typical example of total output power spectral density, along with the inband signal-correlated and uncorrelated distortion components.

Because signal-correlated components' power spectral density function of (2.29b) follows exactly the shape of the input power spectral density, it is obvious that only the second term of (2.29b) contributes to observed distortion within the notch measurement window. So, an NPR test, being a measure of signal-to-distortion noise ratio, made at frequency ω_T leads to an $NPR(\omega_T)$ value of [11]

$$NPR(\omega_T) = \frac{a_1^2\left(\frac{N_0}{2}\right) + 6a_1a_3\frac{N_0^2}{2}Bw + 9a_3^2\frac{N_0^3}{2}Bw^2}{18a_3^2\left(\frac{N_0}{2}\right)^3\left[-\omega_T^2 + (\omega_l + \omega_h)\omega_T + \frac{1}{2}Bw^2 - \omega_l\omega_h\right]} + 1 \tag{2.30}$$

where, in accordance to the definition given in (2.24), NPR actually becomes a measure of the ratio of total fundamental signal plus distortion to distortion noise (the conceptual NPR + 1), and not of signal to distortion noise only (the conceptual NPR).

If, on the other hand, signal-correlated third-order components were also considered as a form of perturbation to the signal (in this case naturally assumed

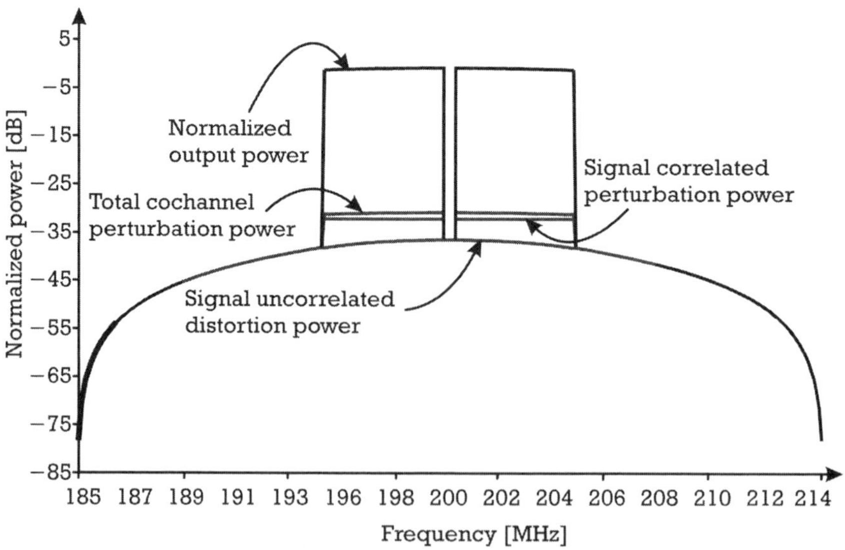

Figure 2.19 Power spectral density functions of the input, output fundamental, and inband signal-correlated and uncorrelated distortion components, resulting from a typical NPR test.

as being exclusively the linear output part, and not the whole signal-correlated components, as is the standard in telecommunication systems), the ratio of fundamental signal to this newly defined total cochannel perturbation power spectral densities would be

$$SCDR(\omega_T) = \frac{a_1^2 \left(\frac{N_0}{2}\right)}{\frac{9}{2} a_3^2 N_0^3 Bw^2 + 18 a_3^2 \left(\frac{N_0}{2}\right)^3 \left[-\omega_T^2 + 2\omega_0 \omega_T + \frac{3}{4} Bw^2 - \omega_0^2\right]} \tag{2.31}$$

Since this $SCDR(\omega_T)$ assumes as distortion also the signal-correlated components, it, in a certain sense, also evaluates signal level induced gain changes as AM-AM conversion.

Similarly to this PSD linear signal-to-distortion ratio, an alternative cochannel distortion figure of merit—*cochannel power ratio* ($CCPR$)—was specifically introduced to quantify the ratio of fundamental signal to total cochannel perturbation power.

This cochannel power ratio shares the same objectives of noise power ratio, in characterizing cochannel perturbation, except that it accounts for all signal-correlated and uncorrelated components. That is, it evaluates all deviations of the actual output from the ideal one, which would be obtained if the system were purely linear. Therefore, $CCPR$ not only measures signal-uncorrelated (or nonlinear distortion noise), like NPR, as it also evaluates other forms of signal-correlated perturbation, like gain compression or expansion and phase variation. And, although some of these may simply report a gain deviation from linearity, uniquely determined by the input average power level, as happens in memoryless systems, some others represent dynamic amplitude and phase deviations. These, arising in systems presenting memory to the baseband, are no longer only dependent on the signal average power, but also on the particular shape of the operating signal PSD function. Consequently, they represent a nonlinear contribution that varies in amplitude and phase at each cochannel frequency spot, being thus much more difficult to be accounted for, and so actually behaving as (in practice, unpredictable) nonlinear signal perturbation.

A $CCPR$ test setup is, indeed, very similar to NPR, except that now the fundamentals are not deleted from the input, but from the DUT's output. This guarantees the required generation of all forms of distortion in the nonlinearity, while still allowing unperturbed distortion observation.

Accordingly, $CCPR$ was conceived in accordance to the adjacent-channel and noise power ratios, and is defined as the ratio of integrated output power measured in the fundamental zone, P_o, to total integrated cochannel perturbation, P_{CC}:

$$CCPR \equiv \frac{P_O}{P_{CC}} = \frac{\int_{\omega_{L_2}}^{\omega_{U_1}} S_O(\omega)\, d\omega}{\int_{\omega_{L_2}}^{\omega_{U_1}} S_{yy3}(\omega)\, d\omega} \tag{2.32}$$

2.4.5 Multitone Characterization Setups

As depicted in Figure 2.20, multitone measurement setups share a common architecture with the other one-tone or two-tone tests, although their excitation source is more involved.

Depending on the objective of the characterization, this source may be a multitone generator, a noise-like continuous spectrum source, or even a specific real telecommunications signal.

If the excitation is a real or noise-like signal, its source can be either a special generator with the required modulation format capabilities, or simply part of a telecommunications network.

Except for special cases of very small number of tones, where they can be created independently and then added with a power combiner, RF multitone excitations are generally built by upconverting a baseband multitone to the desired spectral location. Unfortunately, because real signals have spectral representations whose negative frequency lines are simply the complex conjugate of positive ones, all RF multitones built this way have spectra that show complex conjugate symmetry with respect to the carrier. And this is a severe limitation, not only in selecting the power of each tone, but also in choosing the desired phase distribution. To overcome that restriction, a complex baseband must be used. For that, what will be its real and the symmetrical of the imaginary parts are digitally created in two independent arbitrary waveform generators, and then fed to an I/Q modulator. That is, the real part modulates the desired carrier, while the symmetrical of the imaginary part

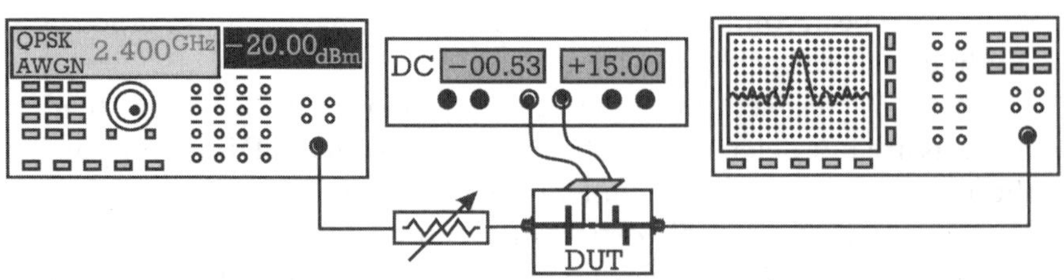

Figure 2.20 Laboratory measurement setup for the evaluation of various multitone distortion figures of merit.

modulates a quadrature replica of that same carrier. Then, these two modulated signals are added. Furthermore, any time a baseband is upconverted, a sufficient mixer rejection must be guaranteed or the leaking carrier may perturb observation of very weak distortion components. This is especially true in NPR tests whenever the carrier appears exactly within the notch position.

Since adjacent-channel distortion tests can directly use the setup of Figure 2.20, let us concentrate on the two different tests proposed for evaluating cochannel distortion performance.

An NPR test may be performed with a digitally produced multitone or noise signal. In the first situation, a multitone is made from an arbitrary waveform generator, at the desired RF center frequency. The notch is made by selective elimination of one of these tones. Alternatively, if the NPR test is to be performed with band-limited noise, then a narrowband notch filter must be provided between the RF noise generator and the DUT's input.

Measuring cochannel power ratio demands for a more complex setup. Indeed, its main difficulty comes from the necessity of eliminating the fundamentals from the output, without significantly perturbing cochannel distortion.

A block diagram of such a measurement setup is depicted in Figure 2.21. It is composed of a variable multitone or noise spectrum generator followed by a bridge setup similar to the one already proposed in Figure 2.12, and a spectrum analyzer for visualization.

The heart of this setup is the bridge network, which was inspired in the signal cancellation loop of a feed-forward linearizer system [15].

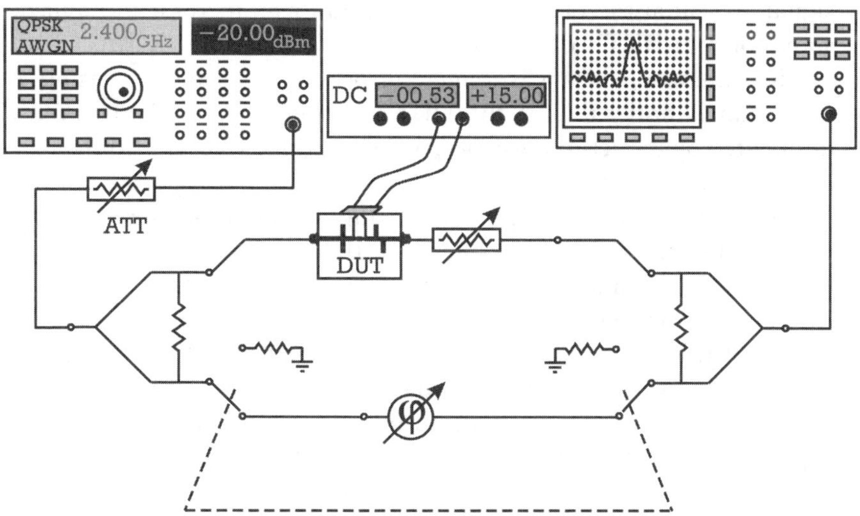

Figure 2.21 Illustrative cochannel power ratio measurement setup.

Its upper branch includes the device under test, while the lower is simply a phase adjustment network intended to replicate the DUT's output linear components. Alternatively, the auxiliary branch can be deactivated (by switching the power divider and combiner ports to the matched loads), for complete fundamental and distortion output readings.

The desired output signal subtraction is performed in the final power combiner, whose output is then observed with a spectrum analyzer.

Finally, the calibrated attenuator (ATT) is required to control the DUT's input excitation level.

Although variants of this setup were also proposed for cochannel distortion evaluation of lossy devices, or even frequency translating DUTs like mixers [16], the particular implementation depicted in Figure 2.21 was conceived for DUTs presenting net gain as usual microwave and wireless power amplifiers. That is why the DUT is followed by an attenuator.

The measurement process is based on the knowledge that the DUT's output can be modeled as the sum of a strong linear component, $a_1 x(t)$, plus higher order distortion products, $a_2 x(t)^2 + a_3 x(t)^3 + \ldots$. Accordingly, for small enough input drive levels, the linear fundamentals dominate output distortion, which, in turn, is determined by third-order components. Therefore, signal-to-distortion ratio will increase 2 dB for each decibel of driving level reduction. That is, provided the DUT's input power is sufficiently backed-off, the DUT will behave as if it were linear. Let us refer this reduced input power level as Pin_0.

Assuming all elements comprised in the signal cancellation loop, except the DUT, are linear, a bridge adjustment at this Pin_0 determines the elimination of any DUT's output linear component, regardless of drive level. Therefore, any error signal present at the bridge output, caused by an increase in input power, must be some form of DUT's distortion.

In summary, the measurement process begins by increasing ATT, until no distortion is noticeable at the DUT's output (i.e., until observed sidebands are dominated by the setup noise floor). At this drive level (Pin_0) the bridge is adjusted for a full signal cancellation.

After this calibration procedure, input amplitude is resumed to its nominal value, reducing ATT. The error signal then displayed in the spectrum analyzer is the desired cochannel and adjacent-channel distortion.

Although we have concentrated on the cochannel distortion observation capabilities of this setup, it is by no means restricted to those distortion components. It also allows direct adjacent-channel distortion measurements for $ACPR_T$, $ACPR_{L/U}$, and ACP_{SP} evaluation, or even NPR tests, if the appropriate input is supplied. However, if the signal cancellation loop were not adjusted for very small-signal, but for nominal operating power level, a major part of signal-correlated perturbation (in fact total signal-correlated perturbation in bandpass memoryless systems) would also be canceled turning this CCPR arrangement an alternative measurement setup of NPR that does not need a notch in the signal spectrum [17].

A final remark on these multitone or continuous spectra distortion tests should address the problem of measuring power with a spectrum analyzer. Measuring total power within a user-defined bandwidth requires the integration of the power spectral density function, which, in turn, can be obtained dividing the spectrum analyzer power spot readings by the utilized resolution bandwidth, RBw. Alternatively, total power can be simply measured dividing the sought bandwidth in the correspondent number of bandwidth segments of RBw width, and then adding the spectrum analyzer power readings one by one. This calculation can be automatically performed by spectrum analyzers that have special firmware functions that implement the described algorithm. Beyond that, special attention must be paid to the correct definition of fundamental and distortion band-limits, as indicated in Figure 2.22. Indeed, since band-limited continuous spectra always present a roll-off of finite slope, fundamental signal and adjacent-channel edge definition is of primary importance to prevent much higher fundamentals from being misread as distortion components. Because ACP_{SP} definition assumes predetermined distortion measurement band-limits, which are offset from the fundamentals, it does not suffer from this problem.

2.4.6 Relation Between Multitone and Two-Tone Test Results

Before closing the analysis of multitone tests, it is interesting to relate their figures of merit with the ones previously derived for the two-tone stimulus. In fact, since two-tone tests still represent the most widely used nonlinear distortion characterization method, there are many situations where a certain device comes specified with the standard IP_3, and we need to estimate its impact in a real multitone or continuous spectrum application. Or, conversely, we may want to specify such a device in terms of its IP_3, or *IMR,* from the knowledge of its admissible distortion under the actual multitone or continuous spectrum environment.

Unfortunately, this task is mathematically very involved, being intractable for all but a very few special cases. From these, the one providing most useful results,

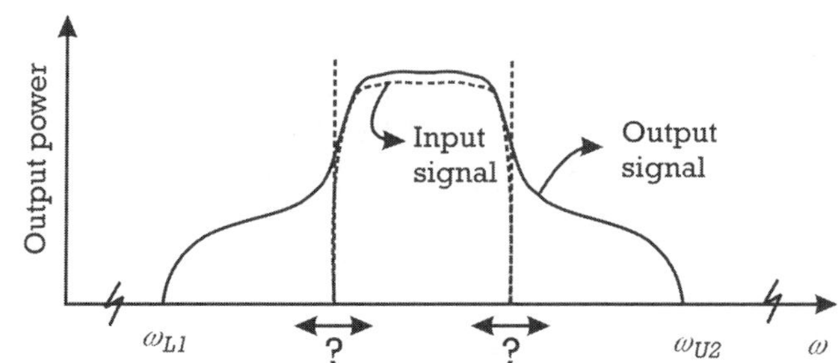

Figure 2.22 Cochannel and adjacent-channel border definition for accurate distortion measurements under excitations of finite slope roll-off.

from a practical point of view, assumes the system is a third-order memoryless nonlinearity, excited by Q evenly spaced but uncorrelated tones of equal amplitude.

Third-order mixing products of these tones produce spectral regrowth components whose new frequency positions are given by $\omega_r = \omega_{q_1} + \omega_{q_2} - \omega_{q_3}$, in which $q_1 \neq q_2 \neq q_3$ (products from now on named as Type A), $q_1 = q_2 \neq q_3$ (Type B); and also generate products falling on the same positions of the input, in which $q_1 \neq q_2 = q_3$ (Type C) and $q_1 = q_2 = q_3$ (Type D).

Now, to calculate the magnitude of adjacent or cochannel distortion components shown in Figure 2.23, we need to calculate the number of different mixing products appearing in each frequency position.

For that, we begin by first determining the number of different ways the set of three input frequencies, $\omega_r = \omega_{q_1} + \omega_{q_2} - \omega_{q_3}$, can be grouped to produce a certain mixing component. That number is the multinomial coefficient of the mixing product [given by (1.23)], and values 6 for Type A products, 3 for type B, 6 for Type C, and finally 3 for type D.

The second step consists of calculating the number of possible combinations of input tones that produce mixing products at the same frequency position, ω_r. That derivation was described in [11] and used straightforward, although quite laborious, combinatory calculus. Its results for the number of mixing products located at the adjacent-channel, $Q + 1 \leq r \leq 2Q - 1$, in Figure 2.23(a) were

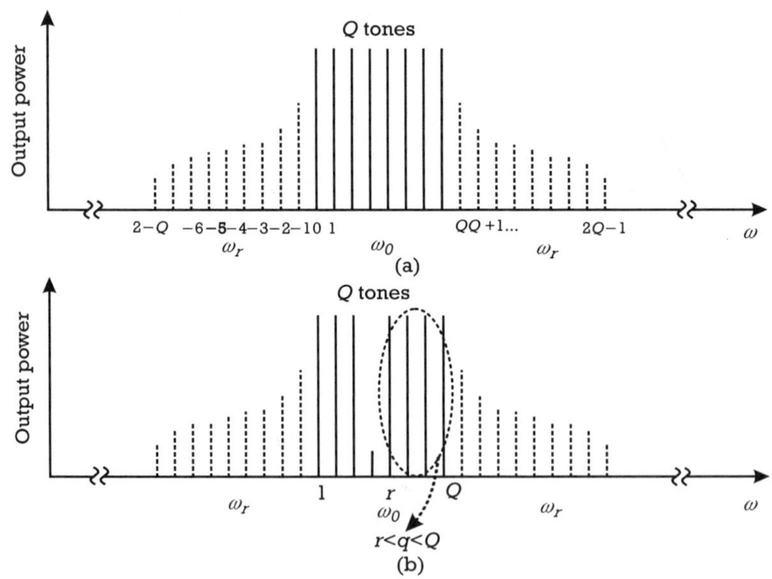

Figure 2.23 Identification of the output spectrum's frequency positions corresponding to (a) adjacent-channel distortion and (b) cochannel distortion.

$$\text{Type A: } q_1 \neq q_2 \neq q_3\text{: } N_A(Q, r) = 6\left[\left(\frac{2K - r}{2}\right)^2 - \frac{\epsilon}{4}\right] \tag{2.33}$$

$$\text{Type B: } q_1 = q_2 \neq q_3\text{: } N_B(Q, r) = 3\left[\left(\frac{2Q - r}{2}\right) + \frac{\epsilon}{2}\right] \tag{2.34}$$

where $\epsilon = \text{mod}[r/2]$, and $\text{mod}(m/n)$ is the remainder of m/n.

Following the same reasoning, the number of mixing products located at cochannel positions, $\text{round}[(Q + 1)/2] \leq r \leq Q$, in Figure 2.23(b) was found to be

Type A_1: $q_1 \neq q_2 \neq q_3$ and $1 \leq q_1, q_2, q_3 < r$:

$$N_{A_1}(Q, r) = 6\left[\left(\frac{r - 2}{2}\right)^2 - \frac{\epsilon_1}{4}\right] \tag{2.35}$$

Type A_2: $q_1 \neq q_2 \neq q_3$ and $r < q_1, q_2, q_3 \leq Q$:

$$N_{A_2}(Q, r) = 6\left[\left(\frac{Q - r - 1}{2}\right)^2 - \frac{\epsilon_2}{4}\right] \tag{2.36}$$

Type A_3: $q_1 \neq q_2 \neq q_3$ and $1 \leq q_1 < r$, $r < q_2 \leq Q$, $q_2 \leq q_3 < q_1$:

$$N_{A_3}(Q, r) = 6(Q - r)(r - 1) \tag{2.37}$$

Then, adding these three partial contributions, gives

$$N_A(Q, r) = 6\left[\left(\frac{r - 2}{2}\right)^2 - \frac{\epsilon_1}{2} + \left(\frac{Q - r - 1}{2}\right)^2 - \frac{\epsilon_2}{4} + (Q - r)(r - 1)\right] \tag{2.38}$$

where $\epsilon_1 = \text{mod}[(2Q - r)/2]$, and $\epsilon_2 = \text{mod}[(Q - r + 1)/2]$.

$$\text{Type } B_1\text{: } q_1 \neq q_2 \text{ and } 1 \leq q_1, q_2 < r\text{: } N_{B_1}(Q, r) = 3\left[\left(\frac{r - 2}{2}\right) + \frac{\epsilon_1}{2}\right] \tag{2.39}$$

$$\text{Type } B_2\text{: } q_1 \neq q_2 \text{ and } r < q_1, q_2 < Q\text{: } N_{B_2}(Q, r) = 3\left[\left(\frac{Q - r - 1}{2}\right) + \frac{\epsilon_2}{2}\right] \tag{2.40}$$

Then,

$$N_B(Q,r) = 3\left[\left(\frac{r-2}{2}\right) + \frac{\epsilon_1}{2} + \left(\frac{Q-r-1}{2}\right) + \frac{\epsilon_2}{2}\right] \tag{2.41}$$

where $\epsilon_1 = \text{mod}[(2Q-r)/2]$, and $\epsilon_2 = \text{mod}[(Q-r+1)/2]$.

$$\text{Type C: } q_1 \neq q_2 \text{ and } q_1 = r,\ q_2 \leq Q\text{: } N_C(Q) = 6(Q-1) \tag{2.42}$$

and

$$\text{Type D: } q_1 = q_2 = r\text{: } N_D = 3 \tag{2.43}$$

Remembering that products of Type A or B are uncorrelated in phase, and so they must add in power, while the ones of Type C or D are correlated in phase, therefore adding linearly, we are now in condition to derive formulas for approximate small-signal level *M-IMR*, $ACPR_{L/U}$, *NPR*, and *CCPR* as a function of the number of tones Q, and the two-tone *IMR*. Those expressions are presented in Tables 2.2 to 2.5, where *IMR* stands for the signal-to-intermodulation distortion

Table 2.2 Relations Between Small-Signal Q-Tone and Band-Limited White Gaussian Noise *M-IMR* and Two-Tone *IMR*

Q-tone *M-IMR*	$M-IMR(Q,r) = \frac{3}{4}\frac{Q^2}{2N_A(Q,r)+N_B(Q,r)} IMR$
Noise *M-IMR* ($Q \to \infty$)	$M-IMR_{\text{noise}}(\omega_T) = \frac{1}{8}\frac{B_w^2}{\frac{\omega_T^2}{2} - (B_w+\omega_b)\omega_T + \frac{(B_w+\omega_b)^2}{2}} IMR$

Table 2.3 Relations Between Small-Signal Q-Tone and Band-Limited White Gaussian Noise $ACPR_{L/U}$ and Two-Tone *IMR*

Q-tone $ACPR_{L/U}$	$ACPR_{L/U}(Q) = \frac{3Q^3}{4Q^3 - 3Q^2 - 4Q - 3\,\text{mod}(Q/2)} IMR$
Noise $ACPR_{L/U}$ ($Q \to \infty$)	$ACPR_{L/U\text{-Noise}} = \frac{3}{4} IMR$

Table 2.4 Relations Between Small-Signal Q-Tone and Band-Limited White Gaussian Noise *NPR* and Two-Tone *IMR*

Q-tone *NPR*	$NPR(Q,r) = \frac{Q^2}{4Q^2 - 8r^2 + 8Qr - 38Q + 24r + 14 - 2(\epsilon_1+\epsilon_2)} IMR$
Noise *NPR* ($Q \to \infty$)	$NPR_{\text{noise}}(\omega_T) = \frac{B_w^2}{-8\omega_T^2 + 8(\omega_l+\omega_h)\omega_T + 4B_w^2 - 8\omega_h\omega_l} IMR$

Table 2.5 Relations Between Small-Signal *Q*-Tone and Band-Limited White Gaussian Noise *CCPR* and Two-Tone *IMR*

Q-tone $CCPR$	$CCPR(Q) = \dfrac{3Q^3}{64Q^3 - 102Q^2 + 56Q + 6\,\mathrm{mod}\left(\frac{Q}{2}\right)} IMR$
Noise $CCPR$ $(Q \rightarrow \infty)$	$CCPR = \dfrac{3}{64} IMR$

ratio that would be measured in the same device, when subject to a two-tone excitation having the same average input power as the considered uncorrelated Q-tones. Since IMR and IP_3 were already related by (2.15) and (2.16), for a given output power, expressing multitone results in terms of IP_3 is now straightforward.

Since the relations presented in Table 2.2 to Table 2.5 were exclusively derived under small-signal regime (imposed by the definition of IP_3), analytical simplicity justified neglecting third order perturbation components in numerators, in comparison to much stronger linear ones.

For completeness, Tables 2.2 to 2.5 also include results derived from (2.29) when the excitation is a band-limited white Gaussian noise, spanning from $\omega_l = \omega_0 - Bw/2$ to $\omega_h = \omega_0 + Bw/2$, and keeping the same input power level. These can also be interpreted as the limit results that would be obtained for the multitone case if the number of input spectral lines were increased indefinitely, but total average power and bandwidth were kept constant.

Figure 2.24 summarizes these results by showing plots of the various two-tone IMR to $M\text{-}IMR[Q, (Q+1)]$, $ACPR_{L/U}(Q)$, $NPR\{Q, \mathrm{round}[(Q+1)/2]\}$ and $CCPR(Q)$ ratios, as a function of the number of input tones Q.

As shown in that figure, the limit of an infinite number of tones (or white Gaussian noise) is almost reached for Q greater than about 10. This serves as an indication of the statistical properties of an uncorrelated multitone signal. Furthermore, this figure also shows that the referred limits are 6 dB for the $IMR/M\text{-}IMR(\omega_h)$ ratio, 1.25 dB for $IMR/ACPR_{L/U}$, 7.78 dB for $IMR/NPR(\omega_0)$, and, 13.29 dB for the $IMR/CCPR$ ratio.

2.5 Illustration Examples of Nonlinear Distortion Characterization

In order to illustrate the distortion characterization techniques detailed in previous sections, we will now present a sample of such results obtained in the laboratory. For that, a typical multistage low-power microwave amplifier was used as the device under test.

2.5.1 One-Tone Characterization Results

Starting with one-tone tests, the setup used to evaluate AM-AM and AM-PM was a manually controlled version of the one presented in Figure 2.4, while THD was

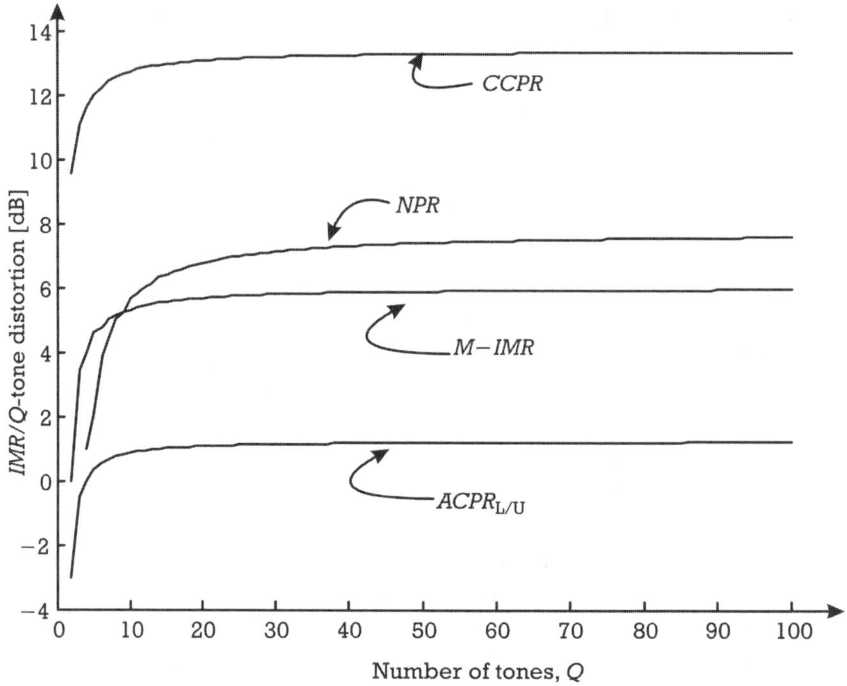

Figure 2.24 Ratios of small-signal two-tone *IMR* to *M-IMR*[Q, (Q + 1)], $ACPR_{L/U}(Q)$, $NPR\{Q$, round[(Q + 1)/2]$\}$ and $CCPR(Q)$ as a function of the number of input tones Q.

computed from the spectrum analyzer readings obtained from a setup similar to the one of Figure 2.6.

Figure 2.25 and Figure 2.26 represent measured AM-AM and AM-PM conversion results, while Figure 2.27 is an illustration of the observed output spectrum. A 1-dB compression point of P_{1dB} = 13 dBm and a total harmonic distortion of *THD* = –19.7 dBc, at this P_{1dB} level, were deduced from Figure 2.25 and Figure 2.27, respectively.

2.5.2 Two-Tone Characterization Results

Two-tone tests were then performed on our low-power amplifier circuit, using the setup illustrated in Figure 2.10. Excitation frequencies were set to f_1 = 1,901 MHz and f_2 = 1,899 GHz, producing the output fundamental power per tone and single sideband *IMD* power results shown in Figure 2.28(a). A direct reading of this plot immediately provides *IMR* as a function of input drive level [shown in Figure 2.28(b)], while the extrapolation of output fundamental and *IMD* power from the small-signal regime, leads to a third-order intercept point of nearly IP_3 = 21.2 dBm.

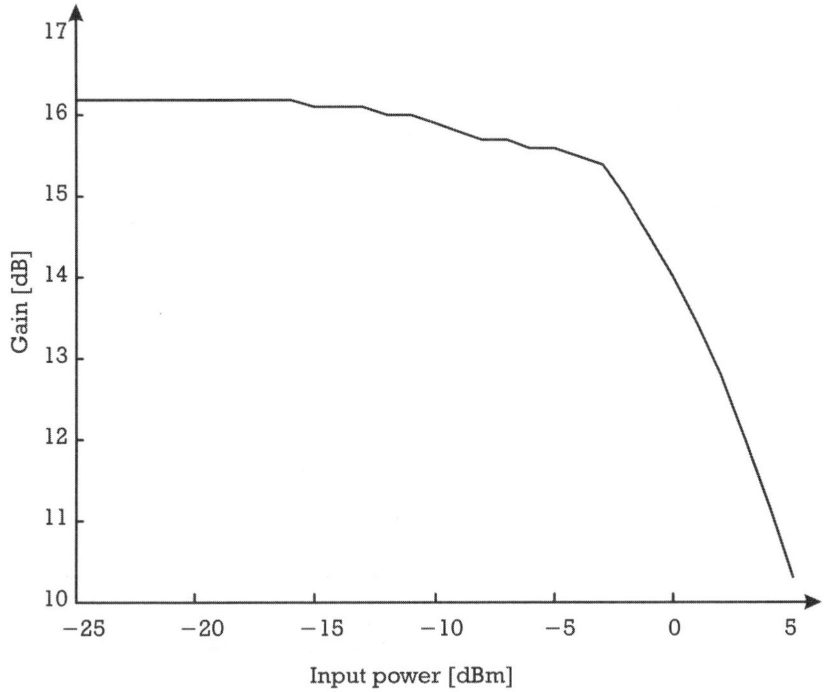

Figure 2.25 Plot of measured power gain versus input drive level, for illustrating AM-AM characterization.

As discussed in previous sections, optimized accuracy in small-signal IMD power readings required that the spectrum analyzer configuration was set to an input attenuation of 20 dB, a frequency span of 10 KHz, and a resolution bandwidth of 300 Hz.

Finally, Figure 2.29 shows one of the observed output power spectra, obtained when the DUT was already under a large-signal regime.

2.5.3 Noise Characterization Results

After these two-tone tests, our amplifier prototype was subject to a continuous spectrum noise signal. The input excitation was created by a *white Gaussian noise* (WGN) generator centered at 1,900 MHz and with a 1-MHz bandwidth.

The next figures show the spectra observed from a CCPR measurement performed with the setup of Figure 2.21. Figure 2.30 shows the inband portion of the output spectrum.[2]

2. Following the CCPR measurement procedure explained in Section 2.4.5, the PSD values presented in Figures 2.30 to 2.32 are the values collected from the output of the CCPR measurement setup loop. That

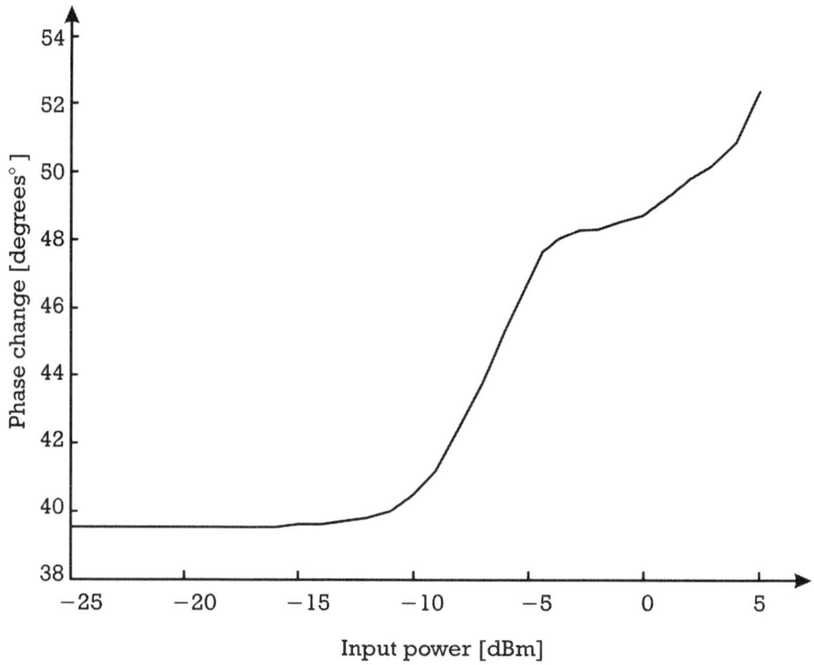

Figure 2.26 Plot of measured excess phase-shift versus input drive level, for illustrating AM-PM characterization.

Figure 2.30 also illustrates the problem of defining the border between the cochannel and adjacent-channel, for accurate $ACPR_T$ measurements. One practical way to circumvent that difficulty consists of saving the input spectrum trace, and then using it as a mask for channel border identification. Doing that, we got an $ACPR_T$ value of 42 dB.

Figure 2.31 depicts the bridge output after loop calibration (DUT's linear operation under very small-signal level) along with the input excitation. Cochannel distortion became evident, when DUT's input power was reset to its nominal value. That spectrum is depicted in Figure 2.32, from which a measured *CCPR* value of about 30.5 dB was obtained.

explains why the input signal has the same magnitude value of the output signal. If the real magnitude values of the output signal were desired, a 26.2-dB gain should be added in order to account for the upper arm attenuator. Accordingly, the input loop PSD should be scaled by 14-dB gain due to the lower arm attenuator.

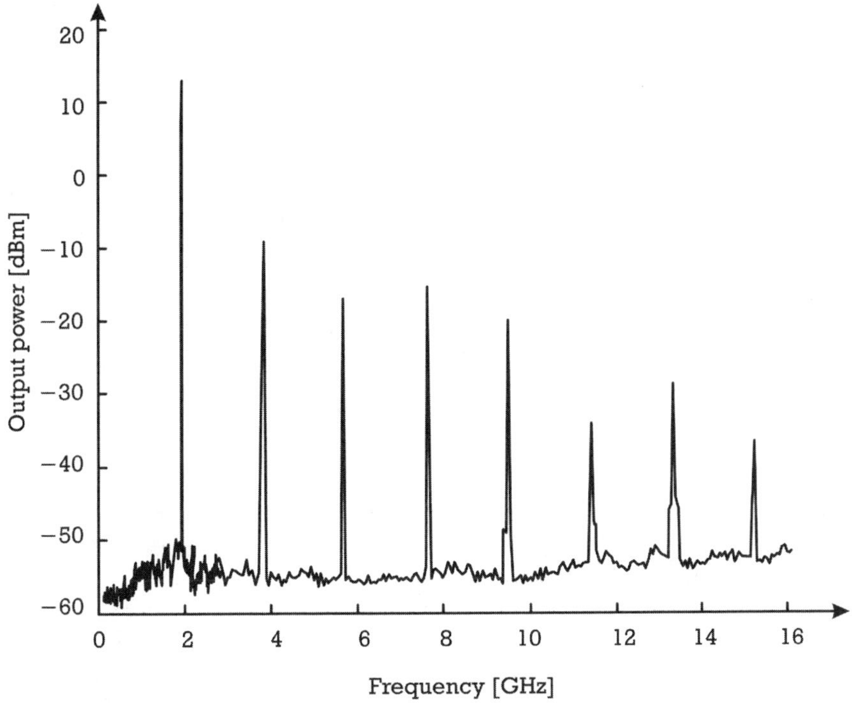

Figure 2.27 Particular view of the DUT's output power spectrum when subject to a one-tone excitation of P_{1dB} (13 dBm) output power level. *THD* deduced from this measurement result was about *THD* = –19.7 dBc.

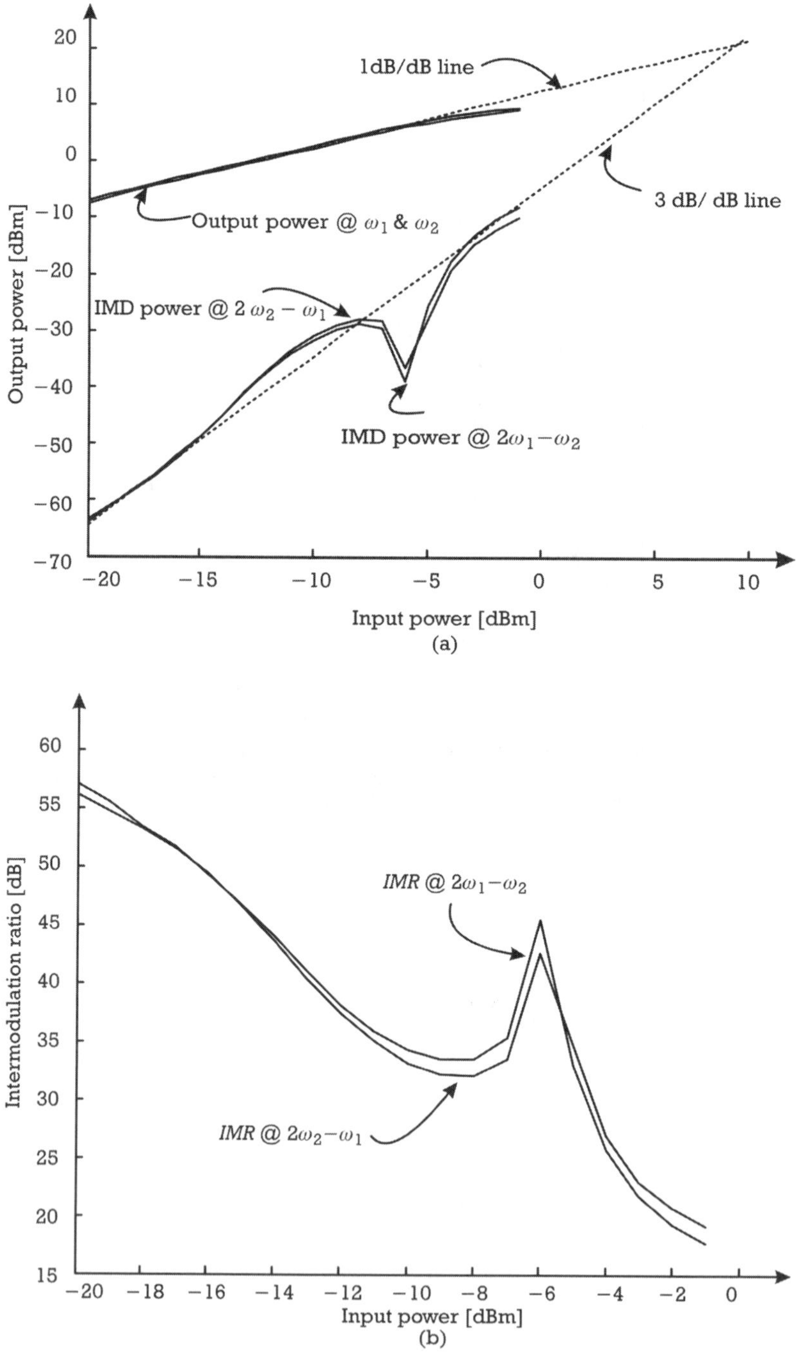

Figure 2.28 (a) Fundamental and IMD output power per tone, and (b) inferred two-tone IMR, as a function of DUT's input drive level per tone. Extrapolating small-signal behavior leads to an $IP_3 = 21.2$ dBm.

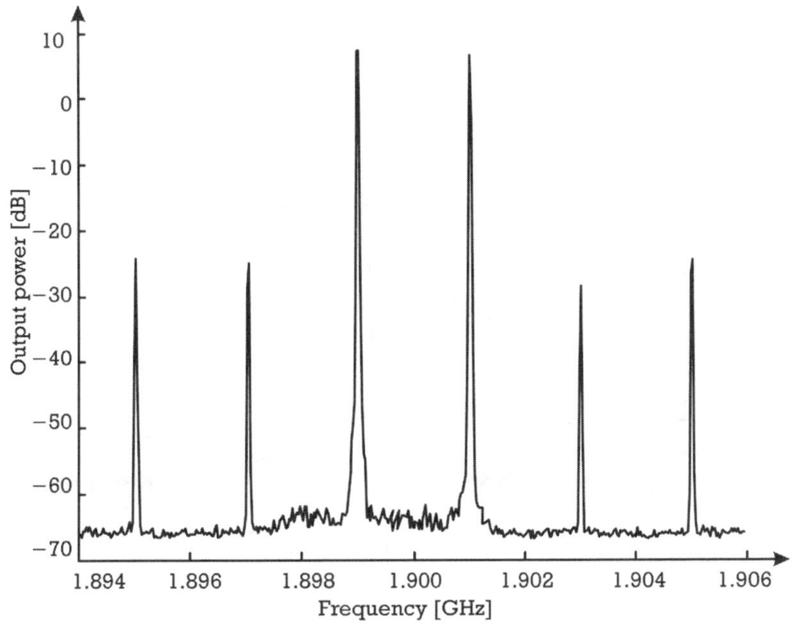

Figure 2.29 Typical output power spectrum observed when the DUT was driven close to saturation.

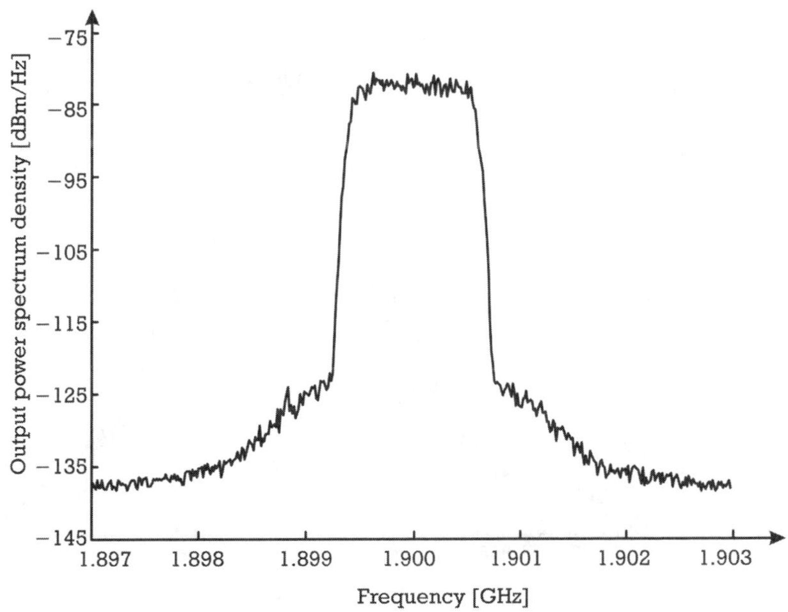

Figure 2.30 Inband output spectrum of our sample amplifier under a WGN spectrum excitation, measured when the lower arm of the CCPR setup is disconnected.

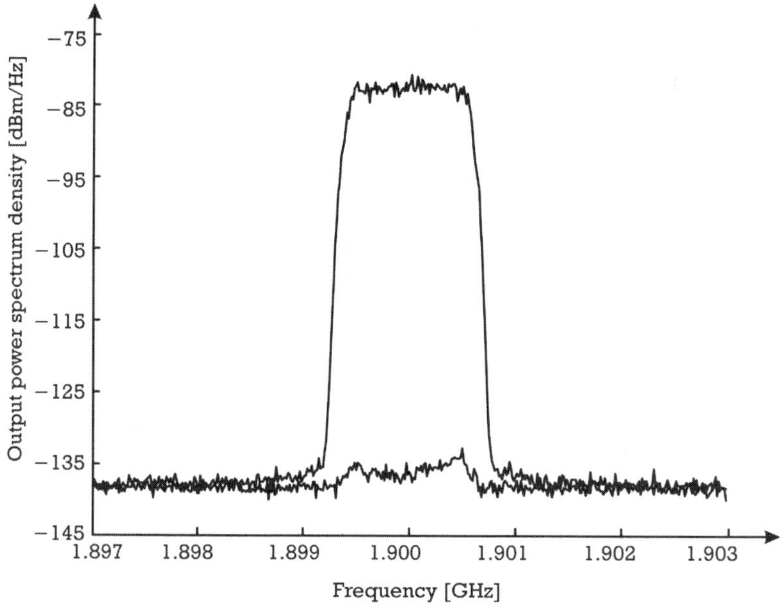

Figure 2.31 Continuous noise spectrum of DUT's input excitation, measured when the upper arm of the CCPR setup is disconnected, and bridge output after setup calibration.

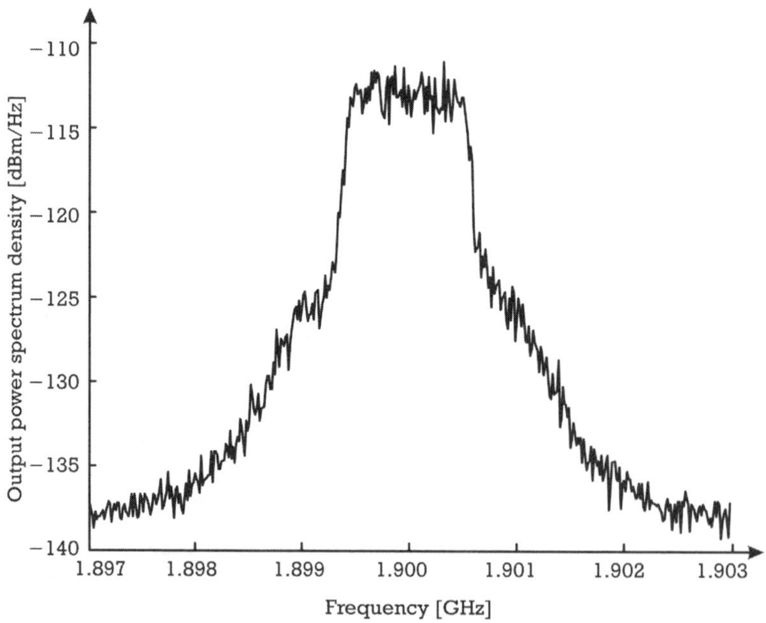

Figure 2.32 Continuous noise spectrum of DUT's bridge output, when stimulus level was reset to its nominal value.

References

[1] "How to Characterize CATV Amplifiers Effectively," in *Hewlett Packard Application Note,* AN 1288-4, Hewlett Packard, 1997.

[2] Leke, A., and J. S. Kenney, "Behavioral Modeling of Narrowband Microwave Power Amplifiers with Applications in Simulating Spectral Regrowth," *Proc. 1996 IEEE International Microwave Theory and Tech. Symposium Digest,* 1996, San Francisco, CA, pp. 1385–1388.

[3] Saleh, A., "Frequency-Independent and Frequency-Dependent Nonlinear Models of TWT Amplifiers," *IEEE Transactions on Communications,* Vol. 29, No. 11, 1981, pp. 1715–1720.

[4] Gannouchi, F. M., H. Wakana, and M. Tanaka, "A New Unequal Three-Tone Signal Method for AM-AM and AM-PM Distortion Measurements Suitable for Characterization of Satellite Communication Transmitters/Transponders," *IEEE Transactions on Microwave Theory and Techniques,* Vol. 48, No. 8, 2000, pp. 1404–1407.

[5] "Network Analyzer Measurements: Filter and Amplifier Examples," in *Application Note Agilent,* AN 1287-4, Agilent Technologies, 1997.

[6] Schubert, Jr., T., and E. Kim, *Active and Non-linear Electronics,* New York: John Wiley & Sons, Inc., 1996.

[7] Carvalho, N. B., and J. C. Pedro, "A Comprehensive Explanation of Distortion Sideband Asymmetries," *IEEE Transactions on Microwave Theory and Techniques,* Vol. 50, No. 9, 2002, pp. 2090–2101.

[8] "Optimizing RF and Microwave Spectrum Analyzer Dynamic Range," in *Application Note Agilent,* AN 1315, Agilent Technologies, 1999.

[9] Krikorian, N., "Bridge Method for Measuring Amplitude Intermodulation Distortion," *RF Design,* March, 1995, pp. 30–34.

[10] Kenington, P., *High-Linearity RF Amplifier Design,* Norwood, MA: Artech House, 2000.

[11] Pedro, J. C., and N. B. Carvalho, "On the Use of Multi-Tone Techniques for Assessing RF Components' Intermodulation Distortion," *IEEE Transactions on Microwave Theory and Techniques,* Vol. 47, No. 12, 1999, pp. 2393–2402.

[12] Pintelon, R., and J. Schoukens, *System Identification—A Frequency Domain Approach,* New York: IEEE Press, 2001.

[13] Carvalho, N. B., and J. C. Pedro, "Multi-Tone Frequency Domain Simulation of Nonlinear Circuits in Large and Small Signal Regimes," *IEEE Transactions on Microwave Theory and Techniques,* Vol. 46, No. 12, 1998, pp. 2016–2024.

[14] Staudinger, J., "Specifying Power Amplifier Linearity via Intermodulation Distortion and Channel Spectral Regrowth," *Applied Microwave & Wireless,* Vol. 9, No. 4, 1997, pp. 62–78.

[15] Seidel, H., "A Feedforward Experiment Applied to an L-4 Carrier System Amplifier," *IEEE Transactions on Communications,* Vol. 19, No. 3, 1971, pp. 320–325.

[16] Pedro, J. C., and N. B. Carvalho, "Analysis and Measurement of Multi-Tone Intermodulation Distortion of Microwave Frequency Converters," *Proc. 2001 IEEE International Microwave Theory and Tech. Symposium Digest,* Phoenix, AZ, May 2001, pp. 1671–1674.

[17] Ku, H., W. Woo, and J. S. Kenney, "Carrier-to-Interference Ratio Prediction of Nonlinear RF Devices," *Microwave Journal,* Vol. 44, No. 2, 2001, pp. 154–164.

CHAPTER 3

Nonlinear Analysis Techniques for Distortion Prediction

3.1 Introduction

The main goal of this chapter is to provide the reader with the minimum background necessary for understanding, and using, nonlinear circuit analysis techniques in distortion prediction.

Unfortunately, nonlinear circuit analysis is far beyond the curricula of most electronics and telecommunication engineering degrees, which makes the subjects treated in this chapter quite new for the major part of our readers. Because of that, we have decided to begin the text with an introductory review of system classification, and to give an explanation of the ideas subjacent to each of the presented methods. Anyway, if the reader is, at this time, not interested in the operating details of the various nonlinear analysis techniques, but only on the simulation possibilities they offer, the authors suggest a bypass of Sections 3.2 to 3.4. Section 3.5 gives a brief overview of the most used methods.

The presentation of the various nonlinear circuit analysis methods was organized by dividing them into three groups.

The first one, the Volterra series method, is an analytical procedure capable of describing the response of a certain class of nonlinear systems by closed form expressions. Since it can represent the system by a set of nonlinear operators independent of the excitation, it can be applicable, not only to circuit or system analysis, but also to design. That is the reason why it plays the most important role among all nonlinear circuit analysis methods in the field of distortion studies and therefore deserves special attention in the following sections.

Unfortunately, the major drawback of Volterra series resides on its inability to handle large-signal distortion problems. Consequently, we have introduced a second frequency-domain nonlinear analysis tool: the harmonic balance technique. At present, it is the most spread simulation method in the RF and microwave community. This is due to the fact that, unlike the more traditional time-step integration methods, harmonic balance handles the circuit, its excitation, and its output in the frequency-domain, the format normally adopted by RF circuit designers. Because of that, it also benefits from allowing the direct inclusion of distributed devices (like dispersive transmission lines and their discontinuities), circuit elements that do not have an exact time-domain representation. For didactic purposes, the presentation of the harmonic balance technique will be restricted to its Harmonic-Newton algorithm (the one also encountered in modern nonlinear RF circuit simulators) because it can be interpreted as an extension of the former Volterra series techniques to large-signal regimes. Harmonic balance is nowadays a mature computer simulation tool with a large number of different implementations. So, space constrains of this text determined that only the features of major impact on the distortion prediction would be treated. That was also the main driving force for having sometimes traded algorithm efficiency for simplicity and clarity of presentation. The reader who wants to get a view of this technique in greater detail is invited to consult some of the many published works on this subject [1–3].

The third group of nonlinear analysis methods addressed is the one based on time-step integration. Although they are not as appropriate as the Volterra series or harmonic balance techniques for multitone distortion prediction, they have been included in this book as they are the analysis methods more spread in the electronics circuits design community. As those methods simply try to numerically solve the nonlinear differential equations that arise by applying Kirchoff laws (and the constitutive element relations) to the circuit, they are the most "natural" way of circuit simulation. And so, they have also been the ones first introduced and for which there is a more intuitive interpretation.

3.1.1 System Classification

The Concept of Signal

A *signal* is an entity capable of carrying on some kind of information. Because it is assumed as evolving with time in a continuous manner, in mathematical terms, it is represented, or *modeled,* as a continuous real function of time, $x_I(t)$ [i.e., a function describing the dependence of the real variable x_I on the real variable $t \to x_I = f(t)$: $x_I, t \in R$]. In electrical circuits, this function models the variation of time of some measurable electrical quantity as a voltage, $v(t)$, current, $i(t)$, power, etc. Common examples of signals vary from the simple sinusoids to more involved modulated signals.

For example, the pure cosine

$$x_I(t) = A \cos \omega_c t \tag{3.1}$$

cannot actually carry any information, as its periodicity immediately tells us that, by knowing only one period $T = 2\pi/\omega_c$, we automatically know the function at all time. On the other hand, a modulated sinusoid can carry information either as a varying amplitude, a varying phase, or both. That is the case of the new $x_I(t)$

$$x_I(t) = A(t)\cos[\omega_c t + \theta(t)] \tag{3.2}$$

in which the radio-frequency tone, or carrier, of (3.1) has been amplitude and phase modulated by an *information envelope* of $A(t)$ and $\theta(t)$, respectively.

Systems as Signal Operators

An *analog system,* or simply a *system,* is the entity that processes those continuous signals. It is also a relationship description, but now between two signals, $x_I(t) \to y_O(t)$. The signal, $x_I(t)$, to be processed is called the *input, excitation,* or *stimulus,* while the resulting one, $y_O(t)$, is called the *output,* or *response.* Because a system no longer relates two real variables, but two real functions, its mathematical *model* is no longer a function but an operator or rule—$y_O(t) = S[x_I(t)]$.

Memoryless and Dynamic Systems

A system for which the output reacts instantaneously to its input [i.e., where the response at any time instant, $y_O(t_1)$, is only dependent on the input at that time instant, $x_I(t_1)$] is called a *zero memory system* or *memoryless system.* On the contrary, a system in which the output $y_O(t_1)$ is dependent on the input $x_I(t_1)$ but also to the past of that input, $x_I(t < t_1)$, is said to have memory, and so it is named a *system with memory* or *dynamic system.* This ability to keep memory of the past is usually modeled by a set of system's internal variables called the *system state.* So, while this system state concept has no meaning in memoryless systems, it is very useful in dynamic systems since it represents the integrated system's past. In this way, we can also say that, while the response of a memoryless system only depends on the input at that time instant, $y_O(t) = f[x_I(t)]$, the response of a dynamic system depends on the input $x(t)$, but also on the system state, $s(t)$, at that time, $y_O(t) = f[x_I(t), s(t)]$.

For example, a linear (but also a nonlinear) conductance is a memoryless system whose output current depends instantaneously on the input voltage, $i(t) = Gv(t)$. A similar, but dynamic system, is the linear capacitance, where the output current, $i(t)$, depends not only on the applied voltage, $v(t)$, but also on the past voltages or the accumulated charge (the capacitance state):

$$i(t) = \lim_{\Delta t \to 0} \frac{q(t) - q(t - \Delta t)}{\Delta t} = C \lim_{\Delta t \to 0} \frac{v(t) - v(t - \Delta t)}{\Delta t} = C\frac{dv(t)}{dt} \tag{3.3}$$

Actually, it is exactly the memory formulation of (3.3) that leads us to the conclusion that, while memoryless systems can be represented by input/output algebraic rules, dynamic systems have to be represented by ordinary differential equations of time.

In electronic circuits, memory is associated with electric charge storage, magnetic flux storage, and delay effects. Therefore, any circuit having capacitors, inductors, time delays, or distributed elements will exhibit memory.

Time-Varying and Time-Invariant Systems

Dynamic systems should not be confused with time-varying systems. A system is said to be *time-invariant* when its input/output relationship is constant no matter the time instant where the system is observed. On the contrary, a system whose input/output rule varies with time is a *time-varying system.*

Note that this by no means signifies that the response of a time-invariant system cannot vary with time. It only states that the system's operator is independent of time. Also note that any time-invariant system that has two or more inputs, but whose output is modeled as being dependent on only one of these stimuli, may appear as a time-varying system. This is the property that is normally used to model a time-invariant double-input/single-output multiplier, $y_O(t) = S_{TI}[x_I(t), z_I(t)] = Kx_I(t)\, z_I(t)$, as a time-varying single-input/single-output system: $y_O(t) = S_{TV}[x_I(t), t] = k(t)x_I(t)$, where $k(t) = Kz(t)$.

In mathematical terms, a time-invariant system is one in which the response to $x_I(t+\tau)$ is $y_O(t+\tau) = S[x_I(t+\tau)]$ (when $y_O(t) = S[x_I(t)]$), since the operator does not vary with time. A time-varying system is any system that does not obey this property.

To illustrate this important concept with a practical example, consider a simple instrumentation chopper amplifier (or square-wave amplitude modulator) in which the information signal is $x_I(t)$, the chopping signal is a zero mean square-wave of fundamental frequency ω_0 and amplitude 1:

$$c(t) = \text{Sign}[\cos(\omega_0 t)] \tag{3.4}$$

and the output is the product of these two:

$$y_O(t) = c(t)x_I(t) \quad \text{or} \quad y_O(t) = S[x_I(t), c(t)] \tag{3.5}$$

This chopper amplifier is a dual-input single-output memoryless time-invariant nonlinear system whose analysis is quite difficult. However, if the chopping signal, $c(t)$, is not correlated with the information signal, $x_I(t)$, the output can be rewritten as if it were dependent on two distinct time variables, t and τ, such that

$$y_O(t, \tau) = c(\tau)x_I(t) \quad \text{or} \quad y_O(t, \tau) = S[x_I(t), \tau] \tag{3.6}$$

Now, the circuit is being treated as a single-input single-output linear memoryless time-varying system, in which the excitation is $x_I(t)$ and the response is $y_O(t)$.

The output is simply given by the input multiplied by a varying (in time) gain, $c(\tau)$, of +1 or −1. As we shall see later, this new interpretation is advantageous because of the analysis simplicity offered by regained superposition. So, as a common procedure, all nonlinear systems where there is one large amplitude input (for which system's nonlinearity is unavoidable) and another one, uncorrelated with the former, whose amplitude is so small that linearity would apply if it were the sole excitation, are usually treated as linear time-varying systems. Examples of these are parametric amplifiers, RF mixers, modulators, samplers, and switched-capacitor filters.

Linear and Nonlinear Systems

A general single-input, $x_I(t)$, single-output, $y_O(t)$, system, $y_O(t) = S[x_I(t)]$, is said to be linear if it complies with superposition:

If

$$y_{O_1}(t) \equiv S[x_{I_1}(t)] \quad \text{and} \quad y_{O_2}(t) \equiv S[x_{I_2}(t)] \tag{3.7}$$

then

$$y_O(t) \equiv S[k_1 x_{I_1}(t) + k_2 x_{I_2}(t)] = k_1 y_{O_1}(t) + k_2 y_{O_2}(t) \tag{3.8}$$

In the opposite case [i.e., if $y_O(t) \neq k_1 y_{O_1}(t) + k_2 y_{O_2}(t)$], the system is nonlinear.

As linear systems respond to a sinusoid with a sinusoid of equal frequency, and obey superposition, they respond to a sum of sinusoids with the same frequency components content. Only their relative amplitude and phase can be varied. Therefore, nonlinear systems are the sole systems that perform qualitative signal spectrum transformations (add or eliminate certain spectral components), contributing with nonlinear distortion. They are, consequently, the object of this text.

For example, any system in which the output can be expressed by the following algebraic relation of the input,

$$y_O(t) \equiv S[x_I(t)] = k x_I(t) \tag{3.9}$$

or any linear differential equation, like

$$b_n \frac{d^n y_O(t)}{dt^n} + \ldots + b_1 \frac{d y_O(t)}{dt} + b_0 y_O(t) = c_m \frac{d^m x_I(t)}{dt^m} + \ldots + c_1 \frac{d x_I(t)}{dt} + c_0 x_I(t) \tag{3.10}$$

[in which the multiplying constants can, eventually, vary with time, but not with $x_I(t)$ or $y_O(t)$] is linear. Any other is nonlinear. In electronic circuits, nonlinearity is usually expected from electron devices, while all other electrical elements, like resistors, capacitors, inductors, and transmission lines are normally approximately modeled by linear relations. Nevertheless, there are some special situations where even the nonlinearity of these elements has to be considered.

The fundamental properties of nonlinearity have been already described in Chapter 1. In the present chapter we will discuss some analysis techniques amenable to determine the responses of nonlinear electronic circuits usually encountered in microwave and wireless systems.

3.1.2 Nonlinear Circuit Example

In order to use a common case study for the various nonlinear analysis techniques throughout this chapter, we propose the single node circuit of Figure 3.1.

The circuit is composed of a linear conductance G connected across a port of nonlinear capacitance and current. These nonlinearities are assumed as quasistatic and are thus described by algebraic constitutive relations of voltage-dependent current and charge.

For the current, we considered a nonlinear voltage-dependent current source, representing a saturating velocity-field resistor, as the ones usually encountered in doped semiconductors:

$$i_{NL}[v_O(t)] = I_0 \tanh[\alpha v_O(t)] \tag{3.11}$$

The form of $i_{NL}[v_O(t)]$ and its first-order derivative (the corresponding nonlinear conductance) are plotted in Figure 3.2.

The nonlinear capacitance behaves as a junction diffusion capacitance, in which the storage charge is proportional to the conductive current $i_{NL}[v_O(t)]$,

$$q_{NL}[v_O(t)] = \tau_F \, i_{NL}[v_O(t)] \tag{3.12}$$

and thus,

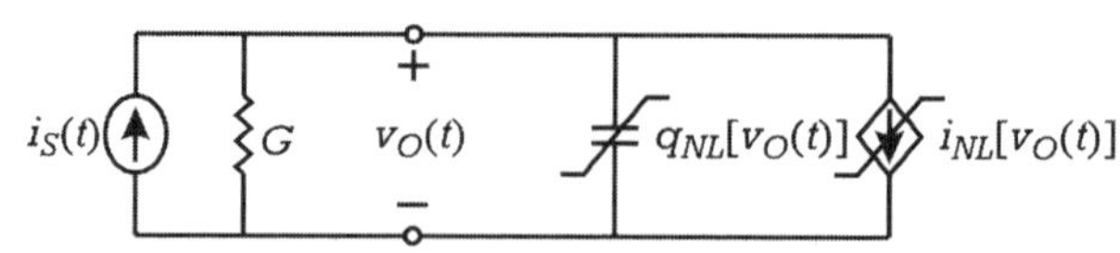

Figure 3.1 Schematic diagram of the circuit used in nonlinear analysis.

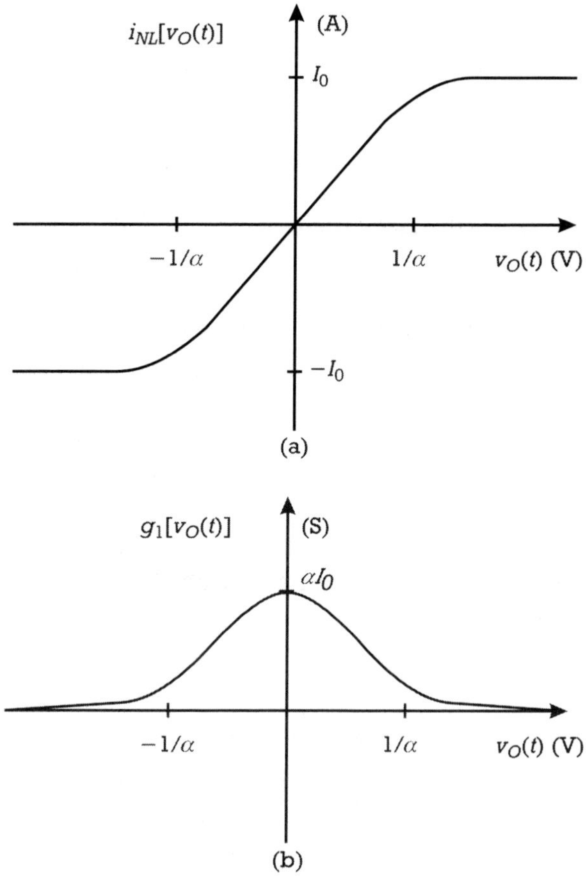

Figure 3.2 (a) Nonlinear current and (b) its first-order derivative used in our example circuit.

$$C[v_O(t)] \equiv \frac{dq_{NL}[v_O(t)]}{dv_O(t)} = \tau_F \frac{di_{NL}[v_O(t)]}{dv_O(t)} = \tau_F I_0 \alpha \operatorname{sech}^2[\alpha v_O(t)] \tag{3.13}$$

The nodal analysis of the circuit of Figure 3.1 leads to the following nonlinear *ordinary differential equation* (ODE):

$$G v_O(t) + \frac{dq_{NL}[v_O(t)]}{dt} + i_{NL}[v_O(t)] = i_S(t) \tag{3.14}$$

typical of a forced dynamic system of input $i_S(t)$ and output $v_O(t)$. Rewriting (3.14) gives

$$Gv_O(t) + \frac{d[\tau_F I_0 \tanh(\alpha v_O)]}{dv_O}\frac{dv_O(t)}{dt} + I_0 \tanh[\alpha v_O(t)] = i_S(t) \quad (3.15)$$

which is also a convenient circuit model alternative to be used through this chapter.

3.2 Frequency-Domain Techniques for Small-Signal Distortion Analysis

3.2.1 Volterra Series Model of Weakly Nonlinear Systems

In the following sections we will discuss two powerful analysis techniques that are of paramount importance to intermodulation problems: *power series analysis* and *Volterra series analysis*. Their merit comes from the fact that they can be conceived as a direct extension of the widely known linear techniques. They have a precise mathematical foundation, provide closed-form solutions for nonlinear system responses, and can be directly used in frequency-domain. These properties are consequences of the restriction imposed on the system's nonlinearities: they must be approximated by a power series (a polynomial) if they are memoryless, or by a generalization of this (a Volterra series) if they are dynamic.

3.2.1.1 Memoryless Systems' Representation

Like any other model, or scientific theory, power series is a representation of nature, in the sense that it cannot be said to be true or false, but only that it may be simply useful or not. A power series is a useful model for, at least, two different orders of reasons.

In the first place, a power series is a simple mathematical representation that has the benefit of allowing the direct response computation of a nonlinear device, circuit or system, in the frequency-domain. That is, contrary to all other natural "time-domain" models, we need not convert our frequency-domain excitations to their time-domain representation (usually by an appropriate inverse Fourier transform), calculate the model response, and then go back to the frequency-domain. With a power series, we simply have to make multiple convolutions of signals' spectra. In fact, since a power series is nothing more than the addition of several time-domain product terms, one can directly compute them by the spectral addition of the correspondent frequency-domain convolutions.

For the system defined by $y_O(t) \equiv S[x_I(t)]$, if in a limited range of $x_I(t)$ amplitude, $y_O(t)$ can be approximated by

$$y_O(t) \approx a_1 x_I(t) + a_2 x_I(t)^2 + a_3 x_I(t)^3 + \ldots \quad (3.16)$$

then, in the frequency-domain,

$$Y_O(\omega) \approx a_1 X_i(\omega) + a_2 X_i(\omega) * X_i(\omega) + a_3 X_i(\omega) * X_i(\omega) * X_i(\omega) + \ldots \tag{3.17}$$

Because most analog, RF and microwave circuit designers generally deal with signals represented in the frequency-domain as a sum of a small number of discrete tones—and repeated convolutions of those signals are very easily calculated—this property of a power series model becomes a very attractive advantage.

In the second place, there is a rigorous mathematical foundation that provides certain power series models with two other important advantages. If we restrict our power series to be a Taylor series expansion around a predetermined quiescent point (usually the dc bias point), we immediately gain a *systematic parameter extraction procedure* and *model consistency.*

The former refers to the fact that each of the model coefficients can be easily extracted from the nth-order device's derivatives:

$$a_1 \equiv \left.\frac{\mathrm{d}S(x_I)}{\mathrm{d}x_I}\right|_{x_I=X_I}; \; a_2 \equiv \frac{1}{2}\left.\frac{\mathrm{d}^2S(x_I)}{\mathrm{d}x_I^2}\right|_{x_I=X_I}; \ldots; \; a_n \equiv \frac{1}{n!}\left.\frac{\mathrm{d}^nS(x_I)}{\mathrm{d}x_I^n}\right|_{x_I=X_I} \tag{3.18}$$

where X_I is the referred quiescent point.

The other Taylor series intrinsic property we mentioned is consistency. This means that, even though our power series model was derived to predict moderate signal level nonlinear effects, it inherently represents the device's small-signal behavior. In electronic device terms, this corresponds to saying the model is able to accurately predict the circuits' weakly nonlinear behavior, while it nicely converges to the small-signal [Y], [Z], [S], etc., parameters, if input excitation level is decreased. This is a consequence of the fact that the Taylor series representation of $S[.]$ around the bias point (X_I, Y_O), with an input signal $x_i(t) \equiv x_I(t) - X_I$ (defined as the dynamic deviation of the control variable $x_I(t)$ from its quiescent value X_I), is

$$y_o(t) \equiv y_O(t) - Y_O = \left.\frac{\mathrm{d}S(x_I)}{\mathrm{d}x_I}\right|_{x_I=X_I}[x_I(t) - X_I] + \ldots \tag{3.19}$$

$$+ \frac{1}{n!}\left.\frac{\mathrm{d}^nS[x_I]}{\mathrm{d}x_I^n}\right|_{x_I=X_I}[x_I(t) - X_I]^n + \ldots$$

or, in our power series model form,

$$y_o(t) = a_1 x_i(t) + a_2 x_i(t)^2 + a_3 x_i(t)^3 + \ldots + a_n x_i(t)^n + \ldots \tag{3.20}$$

Thus, if $x_i(t)$ is very small [$x_I(t)$ tends to X_I], the higher nth-order terms rapidly become negligible compared to $a_1 x_i(t)$, and the model automatically behaves as a linear one. In this sense, the Taylor series is what one could ever think as the simplest nonlinear extension of a linear memoryless, or algebraic, model.

By the way, this explanation also gives an insight onto the Taylor series model validity. It gets useless (or, in other words, hopelessly inaccurate) whenever the device excitation is so hard that other higher order terms we have not initially considered become important.

This rather small validity domain, which restricts power series analysis to small-signal nonlinear distortion studies (or weak nonlinearities), is one of its two major disadvantages. The other is the absence of memory.

Although it is not possible to represent a general dynamic system by a power series model, this kind of representation can still be used in cases where the system can be described by several noninteracting subsystems, and where the nonlinearities are memoryless. An illustrative example is depicted in Figure 3.3.

Indeed, if the input and output subsystems are both linear, defined by $y_i(t) = S_i[x_i(t)]$, $y_o(t) = S_o[x_o(t)]$, and characterized by frequency-domain transfer functions $H_i(\omega)$, $H_o(\omega)$, while the inner one is a memoryless nonlinear system represented by a power series like (3.20), the output to any frequency-domain excitation can be easily computed using the simple relations of (3.17). For example, if $x_i(t)$ is given by a sum of Q complex exponentials,

$$x_i(t) = \sum_{q=1}^{Q} X_{iq} e^{j\omega_q t} \tag{3.21}$$

the linear output can be calculated by

$$y_{o1}(t) = \sum_{q=1}^{Q} H_o(\omega_q) a_1 H_i(\omega_q) X_{iq} e^{j\omega_q t} \tag{3.22}$$

the second-order components' response by

$$y_{o2}(t) = \sum_{q_1=1}^{Q} \sum_{q_2=1}^{Q} H_o(\omega_{q_1} + \omega_{q_2}) a_2 H_i(\omega_{q_1}) H_i(\omega_{q_2}) X_{iq_1} X_{iq_2} e^{j(\omega_{q_1} + \omega_{q_2})t} \tag{3.23}$$

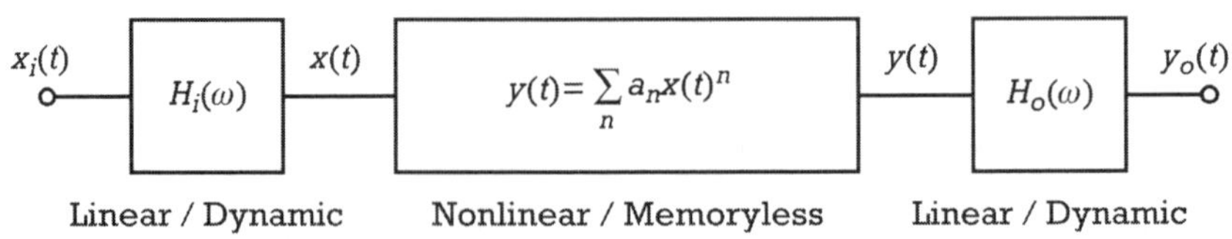

Figure 3.3 Power series model system's representation.

the third-order ones by

$$y_{o3}(t) = \sum_{q_1=1}^{Q} \sum_{q_2=1}^{Q} \sum_{q_3=1}^{Q} H_o(\omega_{q_1} + \omega_{q_2} + \omega_{q_3}) \cdot a_3 H_i(\omega_{q_1}) H_i(\omega_{q_2}) H_i(\omega_{q_3}) X_{iq_1} X_{iq_2} X_{iq_3} e^{j(\omega_{q_1} + \omega_{q_2} + \omega_{q_3})t} \quad (3.24)$$

and so on.

Unfortunately, if the blocks interact with each other, or the system cannot be described by the simple cascade connection of Figure 3.3, as is the situation presented in Figure 3.4, then the straightforward calculation just performed is no longer possible, and the analysis demands for the true nonlinear dynamic representation of Volterra series.

3.2.1.2 Dynamic Systems' Representation

The main difference between power series and Volterra series models is the ability of the latter to represent true nonlinear dynamic systems. A Volterra series is, in fact, nothing more than a Taylor series with memory. Hence, it can also be interpreted as the extension of linear, or first-order, dynamic systems. And so, to introduce its foundations, we will begin by recalling the derivation of the convolutive response of a time-invariant linear dynamic system. The explanation follows the one presented by [4].

Let us begin by considering again our general single-input single-output system $S[.]$ whose signal response $y_o(t)$ to an input signal $x_i(t)$ can be expressed as $y_o(t) \equiv S[x_i(t)]$. As is seen in Figure 3.5, $x_i(t)$ may be approximated by an appropriate

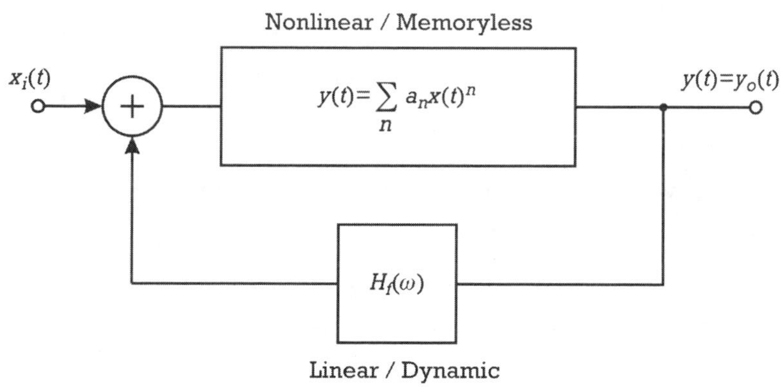

Figure 3.4 Example of a nonlinear dynamic system for which the power series model is no longer valid.

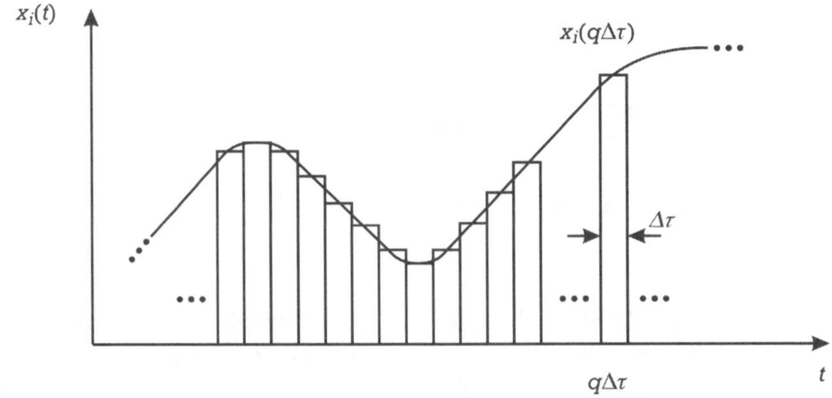

Figure 3.5 Ladder function approximation of the system's excitation.

ladder function in the domain $-T < t < T$ composed by a sum of $2Q + 1$ rectangular pulses, $p(t)$, of $\Delta\tau$ duration and $1/\Delta\tau$ amplitude:

$$x_i(t) \approx \sum_{q=-Q}^{Q} x_i(q\Delta\tau)p(t - q\Delta\tau)\,\Delta\tau \tag{3.25}$$

Assuming $S(t, q\Delta\tau)$ is the response of $S[.]$ to the rectangular pulse located at $q\Delta\tau$,

$$S(t, q\Delta\tau) \equiv S[p(t - q\Delta\tau)] \tag{3.26}$$

$y_o(t)$ may be also approximated by

$$y_o(t) \approx S\left[\sum_{q=-Q}^{Q} x_i(q\Delta\tau)p(t - q\Delta\tau)\,\Delta\tau\right] \tag{3.27}$$

If $S[.]$ were a linear dynamic system, superposition would apply, and thus,

$$y_o(t) \approx \sum_{q=-Q}^{Q} S[x_i(q\Delta\tau)p(t - q\Delta\tau)\,\Delta\tau] = \sum_{q=-Q}^{Q} x_i(q\Delta\tau)S(t, q\Delta\tau)\,\Delta\tau \tag{3.28}$$

In a time-invariant system $S(t, q\Delta\tau) = S(t - q\Delta\tau)$, and $y_o(t)$ can be approximated by

$$y_o(t) \approx \sum_{q=-Q}^{Q} x_i(q\Delta\tau)S(t - q\Delta\tau)\,\Delta\tau \tag{3.29}$$

as is depicted in Figure 3.6.

The approximations assumed for both $x_i(t)$, (3.25), and $y_o(t)$, (3.29), improve their accuracy when the pulses' duration, $\Delta\tau$, is reduced. In the limit where $\Delta\tau$ tends to zero, (3.25) tends to an infinite sum of Dirac delta functions, $\delta(t - \tau)$, and $x_i(t)$ can be represented in its whole domain $]-\infty, +\infty[$ by

$$x_i(t) = \int_{-\infty}^{\infty} x_i(\tau)\,\delta(t - \tau)\,d\tau \tag{3.30}$$

If $h(t - \tau)$ is now defined as the response of $S[.]$ to the Dirac impulse located at τ, (3.29) turns into the convolution integral

$$y_o(t) = \int_{-\infty}^{\infty} x_i(\tau)h(t - \tau)\,d\tau = \int_{-\infty}^{\infty} h(\tau)x_i(t - \tau)\,d\tau \tag{3.31}$$

Equation (3.31) is the usual time-domain response representation of a general linear dynamic time-invariant system.

Any direct attempt of extending that theory to nonlinear systems would fail, because superposition assumed in (3.28) no longer applies.

A convenient way of circumventing that difficulty consists of expanding a wide class of nonlinearities into Taylor series around some quiescent point (null

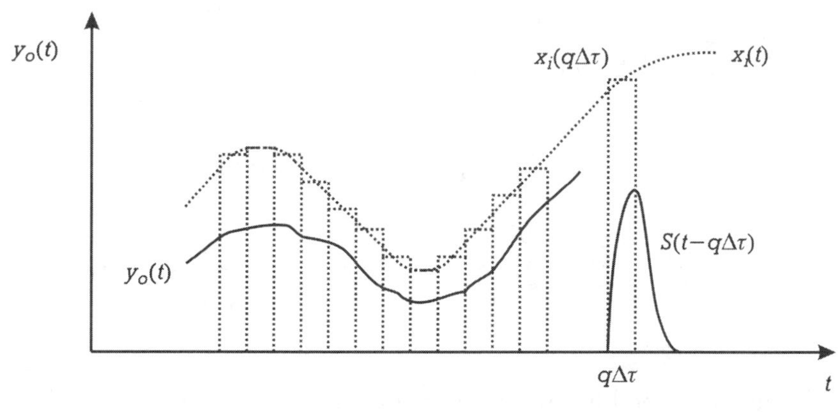

Figure 3.6 Linear system's output approximation as the response of the ladder function.

excitation). In that way, the response of $S[.]$ to the generic rectangular pulse centered at $q_1\Delta\tau$ may be expressed as

$$\begin{aligned} S[x_i(q_1\Delta\tau)p(t-q_1\Delta\tau)\,\Delta\tau] = S_0 &+ S_1(t-q_1\Delta\tau)x_{i1}\Delta\tau \\ &+ S_2(t-q_1\Delta\tau)x_{i1}^2\Delta\tau^2 \\ &+ S_3(t-q_1\Delta\tau)x_{i1}^3\Delta\tau^3 + \ldots \end{aligned} \tag{3.32}$$

The response to a sum of two rectangular pulses $x_i(q_1\Delta\tau)p(t-q_1\Delta\tau)\Delta\tau + x_i(q_2\Delta\tau)p(t-q_2\Delta\tau)\Delta\tau$, would then be

$$\begin{aligned} &S[x_i(q_1\Delta\tau)p(t-q_1\Delta\tau)\Delta\tau + x_i(q_2\Delta\tau)p(t-q_2\Delta\tau)\Delta\tau] \\ &= S_0 + S_1(t-q_1\Delta\tau)x_{i1}\Delta\tau + S_1(t-q_2\Delta\tau)x_{i2}\Delta\tau \\ &\quad + S_2(t-q_1\Delta\tau, t-q_1\Delta\tau)x_{i1}^2\Delta\tau^2 \\ &\quad + 2S_2(t-q_1\Delta\tau, t-q_2\Delta\tau)x_{i1}x_{i2}\Delta\tau^2 \\ &\quad + S_2(t-q_2\Delta\tau, t-q_2\Delta\tau)x_{i2}^2\Delta\tau^2 \\ &\quad + S_3(t-q_1\Delta\tau, t-q_1\Delta\tau, t-q_1\Delta\tau)x_{i1}^3\Delta\tau^3 + \ldots \\ &\quad + S_3(t-q_2\Delta\tau, t-q_2\Delta\tau, t-q_2\Delta\tau)x_{i2}^3\Delta\tau^3 + \ldots \end{aligned} \tag{3.33}$$

In general, when the input is a sum of $2Q+1$ pulses, $y_o(t)$ may be given by

$$\begin{aligned} y_o(t) = &\sum_{q_1=-Q}^{Q} S_1(t-q_1\Delta\tau)x_{i1}\Delta\tau \\ &+ \sum_{q_1=-Q}^{Q}\sum_{q_2=-Q}^{Q} S_2(t-q_1\Delta\tau, t-q_2\Delta\tau)x_{i1}x_{i2}\Delta\tau^2 \\ &+ \sum_{q_1=-Q}^{Q}\sum_{q_2=-Q}^{Q}\sum_{q_3=-Q}^{Q} S_3(t-q_1\Delta\tau, t-q_2\Delta\tau, t-q_3\Delta\tau)x_{i1}x_{i2}x_{i3}\Delta\tau^3 + \ldots \end{aligned} \tag{3.34}$$

Again, in the limit where $\Delta\tau$ tends to zero, and the rectangular pulses tend to Dirac impulses, we have

$$y_o(t) = \int_{-\infty}^{\infty} h_1(t - \tau_1)x_i(\tau_1)\, d\tau_1$$

$$+ \int_{-\infty}^{\infty}\int_{-\infty}^{\infty} h_2(t - \tau_1, t - \tau_2)x_i(\tau_1)x_i(\tau_2)\, d\tau_1\, d\tau_2 \tag{3.35}$$

$$+ \int_{-\infty}^{\infty}\int_{-\infty}^{\infty}\int_{-\infty}^{\infty} h_3(t - \tau_1, t - \tau_2, t - \tau_3)x_i(\tau_1)x_i(\tau_2)x_i(\tau_3)\, d\tau_1\, d\tau_2\, d\tau_3 + \ldots$$

which, rewritten in the compact form,

$$y_o(t) = \sum_{n=1}^{\infty} y_{on}(t) \tag{3.36a}$$

with

$$y_{on}(t) \equiv \int_{-\infty}^{\infty} \ldots \int_{-\infty}^{\infty} h_n(\tau_1, \ldots, \tau_n)x_i(t - \tau_1) \ldots x_i(t - \tau_n)\, d\tau_1 \ldots d\tau_n \tag{3.36b}$$

results in the wanted Volterra series expansion of the nonlinear dynamic system's response, $y_o(t)$, to a general input $x_i(t)$.

As an extension to the linear case, $h_n(\tau_1, \ldots, \tau_n)$ is called the *nth-order impulse response,* or *nth-order Volterra kernel.*

The above deduction was made assuming the system, its input, and output were represented in their natural domain: time-domain. However, we usually have $x_i(t)$ described in the frequency-domain by some spectral representation $X_i(\omega)$ and would like to directly compute $y_o(t)$ in the same domain [i.e., $Y_o(\omega)$]. To see how we can do that using the Volterra series model, we will assume that the input can be expressed as a finite sum of sinusoidal functions, or elementary complex exponentials:

$$x_i(t) = \frac{1}{2}\sum_{q=-Q}^{Q} X_{iq}e^{j\omega_q t} \tag{3.37}$$

in which no dc component is expected (i.e., $q \neq 0$). (The dc term is really already embedded in the Taylor series expansion of the nonlinearity, as its quiescent point.)

Substituting that input into the generic nth order $S[.]$ response, (3.36), we can obtain, after some algebraic manipulation,

$$y_{on}(t) = \frac{1}{2^n} \sum_{q_1=-Q}^{Q} \dots \sum_{q_n=-Q}^{Q} X_{iq_1} \dots X_{iq_n} e^{j(\omega_{q_1} + \dots + \omega_{q_n})t} \tag{3.38}$$

$$\cdot \int_{-\infty}^{\infty} \dots \int_{-\infty}^{\infty} h_n(\tau_1, \dots, \tau_n) e^{-j(\omega_{q_1}\tau_1 + \dots + \omega_{q_n}\tau_n)} d\tau_1 \dots d\tau_n$$

The integral part of (3.38) is a generalization of the conventional Fourier transform, known as the multidimensional Fourier transform:

$$H_n(\omega_{q_1}, \dots, \omega_{q_n}) \equiv \int_{-\infty}^{\infty} \dots \int_{-\infty}^{\infty} h_n(\tau_1, \dots, \tau_n) e^{-j(\omega_{q_1}\tau_1 + \dots + \omega_{q_n}\tau_n)} d\tau_1 \dots d\tau_n \tag{3.39}$$

and the $H_n(\omega_{q_1}, \dots, \omega_{q_n})$ is called the *nth-order nonlinear transfer function* (NLTF). In that way, the system's response $y_o(t)$ can be finally given by

$$y_o(t) = \sum_{n=1}^{\infty} \frac{1}{2^n} \sum_{q_1=-Q}^{Q} \dots \sum_{q_n=-Q}^{Q} X_{iq_1} \dots X_{iq_n} H_n(\omega_{q_1}, \dots, \omega_{q_n}) e^{j(\omega_{q_1} + \dots + \omega_{q_n})t} \tag{3.40}$$

from which it is possible to derive the various $Y_o(\omega)$ components. Therefore, (3.40) constitutes the basis for the frequency-domain analysis of the steady-state response of all mildly nonlinear systems.

Using this Volterra series formalism in real analog and RF circuits resumes to determining each of the nonlinear transfer functions. For that, the *harmonic input method* and the *nonlinear currents method* were proposed, which are the object of subsequent sections.

3.2.2 Volterra Series Analysis of Time-Invariant Circuits

The two methods of nonlinear circuit analysis under the theoretical framework of Volterra series are the harmonic input method (also known as the probing method) and the nonlinear currents method. The former provides a direct way of determining the system's nonlinear transfer functions, while the latter is intended for evaluating the system's response to a certain excitation. So, as we shall see later, the nonlinear currents method can also be used for NLTF identification, if appropriate forcing functions are used.

For a general presentation of both methods, let us go back to the circuit of Section 3.1.2 whose ODE was given by (3.15) and is here reproduced for convenience:

$$Gv_O(t) + \frac{d[\tau_F I_0 \tanh(\alpha v_O)]}{dv_O}\frac{dv_O(t)}{dt} + I_0 \tanh[\alpha v_O(t)] = i_S(t) \quad (3.41)$$

Since we want to analyze this system by Volterra series techniques, we first express the complete excitation $i_S(t)$ as a signal component $i_s(t)$ superimposed on a constant bias I_S:

$$i_S(t) = I_S + i_s(t) \quad (3.42)$$

which produces an output of the same form:

$$v_O(t) = V_O + v_o(t) \quad (3.43)$$

Therefore, $q_{NL}(t)$ and $i_{NL}(t)$ will also be given by

$$q_{NL}(t) = Q_{NL} + q_{nl}(t) \quad (3.44)$$

and

$$i_{NL}(t) = I_{NL} + i_{nl}(t) \quad (3.45)$$

which define the correspondent quiescent points of (Q_{NL}, V_O) and (I_{NL}, V_O) for the nonlinear charge and current, respectively.

The constitutive relations are now approximated by Taylor series expansions around those bias points, leading to

$$\begin{aligned} q_{NL}(v_O) = Q_{NL} &+ c_1[v_O(t) - V_O] + c_2[v_O(t) - V_O]^2 \\ &+ c_3[v_O(t) - V_O]^3 + \ldots \end{aligned} \quad (3.46)$$

where

$$Q_{NL} \equiv q_{NL}(V_O) = \tau_F I_0 \tanh(\alpha V_O) \quad (3.47)$$

$$c_1 \equiv \left.\frac{dq_{NL}(v_O)}{dv_O}\right|_{v_O = V_O} = \tau_F I_0 \alpha \operatorname{sech}^2(\alpha V_O) \quad (3.48)$$

$$c_2 \equiv \frac{1}{2!} \frac{\mathrm{d}^2 q_{NL}(v_O)}{\mathrm{d}v_O^2}\bigg|_{v_O = V_O} = -\tau_F I_0 \alpha^2 \tanh(\alpha V_O) \operatorname{sech}^2(\alpha V_O) \quad (3.49)$$

$$c_3 \equiv \frac{1}{3!} \frac{\mathrm{d}^3 q_{NL}(v_O)}{\mathrm{d}v_O^3}\bigg|_{v_O = V_O} = \frac{1}{3} \tau_F I_0 \alpha^3 \frac{2 \sinh^2(\alpha V_O) - 1}{\cosh^4(\alpha V_O)} \quad (3.50)$$

and

$$i_{NL}(v_O) = I_{NL} + g_1[v_O(t) - V_O] + g_2[v_O(t) - V_O]^2 + g_3[v_O(t) - V_O]^3 + \dots \quad (3.51)$$

where

$$I_{NL} \equiv i_{NL}(V_O) = I_0 \tanh(\alpha V_O) \quad (3.52)$$

$$g_1 \equiv \frac{\mathrm{d}i_{NL}(v_O)}{\mathrm{d}v_O}\bigg|_{v_O = V_O} = I_0 \alpha \operatorname{sech}^2(\alpha V_O) \quad (3.53)$$

$$g_2 \equiv \frac{1}{2!} \frac{\mathrm{d}^2 i_{NL}(v_O)}{\mathrm{d}v_O^2}\bigg|_{v_O = V_O} = -I_0 \alpha^2 \tanh(\alpha V_O) \operatorname{sech}^2(\alpha V_O) \quad (3.54)$$

$$g_3 \equiv \frac{1}{3!} \frac{\mathrm{d}^3 i_{NL}(v_O)}{\mathrm{d}v_O^3}\bigg|_{v_O = V_O} = \frac{1}{3} I_0 \alpha^3 \frac{2 \sinh^2(\alpha V_O) - 1}{\cosh^4(\alpha V_O)} \quad (3.55)$$

Substituting (3.46) and (3.51) into the system's nonlinear differential equation (3.41), and retaining only the dynamic signal components up to third order, we get

$$[c_1 + 2c_2 v_o(t) + 3c_3 v_o(t)^2] \frac{\mathrm{d}v_o(t)}{\mathrm{d}t} + (G + g_1) v_o(t) + g_2 v_o(t)^2 + g_3 v_o(t)^3 = i_s(t) \quad (3.56)$$

Note that even though the terms of c_2 and c_3 only involve $v_o(t)$ and its square, respectively, they really produce components of second and third order because the dynamic charge is multiplied by $\mathrm{d}v_o(t)/\mathrm{d}t$.

3.2.2.1 Nonlinear Currents Method

The process of deriving the solution, $v_o(t)$, of (3.56) for a certain input excitation $i_s(t)$ comes from the following property of Volterra series [4]. If

$$v_o(t) = \sum_{n=1}^{\infty} v_{on}(t) \tag{3.57a}$$

$$v_{on}(t) = \int_{-\infty}^{\infty} \ldots \int_{-\infty}^{\infty} h_n(\tau_1, \ldots, \tau_n) i_s(t-\tau_1) \ldots i_s(t-\tau_n)\, d\tau_1 \ldots d\tau_n \tag{3.57b}$$

is the solution of (3.56) for $i_s(t)$, then

$$v_o(t)' = \sum_{n=1}^{\infty} C^n v_{on}(t) \tag{3.58}$$

will be the solution of (3.56) for the new forcing function $i_s(t)' = Ci_s(t)$, for every constant C. This means that $v_o(t)'$ and $i_s(t)'$ must verify (3.56), and thus,

$$\begin{aligned} &\left[c_1 + 2c_2 \sum_{n=1}^{\infty} C^n v_{on}(t) + 3c_3 \sum_{n_1=1}^{\infty} \sum_{n_2=1}^{\infty} C^{n_1+n_2} v_{on_1}(t) v_{on_2}(t)\right]\left[\sum_{n=1}^{\infty} C^n \frac{dv_{on}(t)}{dt}\right] \\ &+ (G+g_1) \sum_{n=1}^{\infty} C^n v_{on}(t) + g_2 \sum_{n_1=1}^{\infty} \sum_{n_2=1}^{\infty} C^{n_1+n_2} v_{on_1}(t) v_{on_2}(t) \\ &+ g_3 \sum_{n_1=1}^{\infty} \sum_{n_2=1}^{\infty} \sum_{n_3=1}^{\infty} C^{n_1+n_2+n_3} v_{on_1}(t) v_{on_2}(t) v_{on_3}(t) = Ci_s(t) \end{aligned} \tag{3.59}$$

To determine $v_o(t)$, one must calculate each of the nth-order partial solutions, $v_{on}(t)$. These can be easily obtained from (3.59), recognizing that (3.59) can only be verified for a general C if and only if any of its nonlinear differential equations, obtained from equating equal powers of C, are verified. For example, for the first degree, C, we will have

$$c_1 C \frac{dv_{o1}(t)}{dt} + (G+g_1) C v_{o1}(t) = Ci_s(t) \tag{3.60a}$$

or

$$c_1 \frac{dv_{o1}(t)}{dt} + (G+g_1) v_{o1}(t) = i_s(t) \tag{3.60b}$$

This is a linear ODE of constant coefficients that can be represented in compact form by

$$\mathcal{L}[v_{o1}(t)] = i_s(t) \tag{3.61}$$

where $\mathcal{L}[.]$ stands for the linear time-invariant dynamic operator of (3.60).

Therefore, the first-order output component, $v_{o1}(t)$, can be derived from (3.60) using any of the usual methods of linear time-invariant systems. Clearly, the best one is to solve (3.60) in the frequency-domain, using Fourier or Laplace transforms. Let us represent this solution using the inverse of the linear operator $\mathcal{L}[.]$:

$$v_{o1}(t) = \mathcal{L}^{-1}[i_s(t)] \tag{3.62}$$

The second-order component, $v_{o2}(t)$, is derived in a similar way, equating all terms of second degree of C, C^2. Note that the excitation generates no contribution to C^n if $n \neq 1$.

$$c_1 C^2 \frac{dv_{o2}(t)}{dt} + 2c_2 C^2 v_{o1}(t) \frac{dv_{o1}(t)}{dt} + (G + g_1) C^2 v_{o2}(t) + g_2 C^2 v_{o1}(t)^2 = 0 \tag{3.63a}$$

or

$$c_1 \frac{dv_{o2}(t)}{dt} + 2c_2 v_{o1}(t) \frac{dv_{o1}(t)}{dt} + (G + g_1) v_{o2}(t) + g_2 v_{o1}(t)^2 = 0 \tag{3.63b}$$

At this time it should be noted that (3.63) is an equation in the single unknown $v_{o2}(t)$, as $v_{o1}(t)$ was already determined from (3.62). Therefore, the terms involving only $v_{o1}(t)$ can be treated as new forcing functions, which may be passed to the right side of (3.63), leading to

$$c_1 \frac{dv_{o2}(t)}{dt} + (G + g_1) v_{o2}(t) = -i_{c2}(t) - i_{nl2}(t) \tag{3.64a}$$

where

$$i_{c2}(t) \equiv \frac{d}{dt}[c_2 v_{o1}(t)^2] \tag{3.64b}$$

and

$$i_{nl2}(t) \equiv g_2 v_{o1}(t)^2 \tag{3.64c}$$

are the *second-order nonlinear currents* of the nonlinear capacitance and conductance, respectively.

Comparing (3.64) and (3.60) we can conclude that (3.64) is the same linear ODE of (3.60) except that now the forcing function is no longer $i_s(t)$ but $-i_{c2}(t) - i_{nl2}(t)$. So, $v_{o2}(t)$ can be again calculated by

$$v_{o2}(t) = \mathcal{L}^{-1}[-i_{c2}(t) - i_{nl2}(t)] \tag{3.65}$$

The third-order output component, $v_{o3}(t)$, can be obtained in just the same manner, retaining only terms of C^3 in (3.59).

$$\begin{aligned} &c_1 C^3 \frac{dv_{o3}(t)}{dt} + 2c_2 C^3 \left[v_{o1}(t) \frac{dv_{o2}(t)}{dt} + v_{o2}(t) \frac{dv_{o1}(t)}{dt} \right] \\ &+ 3c_3 C^3 \left[v_{o1}(t)^2 \frac{dv_{o1}(t)}{dt} \right] \\ &+ (G + g_1) C^3 v_{o3}(t) + 2g_2 C^3 v_{o1}(t) v_{o2}(t) + g_3 C^3 v_{o1}(t)^3 = 0 \end{aligned} \tag{3.66}$$

Now, the forcing function is composed by the terms involving the already-known $v_{o1}(t)$ and $v_{o2}(t)$, while the unknown is $v_{o3}(t)$, and thus (3.66) can again be rewritten as

$$c_1 \frac{dv_{o3}(t)}{dt} + (G + g_1) v_{o3}(t) = -i_{c3}(t) - i_{nl3}(t) \tag{3.67a}$$

where now the *third-order nonlinear currents* of the capacitance and conductance are

$$i_{c3}(t) \equiv \frac{d}{dt}[2c_2 v_{o1}(t) v_{o2}(t)] + \frac{d}{dt}[c_3 v_{o1}(t)^3] \tag{3.67b}$$

and

$$i_{nl3}(t) \equiv 2g_2 v_{o1}(t) v_{o2}(t) + g_3 v_{o1}(t)^3 \tag{3.67c}$$

$v_{o3}(t)$ can, once again, be obtained from

$$v_{o3}(t) = \mathcal{L}^{-1}[-i_{c3}(t) - i_{nl3}(t)] \tag{3.68}$$

If the system's nonlinearities were expanded in Taylor series up to order n, this procedure could be generalized to that order, giving

$$v_{on}(t) = \mathcal{L}^{-1}\{-i_{cn}[v_{o1}(t), \ldots, v_{on-1}(t)] - i_{nln}[v_{o1}(t), \ldots, v_{on-1}(t)]\} \tag{3.69}$$

Equation (3.69) summarizes two important conclusions that have to be drawn from the above derivations.

The first one can be stated in the following manner:

> Determining the Volterra series solution of a nonlinear ODE up to order n, can be done by solving n times the linearized ODE with the appropriate forcing functions.

The second conclusion refers to these forcing functions, and can be stated as:

> The first-order forcing function [or the one which is applied to the first linearized ODE needed to determine $v_{o1}(t)$] is the system's excitation, while the one of general order $n > 1$ is composed by the nth-order nonlinear controlled variables corresponding to all system's nonlinearities. These nth-order controlled variables can be calculated by substituting the controlling variable components of order 1 to $n - 1$ in the Taylor series terms of degree 2 to n.

The former of these conclusions is really the reason for one of the Volterra series' greatest advantages: it provides an analytical (although approximate) solution to a mildly nonlinear ordinary differential equation which otherwise could only be solved by numerical techniques. Volterra series enables, therefore, drawing qualitative conclusions about the system, and this is of paramount importance to system design.

Unfortunately, the latter statement goes right in the opposite direction. It implies that, although the response of any system (which is stable, continuous, and infinitely differentiable) can be obtained with any desired small amount of error, by simply increasing the maximum order of the series' expansion, in practice Volterra series suffers from convergence problems [4], and becomes hopelessly useless for systems requiring orders higher than about five. In fact, since the nth-order forcing functions are dependent on the combinations of all the first to $(n - 1)$th order solutions, they become extremely laborious to find, as the number of possible different combinations rapidly increases with n.

Finally, note that, even though this analysis technique was named nonlinear currents method—because the nonlinearities were considered as voltage dependent current sources—it is general in nature, since it can be applied to any ODE.

Nonlinear Currents Method Applied to Circuit Analysis

In this section we will show how the above procedure can be reflected at the circuit analysis level. Since the method reduces to repeatedly determining the solution of a linear ODE of constant coefficients, it is better to do it in the frequency-domain.

Therefore, it is assumed that $i_S(t)$ and $v_O(t)$ are given as sums of phasors I_{sq}, V_{ok}, plus their respective quiescent values I_S, V_O:

$$i_S(t) \equiv I_S + i_s(t) = I_S + \sum_{\substack{q=-Q \\ q \neq 0}}^{Q} I_{sq} e^{j\omega_q t} \tag{3.70}$$

and

$$v_O(t) \equiv V_O + v_o(t) = V_O + \sum_{\substack{k=-K \\ k \neq 0}}^{K} V_{ok} e^{j\omega_k t} \tag{3.71}$$

As the various linear ODE to be solved are derived from the linearization of the circuit in the quiescent point, the analysis process begins by calculating these quiescent voltage and current values. Contrary to what was done to the dynamic signal components, considered small perturbations of the dc magnitudes, and thus enabling the Taylor series expansions of the nonlinearities, the quiescent values are, themselves, large-signal components. Therefore, the dc analysis has to be performed using the full nonlinearity expressions, and for which there is, in general, no analytical solution. The way normally used to obtain these quiescent values is the Newton-Raphson iteration scheme. In our example, the algebraic equation to be analyzed is, from (3.41),

$$GV_O + I_0 \tanh(\alpha V_O) = I_S \tag{3.72}$$

Assuming an initial estimate for the solution, ${}^0V_O$, and expanding the nonlinearity into a Taylor series of first-order around this ${}^0V_O$, we obtain

$$G\,{}^0V_O + I_0 \tanh(\alpha\,{}^0V_O) + \left.\frac{dI_0 \tanh(\alpha V_O)}{dV_O}\right|_{V_O={}^0V_O} ({}^1V_O - {}^0V_O) - I_S = 0 \tag{3.73}$$

from which we get a refined estimate as

$$ {}^1V_O = {}^0V_O + \left[\left.\frac{dI_0 \tanh(\alpha V_O)}{dV_O}\right|_{V_O={}^0V_O}\right]^{-1} [I_S - G\,{}^0V_O + I_0 \tanh(\alpha\,{}^0V_O)] \tag{3.74a}$$

or

$$^1V_O = {^0V_O} + \frac{1}{\alpha I_0}\cosh^2(\alpha\, {^0V_O})[I_S - G\, {^0V_O} - I_0 \tanh(\alpha\, {^0V_O})] \tag{3.74b}$$

Since the hyperbolic tangent was substituted by a rough first-order approximation, it is expected that ^1V_O does not exactly verify (3.72). In fact,

$$G\, {^1V_O} + I_0 \tanh(\alpha\, {^1V_O}) - I_S = \epsilon \quad \text{where } \epsilon \neq 0 \tag{3.75}$$

If $|\epsilon|$ is less than an acceptable amount of error δ, then ^1V_O can be taken as a good approximation to the solution. If not, ^1V_O should be considered a new estimate, and the process repeated until

$$|G\, {^fV_O} + I_0 \tanh(\alpha\, {^fV_O}) - I_S| \leq \delta \tag{3.76}$$

This $^fV_O \approx V_O$ is the sought quiescent solution of (3.41) for the dc excitation I_S, and $[Q_{NL} = q_{NL}(^fV_O), I_{NL} = i_{NL}(^fV_O)]$ its correspondent nonlinear charge and current quiescent values.

The second step in the nonlinear currents method consists of redrawing the original circuit in such a way that the linear and nonlinear components of the nonlinearity are separated. Since these circuit elements are modeled as the series of (3.46) and (3.51), their dynamic current components can be given by

$$i_c(t) \equiv \frac{dq_{nl}(t)}{dt} = i_{c1}(t) + i_{c2}(t) + i_{c3}(t) + \ldots \tag{3.77}$$

and

$$i_{nl}(t) \equiv i_{nl1}(t) + i_{nl2}(t) + i_{nl3}(t) + \ldots \tag{3.78}$$

Since the linear, or first-order, current components pertain to the linear dynamic operator $\mathscr{L}[.]$, they must be incorporated in the linear subcircuit as linear voltage-controlled current sources. The nonlinear components, instead, behave as forcing functions, and thus must be represented as independent current sources. They should not be present all at the same time, but connected, one by one, each time its corresponding order term of the output voltage is being determined. The circuit of Figure 3.1 should then be redrawn as the one of Figure 3.7.

First-Order Output Components Determination. For the calculation of $v_{o1}(t)$, the circuit includes only the linear current components $i_{c1}(t)$ and $i_{nl1}(t)$, and $i_s(t)$ as its driving source. A frequency-domain version of this circuit is shown in Figure 3.8. Herein, it will be called the first-order circuit of Figure 3.7.

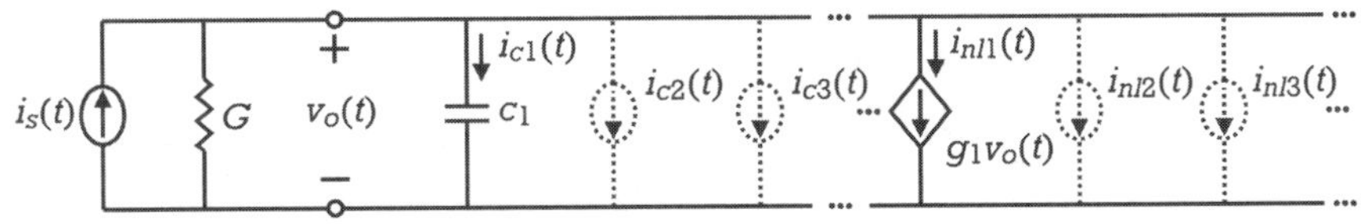

Figure 3.7 Circuit schematic redrawn for nonlinear currents method application.

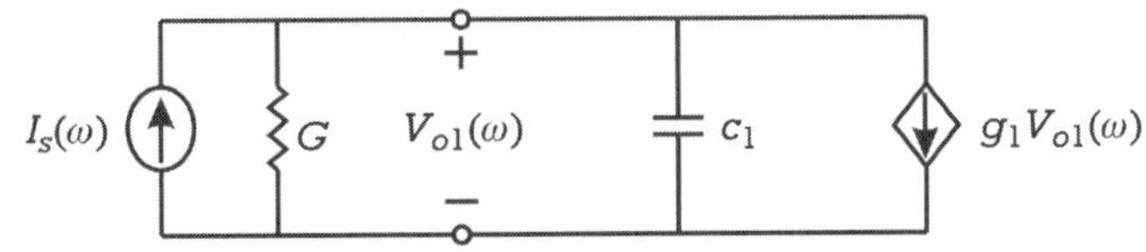

Figure 3.8 First-order circuit schematic diagram.

From Figure 3.8, $V_{o1}(\omega)$ can be given by

$$V_{o1}(\omega) = \frac{I_s(\omega)}{G + g_1 + j\omega c_1} \tag{3.79}$$

or

$$v_{o1}(t) = \sum_{q=-Q}^{Q} V_{o1_q} e^{j\omega_q t} \tag{3.80}$$

Because the nonlinear current components, $I_{c2}(\omega)$ and $I_{nl2}(\omega)$ or $I_{c3}(\omega)$ and $I_{nl3}(\omega)$, depend on the correspondent nonlinearities' control variable, it is convenient to derive the transfer functions that relate these control voltages to the driving source $I_s(\omega)$. In our circuit example, the two nonlinearities share the same control variable, which also coincides with the output voltage. Therefore, in this case, (3.79) is sufficient for providing all these relations.

Second-Order Output Components Determination. The second-order output components determination begins by calculating the second-order nonlinear currents:

$$i_{c2}(t) = 2c_2 v_{o1}(t)\frac{dv_{o1}(t)}{dt} \tag{3.81}$$

$$= 2c_2 \sum_{q_1=-Q}^{Q} \sum_{q_2=-Q}^{Q} j\omega_{q_2} V_{o1_{q_1}} V_{o1_{q_2}} e^{j(\omega_{q_1}+\omega_{q_2})t} = \sum_{r=-R}^{R} I_{c2_r} e^{j\omega_r t}$$

$$i_{nl2}(t) = g_2 v_{o1}(t)^2 \tag{3.82}$$
$$= g_2 \sum_{q_1=-Q}^{Q} \sum_{q_2=-Q}^{Q} V_{o1_{q_1}} V_{o1_{q_2}} e^{j(\omega_{q_1} + \omega_{q_2})t} = \sum_{r=-R}^{R} I_{nl2_r} e^{j\omega_r t}$$

According to what was stated above, the second-order circuit is drawn as in Figure 3.9.

Analyzing the same linear circuit with $I_{c2}(\omega)$ and $I_{nl2}(\omega)$ as its driving sources, we find

$$V_{o2}(\omega) = -\frac{I_{c2}(\omega) + I_{nl2}(\omega)}{G + g_1 + j\omega c_1} \tag{3.83}$$

or

$$v_{o2}(t) = \sum_{r=-R}^{R} V_{o2_r} e^{j\omega_r t} \tag{3.84}$$

Again, (3.83) also provides the nonlinearities' control voltage as function of the nonlinear current sources.

Third-Order Output Components Determination. Now, the third-order nonlinear currents are

$$i_{c3}(t) = 2c_2 \left[v_{o1}(t) \frac{dv_{o2}(t)}{dt} + v_{o2}(t) \frac{dv_{o1}(t)}{dt} \right] + 3c_3 v_{o1}(t)^2 \frac{dv_{o1}(t)}{dt}$$
$$= 2c_2 \left[\sum_{q=-Q}^{Q} \sum_{r=-R}^{R} j\omega_r V_{o1_q} V_{o2_r} e^{j(\omega_q + \omega_r)t} \right.$$
$$\left. + \sum_{r=-R}^{R} \sum_{q=-Q}^{Q} j\omega_q V_{o1_q} V_{o2_r} e^{j(\omega_q + \omega_r)t} \right] \tag{3.85}$$
$$+ 3c_3 \sum_{q_1=-Q}^{Q} \sum_{q_2=-Q}^{Q} \sum_{q_3=-Q}^{Q} j\omega_{q_3} V_{o1_{q_1}} V_{o1_{q_2}} V_{o1_{q_3}} e^{j(\omega_{q_1} + \omega_{q_2} + \omega_{q_3})t}$$
$$= \sum_{k=-K}^{K} I_{c3_k} e^{j\omega_k t}$$

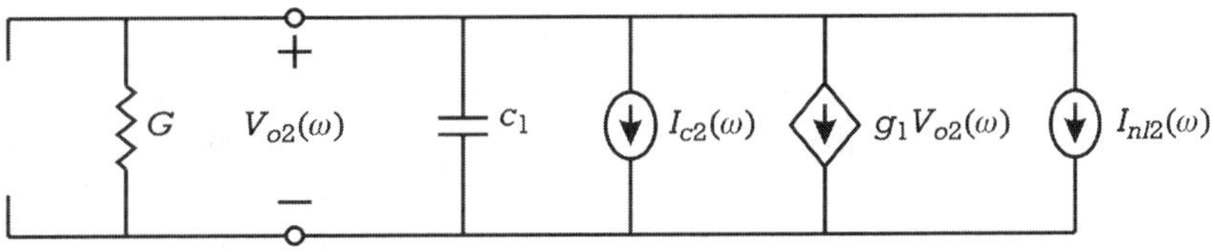

Figure 3.9 Second-order circuit schematic diagram.

$$\begin{aligned} i_{nl3}(t) &= 2g_2 v_{o1}(t) v_{o2}(t) + g_3 v_{o1}(t)^3 \\ &= 2g_2 \sum_{q=-Q}^{Q} \sum_{r=-R}^{R} V_{o1_q} V_{o2_r} e^{j(\omega_q + \omega_r)t} \\ &\quad + g_3 \sum_{q_1=-Q}^{Q} \sum_{q_2=-Q}^{Q} \sum_{q_3=-Q}^{Q} V_{o1_{q_1}} V_{o1_{q_2}} V_{o1_{q_3}} e^{j(\omega_{q_1} + \omega_{q_2} + \omega_{q_3})t} \\ &= \sum_{k=-K}^{K} I_{nl3_k} e^{j\omega_k t} \end{aligned} \tag{3.86}$$

Accordingly, the third-order circuit is drawn in Figure 3.10. The linear analysis of this circuit gives $V_{o3}(\omega)$ as

$$V_{o3}(\omega) = -\frac{I_{c3}(\omega) + I_{nl3}(\omega)}{G + g_1 + j\omega c_1} \tag{3.87}$$

or

$$v_{o3}(t) = \sum_{k=-K}^{K} V_{o3_k} e^{j\omega_k t} \tag{3.88}$$

which completes the analysis up to order three.

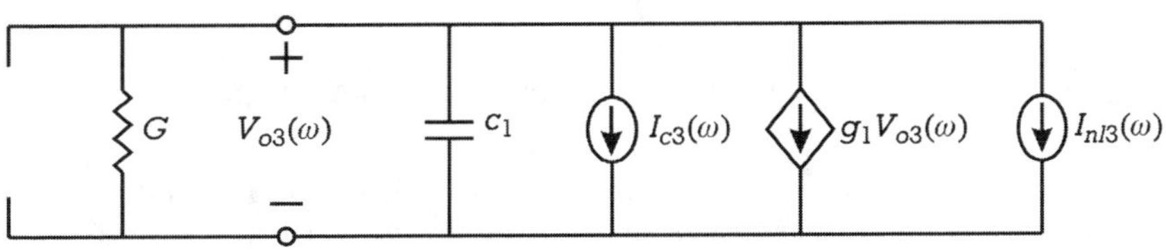

Figure 3.10 Third-order circuit schematic diagram.

Nonlinear Currents Method Applied to Network Analysis

The generalization of the above analysis process to a large network is straightforward. To exemplify, let us consider the mildly nonlinear network depicted in Figure 3.11.

This network is assumed to have a driving source $i_s(t)$, $I_s(\omega)$, and an output variable $v_O(t)$, $V_O(\omega)$, beyond the J dc bias voltages supplies $V_{DC_1}, \ldots, V_{DC_J}$. It also includes M mildly nonlinear voltage-dependent current sources, whose controlled variables are $i_{NL}^{(1)}(t), \ldots, i_{NL}^{(M)}(t)$. These M nonlinearities are dependent on L controlling voltages, $v^{(1)}(t), \ldots, v^{(L)}(t)$, such that $i_{NL}^{(1)}(t) = f_{NL}^{(1)}(t)[v^{(1)}(t), \ldots, v^{(L)}(t)], \ldots, i_{NL}^{(M)}(t) = f_{NL}^{(M)}(t)[v^{(1)}(t), \ldots, v^{(L)}(t)]$.

Normally, since most used electron devices are two or three terminal elements, these $i_{NL}^{(m)}(t)$ are dependent on a single or two controlling voltages.[1] They can represent conductive or capacitive nonlinearities, which can be dependent on local voltages (e.g., nonlinear conductances or capacitances) or remote voltages (e.g., nonlinear transconductances or transcapacitances). In the case of capacitive nonlinearities, $i_{NL}^{(m)}(t)$ must be computed as the time derivative of a nonlinear voltage-dependent charge. Although mildly nonlinear controlled-voltage sources, or nonlinear inductors, could also be considered, they were not included since they are generally not used for nonlinear electron device modeling.

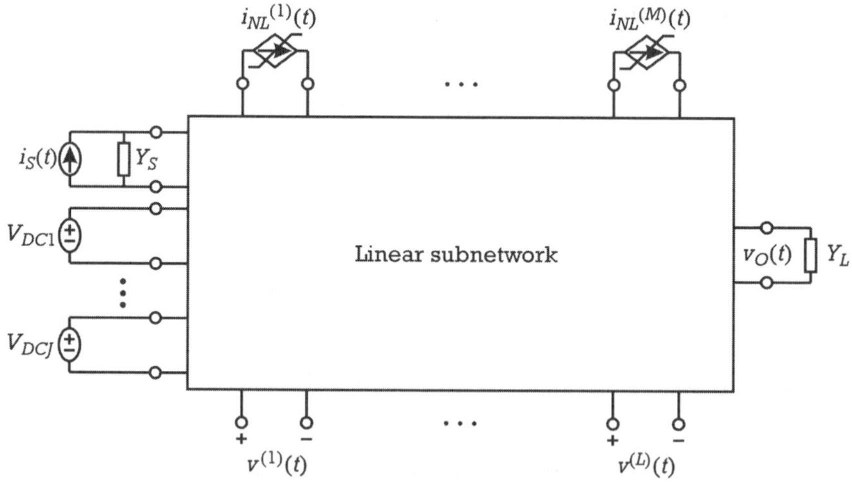

Figure 3.11 Network example for nonlinear currents method application.

1. Nonlinearities dependent on three or more controlling voltages are rare, although they are sometimes encountered. An example is the drain-source current of a MOSFET device which can be expressed as a function of three independent voltages, referred to the substrate potential: source voltage, gate voltage, and drain voltage.

Again, the analysis procedure begins by a dc calculation to find the quiescent point. It can be done by a simple nonlinear nodal analysis of the static subcircuit, creating a system of nonlinear algebraic equations. This nonlinear system is then numerically solved by a multidimensional Newton-Raphson iteration scheme, similar to the one above explained.

The various nonlinearities are then expanded in Taylor series that may be one-dimensional or multidimensional, depending on the number of controlling variables. For example, if $i_{NL}^{(m_1)}(t)$ were only dependent on $v^{(l_1)}(t)$, and $i_{NL}^{(m_2)}(t)$ were dependent on $v^{(l_2)}(t)$ and $v^{(l_3)}(t)$, we would have

$$\begin{aligned} i_{NL}^{(m_1)}[v^{(l_1)}] &= I_{NL}^{(m_1)} + \sum_{n=1}^{\infty} \frac{1}{n!} \frac{\mathrm{d}^n i_{NL}^{(m_1)}}{\mathrm{d}v^{(l_1)^n}}\bigg|_{v^{(l_1)} = V^{(l_1)}} \left[v^{(l_1)} - V^{(l_1)}\right]^n \\ &= I_{NL}^{(m_1)} + \sum_{n=1}^{\infty} g_n^{(m_1)}\left[v^{(l_1)} - V^{(l_1)}\right]^n \end{aligned} \tag{3.89}$$

and

$$\begin{aligned} &i_{NL}^{(m_2)}\left[v^{(l_2)}, v^{(l_3)}\right] \\ &= I_{NL}^{(m_2)} \\ &+ \sum_{n_1=0}^{\infty}\sum_{n_2=0}^{\infty} \frac{1}{n_1!}\frac{1}{n_2!} \frac{\partial^{(n_1+n_2)} i_{NL}^{(m_2)}}{\partial v^{(l_2)^{n_1}} \partial v^{(l_3)^{n_2}}}\bigg|_{\substack{v^{(l_2)} = V^{(l_2)} \\ v^{(l_3)} = V^{(l_3)}}} \left[v^{(l_2)} - V^{(l_2)}\right]^{n_1}\left[v^{(l_3)} - V^{(l_3)}\right]^{n_2} \\ &= I_{NL}^{(m_2)} \\ &+ \sum_{n_1=0}^{\infty}\sum_{n_2=0}^{\infty} g_{n_1 n_2}^{(m_2)}\left[v^{(l_2)} - V^{(l_2)}\right]^{n_1}\left[v^{(l_3)} - V^{(l_3)}\right]^{n_2} \end{aligned} \tag{3.90}$$

in which n_1 and n_2 can never be simultaneously zero.

As was seen, the terms of degree one in (3.89) and (3.90) are the ones responsible for the first-order current components and must be incorporated in the linear subnetwork. The terms of degree $n > 1$ produce current components of order equal or greater then n and become the forcing functions of the corresponding order subcircuit. So, the network of Figure 3.11 can now be redrawn as in Figure 3.12.

This new network is really a set of linear subcircuits, each one valid for a certain order n, according to the driving sources: $i_s(t)$ for $n = 1$ or $i_{nln}^{(1)}(t), \ldots,$ $i_{nln}^{(M)}(t)$, otherwise. Any of these nth-order linear subcircuits are networks of $(M + L + 2)$ ports, whose currents and voltages are identified by:

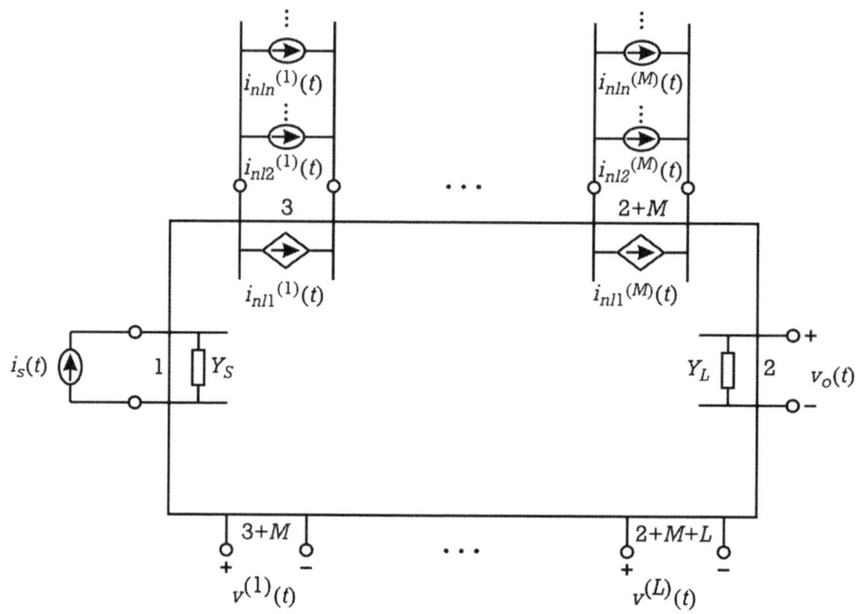

Figure 3.12 Network schematic redrawn for nonlinear currents method application.

$$
\begin{array}{lll}
i_{1n}(t) = i_s(t) & (n = 1 \text{ or zero if } n > 1); & v_{1n}(t) \\
i_{2n}(t) = 0 & ; & v_{2n}(t) = v_{on}(t) \\
i_{3n}(t) = -i_{nln}^{(1)}(t) & (n > 1 \text{ or zero if } n = 1); & v_{3n}(t) \\
\vdots & \vdots & \vdots \\
i_{(m+2)n}(t) = -i_{nln}^{(m)}(t) & (n > 1 \text{ or zero if } n = 1); & v_{(m+2)n}(t) \\
\vdots & \vdots & \vdots \\
i_{(M+2)n}(t) = -i_{nln}^{(M)}(t) & (n > 1 \text{ or zero if } n = 1); & v_{(M+2)n}(t) \\
i_{(M+3)n}(t) = 0 & ; & v_{(M+3)n}(t) = v_n^{(1)}(t) \\
\vdots & \vdots & \vdots \\
i_{(M+l+2)n}(t) = 0 & ; & v_{(M+l+2)n}(t) = v_n^{(l)}(t) \\
\vdots & \vdots & \vdots \\
i_{(M+L+2)n}(t) = 0 & ; & v_{(M+L+2)n}(t) = v_n^{(L)}(t)
\end{array}
$$

and may be analyzed in the frequency-domain by the following set of $(M + L + 2)$ equations:

$$\begin{bmatrix} Y_{11} & \cdots & Y_{1j} & \cdots & Y_{1(M+L+2)} \\ & & \vdots & & \\ \vdots & & Y_{ij} & & \vdots \\ & & \vdots & & \\ Y_{(M+L+2)1} & \cdots & Y_{(M+L+2)j} & \cdots & Y_{(M+L+2)(M+L+2)} \end{bmatrix} \begin{bmatrix} V_{1n}(\omega) \\ \vdots \\ V_{jn}(\omega) \\ \vdots \\ V_{(M+L+2)n}(\omega) \end{bmatrix} = \begin{bmatrix} I_{1n}(\omega) \\ \vdots \\ I_{jn}(\omega) \\ \vdots \\ I_{(M+L+2)n}(\omega) \end{bmatrix} \tag{3.91}$$

As was explained for the single node circuit, several transimpedance gain factors should be derived for calculating the output voltage component $v_{on}(t)$ and each one of the controlling voltages $v^{(1)}(t), \ldots, v^{(L)}(t)$, from the successive driving current sources. These gains can be defined by

$$V_{on}(\omega) = Z_{21}(\omega) I_{1n}(\omega) + \sum_{j=3}^{M+2} Z_{2j}(\omega) I_{jn}(\omega) \tag{3.92}$$

and

$$\begin{aligned} V_n^{(1)}(\omega) &= Z_{(M+3)1}(\omega) I_{1n}(\omega) + \sum_{j=3}^{M+2} Z_{(M+3)j}(\omega) I_{jn}(\omega) \\ \vdots \quad & \qquad \vdots \\ V_n^{(l)}(\omega) &= Z_{(M+l+2)1}(\omega) I_{1n}(\omega) + \sum_{j=3}^{M+2} Z_{(M+l+2)j}(\omega) I_{jn}(\omega) \\ \vdots \quad & \qquad \vdots \\ V_n^{(L)}(\omega) &= Z_{(M+L+2)1}(\omega) I_{1n}(\omega) + \sum_{j=3}^{M+2} Z_{(M+L+2)j}(\omega) I_{jn}(\omega) \end{aligned} \tag{3.93}$$

Since $I_{2n}(\omega) = I_{(M+3)n}(\omega) = \ldots = I_{(M+L+2)n}(\omega) = 0$, these $Z_{ij}(\omega)$ are the impedance parameters of the linear $(M + L + 2)$-port network which are related to the previous admittance matrix by

$$[Z_{ij}] = [Y_{ij}]^{-1} \tag{3.94}$$

The calculation of the various terms of $v_o(t)$ can now be performed in the frequency-domain as follows.

The frequency-domain first-order output voltage component is directly given by (3.92) as

$$V_{o1}(\omega) = Z_{21}(\omega) I_s(\omega) \tag{3.95}$$

Now, for calculating the second-order component, we first proceed to the determination of the controlling voltages' first-order components. By (3.93):

$$V_1^{(l)}(\omega) = Z_{(M+l+2)1}(\omega) I_s(\omega) \quad (l = 1, \ldots, L) \tag{3.96}$$

These first-order control voltages produce second-order nonlinear currents which have to be calculated by substituting (3.96) into the second-degree terms of the Taylor series expansions of $i_{NL}^{(m)}(t)$. For example, the substitution of (3.96) into (3.89) would lead to

$$i_{nl2}^{(m_1)}(t) = g_2^{(m_1)} \sum_{q_1=-Q}^{Q} \sum_{q_2=-Q}^{Q} V_{1_{q_1}}^{(l_1)} V_{1_{q_2}}^{(l_1)} e^{j(\omega_{q_1} + \omega_{q_2})t} = \sum_{r=-R}^{R} I_{nl2_r}^{(m_1)} e^{j\omega_r t} \tag{3.97}$$

If (3.96) were to be substituted into a bidimensional Taylor series like the one of (3.90), then the second-order nonlinear current would be

$$\begin{aligned} i_{nl2}^{(m_2)}(t) &= g_{20}^{(m_2)} \sum_{q_1=-Q}^{Q} \sum_{q_2=-Q}^{Q} V_{1_{q_1}}^{(l_2)} V_{1_{q_2}}^{(l_2)} e^{j(\omega_{q_1} + \omega_{q_2})t} \\ &\quad + g_{11}^{(m_2)} \sum_{q_1=-Q}^{Q} \sum_{q_2=-Q}^{Q} V_{1_{q_1}}^{(l_2)} V_{1_{q_2}}^{(l_3)} e^{j(\omega_{q_1} + \omega_{q_2})t} \\ &\quad + g_{02}^{(m_2)} \sum_{q_1=-Q}^{Q} \sum_{q_2=-Q}^{Q} V_{1_{q_1}}^{(l_3)} V_{1_{q_2}}^{(l_3)} e^{j(\omega_{q_1} + \omega_{q_2})t} \\ &= \sum_{r=-R}^{R} I_{nl2_r}^{(m_2)} e^{j\omega_r t} \end{aligned} \tag{3.98}$$

After calculating all second-order nonlinear current components, the second-order output voltage, $V_{o2}(\omega)$, becomes [by (3.92)]:

$$V_{o2}(\omega) = \sum_{j=3}^{M+2} Z_{2j}(\omega) I_{j2}(\omega) \tag{3.99}$$

The process is now repeated to the third-order component $V_{o3}(\omega)$, by first calculating second-order control voltages:

$$V_2^{(l)}(\omega) = \sum_{j=3}^{M+2} Z_{(M+l+2)j}(\omega) I_{j2}(\omega) \quad (l = 1, \ldots, L) \tag{3.100}$$

First and second-order control voltages are then substituted into second and third-degree terms of the Taylor series expansions of every $i_{NL}^{(m)}[v^{(l_1)}, \ldots, v^{(L)}]$, to determine third-order nonlinear currents' components. Following the example of (3.89) and (3.90), we would get

$$\begin{aligned}
i_{nl3}^{(m_1)}(t) &= 2g_2^{(m_1)} \sum_{q=-Q}^{Q} \sum_{r=-R}^{R} V_{1_q}^{(l_1)} V_{2_r}^{(l_1)} e^{j(\omega_q + \omega_r)t} \\
&\quad + g_3^{(m_1)} \sum_{q_1=-Q}^{Q} \sum_{q_2=-Q}^{Q} \sum_{q_3=-Q}^{Q} V_{1_{q_1}}^{(l_1)} V_{1_{q_2}}^{(l_1)} V_{1_{q_3}}^{(l_1)} e^{j(\omega_{q_1} + \omega_{q_2} + \omega_{q_3})t} \\
&= \sum_{k=-K}^{K} I_{nl3_k}^{(m_1)} e^{j\omega_k t}
\end{aligned} \tag{3.101}$$

for the one-dimensional Taylor series of (3.89), and

$$\begin{aligned}
i_{nl3}^{(m_2)}(t) &= 2g_{20}^{(m_2)} \sum_{q=-Q}^{Q} \sum_{r=-R}^{R} V_{1_q}^{(l_2)} V_{2_r}^{(l_2)} e^{j(\omega_q + \omega_r)t} \\
&\quad + g_{11}^{(m_2)} \sum_{q=-Q}^{Q} \sum_{r=-R}^{R} V_{1_q}^{(l_2)} V_{2_r}^{(l_3)} e^{j(\omega_q + \omega_r)t} \\
&\quad + g_{11}^{(m_2)} \sum_{q=-Q}^{Q} \sum_{r=-R}^{R} V_{1_q}^{(l_3)} V_{2_r}^{(l_2)} e^{j(\omega_q + \omega_r)t} \\
&\quad + 2g_{02}^{(m_2)} \sum_{q=Q}^{Q} \sum_{r=-R}^{R} V_{1_q}^{(l_3)} V_{2_r}^{(l_3)} e^{j(\omega_q + \omega_r)t} \\
&\quad + g_{30}^{(m_2)} \sum_{q_1=-Q}^{Q} \sum_{q_2=-Q}^{Q} \sum_{q_3=-Q}^{Q} V_{1_{q_1}}^{(l_2)} V_{1_{q_2}}^{(l_2)} V_{1_{q_3}}^{(l_2)} e^{j(\omega_{q_1} + \omega_{q_2} + \omega_{q_3})t} \\
&\quad + g_{21}^{(m_2)} \sum_{q_1=-Q}^{Q} \sum_{q_2=-Q}^{Q} \sum_{q_3=-Q}^{Q} V_{1_{q_1}}^{(l_2)} V_{1_{q_2}}^{(l_2)} V_{1_{q_3}}^{(l_3)} e^{j(\omega_{q_1} + \omega_{q_2} + \omega_{q_3})t} \\
&\quad + g_{12}^{(m_2)} \sum_{q_1=-Q}^{Q} \sum_{q_2=-Q}^{Q} \sum_{q_3=-Q}^{Q} V_{1_{q_1}}^{(l_2)} V_{1_{q_2}}^{(l_3)} V_{1_{q_3}}^{(l_3)} e^{j(\omega_{q_1} + \omega_{q_2} + \omega_{q_3})t}
\end{aligned}$$

$$+ g_{03}^{(m_2)} \sum_{q_1=-Q}^{Q} \sum_{q_2=-Q}^{Q} \sum_{q_3=-Q}^{Q} V1_{q_1}^{(l_3)} V1_{q_2}^{(l_3)} V1_{q_3}^{(l_3)} e^{j(\omega_{q_1} + \omega_{q_2} + \omega_{q_3})t}$$

$$= \sum_{k=-K}^{K} I_{nl3_k}^{(m_2)} e^{j\omega_k t} \tag{3.102}$$

for the bidimensional Taylor series of (3.90).

Now, $V_{o3}(\omega)$ comes [from (3.92)] as

$$V_{o3}(\omega) = \sum_{j=3}^{M+2} Z_{2j}(\omega) I_{j3}(\omega) \tag{3.103}$$

This process should then be repeated up to the desired order $V_{on}(\omega)$.

3.2.2.2 Harmonic Input Method

This section is devoted to the calculation of the various NLTFs using the harmonic input method (or probing method). The frequency-domain representation of the Volterra kernels is preferred against their time-domain version, because it is more appropriate for the analysis and design of RF and microwave circuits.

The technique is a generalization of the linear system's harmonic input method, which is based on the calculation of the system's response to a harmonic input (a cosine or complex exponential). Because the time-domain representation of a complex exponential, $e^{-j\omega t}$, is a Dirac delta function $\delta(t - \tau)$, we are, in fact, determining the system's impulse response, or the first-order Volterra kernel. To proceed with the calculation directly in the frequency-domain, we use the following property.

If a linear time-invariant system of input $x_i(t)$ and output $y_o(t)$, characterized by its impulse response $h_1(\tau)$

$$y_o(t) = \int_{-\infty}^{\infty} h_1(\tau) x_i(t - \tau)\, d\tau \tag{3.104}$$

is excited by an elementary complex exponential

$$x_i(t) = e^{j\omega t} \tag{3.105}$$

then its output will be

$$y_o(t) = \int_{-\infty}^{\infty} h_1(\tau) e^{j\omega t} e^{-j\omega\tau}\, d\tau = e^{j\omega t} \int_{-\infty}^{\infty} h_1(\tau) e^{-j\omega\tau}\, d\tau = H_1(\omega) e^{j\omega t} \tag{3.106}$$

That is, the output of a linear system, excited by an elementary complex exponential, is given by the product of the input by its linear transfer function. Therefore, this transfer function can be determined by dividing the calculated system's output, by the elementary complex exponential excitation.

For generalizing that conclusion to the second-order nonlinear transfer function we should realize that a second-order system requires an input with two degrees of freedom, either two independent time delays for the time-domain kernel $h_2(\tau_1, \tau_2)$, or two independent frequencies for its bidimensional Fourier transform $H_2(\omega_1, \omega_2)$. And so, the elementary input should be

$$x_i(t) = e^{j\omega_1 t} + e^{j\omega_2 t} \tag{3.107}$$

Substituting (3.107) into the second-order response expression gives

$$y_{o2}(t) = \int_{-\infty}^{\infty}\int_{-\infty}^{\infty} h_2(\tau_1, \tau_2)\left(e^{j\omega_1(t-\tau_1)} + e^{j\omega_2(t-\tau_1)}\right)\left(e^{j\omega_1(t-\tau_2)} + e^{j\omega_2(t-\tau_2)}\right) d\tau_1\, d\tau_2 \tag{3.108}$$

Since $h_2(\tau_1, \tau_2)$ and $H_2(\omega_1, \omega_2)$ are symmetric in their arguments, (3.108) can be simplified to

$$\begin{aligned} y_{o2}(t) &= e^{j2\omega_1 t}\int_{-\infty}^{\infty}\int_{-\infty}^{\infty} h_2(\tau_1, \tau_2)e^{-j\omega_1(\tau_1+\tau_2)}\, d\tau_1\, d\tau_2 \\ &\quad + 2e^{j(\omega_1+\tau_2)t}\int_{-\infty}^{\infty}\int_{-\infty}^{\infty} h_2(\tau_1, \tau_2)e^{-j(\omega_1\tau_1+\omega_2\tau_2)}\, d\tau_1\, d\tau_2 \\ &\quad + e^{j2\omega_2 t}\int_{-\infty}^{\infty}\int_{-\infty}^{\infty} h_2(\tau_1, \tau_2)e^{-j\omega_2(\tau_1+\tau_2)}\, d\tau_1\, d\tau_2 \\ &= H_2(\omega_1, \omega_2)e^{j2\omega_1 t} + 2H_2(\omega_1, \omega_2)e^{j(\omega_1+\omega_2)t} + H_2(\omega_2, \omega_2)e^{j2\omega_2 t} \end{aligned} \tag{3.109}$$

which shows that the second-order nonlinear transfer function can be calculated by dividing the output component at the sum frequency by $2e^{j(\omega_1+\tau_2)t}$.

The generalization of this process for determining the nth-order NLTF, $H_n(\omega_1, \ldots, \omega_n)$ would require an elementary excitation of the form

$$x_i(t) = \sum_{q=1}^{n} e^{j\omega_q t} \tag{3.110}$$

which produces an nth-order output given by

$$y_{on}(t) = \int_{-\infty}^{\infty} \dots \int_{-\infty}^{\infty} h_n(\tau_1, \dots, \tau_n) \sum_{q_1=1}^{n} \dots \tag{3.111}$$

$$\sum_{q_n=1}^{n} e^{j(\omega_{q_1} + \dots + \omega_{q_n})t} e^{-j(\omega_{q_1}\tau_1 + \dots + \omega_{q_n}\tau_n)} d\tau_1 \dots d\tau_n$$

Again, $y_{on}(t)$ includes components at all possible beat frequencies $m_1\omega_1 + \dots + m_n\omega_n$ ($m_q \in \{1, 2, \dots, n\}$ and $\Sigma_{q=1}^{n} m_q = n$). Looking only into the component at $\omega_1 + \dots + \omega_n$, we will have

$$n! e^{j(\omega_1 + \dots + \omega_n)t} \int_{-\infty}^{\infty} \dots \int_{-\infty}^{\infty} h_n(\tau_1, \dots, \tau_n) e^{-j(\omega_1\tau_1 + \dots + \omega_n\tau_n)} d\tau_1 \dots d\tau_n$$

$$= n! H_n(\omega_1, \dots, \omega_n) e^{j(\omega_1 + \dots + \omega_n)t} \tag{3.112}$$

which shows that the nth-order NLTF, $H_n(\omega_1, \dots, \omega_n)$, can be obtained from the system's response to an excitation of the form of (3.110), dividing the output component at $(\omega_1 + \dots + \omega_n)$ by $n! e^{j(\omega_1 + \dots + \omega_n)t}$.

Harmonic Input Method Applied to Circuit Analysis

The main step of the harmonic input method consists of determining the circuit's response to a sum of elementary exponentials. The nonlinear currents method can thus be used for this task, as it will be next illustrated for our example circuit.

For the application of the nonlinear currents method, the circuit of Figure 3.1 is redrawn as in Figure 3.7. We begin by determining the first-order NLTF, $H_1(\omega)$, for which the excitation is

$$i_s(t) = e^{j\omega t} \tag{3.113}$$

The first-order output voltage $V_{o1}(\omega)$ was given by (3.79), and thus,

$$H_1(\omega) = \frac{1}{G + g_1 + j\omega c_1} \tag{3.114}$$

For the second-order NLTF we assume

$$i_s(t) = e^{j\omega_1 t} + e^{j\omega_2 t} \tag{3.115}$$

and look for the output voltage component at the sum frequency, $V_{o2}(\omega_1 + \omega_2)$. From (3.81) and (3.82) we see that the $\omega_1 + \omega_2$ components of the nonlinear currents are

$$\begin{aligned} &2c_2[j\omega_1 H_1(\omega_1)H_1(\omega_2) + j\omega_2 H_1(\omega_2)H_1(\omega_1)]e^{j(\omega_1+\omega_2)t} \\ &= j(\omega_1 + \omega_2)2c_2 H_1(\omega_1)H_1(\omega_2)e^{j(\omega_1+\omega_2)t} \end{aligned} \tag{3.116}$$

and

$$g_2[H_1(\omega_1)H_1(\omega_2) + H_1(\omega_2)H_1(\omega_1)]e^{j(\omega_1+\omega_2)t} = 2g_2 H_1(\omega_1)H_1(\omega_2)e^{j(\omega_1+\omega_2)t} \tag{3.117}$$

and from (3.83) $H_2(\omega_1, \omega_2)$ is given by

$$H_2(\omega_1, \omega_2) = -[j(\omega_1 + \omega_2)c_2 + g_2]H_1(\omega_1)H_1(\omega_2)H_1(\omega_1 + \omega_2) \tag{3.118}$$

The third-order NLTF derivation requires an input of the form

$$i_s(t) = e^{j\omega_1 t} + e^{j\omega_2 t} + e^{j\omega_3 t} \tag{3.119}$$

which leads to the following nonlinear current components at $(\omega_1 + \omega_2 + \omega_3)$ [see (3.85) and (3.86)]:

$$\begin{aligned} &j(\omega_1 + \omega_2 + \omega_3)\{4c_2[H_1(\omega_1)H_2(\omega_2, \omega_3) + H_1(\omega_2)H_2(\omega_1, \omega_3) \\ &+ H_1(\omega_3)H_2(\omega_1, \omega_2)] + 6c_3 H_1(\omega_1)H_1(\omega_2)H_1(\omega_3)\}e^{j(\omega_1+\omega_2+\omega_3)t} \\ &= I_{c3}(\omega_1 + \omega_2 + \omega_3)e^{j(\omega_1+\omega_2+\omega_3)t} \end{aligned} \tag{3.120}$$

and

$$\begin{aligned} &\{4g_2[H_1(\omega_1)H_2(\omega_2, \omega_3) + H_1(\omega_2)H_2(\omega_1, \omega_3) + H_1(\omega_3)H_2(\omega_1, \omega_2)] \\ &+ 6g_3 H_1(\omega_1)H_1(\omega_2)H_1(\omega_3)\}e^{j(\omega_1+\omega_2+\omega_3)t} \\ &= I_{nl3}(\omega_1 + \omega_2 + \omega_3)e^{j(\omega_1+\omega_2+\omega_3)t} \end{aligned} \tag{3.121}$$

Finally, (3.87) allows the calculation of $H_3(\omega_1, \omega_2, \omega_3)$ as

$$H_3(\omega_1, \omega_2, \omega_3) = -\frac{1}{3!} H_1(\omega_1 + \omega_2 + \omega_3)[I_{c3}(\omega_1 + \omega_2 + \omega_3) + I_{nl3}(\omega_1 + \omega_2 + \omega_3)] \quad (3.122)$$

To close this study, an important property of the NLTFs is worth noting. Because of the recursivity already noted in the nonlinear currents method, the NLTFs are also recursive in nature. That is, in general, the nth-order NLTF, $H_n(\omega_1, \ldots, \omega_n)$, depends on all NLTFs of lower order, $H_1(\omega)$, $H_2(\omega_1, \omega_2), \ldots,$ $H_{n-1}(\omega_1, \ldots, \omega_{n-1})$ and, thus, cannot be calculated before all lower order NLTFs are previously determined.

3.2.3 Volterra Series Analysis of Time-Varying Circuits

In the same way as Volterra series analysis of mildly nonlinear time-invariant circuits was an extension, to nth order, of the traditional linear (or first-order) circuit analysis methods, Volterra series analysis of mildly nonlinear time-varying circuits is an extension of linear time-varying ones. Its development is thus devoted to intermodulation phenomena in many types of analog, RF and microwave circuits, switched capacitor filters, sampler circuits, frequency converters, analog switches, parametric amplifiers, and modulators. The method that follows is undertaken in the frequency-domain and is an extension of the *Conversion Matrix* formalism [2, 5]. So, for the reader who is not familiar with that framework, we will begin by an introductory explanation of linear mixer analysis techniques.

Let us recall the illustrative nonlinear circuit of Figure 3.1. Since we want to perform a small-signal analysis of this circuit, we assume we have already calculated the quiescent point, so that the signal node voltages or branch currents behave as small perturbations to that fixed point. In the mixer case, for example, this quiescent point is composed by the voltages and currents forced by the large local oscillator, or pumping signal, plus any possible dc value. In previous sections, we have seen that the quiescent point could be calculated by a suitable nonlinear numerical method like the Newton-Raphson iteration. Thus, a similar task has to be done here, with the only difference that, now, the quiescent point is no longer a constant dc, but a time-varying (usually periodic) generalized one. An appropriate method for this task is the harmonic balance algorithm, explained in detail in Section 3.3.2.

If we have a time-dependent quiescent point, the Taylor series expansions of the current and charge nonlinearities should be given by

$$q_{NL}(t) = Q_{NL}(t) + c_1(t)[v_O(t) - V_O(t)] + c_2(t)[v_O(t) - V_O(t)]^2 + c_3(t)[v_O(t) - V_O(t)]^3 + \ldots \quad (3.123)$$

where $c_n(t)$ $(n = 1, 2, 3, \ldots)$ is the nth-order derivative, now evaluated in the time-varying quiescent point $V_O(t)$:

$$c_n(t) \equiv \frac{1}{n!}\frac{d^n q_{NL}(v_O)}{dv_O^n}\bigg|_{v_O = V_O(t)} \tag{3.124}$$

and

$$i_{NL}(t) = I_{NL}(t) + g_1(t)[v_O(t) - V_O(t)] + g_2(t)[v_O(t) - V_O(t)]^2 + g_3(t)[v_O(t) - V_O(t)]^3 + \ldots \tag{3.125}$$

where, accordingly,

$$g_n(t) \equiv \frac{1}{n!}\frac{d^n i_{NL}(v_O)}{dv_O^n}\bigg|_{v_O = V_O(t)} \tag{3.126}$$

The mildly nonlinear differential equation that models the dynamic circuit then becomes

$$\frac{d}{dt}[c_1(t)v_o(t) + c_2(t)v_o(t)^2 + c_3(t)v_o(t)^3] + [G + g_1(t)]v_o(t) + g_2(t)v_o(t)^2 + g_3(t)v_o(t)^3 = i_s(t) \tag{3.127}$$

and is supposed to admit a solution, $v_o(t)$, described by the following time-varying Volterra series:

$$v_o(t) = \sum_{n=1}^{\infty} v_{on}(t) \tag{3.128a}$$

$$v_{on}(t) = \int_{-\infty}^{\infty} \cdots \int_{-\infty}^{\infty} h_n(t, \tau_1, \ldots, \tau_n) i_s(t - \tau_1) \ldots i_s(t - \tau_n)\, d\tau_1 \ldots d\tau_n \tag{3.128b}$$

For expressing (3.128) in frequency-domain, we again consider $i_s(t)$ as a sum of Q sinusoids:

$$i_s(t) = \frac{1}{2} \sum_{\substack{q=-Q \\ (q \neq 0)}}^{Q} I_{s_q} e^{j\omega_q t} \tag{3.129}$$

which leads to

$$v_{on}(t) = \frac{1}{2^n} \sum_{q_1=-Q}^{Q} \ldots \sum_{q_n=-Q}^{Q} I_{s_{q_1}} \ldots I_{s_{q_n}} H_n(t, \omega_{q_1}, \ldots, \omega_{q_n}) e^{j(\omega_{q_1}+\ldots+\omega_{q_n})t} \tag{3.130}$$

For further describing $H_n(t, \omega_1, \ldots, \omega_n)$ entirely in the frequency-domain, we assume that the large pumping is a sinusoid of frequency ω_p, or, more generally, a periodic signal of fundamental frequency ω_p. In any case, every $c_n(t)$ or $g_n(t)$ will be periodic functions of the same fundamental frequency, which may then be represented by the Fourier series:[2]

$$c_n(t) = \sum_{k=-K}^{K} C_{n_k} e^{jk\omega_p t} \tag{3.131}$$

and

$$g_n(t) = \sum_{k=-K}^{K} G_{n_k} e^{jk\omega_p t} \tag{3.132}$$

The product of any of these time-varying Taylor coefficients by a control voltage composed of Q sinusoids generates a current whose components have all possible frequency mixing products: $\omega_{q,k} = k\omega_p + \omega_q$. That current will be then converted again into voltage in any circuit impedance. So, the pumping signal translates the input spectrum at ω_q into a large number of frequency clusters located around every pumping harmonic of ω_p plus dc. An example of such a spectrum, where $Q = 2$, is depicted in Figure 3.13.

A spectrum arrangement like the one of Figure 3.13 can be either created by mixing ω_p and its harmonics with $\pm\omega_1$ and $\pm\omega_2$, or by mixing the same ω_p with an input composed by the base-band, $\pm\omega_{b_1}$ and $\pm\omega_{b_2}$. In this case, the mixing components would be given by $\omega_{b,k'} = k'\omega_p + \omega_b$ $(b = -2, -1, 1, 2)$, where the new k' can be related to the previous k by $k' = k + 1$ or $k' = k - 1$, whether $\omega_b > 0$ or $\omega_b < 0$, respectively. This alternative description of the frequency components has a much more intuitive index k' distribution, as can be verified in Figure 3.14, and is thus preferred against the original $\omega_{k,p}$.

2. Note that although (3.129) and (3.131) look like similar only in (3.131) we are dealing with a periodic function. Therefore, (3.131) is, indeed, a Fourier series, whereas (3.129) is not.

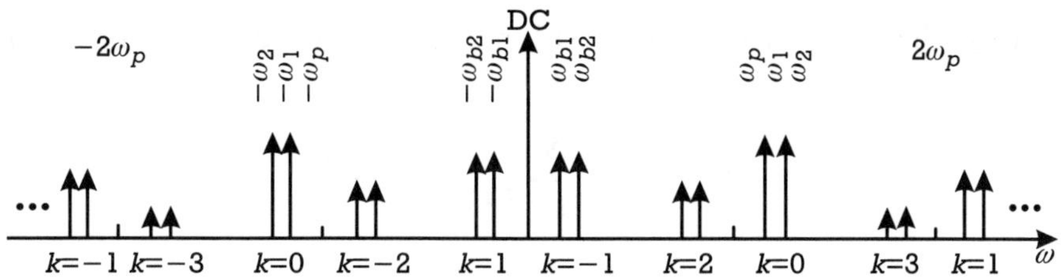

Figure 3.13 Output spectrum components of a mixer driven by a local oscillator and two RF tones.

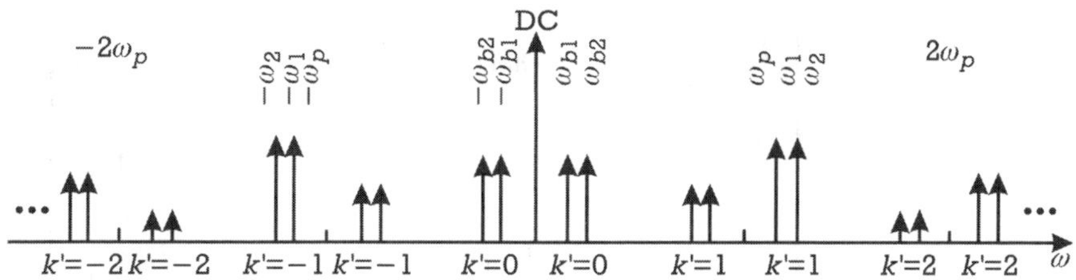

Figure 3.14 Alternative mixer output frequency components' indexing scheme.

Therefore, from now on we will adopt this new indexing scheme, in a way that the mixing product referred to as $\omega_{b,k}$ has a frequency component of $\omega_{b,k} = k\omega_p + \omega_b$. This component corresponds to $k\omega_p + \omega_b$ if ω_b are the excitation signals, or to $(k-1)\omega_p + \omega_q$ in case the inputs are at $\omega_q = \omega_p + \omega_b$, $\omega_q > \omega_p$, or $(k+1)\omega_p - \omega_q$ if the inputs are located at $\omega_q = \omega_p - \omega_b$, $\omega_q < \omega_p$. If k is a negative integer, $\omega_{b,k}$ represents a frequency whose value is symmetric to the one given.

The product of this voltage by another Taylor series coefficient like $g_1(t)$ in (3.132) produces a current $i_1(t)$ with a set of newly generated components given by: $\omega_{b,k_1+k_2} = (k_1 + k_2)\omega_p + \omega_b$. Since k_1 and k_2 are integers varying from $k_{1,2} = -K, \ldots, -1, 0, 1, \ldots, K$, $k_1 + k_2$ spans from $-2K$ to $+2K$. Another product sequence like this one would lead to frequencies ranging from $-3K$ to $+3K$, and so on. So, in practical terms, it is necessary to truncate this spectral regrowth to a certain pumping harmonic $K\omega_p$, determined by desired results' accuracy criteria.[3] The circuit's first-order voltage can then be represented by

$$v_{o1}(t) = \sum_{k=-K}^{K} \sum_{b=-B}^{B} V_{o1_{b,k}} e^{j(k\omega_p + \omega_b)t} \quad (B = Q) \tag{3.133}$$

3. The problem of selecting the highest harmonic order $K\omega_p$ is a very important issue in frequency-domain CAD, generally referred as *spectrum truncation*. It will be addressed in more detail in Section 3.3.

while the first-order current resulting from the product of $g_1(t)$ by $v_{o1}(t)$ is

$$i_{nl1}(t) \equiv g_1(t)v_{o1}(t) = \sum_{k_1=-K}^{K} \sum_{k_2=-K}^{K} \sum_{b=-B}^{B} G1_{k_1} Vo1_{b,k_2} e^{j[(k_1+k_2)\omega_p+\omega_b)]t} \tag{3.134}$$

If the components of $i_{nl1}(t)$ are to be truncated at $K\omega_p$ (i.e., such that $|k_1 + k_2| \le K$), then (3.134) represents the following matrix product:

$$\begin{bmatrix} Inl1_{-B,-K} & \cdots & Inl1_{-1,-K} & Inl1_{1,-K} & \cdots & Inl1_{B,-K} \\ \vdots & & \vdots & \vdots & & \vdots \\ Inl1_{-B,0} & \cdots & Inl1_{-1,0} & Inl1_{1,0} & \cdots & Inl1_{B,0} \\ \vdots & & \vdots & \vdots & & \vdots \\ Inl1_{-B,K} & \cdots & Inl1_{-1,K} & Inl1_{1,K} & \cdots & Inl1_{B,K} \end{bmatrix}$$

$$= \begin{bmatrix} G1_0 & \cdots & \cdots & G1_{-K} & 0 & \cdots & 0 \\ \vdots & & & \vdots & & & \vdots \\ \vdots & & & \vdots & & & 0 \\ G1_K & \cdots & \cdots & G1_0 & \cdots & \cdots & G1_{-K} \\ 0 & & & \vdots & & & \vdots \\ \vdots & & & \vdots & & & \vdots \\ 0 & \cdots & 0 & G1_K & \cdots & \cdots & G1_0 \end{bmatrix} \tag{3.135a}$$

$$\begin{bmatrix} Vo1_{-B,-K} & \cdots & Vo1_{-1,-K} & Vo1_{-1,-K} & \cdots & Vo1_{B,-K} \\ \vdots & & \vdots & \vdots & & \vdots \\ Vo1_{-B,0} & \cdots & Vo1_{-1,0} & Vo1_{1,0} & \cdots & Vo1_{B,0} \\ \vdots & & \vdots & \vdots & & \vdots \\ Vo1_{-B,K} & \cdots & Vo1_{-1,K} & Vo1_{1,K} & \cdots & Vo1_{B,K} \end{bmatrix}$$

or

$$\mathbf{I_{nl1}} = \mathbf{G_1 V_{o1}} \tag{3.135b}$$

$\mathbf{G}_1$ in (3.135) is the so-called *conversion matrix* of the first-order time-varying conductance $g_1(t)$.

Before moving forward with the explanation, it is useful to comment on the form herein adopted for (3.135).

We can recognize a clear regularity in the position of the terms of the conversion matrix $\mathbf{G}_1$. They are all located as if the matrix was filled by simply horizontally shifting to the right the $4K + 1$ vector $\left[0 \ldots 0 \; G1_K \ldots G1_0 \ldots G1_{-K} \; 0 \ldots 0\right]$ and retaining only the middle $2K + 1$ positions. A matrix in which the elements verify the relation $a_{ij} = t_{i-j}$, where the t_{i-j} are the elements of a line vector, as is the case of $\mathbf{G}_1$, is called a Toeplitz matrix, and can be used to represent a linear convolution by a matrix-vector product. Noting also that the vector $[t_{i-j}]$ is nothing more than the inverted vector of the Fourier coefficients of $g_1(t)$, we can conclude that (3.135) is, in fact, the matrix form of the frequency-domain convolution corresponding to the time-domain product $i_{nl1}(t) = g_1(t)v_{o1}(t)$. In this sense, the null positions located next to the Fourier coefficients $\left[G1_{-K} \ldots G1_0 \ldots G1_K\right]$ could be filled by nonzero values expanding the Fourier series from $-2K\omega_p$ up to $+2K\omega_p$. In that case, $\mathbf{G}_1$ would have the more common aspect of [5]

$$\mathbf{G}_1 = \begin{bmatrix} G1_0 & \cdots & G1_{-K} & \cdots & G1_{-2K} \\ \vdots & & \vdots & & \vdots \\ G1_K & \cdots & G1_0 & \cdots & G1_{-K} \\ \vdots & & \vdots & & \vdots \\ G1_{2K} & \cdots & G1_K & \cdots & G1_0 \end{bmatrix} \tag{3.136}$$

[Because of the potential increased accuracy of this formulation in comparison to the one in (3.135), it will be adopted in the mixer studies carried on in Chapter 5.]

Continuing with the mixer analysis, if the intended current was now the one generated in the nonlinear charge,

$$i_{c1}(t) = \frac{\mathrm{d}}{\mathrm{d}t}\left[c_1(t)v_{o1}(t)\right] \tag{3.137}$$

then, in the frequency-domain it would be given by

$$i_{c1}(t) = \sum_{k_1=-K}^{K} \sum_{k_2=-K}^{K} \sum_{b=-B}^{B} j[(k_1 + k_2)\omega_p + \omega_b] C1_{k_1} Vo1_{b,k_2} e^{j[(k_1+k_2)\omega_p + \omega_b]t} \tag{3.138}$$

which can again be described in conversion matrix form as

$$\begin{bmatrix} I_{c1_{-B,-K}} & \cdots & I_{c1_{-1,-K}} & I_{c1_{1,-K}} & \cdots & I_{c1_{B,-K}} \\ \vdots & & \vdots & \vdots & & \vdots \\ I_{c1_{-B,0}} & \cdots & I_{c1_{-1,0}} & I_{c1_{1,0}} & \cdots & I_{c1_{B,0}} \\ \vdots & & \vdots & \vdots & & \vdots \\ I_{c1_{-B,K}} & \cdots & I_{c1_{-1,K}} & I_{c1_{1,K}} & \cdots & I_{c1_{B,K}} \end{bmatrix}$$

$$= j \begin{bmatrix} -K\omega_p - \omega_B & \cdots & -K\omega_p - \omega_1 & -K\omega_p + \omega_1 & \cdots & -K\omega_p + \omega_B \\ \vdots & & \vdots & \vdots & & \vdots \\ -\omega_B & \cdots & -\omega_1 & \omega_1 & \cdots & \omega_B \\ \vdots & & \vdots & \vdots & & \vdots \\ K\omega_p - \omega_B & \cdots & K\omega_p - \omega_1 & K\omega_p + \omega_1 & \cdots & K\omega_p + \omega_B \end{bmatrix}$$

$$.x \begin{bmatrix} C1_0 & \cdots & \cdots & C1_{-K} & 0 & \cdots & 0 \\ \vdots & & & \vdots & & & \vdots \\ \vdots & & & \vdots & & & 0 \\ C1_K & \cdots & \cdots & C1_0 & \cdots & \cdots & C1_{-K} \\ 0 & & & \vdots & & & \vdots \\ \vdots & & & \vdots & & & \vdots \\ 0 & \cdots & 0 & C1_K & \cdots & \cdots & C1_0 \end{bmatrix}$$

$$\begin{bmatrix} V_{o1_{-B,-K}} & \cdots & V_{o1_{-1,-K}} & V_{o1_{1,-K}} & \cdots & V_{o1_{B,-K}} \\ \vdots & & \vdots & \vdots & & \vdots \\ V_{o1_{-B,0}} & \cdots & V_{o1_{-1,0}} & V_{o1_{1,0}} & \cdots & V_{o1_{B,0}} \\ \vdots & & \vdots & \vdots & & \vdots \\ V_{o1_{-B,K}} & \cdots & V_{o1_{-1,K}} & V_{o1_{1,K}} & \cdots & V_{o1_{B,K}} \end{bmatrix} \tag{3.139a}$$

or

$$\mathbf{I}_{c1} = j\boldsymbol{\Omega} \ .x \ \mathbf{C}_1 \mathbf{V}_{o1} \tag{3.139b}$$

where the ".x" operator represents a matrix product on an element by element basis:

$$\mathbf{Z} = \mathbf{X} \ .x \ \mathbf{Y}: \ z_{ij} = x_{ij}\, y_{ij}$$

Finally, the first-order current passing through a time-invariant capacitance, C', or conductance, G, (conventional linear circuit elements) would be equal to

$$\begin{bmatrix} I_{c'_{-B,-K}} & \cdots & I_{c'_{-1,-K}} & I_{c'_{1,-K}} & \cdots & I_{c'_{B,-K}} \\ \vdots & & \vdots & \vdots & & \vdots \\ I_{c'_{-B,0}} & \cdots & I_{c'_{-1,0}} & I_{c'_{1,0}} & \cdots & I_{c'_{B,0}} \\ \vdots & & \vdots & \vdots & & \vdots \\ I_{c'_{-B,K}} & \cdots & I_{c'_{-1,K}} & I_{c'_{1,K}} & \cdots & I_{c'_{B,K}} \end{bmatrix}$$

$$= jC' \begin{bmatrix} -K\omega_p - \omega_B & \cdots & -K\omega_p - \omega_1 & -K\omega_p + \omega_1 & \cdots & -K\omega_p + \omega_B \\ \vdots & & \vdots & \vdots & & \vdots \\ -\omega_B & \cdots & -\omega_1 & \omega_1 & \cdots & \omega_B \\ \vdots & & \vdots & \vdots & & \vdots \\ K\omega_p - \omega_B & \cdots & K\omega_p - \omega_1 & K\omega_p + \omega_1 & \cdots & K\omega_p + \omega_B \end{bmatrix}$$

$$.x \begin{bmatrix} Vo1_{-B,-K} & \cdots & Vo1_{-1,-K} & Vo1_{1,-K} & \cdots & Vo1_{B,-K} \\ \vdots & & \vdots & \vdots & & \vdots \\ Vo1_{-B,0} & \cdots & Vo1_{-1,0} & Vo1_{1,0} & \cdots & Vo1_{B,0} \\ \vdots & & \vdots & \vdots & & \vdots \\ Vo1_{-B,K} & \cdots & Vo1_{-1,K} & Vo1_{1,K} & \cdots & Vo1_{B,K} \end{bmatrix} \quad (3.140a)$$

or

$$\mathbf{I}_{c'} = j\mathbf{\Omega} C' \;.x\; \mathbf{V}_{o1} \quad (3.140b)$$

and

$$\begin{bmatrix} I_{G_{-B,-K}} & \cdots & I_{G_{-1,-K}} & I_{G_{1,-K}} & \cdots & I_{G_{B,-K}} \\ \vdots & & \vdots & \vdots & & \vdots \\ I_{G_{-B,0}} & \cdots & I_{G_{-1,0}} & I_{G_{1,0}} & \cdots & I_{G_{B,0}} \\ \vdots & & \vdots & \vdots & & \vdots \\ I_{G_{-B,K}} & \cdots & I_{G_{-1,K}} & I_{G_{1,K}} & \cdots & I_{G_{B,K}} \end{bmatrix} \quad (3.141a)$$

$$= G \begin{bmatrix} Vo1_{-B,-K} & \cdots & Vo1_{-1,-K} & Vo1_{1,-K} & \cdots & Vo1_{B,-K} \\ \vdots & & \vdots & \vdots & & \vdots \\ Vo1_{-B,0} & \cdots & Vo1_{-1,0} & Vo1_{1,0} & \cdots & Vo1_{B,0} \\ \vdots & & \vdots & \vdots & & \vdots \\ Vo1_{-B,K} & \cdots & Vo1_{-1,K} & Vo1_{1,K} & \cdots & Vo1_{B,K} \end{bmatrix}$$

or

$$I_G = G\,\mathbf{V}_{o1} \tag{3.141b}$$

respectively.

Since a constant-matrix product can be substituted by a matrix-matrix product if the constant is replaced by a diagonal matrix in which all elements are equal to that constant, it is usually accepted that such a diagonal matrix is the conversion matrix representation of time-invariant elements. In this way, for instance, (3.141) can also be expressed by

$$\mathbf{I}_G = \mathbf{G}\mathbf{V}_{o1} \tag{3.141c}$$

where $\mathbf{G}$: $g_{ij} = 0$ if $i \neq j$ and $g_{ij} = G$ if $i = j$.

With the above definitions in mind, it is now possible to use a conversion matrix form for the Kirchoff laws, enabling the analysis of any time-varying linear circuit. For example, the linearized time-varying model equation of (3.127) can be written as

$$j\boldsymbol{\Omega}\ .x\ \mathbf{C}_1\mathbf{V}_{o1} + (\mathbf{G} + \mathbf{G}_1)\mathbf{V}_{o1} = \mathbf{I}_s \tag{3.142}$$

$\mathbf{I}_s$ is a matrix representation of the input, where all elements are zeros except the line of $k = 0$, for the inputs at ω_b (see Figure 3.14), or the lines of $k = -1$ and $k = 1$ for the excitation at ω_q, respectively.

The first-order output voltage $v_{o1}(t)$ can now be derived from (3.142) as

$$\mathbf{V}_{o1} = [\mathbf{G} + \mathbf{G}_1 + j\boldsymbol{\Omega}\ .x\ \mathbf{C}_1]^{-1}\mathbf{I}_s = \mathbf{Z}_1\mathbf{I}_s \tag{3.143}$$

or

$$v_{o1}(t) = \sum_{k_1=-K}^{K}\sum_{k_2=-K}^{K}\sum_{b=-B}^{B} Z_{1_{k_1}} I_{s_{b,k_2}} e^{j[(k_1+k_2)\omega_p+\omega_b]t}$$

$$\Rightarrow v_{o1}(t) = \sum_{k=-K}^{K}\sum_{b=-B}^{B} V_{o1_{b,k}} e^{j(k\omega_p+\omega_b)t} \tag{3.144}$$

as was previously assumed by (3.143).

Using (3.143), it is obvious that the frequency-domain first-order transfer function can be expressed by

$$\mathbf{H}_1(\omega) = \mathbf{Z}_1 \tag{3.145}$$

Proceeding with the nonlinear currents method for next order components, $v_{on}(t)$, we now calculate $v_{o1}(t)^2$:

$$v_{o1}(t)^2 = \sum_{k_1=-K}^{K} \sum_{k_2=-K}^{K} \sum_{b_1=-B}^{B} \sum_{b_2=-B}^{B} V_{o1_{b_1,k_1}} V_{o1_{b_2,k_2}} e^{j[(k_1+k_2)\omega_p + \omega_{b_1} + \omega_{b_2}]t} \tag{3.146}$$

which, again truncated to the Kth ω_p harmonic, gives

$$\begin{aligned} v_{o1}(t)^2 &= \sum_{k=-K}^{K} \sum_{b_1=-B}^{B} \sum_{b_2=-B}^{B} V_{o1_{(b_1+b_2),k}}^{(2)} e^{j(k\omega_p + \omega_{b_1} + \omega_{b_2})t} \\ &= \sum_{k=-K}^{K} \sum_{c=-C}^{C} V_{o1_{c,k}}^{(2)} e^{j(k\omega_p + \omega_c)t} \end{aligned} \tag{3.147}$$

where $V_{o1_{c,k}}^{(2)}$ stands for the frequency-domain representation of the second-order products generated from $v_{o1}(t)$.

Following the conversion matrix notation presented above, $i_{nl2}(t) = g_2(t)v_{o1}(t)^2$ is given by

$$\begin{bmatrix} I_{nl2_{-C,-K}} & \cdots & I_{nl2_{C,-K}} \\ \vdots & & \vdots \\ I_{nl2_{-C,0}} & \cdots & I_{nl2_{C,0}} \\ \vdots & & \vdots \\ I_{nl2_{-C,K}} & \cdots & I_{nl2_{C,K}} \end{bmatrix} = \begin{bmatrix} G_{2_0} & \cdots & G_{2_{-K}} & \cdots & 0 \\ \vdots & & \vdots & & \vdots \\ G_{2_K} & \cdots & G_{2_0} & \cdots & G_{2_{-K}} \\ \vdots & & \vdots & & \vdots \\ 0 & \cdots & G_{2_K} & \cdots & G_{2_0} \end{bmatrix} \begin{bmatrix} V_{o1_{-C,-K}}^{(2)} & \cdots & V_{o1_{C,-K}}^{(2)} \\ \vdots & & \vdots \\ V_{o1_{-C,0}}^{(2)} & \cdots & V_{o1_{C,0}}^{(2)} \\ \vdots & & \vdots \\ V_{o1_{C,K}}^{(2)} & \cdots & V_{o1_{C,K}}^{(2)} \end{bmatrix} \tag{3.148a}$$

or

$$\mathbf{I}_{nl2} = \mathbf{G}_2 \mathbf{V}_{o1}^{(2)} \tag{3.148b}$$

and

$$i_{c2}(t) = \frac{\mathrm{d}}{\mathrm{d}t}[c_2(t)v_{o1}(t)^2] \tag{3.149}$$

$$= \sum_{k_1=-K}^{K} \sum_{k_2=-K}^{K} \sum_{c=-C}^{C} j[(k_1+k_2)\omega_p + \omega_c]\, C_{2_{k_1}} V_{o1_{c,k_2}}^{(2)} e^{j[(k_1+k_2)\omega_p+\omega_c]t}$$

or

$$\begin{bmatrix} I_{c2_{-C,-K}} & \cdots & I_{c2_{C,-K}} \\ \vdots & & \vdots \\ I_{c2_{-C,0}} & \cdots & I_{c2_{C,0}} \\ \vdots & & \vdots \\ I_{c2_{-C,K}} & \cdots & I_{c2_{C,K}} \end{bmatrix} = j \begin{bmatrix} -K\omega_p - \omega_C & \cdots & -K\omega_p + \omega_C \\ \vdots & & \vdots \\ -\omega_C & \cdots & \omega_C \\ \vdots & & \vdots \\ K\omega_p - \omega_C & \cdots & K\omega_p + \omega_C \end{bmatrix}$$

$$.x \begin{bmatrix} C_{2_0} & \cdots & C_{2_{-K}} & \cdots & 0 \\ \vdots & & \vdots & & \vdots \\ C_{2_K} & \cdots & C_{2_0} & \cdots & C_{2_{-K}} \\ \vdots & & \vdots & & \vdots \\ 0 & \cdots & C_{2_K} & \cdots & C_{2_0} \end{bmatrix} \begin{bmatrix} V_{o1_{-C,-K}}^{(2)} & \cdots & V_{o1_{C,-K}}^{(2)} \\ \vdots & & \vdots \\ V_{o1_{-C,0}}^{(2)} & \cdots & V_{o1_{C,0}}^{(2)} \\ \vdots & & \vdots \\ V_{o1_{-C,K}}^{(2)} & \cdots & V_{o1_{C,K}}^{(2)} \end{bmatrix} \tag{3.150a}$$

or even

$$\mathbf{I}_{c2} = j\boldsymbol{\Omega}\ .x\ \mathbf{C}_2 \mathbf{V}_{o1}^{(2)} \tag{3.150b}$$

The second-order linear time-varying equation of our circuit can thus be expressed as

$$j\boldsymbol{\Omega}\ .x\ \mathbf{C}_1\mathbf{V}_{o2} + (\mathbf{G} + \mathbf{G}_1)\mathbf{V}_{o2} = -\mathbf{I}_{c2} - \mathbf{I}_{nl2} \tag{3.151}$$

which gives the second-order output voltage:

$$\mathbf{V}_{o2} = -[\mathbf{G} + \mathbf{G}_1 + j\boldsymbol{\Omega}\ .x\ \mathbf{C}_1]^{-1}(\mathbf{I}_{c2} + \mathbf{I}_{nl2}) = -\mathbf{Z}_1(\mathbf{I}_{c2} + \mathbf{I}_{nl2}) \tag{3.152}$$

Equation (3.152) can be used as a control voltage to determine third-order output components, or to calculate $\mathbf{H}_2(\omega_1, \omega_2)$. For that, an $i_s(t) = e^{j\omega_1 t} + e^{j\omega_2 t}$ excitation is considered, and the output components at the converted frequency corresponding to $\omega_1 + \omega_2$ should be determined from (3.152):

$$v_{o2}(t) = \sum_{k=-K}^{K} \sum_{c=-C}^{C} V_{o2_{c,k}} e^{j(k\omega_p + \omega_c)t} \tag{3.153}$$

Similarly to what was done to second order, third-order output components are determined from $v_{o1}(t)$ and $v_{o2}(t)$ by first calculating the third-order nonlinear excitations:

$$\begin{aligned} i_{nl3}(t) &= 2g_2(t)v_{o1}(t)v_{o2}(t) + g_3(t)v_{o1}(t)^3 \\ &= 2\sum_{k_1=-K}^{K}\sum_{k_2=-K}^{K}\sum_{k_3=-K}^{K}\sum_{b_1=-B}^{B}\sum_{b_2=-B}^{B}\sum_{b_3=-B}^{B} G_{2_{k_1}} V_{o1_{b_1,k_2}} V_{o2_{(b_2+b_3),k_3}} \\ &\quad \cdot e^{j\left[(k_1+k_2+k_3)\omega_p + \omega_{b_1} + \omega_{b_2} + \omega_{b_3}\right]t} \\ &\quad + \sum_{k_1=-K}^{K}\sum_{k_2=-K}^{K}\sum_{k_3=-K}^{K}\sum_{k_4=-K}^{K}\sum_{b_1=-B}^{B}\sum_{b_2=-B}^{B}\sum_{b_3=-B}^{B} G_{3_{k_1}} V_{o1_{b_1,k_2}} V_{o1_{b_2,k_3}} V_{o1_{b_3,k_4}} \\ &\quad \cdot e^{j\left[(k_1+k_2+k_3+k_4)\omega_p + \omega_{b_1} + \omega_{b_2} + \omega_{b_3}\right]t} \end{aligned} \tag{3.154}$$

or

$$\begin{bmatrix} I_{nl3_{-D,-K}} & \cdots & I_{nl3_{D,-K}} \\ \vdots & & \vdots \\ I_{nl3_{-D,0}} & \cdots & I_{nl3_{D,0}} \\ \vdots & & \vdots \\ I_{nl3_{-D,K}} & \cdots & I_{nl3_{D,K}} \end{bmatrix} =$$

$$2\begin{bmatrix} G_{2_0} & \cdots & G_{2_{-K}} & \cdots & 0 \\ \vdots & & \vdots & & \vdots \\ G_{2_K} & \cdots & G_{2_0} & \cdots & G_{2_{-K}} \\ \vdots & & \vdots & & \vdots \\ 0 & \cdots & G_{2_K} & \cdots & G_{2_0} \end{bmatrix} \begin{bmatrix} V_{o12_{-D,-K}}^{(3)} & \cdots & V_{o12_{D,-K}}^{(3)} \\ \vdots & & \vdots \\ V_{o12_{-D,0}}^{(3)} & \cdots & V_{o12_{D,0}}^{(3)} \\ \vdots & & \vdots \\ V_{o12_{-D,K}}^{(3)} & \cdots & V_{o12_{D,K}}^{(3)} \end{bmatrix}$$

$$+\begin{bmatrix} G_{3_0} & \cdots & G_{3_{-K}} & \cdots & 0 \\ \vdots & & \vdots & & \vdots \\ G_{3_K} & \cdots & G_{3_0} & \cdots & G_{3_{-K}} \\ \vdots & & \vdots & & \vdots \\ 0 & \cdots & G_{3_K} & \cdots & G_{3_0} \end{bmatrix} \begin{bmatrix} V_{o1_{-D,-K}}^{(3)} & \cdots & V_{o1_{D,-K}}^{(3)} \\ \vdots & & \vdots \\ V_{o1_{-D,0}}^{(3)} & \cdots & V_{o1_{D,0}}^{(3)} \\ \vdots & & \vdots \\ V_{o1_{-D,K}}^{(3)} & \cdots & V_{o1_{D,K}}^{(3)} \end{bmatrix} \tag{3.155a}$$

or even

$$\mathbf{I}_{\mathrm{nl}3} = 2\mathbf{G}_2\mathbf{V}_{\mathrm{o}12}^{(3)} + \mathbf{G}_3\mathbf{V}_{\mathrm{o}1}^{(3)} \tag{3.155b}$$

and

$$i_{c3}(t) = \frac{\mathrm{d}}{\mathrm{d}t}\left[2c_2(t)v_{o1}(t)v_{o2}(t) + c_3(t)v_{o1}(t)^3\right]$$

$$= 2\sum_{k_1=-K}^{K}\sum_{k_2=-K}^{K}\sum_{k_3=-K}^{K}\sum_{b_1=-B}^{B}\sum_{b_2=-B}^{B}\sum_{b_3=-B}^{B} j[(k_1+k_2+k_3)\omega_p + \omega_{b_1} + \omega_{b_2} + \omega_{b_3}]$$

$$\cdot\, C2_{k_1} Vo1_{b_1,k_2} Vo2_{(b_2+b_3),k_3}\, e^{j[(k_1+k_2+k_3)\omega_p + \omega_{b_1} + \omega_{b_2} + \omega_{b_3}]t}$$

$$+ \sum_{k_1=-K}^{K}\sum_{k_2=-K}^{K}\sum_{k_3=-K}^{K}\sum_{k_4=-K}^{K}\sum_{b_1=-B}^{B}\sum_{b_2=-B}^{B}\sum_{b_3=-B}^{B} j[(k_1+k_2+k_3+k_4)\omega_p + \omega_{b_1} + \omega_{b_2} + \omega_{b_3}]$$

$$\cdot\, C3_{k_1} Vo1_{b_1,k_2} Vo1_{b_2,k_3} Vo1_{b_3,k_4}\, e^{j[(k_1+k_2+k_3+k_4)\omega_p + \omega_{b_1} + \omega_{b_2} + \omega_{b_3}]t} \tag{3.156}$$

or

$$\mathbf{I}_{c3} = 2j\boldsymbol{\Omega}\ .x\ \mathbf{C}_2\mathbf{V}_{\mathrm{o}12}^{(3)} + j\boldsymbol{\Omega}\ .x\ \mathbf{C}_3\mathbf{V}_{\mathrm{o}1}^{(3)} \tag{3.157}$$

Therefore, $v_{o3}(t)$ can be obtained from

$$\mathbf{V}_{\mathrm{o}3} = -[\mathbf{G} + \mathbf{G}_1 + j\boldsymbol{\Omega}\ .x\ \mathbf{C}_1]^{-1}(\mathbf{I}_{c3} + \mathbf{I}_{\mathrm{nl}3}) = -\mathbf{Z}_1(\mathbf{I}_{c3} + \mathbf{I}_{\mathrm{nl}3}) \tag{3.158}$$

Following what was said for second-order $\mathbf{V}_{\mathrm{o}2}$, (3.158) can also be used to derive the third-order nonlinear transfer function $\mathbf{H}_3(\omega_1, \omega_2, \omega_3)$, if an appropriate elementary excitation is assumed for $i_s(t)$.

The extension of the time-varying nonlinear currents method, just explained, to multiport networks is straightforward, although laboriously involved. So, it will not be further discussed.

The main conclusion one should keep in mind is that the formalism of conversion matrix enables the analysis of any linear time-varying circuit in matrix form, in much the same way linear time-invariant networks were already treated. And, since the Volterra series analysis of any weakly nonlinear circuit simply consists on repeatedly analyzing the linearized circuit with the appropriate nth-order excitation, the extension to time-varying networks simply requires the substitution of the circuit variables and elements by their conversion matrix counterparts.

3.2.4 Volterra Series Analysis at the System Level

Since Volterra series is a technique that provides closed-form solutions to mildly nonlinear systems, it can also be used for behavioral system modeling. This is accomplished by a set of time-domain Volterra kernels or frequency-domain nonlinear transfer functions. To exemplify that, we will now derive the NLTFs up to third-order of three different system configurations of practical interest: feedforward (or parallel), cascade, and feedback connections. Since we are interested in directly obtaining the NLTFs, the harmonic input method will be used throughout the following derivations.

In all cases under study it is assumed that the nonlinear subsystems do not interact with each other. This means that the set of NLTFs used to characterize a certain block do not depend on the input and output terminations (the source and load circuit impedances), or, alternatively, the adjacent blocks provide exactly the terminations previously used for NLTF identification. If such conditions cannot be guaranteed, the composite system must be analyzed as a whole at the circuit level using the nonlinear currents method. Beyond that, two other basic assumptions must apply. First, the whole system must be stable. And second, every subsystem, and the composite system, must be well described by a Volterra series with a few number of terms; in the following examples, the number of terms is three.

3.2.4.1 Parallel Connection of Two Subsystems

Let us consider the feedforward, or parallel connection, of two mildly nonlinear subsystems, $S^A[.]$ and $S^B[.]$, as represented in Figure 3.15.

Each of the three nonlinear systems, $S^A[.]$, $S^B[.]$, and the composite feedforward arrangement, $S^C[.]$, are characterized in the frequency-domain by its frequency-domain NLTFs up to third order, such that, for an input

$$x^{A,B,C}(t) = \frac{1}{2} \sum_{q=-Q}^{Q} X_q^{A,B,C} e^{j\omega_q t} \tag{3.159}$$

the systems respond with

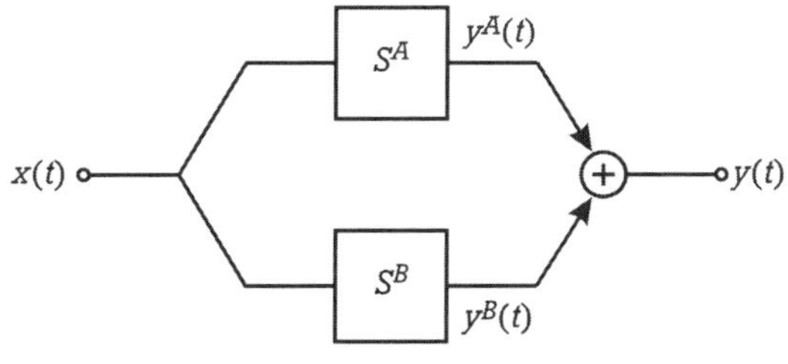

Figure 3.15 Block diagram of a parallel connection of two mildly nonlinear subsystems.

$$y^{A,B,C}(t) = \sum_{n=1}^{3} y_n^{A,B,C}(t) \tag{3.160a}$$

where

$$y_n^{A,B,C}(t) = \frac{1}{2^n} \sum_{q_1=-Q}^{Q} \dots \sum_{q_n=-Q}^{Q} X_{q_1}^{A,B,C} \dots X_{q_n}^{A,B,C} \tag{3.160b}$$

$$\cdot H_n^{A,B,C}(\omega_{q_1}, \dots, \omega_{q_n}) e^{j(\omega_{q_1}+\dots+\omega_{q_n})t}$$

The objective of the following calculations is to derive the first three NLTFs, $H_n(\omega_1, \dots, \omega_n)$ of the composite system, $S^C[.]$, defined by

$$y(t) = y^A(t) + y^B(t) \equiv S^C[x(t)] \tag{3.161a}$$

where

$$x(t) = x^A(t) = x^B(t) \tag{3.161b}$$

First-Order NLTF Derivation

Assuming a first-order elementary input of

$$x(t) = e^{j\omega t} \tag{3.162}$$

substituting it into (3.160) and (3.161) and retaining only components at frequency ω, we obtain

$$y(t) = H_1^C(\omega)e^{j\omega t} + \dots = H_1^A(\omega)e^{j\omega t} + H_1^B(\omega)e^{j\omega t} + \dots \tag{3.163}$$

Thus,

$$H_1^C(\omega) = H_1^A(\omega) + H_1^B(\omega) \tag{3.164}$$

Second-Order NLTF Derivation

The second-order elementary input is

$$x(t) = e^{j\omega_1 t} + e^{j\omega_2 t} \tag{3.165}$$

which, applied to (3.160) and (3.161) gives

$$\begin{aligned} y(t) &= 2!H_2^C(\omega_1, \omega_2)e^{j(\omega_1+\omega_2)t} + \ldots \\ &= 2H_2^A(\omega_1, \omega_2)e^{j(\omega_1+\omega_2)t} + 2H_2^B(\omega_1, \omega_2)e^{j(\omega_1+\omega_2)t} + \ldots \end{aligned} \tag{3.166}$$

and thus,

$$H_2^C(\omega_1, \omega_2) = H_2^A(\omega_1, \omega_2) + H_2^B(\omega_1, \omega_2) \tag{3.167}$$

Third-Order NLTF Derivation

The third-order elementary input is now

$$x(t) = e^{j\omega_1 t} + e^{j\omega_2 t} + e^{j\omega_3 t} \tag{3.168}$$

which substituted into (3.160) and (3.161) gives

$$\begin{aligned} y(t) &= 3!H_3^C(\omega_1, \omega_2, \omega_3)e^{j(\omega_1+\omega_2+\omega_3)t} + \ldots \\ &= 6H_3^A(\omega_1, \omega_2, \omega_3)e^{j(\omega_1+\omega_2+\omega_3)t} \\ &\quad + 6H_3^B(\omega_1, \omega_2, \omega_3)e^{j(\omega_1+\omega_2+\omega_3)t} + \ldots \end{aligned} \tag{3.169}$$

and again,

$$H_3^C(\omega_1, \omega_2, \omega_3) = H_3^A(\omega_1, \omega_2, \omega_3) + H_3^B(\omega_1, \omega_2, \omega_3) \tag{3.170}$$

3.2.4.2 Cascade Connection of Two Subsystems

The cascade arrangement to be analyzed is shown in Figure 3.16.

Using the same assumptions proposed for the parallel arrangement, we now want to determine the composite cascaded system's NLTFs. This arrangement is described by the following equations:

$$x^A(t) = x(t); \quad x^B(t) = y^A(t); \quad y(t) = y^B(t) \tag{3.171}$$

and (3.159) and (3.160).

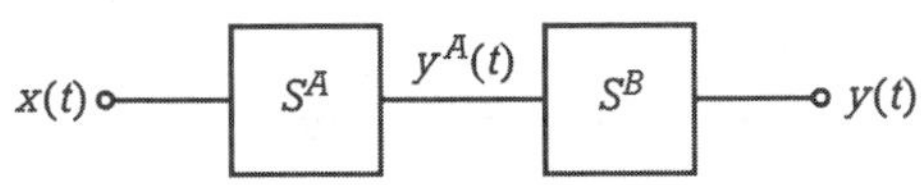

Figure 3.16 Block diagram of a cascade arrangement of two mildly nonlinear subsystems.

First-Order NLTF Derivation

The first-order elementary input is

$$x(t) = e^{j\omega t} \tag{3.172}$$

which substituted into (3.159), (3.160), and (3.161) gives

$$y(t) = H_1^C(\omega)e^{j\omega t} + \ldots = H_1^A(\omega)H_1^B(\omega)e^{j\omega t} + \ldots \tag{3.173}$$

and thus,

$$H_1^C(\omega) = H_1^A(\omega)H_1^B(\omega) \tag{3.174}$$

This is nothing more than the traditional result of the cascaded linear systems' transfer function.

Second-Order NLTF Derivation

The second-order elementary input is

$$x(t) = e^{j\omega_1 t} + e^{j\omega_2 t} \tag{3.175}$$

Thus,

$$y^A(t) = H_1^A(\omega_1)e^{j\omega_1 t} + H_1^A(\omega_2)e^{j\omega_2 t} + 2H_2^A(\omega_1, \omega_2)e^{j(\omega_1+\omega_2)t} + \ldots = x^B(t) \tag{3.176}$$

and

$$\begin{aligned} y(t) &= 2!H_2^C(\omega_1, \omega_2)e^{j(\omega_1+\omega_2)t} + \ldots \\ &= 2H_1^B(\omega_1 + \omega_2)H_2^A(\omega_1, \omega_2)e^{j(\omega_1+\omega_2)t} \\ &\quad + 2H_2^B(\omega_1, \omega_2)H_1^A(\omega_1)H_1^A(\omega_2)e^{j(\omega_1+\omega_2)t} + \ldots \end{aligned} \tag{3.177}$$

Therefore,

$$H_2^C(\omega_1, \omega_2) = H_1^B(\omega_1 + \omega_2)H_2^A(\omega_1, \omega_2) + H_2^B(\omega_1, \omega_2)H_1^A(\omega_1)H_1^A(\omega_2) \tag{3.178}$$

Third-Order NLTF Derivation

The third-order elementary input is now

$$x(t) = e^{j\omega_1 t} + e^{j\omega_2 t} + e^{j\omega_3 t} \tag{3.179}$$

and

$$\begin{aligned} y^A(t) &= H_1^A(\omega_1)e^{j\omega_1 t} + H_1^A(\omega_2)e^{j\omega_2 t} + H_1^A(\omega_3)e^{j\omega_3 t} + 2H_2^A(\omega_1, \omega_2)e^{j(\omega_1+\omega_2)t} \\ &\quad + 2H_2^A(\omega_1, \omega_3)e^{j(\omega_1+\omega_3)t} + 2H_2^A(\omega_2, \omega_3)e^{j(\omega_2+\omega_3)t} \\ &\quad + 6H_3^A(\omega_1, \omega_2, \omega_3)e^{j(\omega_1+\omega_2+\omega_3)t} + \ldots \\ &= x^B(t) \end{aligned} \tag{3.180}$$

which leads to a third-order NLTF of

$$\begin{aligned} H_3^C(\omega_1, \omega_2, \omega_3) &= H_1^B(\omega_1 + \omega_2 + \omega_3)H_3^A(\omega_1, \omega_2, \omega_3) \\ &\quad + \frac{2}{3}H_2^B(\omega_1, \omega_2 + \omega_3)H_1^A(\omega_1)H_2^A(\omega_2, \omega_3) \\ &\quad + \frac{2}{3}H_2^B(\omega_2, \omega_1 + \omega_3)H_1^A(\omega_2)H_2^A(\omega_1, \omega_3) \\ &\quad + \frac{2}{3}H_2^B(\omega_3, \omega_1 + \omega_2)H_1^A(\omega_3)H_2^A(\omega_1, \omega_2) \\ &\quad + H_3^B(\omega_1, \omega_2, \omega_3)H_1^A(\omega_1)H_1^A(\omega_2)H_1^A(\omega_3) \end{aligned} \tag{3.181}$$

3.2.4.3 Feedback Connection of Two Subsystems

As a final example of Volterra series analysis of weakly nonlinear systems, we will now derive the NLTFs of the general (linear or nonlinear) feedback arrangement of Figure 3.17.

The assumptions of (3.159) and (3.160) are again considered, along with the feedback relations,

$$y(t) = y^A(t) = S^A[x^A(t)];\; y^B(t) = S^B[y(t)] \tag{3.182a}$$

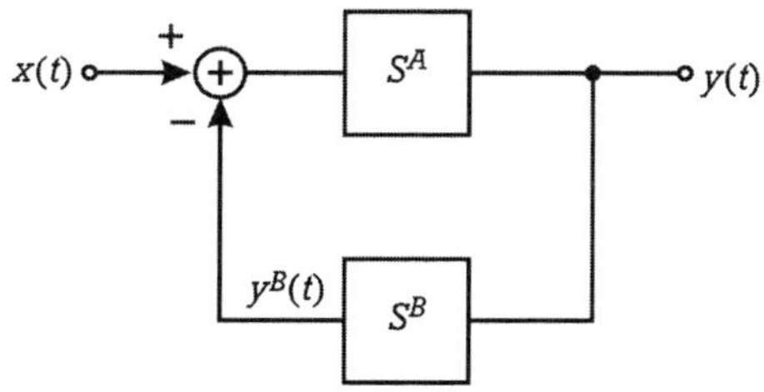

Figure 3.17 Block diagram of a feedback connection of two mildly nonlinear subsystems.

and

$$x^A(t) = x(t) - y^B(t) \tag{3.182b}$$

First-Order NLTF Derivation

Considering again the elementary input $x(t) = e^{j\omega t}$ in (3.159), (3.160), and (3.182), we have

$$y(t) = H_1^C(\omega)e^{j\omega t} + \ldots = H_1^A(\omega)[1 - H_1^B(\omega)H_1^C(\omega)]e^{j\omega t} + \ldots \tag{3.183}$$

or

$$H_1^C(\omega) = \frac{H_1^A(\omega)}{1 + H_1^A(\omega)H_1^B(\omega)} \tag{3.184}$$

the traditional feedback formula of linear control system analysis.

Second-Order NLTF Derivation

For the calculation of the second-order NLTF, a $x(t) = e^{j\omega_1 t} + e^{j\omega_2 t}$ is considered, which, applied to (3.159), (3.160), and (3.182), leads to

$$\begin{aligned} y(t) &= 2!H_2^C(\omega_1, \omega_2)e^{j(\omega_1+\omega_2)t} + \ldots \\ &= -H_1^A(\omega_1 + \omega_2) \\ &\quad \cdot [2H_1^B(\omega_1 + \omega_2)H_2^C(\omega_1, \omega_2) + 2H_2^B(\omega_1, \omega_2)H_1^C(\omega_1)H_1^C(\omega_2)]e^{j(\omega_1+\omega_2)t} \\ &\quad + 2H_2^A(\omega_1, \omega_2)[1 - H_1^B(\omega_1)H_1^C(\omega_1)][1 - H_1^B(\omega_2)H_1^C(\omega_2)]e^{j(\omega_1+\omega_2)t} \\ &\quad + \ldots \end{aligned} \tag{3.185}$$

which is equivalent to

$$\begin{aligned} H_2^C(\omega_1, \omega_2) = &\frac{1}{1 + H_1^A(\omega_1 + \omega_2)H_1^B(\omega_1 + \omega_2)} \\ &\cdot \{H_2^A(\omega_1, \omega_2)[1 - H_1^B(\omega_1)H_1^C(\omega_1)][1 - H_1^B(\omega_2)H_1^C(\omega_2)] \\ &- H_1^A(\omega_1 + \omega_2)H_2^B(\omega_1, \omega_2)H_1^C(\omega_1)H_1^C(\omega_2)\} \end{aligned} \tag{3.186}$$

Third-Order NLTF Derivation

In a similar way, the input is now $x(t) = e^{j\omega_1 t} + e^{j\omega_2 t} + e^{j\omega_3 t}$ and

$$
\begin{aligned}
y(t) &= 3!H_3^C(\omega_1, \omega_2, \omega_3)e^{j(\omega_1+\omega_2+\omega_3)t} + \dots \\
&= -H_1^A(\omega_1 + \omega_2 + \omega_3) \\
&\quad \cdot [6H_1^B(\omega_1 + \omega_2 + \omega_3)H_3^C(\omega_1, \omega_2, \omega_3) \\
&\quad + 4H_2^B(\omega_1, \omega_2 + \omega_3)H_1^C(\omega_1)H_2^C(\omega_2, \omega_3) \\
&\quad + 4H_2^B(\omega_2, \omega_1 + \omega_3)H_1^C(\omega_2)H_2^C(\omega_1, \omega_3) \\
&\quad + 4H_2^B(\omega_3, \omega_1 + \omega_2)H_1^C(\omega_3)H_2^C(\omega_1, \omega_2) \\
&\quad + 6H_3^B(\omega_1, \omega_2, \omega_3)H_1^C(\omega_1)H_1^C(\omega_2)H_1^C(\omega_3)]e^{j(\omega_1+\omega_2+\omega_3)t} \\
&\quad + \{4H_2^A(\omega_1, \omega_2 + \omega_3)[1 - H_1^B(\omega_1)H_1^C(\omega_1)] \\
&\quad \cdot [H_1^B(\omega_2 + \omega_3)H_2^C(\omega_2, \omega_3) + H_2^B(\omega_2, \omega_3)H_1^C(\omega_2)H_1^C(\omega_3)] \\
&\quad + \{4H_2^A(\omega_2, \omega_1 + \omega_3)[1 - H_1^B(\omega_2)H_1^C(\omega_2)] \\
&\quad \cdot [H_1^B(\omega_1 + \omega_3)H_2^C(\omega_1, \omega_3) + H_2^B(\omega_1, \omega_3)H_1^C(\omega_1)H_1^C(\omega_3)] \\
&\quad + \{4H_2^A(\omega_3, \omega_1 + \omega_2)[1 - H_1^B(\omega_3)H_1^C(\omega_3)] \\
&\quad \cdot [H_1^B(\omega_1 + \omega_2)H_2^C(\omega_1, \omega_2) + H_2^B(\omega_1, \omega_2)H_1^C(\omega_1)H_1^C(\omega_2)]\}e^{j(\omega_1+\omega_2+\omega_3)t} \\
&\quad + 6H_3^A(\omega_1, \omega_2, \omega_3)[1 - H_1^B(\omega_1)H_1^C(\omega_1)] \\
&\quad \cdot [1 - H_1^B(\omega_2)H_1^C(\omega_2)][1 - H_1^B(\omega_3)H_1^C(\omega_3)]e^{j(\omega_1+\omega_2+\omega_3)t} + \dots
\end{aligned}
\tag{3.187}
$$

Therefore, $H_3^C(\omega_1, \omega_2, \omega_3)$ can be expressed as

$$H_3^C(\omega_1, \omega_2, \omega_3) = \frac{1}{1 + H_1^A(\omega_1 + \omega_2 + \omega_3) H_1^B(\omega_1 + \omega_2 + \omega_3)}$$

$$\cdot \left\{ -\frac{2}{3} H_1^A(\omega_1 + \omega_2 + \omega_3) C_{ijk} [H_2^B(\omega_i, \omega_j + \omega_k) H_1^C(\omega_i) H_2^C(\omega_j, \omega_k)] \right.$$

$$- H_1^A(\omega_1 + \omega_2 + \omega_3) H_3^B(\omega_1, \omega_2, \omega_3) H_1^C(\omega_1) H_1^C(\omega_2) H_1^C(\omega_3)$$

$$+ \frac{2}{3} C_{ijk} \{ H_2^A(\omega_i, \omega_j + \omega_k) [1 - H_1^B(\omega_i) H_1^C(\omega_i)]$$

$$\cdot [H_1^B(\omega_j + \omega_k) H_2^C(\omega_j, \omega_k) + H_2^B(\omega_j, \omega_k) H_1^C(\omega_j) H_1^C(\omega_k)]\}$$

$$+ H_3^A(\omega_1, \omega_2, \omega_3) [1 - H_1^B(\omega_1) H_1^C(\omega_1)]$$

$$\left. \cdot [1 - H_1^B(\omega_2) H_1^C(\omega_2)] [1 - H_1^B(\omega_3) H_1^C(\omega_3)] \right\} \tag{3.188}$$

in which $C_{ijk}[.]$ stands for the sum of all three possible combinations of (i, j, k), as is detailed in (3.187).

3.2.5 Limitations of Volterra Series Techniques

Although Volterra series techniques are the preferred method for nonlinear distortion modeling, they present some important limitations.

The first problem we should point out refers to the Volterra series as a macromodeling tool: it is the inherent difficulty one faces when measuring the system's Volterra kernels, directly in time-domain, or their correspondent frequency-domain nonlinear transfer functions. Because this book is mainly devoted to analog and RF circuit designers, we will focus this brief discussion in the NLTF's extraction.

To put it in simple and practical terms, let us imagine we would like to measure the first n $H_n(\omega_1, \ldots, \omega_n)$ of the one-node circuit example we have been using. The first-order NLTF of this circuit is nothing but the small-signal frequency-domain impedance seen into the node. Thus, measuring $H_1(\omega)$ in Q frequency points requires Q different tests, which constitutes the usual one-port network analysis.

Extracting any other nth-order NLTF is incomparably more difficult. First of all, we should remember that $H_n(\omega_1, \ldots, \omega_n)$ is a n-dimensional transfer function that can only be uniquely identified by simultaneously exciting the circuit with n sources of distinct (and uncorrelated) frequencies. The experimental setup needs, therefore, n different signal generators. Also, the same requirement of Q frequency points considered above now implies a rapidly impossible-to-handle amount of tests, one for each of the possible distortion products arising from mixing n frequencies from the set of Q.[4]

Beyond this, it is worth mentioning that we would need a quite uncommon, and hard to build, laboratory setup. Indeed, since we are dealing with weakly nonlinear circuits, any distortion product would be easily masked by the much stronger linear responses. Furthermore, the fact that, except for the special case of the harmonics of the input frequencies, the wanted signals are not correlated in phase with any of the sources, obviates, in principle, the use of high dynamic range synchronous receivers.

The second group of problems associated with Voterra series is its inability to handle strong nonlinear circuits. Because this is the most important limitation of the technique, we will explore it in more detail.

First, let us clarify what we mean by the words "weakly," or "mildly nonlinear" and "strongly nonlinear." In the same way we said that a system should be considered as nonlinear if it could not be accurately treated as linear, we now say that a certain system is in a "strong nonlinear regime" if it can not be accurately represented by a Volterra series with a practically small number of terms. And this may happen, either because the system incorporates nonlinear elements, which do not have continuous characteristic functions or derivatives, or because the signal excursion is such that the maximum order considered for the series is no longer enough for the desired results' accuracy. In certain special cases, the Volterra series presents a limited radius of convergence, and is simply not applicable if the input signal exceeds that range [4]. In practical terms, the series loss of accuracy is, by far, the most interesting situation for two orders of reasons. First, we should realize that it is not possible, in general, to say a priori (i.e., without comparing Volterra series results with other simulation means, or measurement data) when the truncated series fails in producing useful results. This indicates that when you simulate a circuit with Volterra series, you will not get any warning, or noticeable strange outcomes, if the excitation is increased up to a level corresponding (in the real world) to a strong nonlinear regime. The second, and probably more omitted in published works, is that you may reach a situation where any practical Volterra series becomes hopelessly inaccurate. That is, the additional range of signal excursion you may gain is insufficient to compensate the increased computation effort required by raising the maximum order of the series.

Figure 3.18 illustrates this idea using our circuit example. It represents the static node current

4. Obviously, these are general remarks valid for any weakly dynamic system. A mildly nonlinear memoryless system can be represented by a power series. Its NLTFs are constants, and thus, a single frequency is enough for their complete characterization.

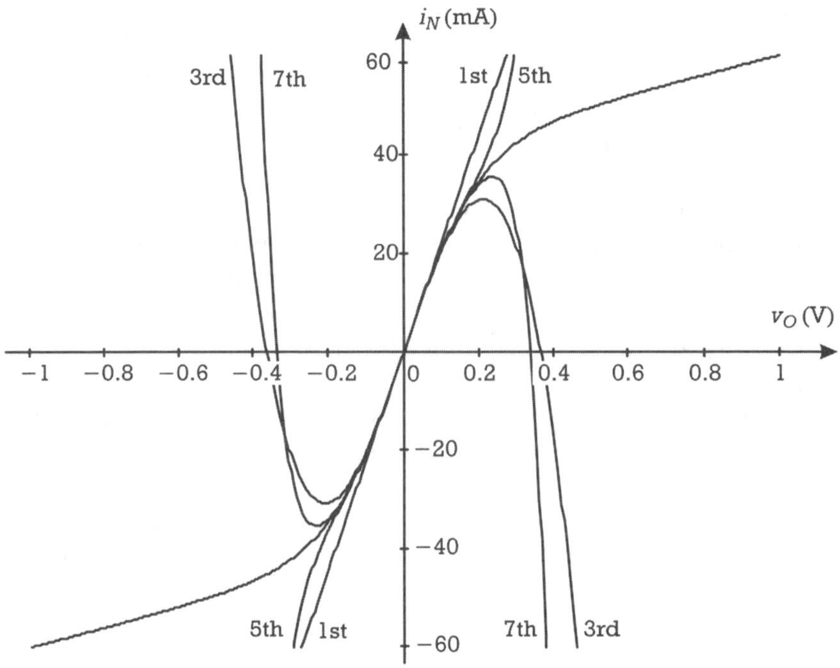

Figure 3.18 Power series approximation of the linear current, $Gv_O(t)$, plus nonlinear voltage dependent current source, $i_{NL}(v_O)$, of our circuit example resistive current.

$$i_N(v_O) = Gv_O + i_{NL}(v_O) = Gv_O + I_0 \tanh(\alpha v_O) \tag{3.189}$$

and its first to seventh-order Taylor series expansions around $V_O = 0$. Having in mind the amount of labor necessary to obtain the output components for orders higher than three or five, a glance onto Figure 3.18 will discourage any attempt to use Volterra series when the input level gets higher than about 35 mA. Indeed, it seems that beyond this limit all orders begin to simultaneously have a nonnegligible effect. In the circuit behavior this corresponds to the observation that every mixing product begins to reflect the presence of higher order contributions. For example, the fundamentals no longer present a linear behavior. Their output power versus input power patterns begin to depart from the small-signal 1 dB per dB slope, and the device reaches its 1-dB compression point. That is why this point is some times used as a rough estimate of the Volterra series utility limit.

Figure 3.19 illustrates the relation between the 1-dB compression point and the validity limit of a Volterra series description of our example circuit. It depicts the fundamental and third-harmonic's output power, in a 50-Ω load, versus source available power, calculated by the third-order Volterra series expansion around V_O = 0V and I_S = 0 mA, and from a large-signal simulator. Note that the odd symmetry of $i_{NL}(v_O)$ around the (0V, 0 mA) quiescent point determines no

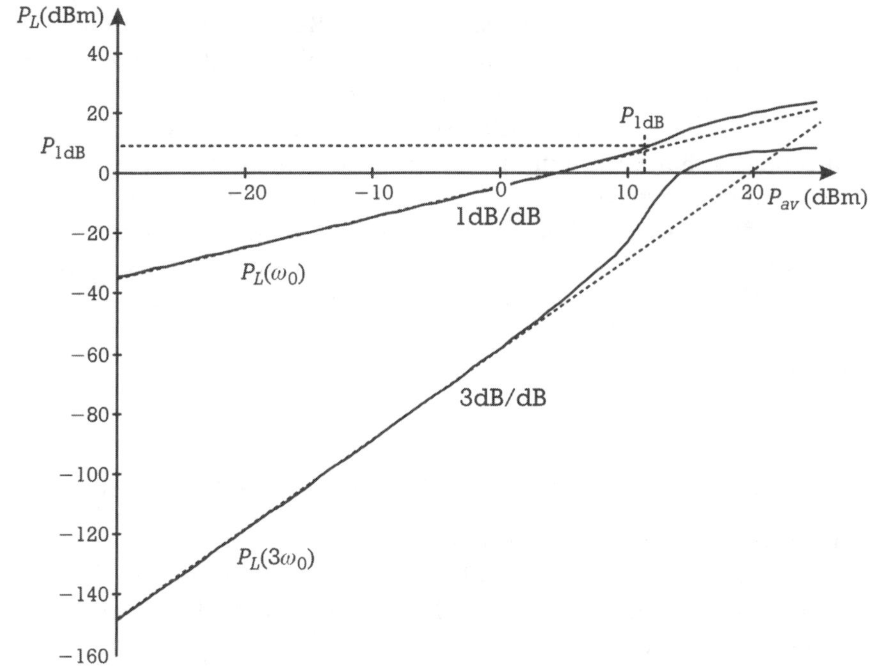

Figure 3.19 Small- and large-signal response of our example circuit.

even-order output voltage components, whereas the compression behavior of i_{NL} current for increased node voltage v_O, is the responsible for the observed gain expansion of the fundamental of v_O when the circuit is excited by i_S.

The validity limit of the third-order Volterra series is clearly associated to the deviation of the output responses from the small-signal 1-dB/dB, and 3-dB/dB straight lines, being related to the onset of $I_{nl}(\omega_0)$ current saturation, as seen by the 1-dB expansion level of $V_o(\omega_0)$.

3.3 Frequency-Domain Techniques for Large-Signal Distortion Analysis

In the following sections we will briefly present the basic ideas behind the most used RF and microwave nonlinear circuits simulation technique, the harmonic balance (HB).

It is not the authors' objective to give a comprehensive view of all particular implementations of HB (which would need an entire book) but only a general description of its underlying concepts, its ability to deal with the nonlinear distortion problem, and major limitations. Contrary to Volterra series, which is a theoretical platform for analyzing nonlinear systems, HB is mainly a set of iterative algorithms for finding an approximate solution of the nonlinear differential equations modeling those systems. So, we will focus our attention in the typical and most used implementations of HB, or in the ones especially able to deal with nonlinear distortion phenomena. Since the aim of this text is to provide the reader with the minimum information required to understand the HB method, we will prefer clarity against algorithmic efficiency. For detailed information on special implementation aspects, the reader is invited to see the abundant literature on this subject [1–3, 6, 7].

Finally, and also contrary to the large majority of previously published works, we will introduce the HB technique as an iterative means of extending Volterra series to strongly nonlinear regimes [8, 9].

3.3.1 Extending Volterra Series' Maximum Excitation Level

In previous sections we saw that the main limitation of Volterra series was its inability to handle strongly nonlinear regimes, a consequence of the adopted Taylor series model for the nonlinearities (Figure 3.18). We will now present two alternative ways of circumventing this disadvantage.

Looking back into Figure 3.18, it is clear that it does not seem practically feasible to use a third-order Volterra series (in that case a Taylor series expansion around $V_O = 0$, $I_{NL} = 0$) to determine the node voltage imposed by any source current higher than some 30 mA. So, determining the node voltage for, let us say, a 100-mA dc current would be completely out of the question. However, there should be no problem in predicting the voltage for $i_S = 25$ mA. In fact, Figure 3.18 indicates that this solution is close to $v_O = 0.15$V. Now, imagine we would use the solution found for 25 mA as another fixed point for Taylor series expansion.

Although we would still be unable to find the solution for $i_S = 100$ mA, we probably could obtain another intermediate solution for some 50 mA, which were impossible before. So, we could expand again $i_N[V_O]$ into a third-order Taylor series around this new quiescent point, and retry the desired solution for $i_S = 100$ mA. This process may be repeated as many times as needed, until the sought solution is found within a certain allowed error. Obviously, if the order of the Taylor expansions is increased, the total number of source samples needed is reduced, as the steps taken for i_S may be widened. If, in the opposite direction, the order of the Taylor series is lowered, the number of total steps is increased. But, this may bring a new advantage: solution simplicity for each of the intermediate steps. In the limit, the selected order for the Taylor expansion is one, the number of total steps is maximized, but we only have to solve linear equations. In fact, Figure 3.20 illustrates exactly that method by starting from an initial value of ${}^0i_S = 0$ mA , ${}^0v_O = 0$V, and then increasing the source current from 0 to 100 mA in 20-mA steps.

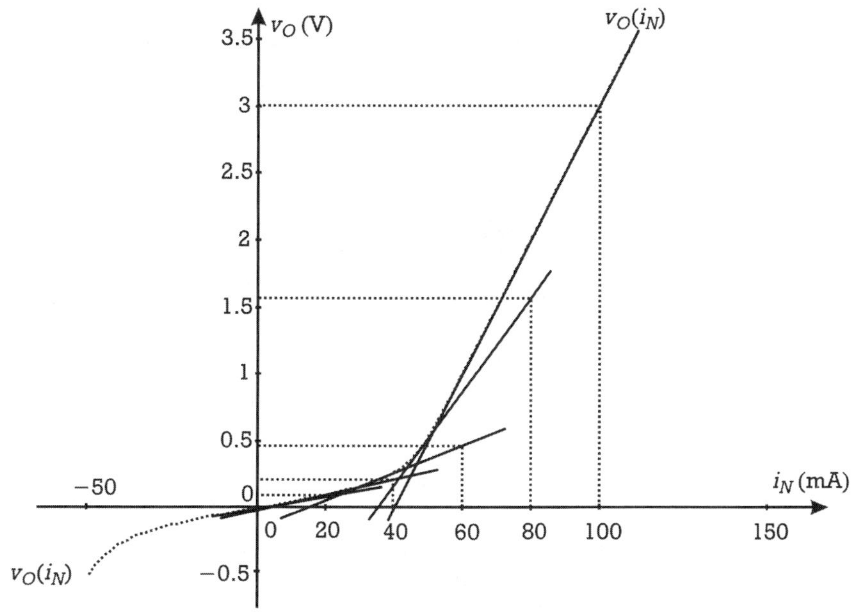

Figure 3.20 Successive expansion point adjustment for improved first-degree Taylor series' approximation range.

The extension of this process to a periodic varying $i_S(t)$ is straightforward, because its time samples, $i_S(t_i)$, may be computed independently as in Figure 3.20. In Figure 3.21 we can see an example where $i_S(t) = 100 \cos \omega t$ mA, the initial excitation was again ${}^0i_S(t) = 0$ mA and the intermediate $i_S(t)$ and $v_O[i_S(t)]$ were computed from increasing the source amplitude in 20-mA steps.

The main difference between the constant and the time-varying input is that, since the intermediate solutions ${}^iv_O(t)$ are functions of time, the Taylor series coefficients are also time-varying:

$$i_N[{}^{i+1}v_O(t)] = i_N[{}^iv_O(t)] + \left.\frac{di_N(v_O)}{dv_O}\right|_{v_O = {}^iv_O(t)} [{}^{i+1}v_O(t) - {}^iv_O(t)] + \dots$$
$$+ \frac{1}{n!} \left.\frac{d^n i_N(v_O)}{dv_O^n}\right|_{v_O = {}^iv_O(t)} [{}^{i+1}v_O(t) - {}^iv_O(t)]^n + \dots \quad (3.190a)$$

or

$${}^{i+1}i_N(t) = {}^ii_N(t) + {}^ig_1(t)\Delta v_O(t) + \dots + {}^ig_n(t)\Delta v_O(t)^n + \dots \quad (3.190b)$$

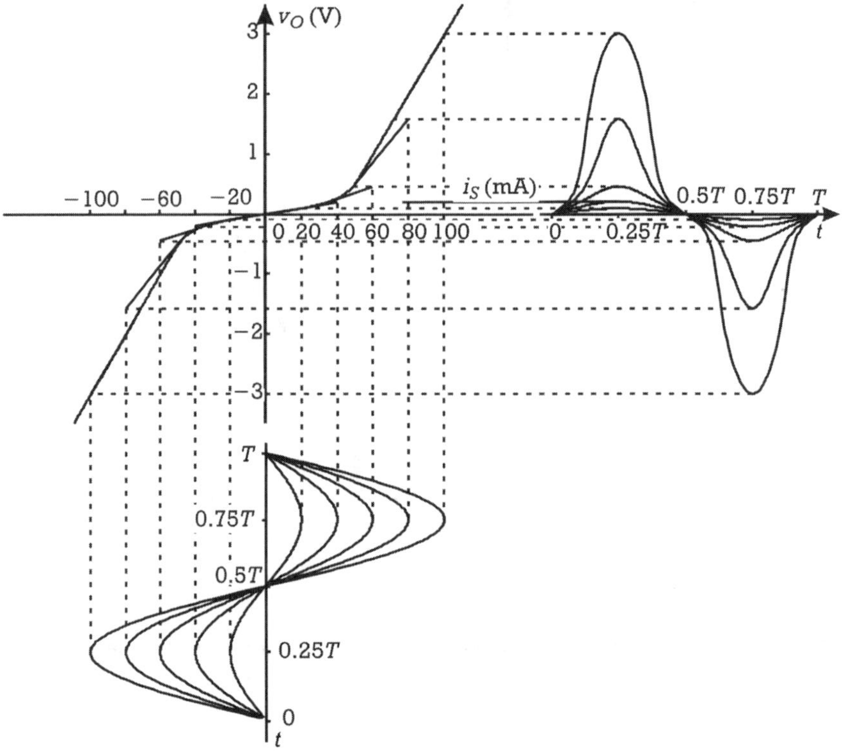

Figure 3.21 Successive time-varying expansion point adjustment for improved Taylor series' approximation range.

If $v_O(t)$ is periodic of fundamental frequency ω_0,

$$v_O(t) = \sum_{k=-\infty}^{\infty} V_{O_k} e^{jk\omega_0 t} \tag{3.191}$$

and the harmonics of ${}^{i+1}i_N(t)$ are truncated at $K\omega_0$, (3.190) can be expressed in conversion matrix form as

$$^{i+1}\begin{bmatrix} I_{n_{-K}} \\ \vdots \\ I_{n_0} \\ \vdots \\ I_{n_K} \end{bmatrix} = {}^{i}\begin{bmatrix} I_{n_{-K}} \\ \vdots \\ I_{n_0} \\ \vdots \\ I_{n_K} \end{bmatrix} + {}^{i}\begin{bmatrix} G1_0 & \cdots & G1_{-K} & \cdots & 0 \\ \vdots & & \vdots & & \vdots \\ G1_K & \cdots & G1_0 & \cdots & G1_{-K} \\ \vdots & & \vdots & & \vdots \\ 0 & \cdots & G1_K & \cdots & G1_0 \end{bmatrix} \begin{bmatrix} \Delta V_{O_{-K}} \\ \vdots \\ \Delta V_{O_0} \\ \vdots \\ \Delta V_{O_K} \end{bmatrix} + \ldots$$

$$+ \begin{bmatrix} G_{n_0} & \cdots & G_{n_{-K}} & \cdots & 0 \\ \vdots & & \vdots & & \vdots \\ G_{n_K} & \cdots & G_{n_0} & \cdots & G_{n_{-K}} \\ \vdots & & \vdots & & \vdots \\ 0 & \cdots & G_{n_K} & \cdots & G_{n_0} \end{bmatrix}^i \begin{bmatrix} \Delta V_{O_{-K}}^{(n)} \\ \vdots \\ \Delta V_{O_0}^{(n)} \\ \vdots \\ \Delta V_{O_K}^{(n)} \end{bmatrix} + \ldots \tag{3.192}$$

where $\Delta V_{O_K}^{(n)}$ represents the kth harmonic component of the voltage $\Delta v_O(t)^n$. That formulation really converted our time-domain problem in a frequency-domain one, since we are no longer calculating each of $v_O(t)$ time points, but its Fourier components. If the function is not algebraic, as $i_N[v_O(t)]$, but dynamic (i.e., represented by a certain nonlinear differential equation), this procedure corresponds to solving the problem using successive time-varying Volterra series calculated in each of the intermediate solutions.

The implementation of this form of *generalized Volterra series* [8, 9] is a complex task. It involves the method of nonlinear currents for time-varying Volterra series applied to each of the source samples determined by the excitation stepping.

Nevertheless, as was already seen, it can be considerably simplified if convergence rate is traded by reduced Volterra series order. The lower end is obtained for first order, where simplicity is maximized. We no longer need to solve any nonlinear system for each of the intermediate solutions, but simply a time-varying linear one. In this case, the inner recurrent loop disappears, and we would end up in a form of harmonic balance based on a *source stepping* procedure.

To exemplify its application, let us use it to solve our circuit when driven by a $i_S(t) = I_0 + I_P \cos \omega_0 t$ (A) sinusoidal current. Since $i_S(t)$ is a sinusoid of frequency ω_0, all circuit voltages and currents are periodic functions of the fundamental ω_0, and may be described by a Fourier series similar to (3.191).

Substituting (3.191) into the nonlinear differential equation describing our circuit

$$G v_O(t) + \frac{dq_{NL}[v_O(t)]}{dt} + i_{NL}[v_O(t)] = i_S(t) \tag{3.193}$$

gives

$$\sum_{k=-K}^{K} G V_{O_k} e^{jk\omega_0 t} + \frac{d}{dt}\left[q_{NL}\left(\sum_k V_{O_k} e^{jk\omega_0 t} \right) \right] + i_{NL}\left(\sum_k V_{O_k} e^{jk\omega_0 t} \right) = \sum_{k=-K}^{K} I_{s_k} e^{jk\omega_0 t} \tag{3.194}$$

where Σ_k stands for a summation in k, spanning from $-K$ to K, and in which all $I_{s_k} = 0$ except the ones for $k = \pm 1$ and $k = 0$ (i.e., $\pm\omega_0$ and dc, respectively).

Since we only know a priori the solution ${}^{0}v_O(t) = 0$ for ${}^{0}i_S(t) = 0$, it is not possible to directly determine the solution for $i_S(t) = I_0 + I_P \cos \omega_0 t$ (A). In fact, it is likely that we can neither obtain a good approximation for $i_S(t) = I_0$ A, based on a first-order expansion at ${}^{0}v_O(t)$, nor even for $i_S(t) = I_0 + I_P \cos \omega_0 t$ (A) from an expansion at $i_S(t) = I_0$ A. At this point we can proceed in three different ways. We can use a source-stepping algorithm for simultaneously increasing the dc and ac part of $i_S(t)$ from zero to its final values. We may, first, increase the ac part, and only after that, the dc part. Or, alternatively, we can begin by solving for the dc solution, and then increase ac excitation. As the problems usually handled in the nonlinear distortion area are such that the dc part of the excitation represents a much stronger component than the time-varying part, it is better to solve for dc first, and only after that, trying to find the complete dc plus ac solution.[5] So, we will begin the analysis by solving the circuit for $i_S(t) = I_0$ A.

At dc, (3.194) is transformed into

$$G V_{O_0} + i_{NL}(V_{O_0}) = I_{s_0} \tag{3.195}$$

As explained above, we will use a source-stepping approach in which the initial expansion point corresponds to the already known solution, ${}^{0}V_{O_0} = 0$ for ${}^{0}I_{s_0} = 0$, and where the intermediate solution ${}^{i+1}V_{O_0}$ for an increased source ${}^{i+1}I_{s_0} = {}^{i}I_{s_0} + \Delta I_{s_0}$ can be obtained from a first-order Taylor series of the $i_N(v_O)$ current:

$$i_N({}^{i+1}V_{O_0}) = G\,{}^{i}V_{O_0} + i_{NL}({}^{i}V_{O_0}) + \left.\frac{d[G v_O + i_{NL}(v_O)]}{dv_O}\right|_{v_O = {}^{i}V_{O_0}} ({}^{i+1}V_{O_0} - {}^{i}V_{O_0}) \tag{3.196}$$

as

$${}^{i+1}V_{O_0} = {}^{i}V_{O_0} - \left[\left.\frac{d[G v_O + i_{NL}(v_O)]}{dv_O}\right|_{v_O = {}^{i}V_{O_0}}\right]^{-1} [G\,{}^{i}V_{O_0} + i_{NL}({}^{i}V_{O_0}) - {}^{i+1}I_{s_0}] \tag{3.197}$$

This procedure continues through all the I_{s_0} steps until the final ${}^{f}V_{O_0}$ is found for the sought I_0 bias point.

This ${}^{f}V_{O_0} = V_{O_0}$ is now used as the first expansion point for the sinusoidal source-stepping algorithm, ${}^{0}v_O(t) = V_{O_0}$ for ${}^{0}i_S(t) = I_{s_0}$. Therefore, for a certain small enough amplitude of the current source

5. Note that in a class C amplifier, for example, where the bias point shifts dramatically from its quiescent value, the best strategy might not be so obvious. Nevertheless, this is the way the majority of nonlinear simulators—either HB or time-domain—work.

$$ {}^{1}i_S(t) = \sum_k {}^{1}I_{s_k} e^{jk\omega_0 t} \tag{3.198} $$

applied to (3.194), we will get

$$ \begin{aligned} & G V_{o_0} + \frac{d}{dt}\left\{ q_{NL}(V_{o_0}) + \left.\frac{dq_{NL}(v_O)}{dv_O}\right|_{v_O = V_{o_0}} \left[\sum_k {}^{1}V_{o_k} e^{jk\omega_0 t} - V_{o_0}\right]\right\} \\ & + i_{NL}(V_{o_0}) + \left[G + \left.\frac{di_{NL}(v_O)}{dv_O}\right|_{v_O = V_{o_0}}\right]\left[\sum_k {}^{1}V_{o_k} e^{jk\omega_0 t} - V_{o_0}\right] \\ & = \sum_{k=-K}^{K} {}^{1}I_{s_k} e^{jk\omega_0 t} \end{aligned} \tag{3.199a} $$

or

$$ G\,{}^{0}\mathbf{V}_o + c_1(V_{o_0})\,j\boldsymbol{\Omega}\,{}^{1}\mathbf{V}_o + \begin{bmatrix} 0 \\ \vdots \\ i_{NL}(V_{o_0}) \\ \vdots \\ 0 \end{bmatrix} + [G + g_1(V_{o_0})] \begin{bmatrix} {}^{1}V_{o_{-K}} \\ \vdots \\ {}^{1}V_{o_0} - V_{o_0} \\ \vdots \\ {}^{1}V_{o_K} \end{bmatrix} = \begin{bmatrix} 0 \\ \vdots \\ {}^{1}I_{s_{-1}} \\ {}^{1}I_{s_0} \\ {}^{1}I_{s_1} \\ \vdots \\ 0 \end{bmatrix} \tag{3.199b} $$

where

$$ {}^{1}\mathbf{V}_o \equiv \begin{bmatrix} {}^{1}V_{o_{-K}} \\ \vdots \\ {}^{1}V_{o_0} \\ \vdots \\ {}^{1}V_{o_K} \end{bmatrix} \text{ and } j\boldsymbol{\Omega} \equiv \begin{bmatrix} -jK\omega_0 & \dots & 0 \\ \vdots & j0 & \vdots \\ 0 & \dots & jK\omega_0 \end{bmatrix} $$

Equation (3.199) represents a set of $(2K + 1)$ linear equations in the $(2K + 1)\,{}^{1}V_{o_k}$ unknowns, whose solution is such that ${}^{1}V_{o_k} = 0$ except for ${}^{1}V_{o_0} = V_{o_0}$ and

$$ {}^{1}V_{o_{\pm 1}} = \frac{{}^{1}I_{s_{\pm 1}}}{G + {}^{0}g_1 \pm j\omega_0\,{}^{0}c_1} \tag{3.200} $$

This small output amplitude ${}^{1}v_O(t)$ is now used as the next expansion point, allowing the solution for a general ${}^{i+1}i_S(t) = {}^{i}i_S(t) + \Delta i_S(t)$, to be obtained from

$$\begin{aligned}
&\sum_k G\,{}^{i}V_{O_k} e^{jk\omega_0 t} \\
&+ \frac{d}{dt}\left\{ q_{NL}\left(\sum_k {}^{i}V_{O_k} e^{jk\omega_0 t}\right) + c_1\left(\sum_k {}^{i}V_{O_k} e^{jk\omega_0 t}\right)\left[\sum_k ({}^{i+1}V_{O_k} - {}^{i}V_{O_k}) e^{jk\omega_0 t}\right]\right\} \\
&+ i_{NL}\left(\sum_k {}^{i}V_{O_k} e^{jk\omega_0 t}\right) + \left[G + g_1\left(\sum_k {}^{i}V_{O_k} e^{jk\omega_0 t}\right)\right]\left[\sum_k ({}^{i+1}V_{O_k} - {}^{i}V_{O_k}) e^{jk\omega_0 t}\right] \\
&= \sum_k {}^{i+1}I_{s_k} e^{jk\omega_0 t}
\end{aligned} \tag{3.201}$$

The problem now arising in solving (3.201) is that we would like to transform it entirely into the frequency-domain to obtain another set of $(2K + 1)$ equations, but we do not know how to compute the Fourier coefficients of $q_{NL}[\cdot]$, $c_1[\cdot]$, $i_{NL}[\cdot]$, and $g_1[\cdot]$. Well, one possible way consists of computing each of these functions in the time-domain (e.g., ${}^{i}q_{NL}(t) = q_{NL}[{}^{i}v_O(t)]$, and then calculate its Fourier coefficients. The time-domain products $c_1[{}^{i}v_O(t)] \cdot [{}^{i+1}v_O(t) - {}^{i}v_O(t)]$ and $g_1[{}^{i}v_O(t)] \cdot [{}^{i+1}v_O(t) - {}^{i}v_O(t)]$ then become spectral convolutions, which can be represented as matrix-vector products using the conversion matrix formalism. Thus, (3.201) can be rewritten as

$$\begin{aligned}
G\,{}^{i}\mathbf{V}_o + j\boldsymbol{\Omega}\,{}^{i}\mathbf{Q}_{nl} + {}^{i}\mathbf{I}_{nl} + j\boldsymbol{\Omega}\,{}^{i}\mathbf{C}_1[{}^{i+1}\mathbf{V}_o - {}^{i}\mathbf{V}_o] \\
+ [G\mathbf{1} + {}^{i}\mathbf{G}_1][{}^{i+1}\mathbf{V}_o - {}^{i}\mathbf{V}_o] - {}^{i+1}\mathbf{I}_s = 0
\end{aligned} \tag{3.202a}$$

(where **1** stands for the identity matrix), or

$$\mathbf{F}({}^{i}\mathbf{V}_o) + \left.\frac{d\mathbf{F}(\mathbf{V}_o)}{d\mathbf{V}_o}\right|_{\mathbf{V}_o = {}^{i}\mathbf{V}_o} ({}^{i+1}\mathbf{V}_o - {}^{i}\mathbf{V}_o) = 0 \tag{3.202b}$$

in which

$$\mathbf{F}(\mathbf{V}_o) \equiv G\mathbf{V}_o + j\boldsymbol{\Omega}\mathbf{Q}_{nl}(\mathbf{V}_o) + \mathbf{I}_{nl}(\mathbf{V}_o) - \mathbf{I}_s = 0 \tag{3.203}$$

is known as the harmonic balance equation.

The composite conversion matrix $d\mathbf{F}(\mathbf{V}_o)/d\mathbf{V}_o|_{\mathbf{V}_o = {}^{i}\mathbf{V}_o} \equiv \mathbf{J}({}^{i}\mathbf{v}_o)$ is known as the Jacobian matrix of $\mathbf{F}(\mathbf{V}_o)$, and its general element of row m and column l is given by

$$J({}^i\mathbf{V}_o)_{m,l} \equiv \left.\frac{\partial F_m(\mathbf{V}_o)}{\partial V_{o_l}}\right|_{\mathbf{V}_o={}^i\mathbf{V}_o} = \frac{1}{T}\int_{-T/2}^{T/2} \left.\frac{df(v_O)}{dv_O}\right|_{v_O={}^iv_O(t)} e^{-j(m-l)\omega_0 t}\,dt \tag{3.204}$$

in which $f[v_O(t)]$ is the time-domain representation of $\mathbf{F}(\mathbf{V}_o)$, and it is assumed that $m, l = 1, \ldots, 2K+1$ and $\mathbf{J}({}^i\mathbf{V}_o)_{m,l} = 0$ if $|m-l| > K$ according to the adopted harmonic truncation.

The next solution ${}^{i+1}\mathbf{V}_o$ then becomes

$${}^{i+1}\mathbf{V}_o = {}^i\mathbf{V}_o - [\mathbf{J}({}^i\mathbf{V}_o)]^{-1}[\mathbf{F}({}^i\mathbf{V}_o)] \tag{3.205}$$

A final remark that refers to both the dc and ac source stepping algorithms, (3.197) and (3.205), regards the magnitude of the source increment. Obviously, larger increments reduce number of intermediate steps but may also compromise desired accuracy. This is exactly the same problem above discussed as the inability of Volterra series in dealing with strong nonlinear regimes. If the source stepping increment $\Delta i_S(t) \equiv {}^{i+1}i_S(t) - {}^i i_S(t)$ is sufficiently small for substituting the nonlinear element descriptions by piecewise linear approximations, then ${}^{i+1}V_{O_0}$ or ${}^{i+1}\mathbf{V}_o$ given by (3.197) or (3.205) is, indeed, a good approximate solution to ${}^{i+1}I_{S_0}$ or ${}^{i+1}i_S(t)$. If not, the HB equation will not be satisfied to the desired accuracy, and (3.197) or (3.205) must be applied to refined source steps until $|F_0({}^fV_{O_0})| < \epsilon$ or $\|\mathbf{F}({}^f\mathbf{V}_o)\| < \epsilon$, where ϵ is an allowed error ceiling, and $\|\mathbf{F}[.]\|$ stands for the norm of $\mathbf{F}[.]$.[6]

In a digital computer, both time- and frequency-domain are represented by discrete quantities. Therefore, the mathematical tool used to perform the conversion between domains is the discrete Fourier transform (DFT), or its fast algorithm, the fast Fourier transform (FFT). The DFT of a signal of $2N+1$ samples, $x(nT_s)$, where T_s is the sampling period, is

$$X(k\omega_0) = \frac{1}{N}\sum_{n=-N}^{N} x(nT_s)e^{-jk\omega_0 nT_s} = \frac{1}{N}\sum_{n=-N}^{N} x(nT_s)\,W^{-kn} \tag{3.206}$$

where ω_0 is the fundamental frequency and $W = \exp(j\omega_0 T_s)$.

Accordingly, the inverse DFT (IDFT) is defined as

$$x(nT_s) = \sum_{k=-K}^{K} X(k\omega_0)e^{jk\omega_0 nT_s} = \sum_{k=-K}^{K} X(k\omega_0)\,W^{+kn} \tag{3.207}$$

6. $\|F({}^fV_o)\| = \left(\Sigma_k |F_k|^2\right)^{1/2}$ is only an example of a number of possible error functions. For instance, if certain $\mathbf{F}_k$ components are found to dominate over other more interesting components, then some different weights could be applied when building the error function.

So, (3.206) and (3.207) can be represented as matrix-vector products like

$$\mathbf{X} = \mathbf{\Gamma x} \text{ and } \mathbf{x} = \mathbf{\Gamma}^{-1}\mathbf{X} \tag{3.208}$$

where

$$\mathbf{\Gamma} \equiv \frac{1}{N}\begin{bmatrix} W^{-KN} & \cdots & W^{Kn} & \cdots & W^{KN} \\ \vdots & & \vdots & & \vdots \\ W^{kN} & \cdots & W^{-kn} & \cdots & W^{-kN} \\ \vdots & & \vdots & & \vdots \\ W^{KN} & \cdots & W^{-Kn} & \cdots & W^{-KN} \end{bmatrix}$$

and

$$\mathbf{\Gamma}^{-1} \equiv \begin{bmatrix} W^{KN} & \cdots & W^{-kN} & \cdots & W^{-KN} \\ \vdots & & \vdots & & \vdots \\ W^{-Kn} & \cdots & W^{kn} & \cdots & W^{Kn} \\ \vdots & & \vdots & & \vdots \\ W^{-KN} & \cdots & W^{kN} & \cdots & W^{KN} \end{bmatrix}$$

In this way, the general Jacobian matrix element of line m and column n can be given by

$$J({}^{i}V_{o})_{m,l} = \frac{\partial F_{m}(\mathbf{V}_{o})}{\partial V_{o_l}} = \frac{1}{N}\sum_{n=-N}^{N} \left.\frac{df(v_O)}{dv_O}\right|_{v_O = v_O(nT_s)} W^{-(m-l)n} \tag{3.209}$$

And the full Jacobian can be computed using

$$\mathbf{J}({}^{i}\mathbf{V}_{o}) = \mathbf{\Gamma}\left[\left.\frac{df(v_O)}{dv_O}\right|\right]\mathbf{\Gamma}^{-1} \tag{3.210}$$

where $[df(v_O)/dv_O]$ is a $(2N + 1) \times (2N + 1)$ diagonal matrix such that $d_{n_1,n_2} = df[v_O]/dv_O\,|_{v_O = v_O(nTs)}$ if $n_1 = n_2 = n$ or $d_{n_1,n_2} = 0$ otherwise.

3.3.2 Harmonic Balance by Newton Iteration

In the following, we will describe an alternative way to extend Volterra series' maximum excitation level.

Consider again the dc problem of finding the V_O output for an $I_S = 100$ mA excitation. Contrary to what was done in the previous section, where the solution for that i_S was calculated from successively solving partial linear problems obtained by increasing the level of a reduced version of the excitation, we will now attempt to solve the circuit for the full excitation amplitude, by successively refining coarse linear solutions. This is illustrated in Figure 3.22, in which the determination of node voltage $v_O[i_N(t)]$ for a $i_S(t) = 100$ mA, based on a first-order Taylor expansion around $i_N[v_O(t)]$ for $v_O(t) = 0$, is sought. A first tentative voltage ${}^1v_O$ is clearly too low, as the first-order approximation does not account for the $i_{NL}(v_O)$ hyperbolic tangent saturation characteristic. As shown in Figure 3.22, this voltage solution can be much better represented if we now change the expansion point to that ${}^1v_O$. Thus, a new ${}^2v_O$ is obtained, which can be used again to further refine the solution, until the desired error level is met.

Passing from dc to the ac problem corresponds to solving successively linear time-varying equations that are dealt, in the frequency-domain, with the appropriate conversion matrices. Therefore, the application of this form of harmonic balance iteration to find the output $v_O(t)$ of our example circuit, when subject to $i_S(t) = I_0 + I_P \cos \omega_0 t$ (A), can be again handled in much the same way as before. We start by computing a better approximation to the dc operation point, and then proceed to calculate the complete dc plus ac solution. The HB equation for dc is

$$GV_{o_0} + i_{NL}(V_{o_0}) - I_{s_0} = 0 \tag{3.211a}$$

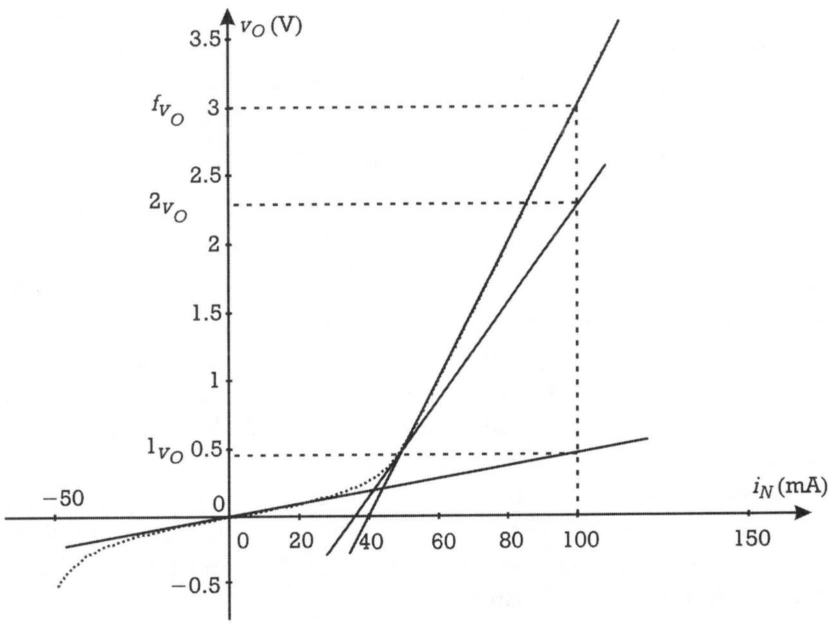

Figure 3.22 Bias point calculation by successive first-order model approximation refinement.

or

$$F_0(V_{o_0}) = 0 \tag{3.211b}$$

which can be solved iteratively using the Newton-Raphson iteration algorithm,

$$^{i+1}V_{o_0} = {}^{i}V_{o_0} - \left[\left. \frac{dF_0(v_O)}{dv_O} \right|_{v_O = {}^{i}V_{o_0}} \right]^{-1} F_0({}^{i}V_{o_0}) \tag{3.212}$$

where the initial solution is again ${}^{0}V_{o_0} = 0$.

After having determined the dc solution up to a desired approximation, ${}^{f}V_{o_0}$, the full dc plus ac HB equation can be iteratively solved for all the considered harmonics. We again use a Newton iteration scheme, in which the starting solution is that dc bias point ${}^{f}V_{o_0}$,

$$\mathbf{F}({}^{i}\mathbf{V}_o) \equiv G\,{}^{i}\mathbf{V}_o + j\boldsymbol{\Omega}\mathbf{Q}_{nl}({}^{i}\mathbf{V}_o) + \mathbf{I}_{nl}({}^{i}\mathbf{V}_o) - \mathbf{I}_s = 0 \tag{3.213}$$

and the next improved solution can be computed as

$$^{i+1}\mathbf{V}_o = {}^{i}\mathbf{V}_o - [\mathbf{J}({}^{i}\mathbf{V}_o)]^{-1}[\mathbf{F}({}^{i}\mathbf{V}_o)] \tag{3.214}$$

where the Jacobian matrix $\mathbf{J}(\mathbf{V}_o)$ has the same conversion matrix form given by (3.204).

An example for iteratively determining the various 20 harmonics of $v_O[i_S(t)]$ for our previous time-varying source of $i_S(t) = 100 \cos \omega_0 t$ mA using this harmonic balance scheme for an error ceiling of $\epsilon = 1$ mA, is shown in Figure 3.23.

3.3.2.1 Concluding Remarks

The similarities between this *harmonic-Newton* algorithm and the previous one are evident, and far from being accidental. In fact, when we first discussed the source-stepping procedure, we mentioned that we could have decided to use bigger steps against the close samples initially considered, at the expense of loosing accuracy. That is, the HB equation could no longer be verified with a single linear solution, for each source increment, and consequently a nonlinear solver like the Newton iteration would be required. The bigger the steps, the larger the number of Newton iterations to be undertaken for a certain allowed error. In the limit, we could try only one step (i.e., attempt to directly determine the solution for the full excitation, but still anchoring our expansion at $V_{o_0} = 0$V) if we accepted the larger number of Newton-Rapson iterations needed. That is exactly the harmonic-Newton scheme just described.

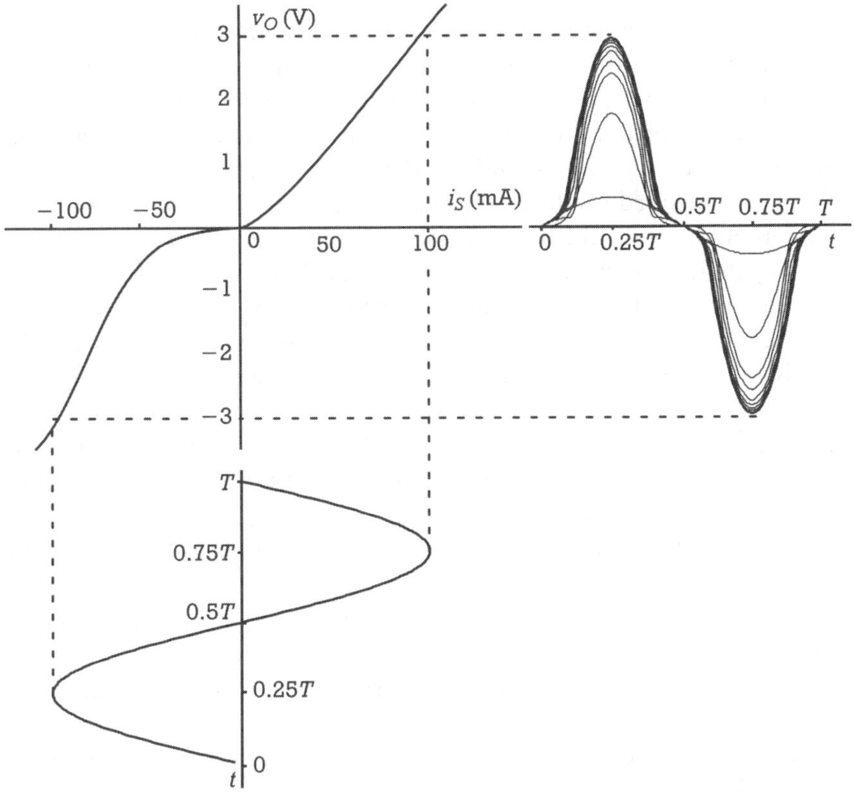

Figure 3.23 Frequency-domain large-signal calculation by successive model approximation refinement.

Whether to use one method or the other depends on the specific problem to be solved. As explained below in more detail, the approach utilized in general-purpose microwave nonlinear circuit simulators is a combination of both. Usually, the problem is attacked with a harmonic-Newton scheme for both the dc and the complete excitation. Nevertheless, if the considered initial condition for the Newton-Raphson iteration is not close enough to the solution, it is likely that convergence problems will be faced. In that case, a source-stepping procedure is used to find a closer initial condition, and the harmonic-Newton is resumed.

Achieving Convergence in the Harmonic-Newton

Unfortunately, the harmonic-Newton is an iterative procedure that has not guaranteed convergence (i.e., in which we can not be sure that we will obtain the desired $\|\mathbf{F}({}^f\mathbf{V}_o)\| < \epsilon$). This may happen because the circuit is unstable or presents a nonunique stable point, the functions describing the nonlinearities are not continuous or show discontinuities in their derivatives, the initial estimate is far from the exact solution, or simply because the selected number of harmonics in the Fourier expansions is insufficient to represent the signals with the required accuracy.

The first type of convergence problems is usually not important when simulating nonlinear distortion phenomena, since the circuits considered are normally stable. An eventual exception could be the analysis of the distortion performance of voltage-controlled oscillator–based FM/PM modulators. In this case, probably the best alternative would be to simulate first the free oscillator with an appropriate HB machine, and then perturbing that stable point with the modulating signal.

The second type of convergence problems must be circumvented by a proper selection of the nonlinear model expressions. As we saw from Volterra series analysis, the ability of the device model in representing well, not only the function, but also its higher order derivatives, is crucial for achieving accurate small-signal intermodulation distortion prediction. Therefore, if one is seeking good IMD simulations, he should not face any harmonic-Newton convergence problems due to nonlinear device model format.

The third origin of convergence failure refers to the initial estimate selection. To understand this, we must remember that the harmonic-Newton algorithm searches the solution of the HB equation by sensing the gradient of its error function $\|\mathbf{F}({}^{i}\mathbf{V}_o)\|$. Thus, it is likely to fail convergence whenever $\|\mathbf{F}({}^{i}\mathbf{V}_o)\|$ does not decay monotonically to zero, in its trajectory from the initial solution ${}^{i}\mathbf{V}_o$ to the final solution of ${}^{f}\mathbf{V}_o$. This is shown in Figure 3.24 for the one-dimensional case, where an initial estimate of ${}^{0}V_o$ would lead to an iteration trajectory of ${}^{0}V_o$, ${}^{1}V_o$, ${}^{2}V_o$, ${}^{3}V_o = {}^{0}V_o, \ldots$ that wanders around the local minimum of $F(V_o)$.

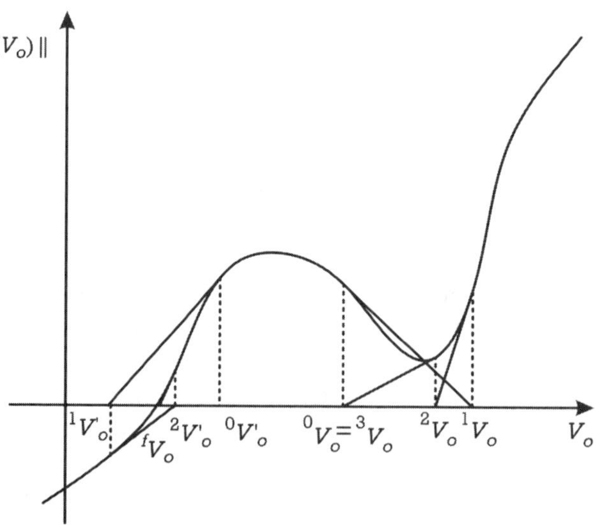

Figure 3.24 Harmonic-Newton convergence sensitivity to initial estimate selection.

The possibility that harmonic-Newton leads to such an anomalous behavior is evident from Figure 3.23. Due to the fact that the initial solution is taken as the linearized system response around the quiescent point to the full excitation, it may be quite far from the desired solution. The harmonic-Newton may even start from a reasonable low error and then temporarily (or definitely) increase the error, if the iterative trajectory passes through a zone of low gradient. In this case, an appropriate way for a successful simulation consists of choosing another initial solution, like the one referred to as ${}^{0}V_0'$ in Figure 3.24. And this may be obtained by switching from the harmonic-Newton to the source-stepping HB. The procedure followed by commercial HB simulators in the presence of such convergence difficulties consists of reducing the excitation to a value where convergence is guaranteed, and using this stimulus level backed-off solution as the alternative harmonic-Newton initial estimate. Obviously, if the simulation already consists of a source power sweep (like in common output power versus input power transfer characteristics) the initial solution to the next amplitude point is taken as the preceding calculated result.

The source-stepping procedure just described pertains to a much broader strategy known as *continuation methods*. Continuation methods rely on varying a circuit parameter from a situation where the solution is easily found, to the desired circuit condition. In principle, any circuit parameter is amenable for use as a continuation parameter, as long as the circuit behavior responds smoothly to its changes. So, beyond the natural excitation level, we could also use the values of some critical circuit components, or even some parameters of the nonlinear device models. Normally, the idea is to convert the nonlinear circuit facing convergence difficulties into another one nearly linear, for which the solution can be obtained in one iteration. This solution is then used as the first estimate to another simulation where the circuit is weakly nonlinear, and the process repeated until the solution is found for the original circuit [1].

Finally, the last referred origin of harmonic-Newton convergence failure is related to an insufficient number of harmonics used to represent the signals. The only remedy for such a situation is to try another simulation run with a more conservative harmonic truncation. In practical RF or microwave circuits, a number of harmonics on the order of eight to ten is generally enough. This is specially true in the nonlinear distortion simulation field, not only because the nonlinearities usually encountered tend to produce small harmonic amplitudes with increasing order, as all circuits behave in an asymptotic lowpass manner when excitation frequency goes to infinity. Also, the package or intrinsic device reactances contribute to soften the impact of higher frequency components.

Summarizing Algorithm of the Harmonic-Newton Method

In order to close this section, we will now summarize the underlying concepts of the harmonic-Newton method by presenting a flow chart of its algorithm. For the sake of generality, it is assumed that our circuit example is now composed of a linear subcircuit that comprises all linear dynamic elements, beyond the previous nonlinear subcircuit including only a memoryless current source and charge models. It all works as if the linear conductance G of our example circuit (see Figure 3.1) were substituted by an admittance matrix, $Y_{cl}(\omega)$, while the nonlinear subcircuit still involves a voltage-dependent current source, $i_{NL}[v_O(t)]$, and a nonlinear capacitance, whose current is given by $d/dt\{q_{NL}[v_O(t)]\}$. So, the harmonic balance equation is formed by imposing Kirchoff's current law to the circuit's node—that is, $\mathbf{F}(\mathbf{V}_o) = \mathbf{I}_{cl} + \mathbf{I}_{nl} + j\Omega\mathbf{Q}_{nl} - \mathbf{I}_s = 0$, where $\mathbf{I}_s(\omega)$ is again our excitation source vector, the linear subcircuit's current is given by $I_{cl}(\omega) = Y_{cl}(\omega)$. $V_o(\omega)$, and the nonlinear currents are computed by Fourier transforming the nonlinear current and charge previously evaluated in time-domain.

As seen in Figure 3.25, the algorithm starts by estimating an initial solution ${}^0\mathbf{V}_o$, which is then used to formulate the harmonic balance equation in that time-varying quiescent point. In the following, a new solution estimate, ${}^1\mathbf{V}_o$, is generated by the Newton-Raphson nonlinear solver, unless the harmonic balance equation is already approximately verified within an error level not greater than a prescribed ϵ. In this case, it is considered that the solution ${}^f\mathbf{V}_o$ has been reached.

3.3.3 Nonlinear Model Representation—Spectral Balance

When formulating the harmonic equation, we faced the problem of determining the spectrum representation of a certain nonlinear response [e.g., $i_{NL}[v_O(t)] = i_{NL}\left[\sum_k V_{o_k} e^{jk\omega_0 t}\right]$ in (3.201)]. At that time, the adopted procedure was to convert the spectral representation of $\mathbf{V}_o$ into the time-domain, $v_O(t)$, compute $i_{NL}[v_O(t)]$ in a time-sample by time-sample basis, and then return to the frequency-domain with the help of the appropriate Fourier transformation. However, that time-domain evaluation of the nonlinearities may not be viable for, at least, two important reasons.

First of all because the signals may not be periodic, thus obviating the use of the DFT (or its fast computation algorithm, the FFT) as the tool to jump between time- and frequency-domains. This is so important in the nonlinear distortion field of problems that it deserves a separate treatment in a special section. For now, just imagine the simple case of trying to simulate a two-tone test where the two frequencies, ω_1 and ω_2, have no common divider (i.e., there are no simultaneous nonzero integers k_1 and k_2 such that $k_1\omega_1 + k_2\omega_2 = 0$).

The second possibility appears whenever the nonlinearity may not be expressed by the cascade of a linear dynamic operator and an algebraic nonlinear function. For example, if the circuit includes one or more electron devices for which quasi-static approximation does not apply. In this case, $i_{NL}[v_O(t)]$ would have memory,

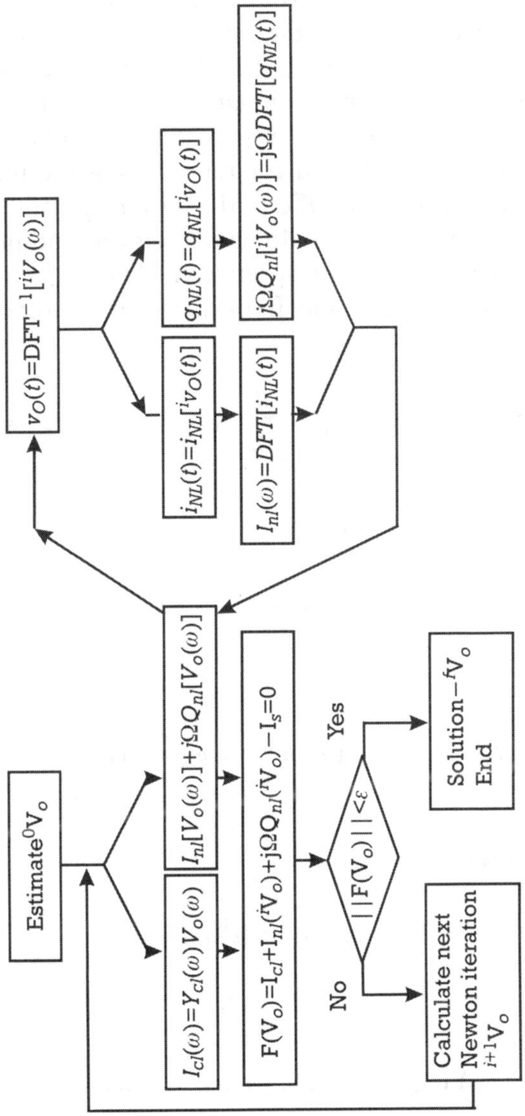

Figure 3.25 Summarizing flow chart of the harmonic-Newton algorithm.

meaning that the time-sample by time-sample calculation is no longer possible [the response at any time also depends on the past $v_O(t)$], and some kind of transient integration is needed. But, this is exactly what we want to obviate when using a steady-state simulation technique like HB.

To cope with these problems, we would like to have the capability of determining the output spectrum directly from the input spectrum. And, this is, in general, not possible, since there are no mathematical tools for analytically solving a nonlinear differential equation subject to any combination of sinusoids. Obviously, the power series or Volterra series are two very special methods to handle that situation, in which the particular format adopted for representing the nonlinearities allowed the desired spectral manipulation. Indeed, the output spectral calculation could be easily done by transforming the time-domain additions and products into spectral additions and pseudoconvolutions, respectively. That is, if $x(t)$, $X(\omega)$; $y(t)$, $Y(\omega)$; $z(t)$, $Z(\omega)$ are the time-domain and frequency-domain representations of three signals such that

$$x(t) = \sum_{k=-K}^{K} X_k e^{j\omega_k t};\ y(t) = \sum_{k=-K}^{K} Y_k e^{j\omega_k t};\ z(t) = \sum_{r=-R}^{R} Z_k e^{j\omega_r t} \tag{3.215}$$

then, the spectrum of the addition

$$z(t) = a_1 x(t) + a_2 y(t) \tag{3.216}$$

will be

$$Z(\omega_r) = a_1 X(\omega_k) + a_2 Y(\omega_k);\ \omega_r = \omega_k = -\omega_K, \ldots, 0, \ldots, \omega_K \tag{3.217}$$

and the spectrum of the product

$$z(t) = x(t)\, y(t) \tag{3.218}$$

will be

$$Z(\omega_r) = X(\omega_{k_1}) Y(\omega_{k_2});\ \omega_r = \omega_{k_1} + \omega_{k_2} = -2\omega_K, \ldots, 0, \ldots, 2\omega_K \tag{3.219a}$$

or

$$\mathbf{Z}(\omega) = \mathbf{X}(\omega) * \mathbf{Y}(\omega) = \mathbf{Y}(\omega) * \mathbf{X}(\omega) \tag{3.219b}$$

where ω_r spans through all possible linear combinations of ω_{k_1} and ω_{k_2}, in a way that $\mathbf{Z}(\omega) = \mathbf{X}(\omega) * \mathbf{Y}(\omega)$ represents true spectral convolution if the various ω_k are multiples of a fundamental ω_0: $\omega_k = k\omega_0$, and another frequency transformation (herein called pseudoconvolution) otherwise. Again, from all mixing products, only the ones falling in the original frequency set are considered, in order to limit the spectral regrowth handled by the simulator. Therefore, it is assumed that many of the $X(\omega_{k_1})$ and $Y(\omega_{k_2})$ are null, and any $\omega_r < -\omega_K$ or $\omega_r > \omega_K$ will be discarded. In this way, that pseudoconvolution corresponds to $(2K + 1)^2$ complex entity products that may be put in matrix form as

$$\begin{bmatrix} Z_{-K} \\ \vdots \\ Z_0 \\ \vdots \\ Z_K \end{bmatrix} = \begin{bmatrix} Tx_{11} & \cdots & Tx_{1(2K+1)} \\ \vdots & & \vdots \\ Tx_{(2K+1)1} & \cdots & Tx_{(2K+1)(2K+1)} \end{bmatrix} \begin{bmatrix} Y_{-K} \\ \vdots \\ Y_0 \\ \vdots \\ Y_K \end{bmatrix} \tag{3.220a}$$

$$= \begin{bmatrix} Ty_{11} & \cdots & Ty_{1(2K+1)} \\ \vdots & & \vdots \\ Ty_{(2K+1)1} & \cdots & Ty_{(2K+1)(2K+1)} \end{bmatrix} \begin{bmatrix} X_{-K} \\ \vdots \\ X_0 \\ \vdots \\ X_K \end{bmatrix}$$

or

$$\mathbf{Z} = \mathbf{TxY} = \mathbf{TyX} \tag{3.220b}$$

in which **Tx** and **Ty** are the *spectrum transform matrices* of the signals $x(t)$ and $y(t)$, respectively.

At this point, it should be noted that if $x(t)$, $y(t)$ and $z(t)$ are periodic ($\omega_k = k\omega_0$), these transform matrices have a regular Toeplitz form such that $Tx_{ij} = X[(i - j)\omega_0]$. The matrix-vector products of (3.220) can be implemented in a much more efficient way in time-domain, if the convolution properties of the FFT are used.[7] Indeed, these order $O[(2K + 1)^2]$ frequency-domain convolutions can be done with only two times $O[(2K+1)\log_2(2K+1)]$ for FFT inversion of $\mathbf{X}(\omega)$ and $\mathbf{Y}(\omega)$, $O[2K + 1]$ for $z(t) = x(t)y(t)$ multiplication and $O[(2K + 1)\log_2(2K + 1)]$ for FFT transformation of $z(t)$ to $\mathbf{Z}(\omega)$, and we end up in the conventional mixed time-domain frequency-domain HB algorithm.

With that definition of the spectrum transform matrix, it is also possible to determine the spectrum mapping imposed by a time-domain division. Consider that the nonlinear function to be evaluated in the frequency-domain is

7. Strictly speaking, the use of the FFT algorithm requires that $(2K + 1)$ be a multiple of two. If it is not, the array should be padded with the necessary number of zeros before processing.

$$z(t) = \frac{y(t)}{x(t)} \tag{3.221}$$

To determine $\mathbf{Z}(\omega)$ from $\mathbf{X}(\omega)$ and $\mathbf{Y}(\omega)$, we should first recognize that if (3.221) holds, then $y(t) = x(t)z(t)$ or $\mathbf{Y}(\omega) = \mathbf{X}(\omega) * \mathbf{Z}(\omega)$ or $\mathbf{Y} = \mathbf{Tx}\,\mathbf{Z}$, and thus

$$\mathbf{Z} = \mathbf{Tx}^{-1}\mathbf{Y} \tag{3.222}$$

where $\mathbf{Tx}^{-1}$ is the inverse matrix of $\mathbf{Tx}$.

In the same manner as the matrix-vector product $\mathbf{Tx}\,\mathbf{Y}$ represented a frequency-domain convolution in the periodic excitation case, now $\mathbf{Tx}^{-1}\mathbf{Y}$ stands for its inverse operation, or deconvolution.

It should be clear by now that since any algebraic function can be approximately expressed as a suitable combination of the four arithmetic operations (addition, subtraction, multiplication, and division), any memoryless nonlinearity can be directly evaluated in frequency-domain. For that, the nonlinearity should be approximated by a convenient power series or rational function (i.e., a ratio between two polynomials).[8] In this way, we can solve the HB equation in the frequency-domain obviating the need for the Fourier transformations. That special version of HB is usually known as *frequency-domain harmonic-balance* or simply *spectral-balance* (SB), to distinguish it from the conventional mixed mode HB.

To solve the spectral-balance equation with the Newton-Raphson method, like in the previous harmonic-Newton, we need to compute the Jacobian matrix. That is straightforward if the spectrum transform matrix is again used for the time-domain products and divisions. If, for example, $x(t)$, $y(t)$, and $z(t)$ are dependent on a signal $s(t)$, and, for example, the derivative of $\mathbf{X}(\omega)$ with respect to the nth component of $\mathbf{S}(\omega)$ is

$$\dot{\mathbf{X}}_n \equiv \begin{bmatrix} \frac{\partial X_{-K}}{\partial S_n} & \cdots & \frac{\partial X_0}{\partial S_n} & \cdots & \frac{\partial X_K}{\partial S_n} \end{bmatrix}^T \tag{3.223}$$

then the Jacobian of the addition (or subtraction) $z(t) = x(t) \pm y(t)$ can be calculated by

$$\dot{\mathbf{Z}}_n = \dot{\mathbf{X}}_n \pm \dot{\mathbf{Y}}_n \tag{3.224}$$

the Jacobian of the product $z(t) = x(t)y(t)$, by

$$\dot{\mathbf{Z}}_n = \mathbf{Tx}\dot{\mathbf{Y}}_n + \mathbf{Ty}\dot{\mathbf{X}}_n \tag{3.225}$$

8. This is only applicable to algebraic nonlinearities. If they include memory, alternative generalized power series could also be used [10].

and the Jacobian of the division $z(t) = y(t)/x(t)$ by

$$\dot{\mathbf{Z}}_n = \mathbf{Tx}^{-1}(\dot{\mathbf{Y}}_n - \mathbf{Tx}^{-1}\mathbf{Ty}\dot{\mathbf{X}}_n) = \mathbf{Tx}^{-1}(\dot{\mathbf{Y}}_n - \mathbf{Tz}\dot{\mathbf{X}}_n) \tag{3.226}$$

3.3.3.1 Comparison of Frequency and Mixed-Mode Harmonic Balance

As previously stated, the main advantages of SB over HB are its ability to handle any type of input spectrum, and dynamic nonlinearities. Although the second argument does not have too strong a practical impact, as common nonlinear devices can generally be described by memoryless currents, charges, or magnetic fluxes, the first one plays an important role on nonlinear distortion simulation. Since no Fourier transformation is required, no additional requirement is imposed to the input spectrum other than that it must be composed of a finite number of discrete points.

However, as the number of different mixing products increases rapidly with the number of nonharmonically related tones, in practice, memory storage and computation time limit SB usage to inputs composed by a few tens of discrete uncommensurated tones.

Another important advantage of SB over HB, which is also very important to the distortion simulations filed, is its higher numerical range. Highly linear systems produce distortion components having amplitudes much lower than the linear components. Therefore, its accurate calculation may be compromised if any error associated to those linear components can be spread over the distortion components. Because Fourier transformations compute any spectral point from all time-samples, and the inverse transformation compute the value of any time-sample from all spectral components, conventional mixed-mode HB is susceptible to the type of inaccuracies just referred, while SB is not. In fact, this is an advantage inherent to all frequency-domain methods, as they handle independently all spectral components.

Unfortunately, these three advantages of SB have a price, one high enough to obviate its widely use in commercial simulators: contrary to mixed-mode HB, which handles any SPICE-like model, SB requires a special device model format. And this implies, really, a twofold problem.

First of all, this requirement imposes another step in circuit simulation: the model substitution by a convenient approximant. It may be a power series or a rational function. Although the second one is generally preferred because of its wider approximation range, it must be selected in a way that it osculates the first n derivatives of the original function, if small-signal distortion up to order n is to be accurately predicted. A good example of that is the Hermite rational [11]. The second criteria that must be observed in selecting these ratios of polynomials is that it must be guaranteed that any possible zero of the denominator within the approximation range must be exactly canceled by a similar zero of the numerator. And this may be some times difficult, due to the finite arithmetic precision used to compute the polynomial coefficients.

The second problem caused by the device model format is excessive computation time. Since we rely on spectral pseudoconvolutions and deconvolutions, it is first necessary to spend time calculating the frequency transform matrices and their inverses. Then, it is necessary to perform many matrix-vector products, which are comparatively much more expensive then evaluating the model in time-domain.

3.3.4 Multitone Harmonic Balance

As was presented in previous sections, HB was initially conceived to compute the steady-state of a nonlinear differential equation when subject to a sinusoidal excitation. So, due to the restrictions imposed by the DFT, this conventional mixed-mode HB is only capable of handling periodic signals. The intermodulation distortion problem usually requires at least two uncommensurated tones, which determines aperiodic (quasiperiodic) signals, demanding for different approaches. One of these alternatives was already presented in the previous section as the spectral balance. In that case, we approximated the nonlinear function with an appropriate power series or rational, to allow its evaluation directly in frequency-domain. In this section we return to the time-domain evaluation of the nonlinearities and explore time-frequency transformations amenable for these quasiperiodic stimulus.

Although most of the techniques to be presented can be generalized to more than two tones, we will assume that simplest case to guarantee explanation clarity. So, our excitation $i_S(t)$ is given by

$$i_S(t) = \sum_{q=-2}^{2} I_{S_q} e^{j\omega_q t} \tag{3.227}$$

and the output variable by

$$v_O(t) = \sum_{k=-K}^{K} V_{O_k} e^{j\omega_k t} \tag{3.228}$$

where $\omega_k = k_1\omega_1 + k_2\omega_2$ for k_1 and k_2 integers.

Two different strategies can be considered for the necessary harmonic truncation, K. *Box truncation* assumes that for a maximum harmonic order, k_1 and k_2 are such that $|k_1| \leq K_1$ and $|k_2| \leq K_2$. *Diamond truncation* considers, instead, that $|k_1| + |k_2| \leq K$. An example of the resulting frequency location for each of these truncation schemes ($K = 3$ and $K_1 = K_2 = 3$) is illustrated in Figure 3.26 and Figure 3.27.

From a practical point of view, diamond truncation is preferable to box truncation since it is more efficient. To understand that, consider two tones of $f_1 = 2.000$ GHz and $f_2 = 2.001$ GHz. If circuit bandwidth limitations determine that all mixing products up to, lets say, about 6 GHz should be considered, minimum

$k_1 \backslash k_2$	−3	−2	−1	0	+1	+2	+3
−3				✓			
−2			✓	✓	✓		
−1		✓	✓	✓	✓	✓	
0	✓	✓	✓	✓	✓	✓	✓
+1		✓	✓	✓	✓	✓	
+2			✓	✓	✓		
+3				✓			

Figure 3.26 Frequency set generated by diamond truncation with $K = 3$ (25 mixing products).

$k_1 \backslash k_2$	−3	−2	−1	0	+1	+2	+3
−3	✓	✓	✓	✓	✓	✓	✓
−2	✓	✓	✓	✓	✓	✓	✓
−1	✓	✓	✓	✓	✓	✓	✓
0	✓	✓	✓	✓	✓	✓	✓
+1	✓	✓	✓	✓	✓	✓	✓
+2	✓	✓	✓	✓	✓	✓	✓
+3	✓	✓	✓	✓	✓	✓	✓

Figure 3.27 Frequency set generated by box truncation with $K_1 = K_2 = 3$ (49 mixing products).

K for diamond truncation must be selected as $K = 3$, as is reported in Figure 3.26, while box truncation would require a much larger frequency set of $K_1 = K_2 = 3$, as shown in Figure 3.27. It can be shown that frequency vector size for box truncation with $k_1 = k_2 = K$ is $(2K + 1)^2 = 4K^2 + 4K + 1$, while it is only $(2K + 1)^2 - 2K(K + 1) = 2K^2 + 2K + 1$ for diamond truncation. Thus, for large K, box truncation frequency vector size tends to double the one of diamond truncation.

A two-tone excitation, like the one considered, can be periodic in T if $x(t + T) - x(t) = 0$ for all t. In that case, the input frequencies are said to be commensurated since there exists a common divider $\omega_0 = 2\pi/T$ such that $\omega_1 = k_1\omega_0$ and $\omega_2 = k_2\omega_0$ where k_1, k_2 are integers.

In the opposite case, it is not possible to find any pair of nonsimultaneous null integers k_1, k_2 that make $k_1\omega_1 + k_2\omega_2 = 0$, and the signal is aperiodic. Although there is no exact period for which the signal repeats itself, we may find an ϵ-almost periodic T such that $|x(t + T) - x(t)| < \epsilon$, and the excitation is said to be *almost periodic* or *quasiperiodic*.

Uniformly sampling an almost periodic signal, and subsequently calculating its spectrum by a DFT is not possible. The problem is that the DFT automatically repeats the sampled data as if the almost period T were a true period, thus creating a new signal with a discontinuity of amplitude ϵ. That discontinuity corresponds to add very high frequency components that are aliased onto the original spectrum.

And this aliasing error, commonly referred to as spectral leakage, is usually so large that masks the desired small distortion components.

But, even if the excitation is periodic, it may not be practical to use a DFT as the means to obtain the signal spectrum representation. For example, the direct application of the DFT to our signal of $f_1 = 2.000$ GHz and f_2 1 MHz apart, and diamond truncation with $K = 3$, demands a sampling frequency of, at least, $f_s = 12.006$ GHz, and a sampling time equal to the period of 1 MHz or $NT_s = 1\mu$s. Therefore, the number of data samples needed to obtain only 25 Fourier components would be $N = 12{,}006$. If frequency separation of tones were reduced to 1 KHz, this number of samples would become nearly 12 million!

3.3.4.1 Almost Periodic Fourier Transform

One reasonable alternative to the DFT is to use what is known as the *almost periodic Fourier transform* (APFT). It is based on the intuitive idea that, in principle, $(K + 1)$ Fourier components V_{o_k} ($2K$ real numbers for all positive frequencies plus one more for dc) can be obtained from $(2K + 1)$ time-samples, t_s, by solving the following linear system of $(2K + 1)$ equations[9]:

$$\begin{bmatrix} v_O(t_1) \\ \vdots \\ v_O(t_s) \\ \vdots \\ v_O(t_{2K+1}) \end{bmatrix} = \begin{bmatrix} 1 & 2\cos(\omega_1 t_1) & -2\,\mathrm{sen}(\omega_1 t_1) & \dots & 2\cos(\omega_K t_1) & -2\,\mathrm{sen}(\omega_K t_1) \\ \vdots & \vdots & \vdots & & \vdots & \vdots \\ 1 & \cos(\omega_1 t_s) & -2\,\mathrm{sen}(\omega_1 t_s) & \dots & 2\cos(\omega_K t_s) & -2\,\mathrm{sen}(\omega_K t_s) \\ \vdots & \vdots & \vdots & & \vdots & \vdots \\ 1 & \cos(\omega_1 t_{2K+1}) & -2\,\mathrm{sen}(\omega_1 t_{2K+1}) & \dots & 2\cos(\omega_k t_{2K+1}) & -2\,\mathrm{sen}(\omega_k t_{2K+1}) \end{bmatrix} \begin{bmatrix} V_{o_0} \\ V_{o_{1_r}} \\ V_{o_{1_i}} \\ \vdots \\ V_{o_{K_r}} \\ V_{o_{K_i}} \end{bmatrix} \tag{3.229a}$$

where $V_{o_k} = V_{o_{K_r}} + jV_{o_{K_i}}$, or

9. For guaranteeing consistency with the previous sinusoidal harmonic-Newton derivations, equation (3.229a) uses a double-sided DFT against the more common single sided definition [1–3]. A comparison between these two possible DFT definitions can be found in [1].

$$\mathbf{v}_O = \mathbf{\Gamma}^{-1}\mathbf{V}_o \tag{3.229b}$$

which gives

$$\mathbf{V}_o = \mathbf{\Gamma}\mathbf{v}_O \tag{3.230}$$

The pair (Γ, Γ^{-1}) is then the wanted APFT and its inverse, respectively. The problem with this approach is that we do not know a priori how to select the $(2K + 1)$ time-samples.

In the sinusoidal excitation case $v_O(t)$ would be periodic and an uniform sampling strategy should be used during one period of $v_O(t)$. The (Γ, Γ^{-1}) thus obtained is the DFT pair. Unfortunately, trying to extend this knowledge to the almost periodic case is, generally, of little use as the system (3.229) becomes ill conditioned. This means that the rows of Γ^{-1} are nearly linearly dependent, the system is close to being undetermined, and the transform result is very sensitive to finite arithmetic round off errors or aliasing. Aliasing errors are particularly important in this context because the spectrum was initially truncated to K [1].

A method proposed to overcome that problem [12] uses twice the minimum number of time-samples theoretically needed (oversampling), randomly distributed over a three-period time duration. This yields two times more rows than unknowns, which then allows a careful selection of the $(2K + 1)$ set that is more linearly independent or orthogonal.

To understand how this can be done, suppose that the row vectors correspondent to time-samples t_{s_1} and t_{s_2}, $\mathbf{r}_1$ and $\mathbf{r}_2$, are not orthogonal. This means that $\mathbf{r}_1^T\mathbf{r}_2 \neq 0$ and that $\mathbf{r}_2$ can be expressed as the sum of an orthogonal component to $\mathbf{r}_1$, $\mathbf{r}_{2_\perp}$, and one linearly dependent on $\mathbf{r}_1$, $a\mathbf{r}_1$:

$$\mathbf{r}_2 = \mathbf{r}_{2_\perp} + a\mathbf{r}_1 \tag{3.231}$$

where a is a real constant. The orthogonal component, $\mathbf{r}_{2_\perp}$, could be obtained from $\mathbf{r}_2$ and $\mathbf{r}_1$, if we knew a. And that constant can be easily determined if we multiply both parts of (3.231) by $\mathbf{r}_1^T$, giving

$$a = \frac{\mathbf{r}_1^T\mathbf{r}_2}{\mathbf{r}_1^T\mathbf{r}_1} \tag{3.232}$$

Therefore,

$$\mathbf{r}_{2_\perp} = \mathbf{r}_2 - \frac{\mathbf{r}_1^T\mathbf{r}_2}{\mathbf{r}_1^T\mathbf{r}_1}\mathbf{r}_1 \tag{3.233}$$

is the desired component of $\mathbf{r}_2$ orthogonal to $\mathbf{r}_1$.

The near orthogonal selection of time-samples is accomplished by selecting one row arbitrarily, say $\mathbf{r}_1$, and using (3.233) to determine the other row in the set which is more orthogonal to it, $\mathbf{r}_s$ (i.e., the one with largest orthogonal component norm $\|\mathbf{r}_{s_\perp}\|$). Those two rows, $\mathbf{r}_1$ and $\mathbf{r}_s$, are then retained for the final $(2K + 1)$ system. Then $\mathbf{r}_1$ is substituted by $\mathbf{r}_s$ in the next iteration to determine another row near-orthogonal to $\mathbf{r}_1$ and $\mathbf{r}_s$. This process is repeated, as many times as needed, to select all the near-orthogonal $(2K + 1)$ rows.

Similarly to what was done for the periodic case, harmonic-Newton implementation based on this APFT requires a Jacobian matrix that may be computed as

$$\mathbf{J} \equiv \frac{d\mathbf{F}(\mathbf{V}_o)}{d\mathbf{V}_o} = \mathbf{\Gamma}\left[\frac{df_m(v_O)}{dv_O}\right]\mathbf{\Gamma}^{-1} \tag{3.234}$$

in which is $[df_m[v_O]/dv_O]$ is again a diagonal matrix such that

$$d_{s_1,s_2} = \left.\frac{df_m(v_O)}{dv_O}\right|_{v_O = v_O(t_s)}$$

if $s_1 = s_2 = s$ and $d_{s_1,s_2} = 0$ otherwise.

3.3.4.2 Multidimensional Fourier Transform

The *multidimensional discrete Fourier transform* (MDFT) is a generalization of the DFT to signals dependent on various parameters. Since the signals we have to deal with in circuit analysis are all dependent on a single parameter, time, MDFT is not directly applicable in this field. However, it can offer an enormous benefit to multitone HB if we note that a two-tone almost-periodic signal like

$$i_S(t) = \sum_{q=-2}^{2} I_{s_q} e^{j\omega_q t} \tag{3.235}$$

can be rewritten as a double periodic two-dimensional signal,

$$i_S(\theta_1, \theta_2) = [I_{s_1} e^{j\theta_1} + I_{s_1}^* e^{-j\theta_1}] + [I_{s_2} e^{j\theta_2} + I_{s_2}^* e^{-j\theta_2}]; \tag{3.236}$$
$$\theta_1 \equiv \omega_1 t,\ \theta_2 \equiv \omega_2 t$$

provided ω_1 and ω_2 are uncommensurated and $i_S(\theta_1, \theta_2) = i_S(\theta_1 + \omega_1 T_1, \theta_2) = i_S(\theta_1, \theta_2 + \omega_2 T_2)$ [13, 14]. If $i_S(t)$ would now be passed through an algebraic nonlinearity, $v_O(t) = f[i_S(t)]$, the output would be composed of multiple combinations of the original base frequencies, ω_1, ω_2, indicating that

$$v_O(t) = \sum_{k_1=-\infty}^{\infty} \sum_{k_2=-\infty}^{\infty} V_{O_{k_1,k_2}} e^{j(k_1\omega_1 t + k_2\omega_2 t)} \tag{3.237}$$

can also be expressed as a double periodic two-dimensional signal:

$$V_o(\theta_1, \theta_2) = \sum_{k_1=-\infty}^{\infty} \sum_{k_2=-\infty}^{\infty} V_{O_{k_1,k_2}} e^{jk_1\theta_1} e^{jk_2\theta_2} \tag{3.238}$$

The two-dimensional Fourier coefficients of a box truncated version of $v_O(t)$,[10] $V_o(k_1, k_2)$, can be determined if a system of $(2K_1 + 1)(2K_2 + 1)$ equations is built from a collection of time-samples of $v_O(t)$ uniformly taken on a rectangular grid defined as

$$t_{s_1} = \frac{2\pi}{(2K_1 + 1)\omega_1} s_1,\ t_{s_2} = \frac{2\pi}{(2K_2 + 1)\omega_2} s_2; \tag{3.239}$$
$$s_1, s_2 = -K_{1,2}, \ldots, 0, \ldots, K_{1,2}$$

This system would then be

$$v_O(s_1, s_2) = \sum_{k_1=-K_1}^{K_1} \sum_{k_2=-K_2}^{K_2} V_o(k_1, k_2) W_1^{k_1 s_1} W_2^{k_2 s_2} \tag{3.240}$$

where $W_{1,2} = \exp[2\pi j/(2K_{1,2} + 1)]$. This is the definition of the inverse 2-DFT. The inversion of (3.240) leads to the two-dimensional DFT:

$$V_o(k_1, k_2) = \frac{1}{(2K_1 + 1)(2K_2 + 1)} \sum_{s_1=-K_1}^{K_1} \sum_{s_2=-K_2}^{K_2} v_O(s_1, s_2) W_1^{-s_1 k_1} W_2^{-s_2 k_2} \tag{3.241}$$

Since $v_O(s_1, s_2)$ is periodic in s_1 and s_2, $V_o(k_1, k_2)$ is usually computed as $(2K_2 + 1)$ one-dimensional FFTs on s_1, followed by $(2K_1 + 1)$ one-dimensional FFTs on s_2, to profit from the computational efficiency provided by that algorithm.

The MDFT pair, thus defined, is used on a multitone harmonic-Newton by evaluating the nonlinear function and their derivatives in the new multidimensional time-domain, and formulating the HB equation in the correspondent multidimensional frequency-domain.

10. Application of MDFT is restricted to box truncation, which imposes an important practical limitation to this technique.

Contrary to the APFT, which is approximate in nature, the MDFT is exact. Therefore, intermodulation distortion calculations made with this technique gain in numerical accuracy. Nevertheless, its algorithm implementation is much more involved and requires a rapidly increasing number of arithmetic operations when the number of base frequencies is greater than two or three.

3.3.4.3 Artificial Frequency Mapping Techniques

In the above sections we have seen that the DFT is only directly applicable to periodic signals. And, even if the signal is periodic, the computational efficiency of this transform is dramatically compromised when the signal's bandwidth is too small compared to its central frequency (i.e., the spectrum is sparse). In this section we will explore some particular features of HB that allow the use of the DFT for solving problems where it would be, a priori, not applicable.

First, we should remember that the only role played by any time-frequency transformation in HB is to permit the evaluation of the nonlinearities in time-domain. Therefore, in this context, any intermediate calculation step should be acceptable, provided the output result is correct. And second, if we restrict the nonlinearities' description to being memoryless, their output spectrum coefficients no longer depend on input absolute frequency values, but only on their relative positions. For example, the output coefficient of any mixing product, identified by (k_1, k_2), of a two-tone signal passed through a third-degree power series is the same whether the input excitations are located at $f_1 = 2$ GHz and $f_2 = 2.001$ GHz, or $f_1 = 2.4573$ KHz and $f_2 = 2.5$ KHz. For this to be true in general, we should also enforce that any mixing product of interest should not coincide in frequency with any other. This means that, either the input base frequencies are uncommensurated, or the spectrum truncation order is sufficiently high to prevent overlapping of frequency mixing components. For example, for $f_1 = 2$ GHz and $f_2 = 2.001$ GHz, no overlapping will be possible under diamond truncation if $K < 2{,}000$.

With these considerations in mind, we could relax the need for the either inaccurate APFT or complex MDFT if we found it possible to convert our mixing component's vector into another one where the original proportions between frequency positions are preserved, and is both dense and harmonically related. This constitutes the basis for the so-called *artificial frequency mapping* (AFM) techniques [14–17].

To exemplify the concept, let us consider the spectrum of Figure 3.28, which resulted from a box truncation of the mixing products of two base frequencies $f_1 = 2$ GHz and $f_2 = 2.001$ GHz, up to $K_1 = 3$ and $K_2 = 4$.

If an AFM spectrum is now constructed such that the base frequency is, let us say, $\lambda_0 = 1$ Hz (this value is, in fact, completely arbitrary), $k_1 f_1$ is transformed into $k_1 \lambda_0$ and $k_2 f_2$ into $(1 + 2K_1) k_2 \lambda_0$, we get the transformation depicted in Figure 3.29.

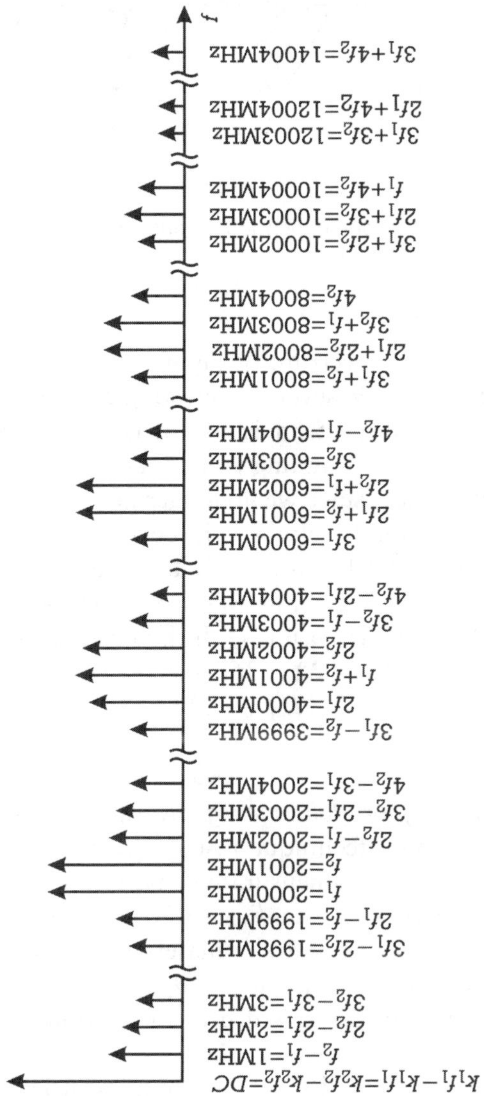

Figure 3.28 Two-tone box truncation spectrum example where $K_1 = 3$ and $K_2 = 4$.

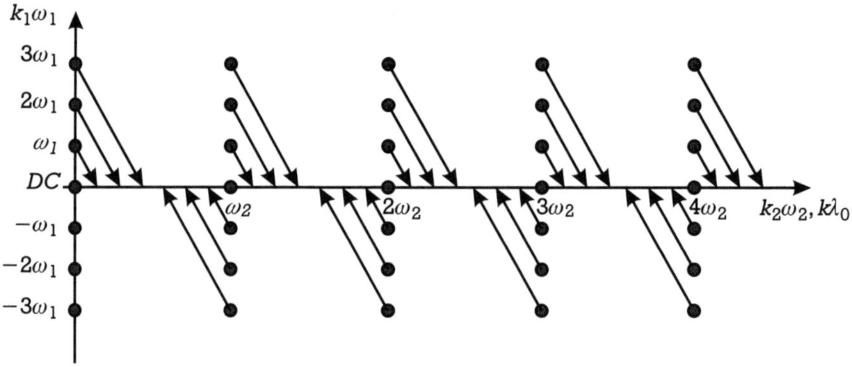

Figure 3.29 Two-dimensional to one-dimensional transformation performed by a two-tone box truncation AFM technique.

Comparing the original spectrum of Figure 3.28 with the one resulting from the transformation, Figure 3.29, several interesting conclusions can be drawn. First, the extremely sparse original spectrum is now dense (i.e., there are no null positions in between the desired mapped mixing components). This corresponds to a large gain in algorithmic efficiency, since we no longer need to handle any null positions. Second, all positions are a multiple of λ_0, the correspondent artificial time-domain signal is periodic, and the DFT can already be used. And finally, although some relative positions were lost (e.g., now $3f_1 = 6$ GHz $\rightarrow 3\lambda_0$ appears at a lower position than $f_2 - f_1 = 1$ MHz $\rightarrow 6\lambda_0$), all the original proportionality relations were preserved: $f_1 \rightarrow \lambda_0$, $2f_1 \rightarrow 2\lambda_0$, $3f_1 \rightarrow 3\lambda_0$, $f_2 \rightarrow 7\lambda_0$, $2f_2 \rightarrow 14\lambda_0$, $3f_2 \rightarrow 21\lambda_0$, and $f_2 - f_1 \rightarrow 6\lambda_0$, $2f_2 - 2f_1 \rightarrow 12\lambda_0$, $3f_2 - 3f_1 \rightarrow 18\lambda_0$, as shown in Figure 3.30.

Artificial Frequency Mapping for Multitone Signals with Box Truncation

To understand the mapping technique just presented, and then to generalize its use to any number, Q, of uncommensurated base frequencies, $\omega_q \in \{\omega_1, \ldots, \omega_Q: k_1\omega_1 + \ldots + k_Q\omega_Q \neq 0,\ k_1, \ldots, k_Q \in Z\}$ we must realize that the relative positions of these base frequencies are immaterial. In fact, when we explained the MDFT we have considered $\omega_1 t \equiv \theta_1$, $\omega_2 t \equiv \theta_2, \ldots, \omega_Q t \equiv \theta_Q$ as Q independent variables. Therefore, we can use this degree of freedom to develop the mapping procedure.

Let us begin by considering that the harmonics of ω_1 were truncated to K_1. Its mixing components are shown in Figure 3.31.

The number of tones is $2K_1 + 1$ $(-K_1\omega_1, \ldots, 0, \ldots, K_1\omega_1)$, which can be mapped onto an artificial frequency-domain, λ, as $\omega_1 \rightarrow \lambda_0$ and $k_1\omega_1 \rightarrow k_1\lambda_0$.

The addition of another tone, ω_2, whose harmonics are truncated to K_2, will produce clusters of mixing products given by $k_1\omega_1 + k_2\omega_2$: $-K_1\omega_1 - K_2\omega_2, \ldots,$

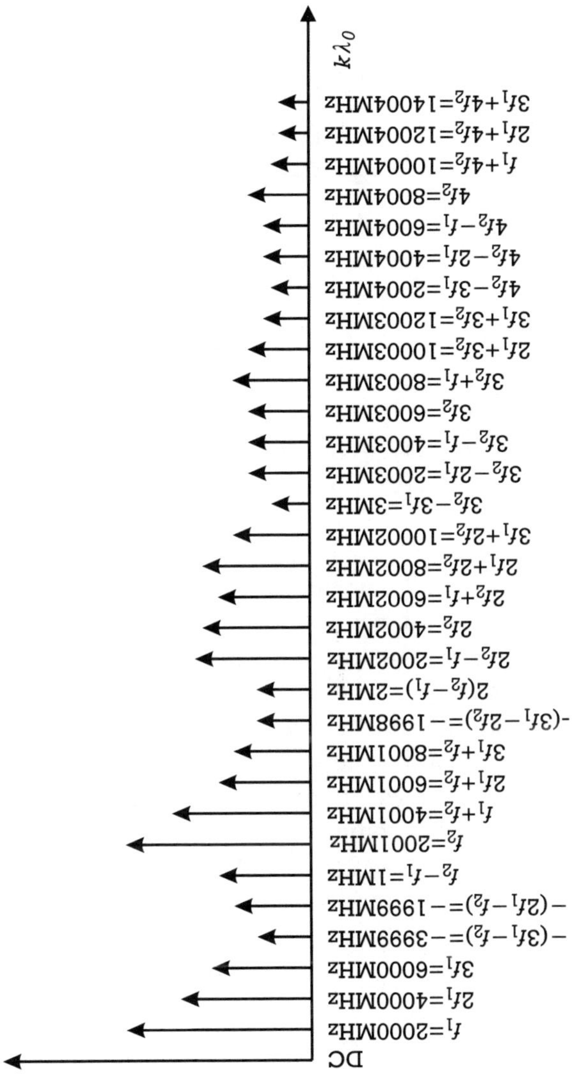

Figure 3.30 Frequency positions of a two-tone box truncated mapped spectrum.

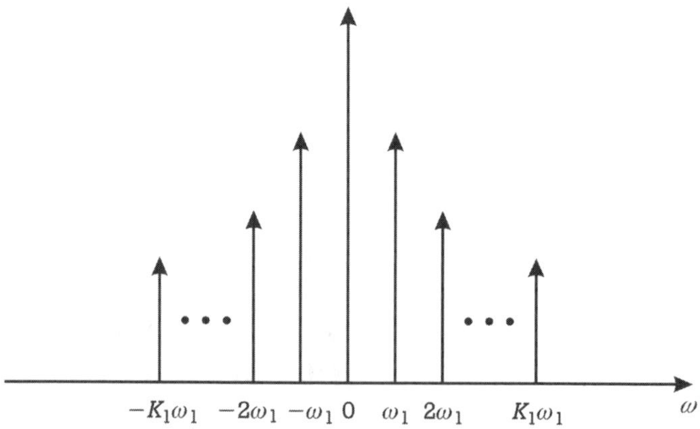

Figure 3.31 Frequency positions for only one input base frequency.

$-K_2\omega_2, \ldots, K_1\omega_1 - K_2\omega_2; \ldots; -K_1\omega_1, \ldots, 0, \ldots, K_1\omega_1; \ldots; -K_1\omega_1 + K_2\omega_2, \ldots, K_2\omega_2, \ldots, K_1\omega_1 + K_2\omega_2$. Depending on the relative positions of ω_1 and ω_2, these clusters may overlap: $\omega_2 < (2K_1 + 1)\omega_1$; or may have vacant bands in between: $\omega_2 > (2K_1 + 1)\omega_1$. As we are free to shift the location of ω_2 with respect to ω_1, making $\omega_2 = (2K_1 + 1)\omega_1$ produces an optimum spectrum that is dense and has no overlapping clusters, as shown in Figure 3.32.

The spectrum of Figure 3.32 includes all possible products generated by combining $k_1\omega_1$ with $k_2\omega_2$, containing a total of $(2K_1 + 1)(2K_2 + 1)$ components. These can be mapped to the artificial frequency-domain such that $k_1\omega_1 \rightarrow k_1\lambda_0$ and $k_2\omega_2 \rightarrow k_2(2K_1 + 1)\lambda_0$.

The addition of a third base frequency, ω_3, can now be handled in much the same way by simply replicating the spectrum of Figure 3.32 in the harmonics of ω_3, $k_3\omega_3$. It will produce a set of components given by $k_1\omega_1 + k_2\omega_2 + k_3\omega_3$. Since the number of frequency components in any of these clusters is $(2K_1 + 1)(2K_2 + 1)$, ω_3 should be located at $\omega_3 = (2K_1 + 1)(2K_2 + 1)\omega_1$ to obtain a nonoverlapping, but dense, spectrum. The new mapping is thus: $k_1\omega_1 \rightarrow k_1\lambda_0$; $k_2\omega_2 \rightarrow k_2(2K_1 + 1)\lambda_0$ and $k_3\omega_3 = k_3(2K_1 + 1)(2K_2 + 1)\lambda_0$.

A generalization of this procedure to Q uncommensurated tones gives a total number of mixing products of

$$\prod_{q=1}^{Q} (2K_q + 1) \tag{3.242}$$

and the following mapping functions:

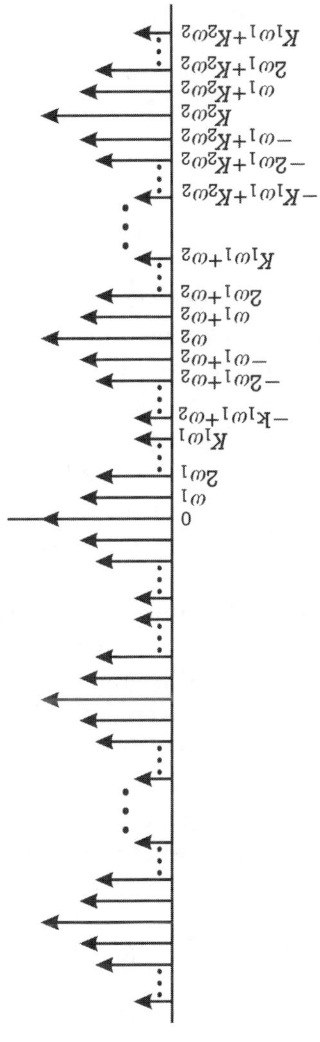

Figure 3.32 Rearranged frequency positions for two input base frequencies with box truncation.

$$\begin{aligned}
&\omega_1 \to \lambda_0 \\
&\omega_2 \to (2K_1 + 1)\lambda_0 \\
&\omega_3 \to (2K_1 + 1)(2K_2 + 1)\lambda_0 \\
&\vdots \\
&\omega_Q \to \left[\prod_{q=1}^{Q-1} (2K_q + 1)\right]\lambda_0
\end{aligned} \tag{3.243a}$$

or

$$\omega_k = \prod_{q=1}^{Q} k_q \omega_q \to \lambda_k = \left\{\prod_{q_1=1}^{Q} k_{q_1}\left[\prod_{q_2=1}^{Q-1} (2K_{q_2} + 1)\right]\right\}\lambda_0 \tag{3.243b}$$

Artificial Frequency Mapping for Two-Tone Signals with Diamond Truncation

Contrary to box truncation, for diamond truncation no AFM producing a dense spectrum is known for any number of base frequencies greater than two. Although it is possible to generate a harmonically related mapped spectrum, it seems difficult to build one without any zero amplitude positions. However, the AFM for the two base frequencies case is simple, and, as we shall see later, can be quite useful.

To explain the way this mapping operates, consider the two-tone spectrum, which was truncated to fifth order ($\omega_k = k_1\omega_1 + k_2\omega_2, |k_1| + |k_2| < 5$), represented in Figure 3.33.

A convenient two-dimensional to one-dimensional mapping for this diamond truncated spectrum is represented in Figure 3.34.

Comparing the positions occupied by the original (Figure 3.33) and mapped frequencies (Figure 3.34) we conclude that this AFM keeps, not only the proportionality between frequency values, but also their relative positions. In fact, the mapping does nothing more than simply removing the zero amplitude spectral positions present in between the various mixing clusters.

The extension of this AFM to two-base frequencies diamond truncated up to order K gives

$$2K^2 + 2K + 1 \tag{3.244}$$

as the total number of mixing products, and the following mapping functions:

$$\begin{aligned}
&\omega_1 \to K\lambda_0 \\
&\omega_2 \to (K + 1)\lambda_0
\end{aligned} \tag{3.245a}$$

or

$$\omega_k = k_1\omega_1 + k_2\omega_2 \to \lambda_k = [k_1 K + k_2(K + 1)]\lambda_0 \tag{3.245b}$$

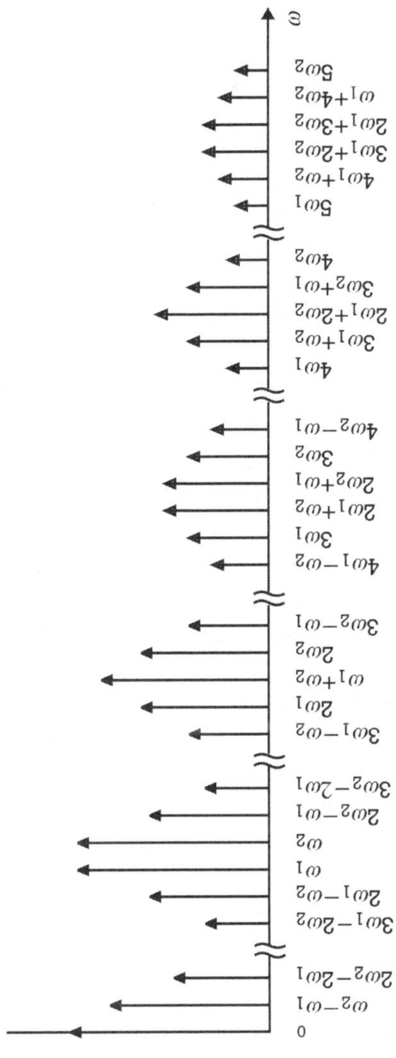

Figure 3.33 Original two-tone spectrum diamond truncated up to fifth order.

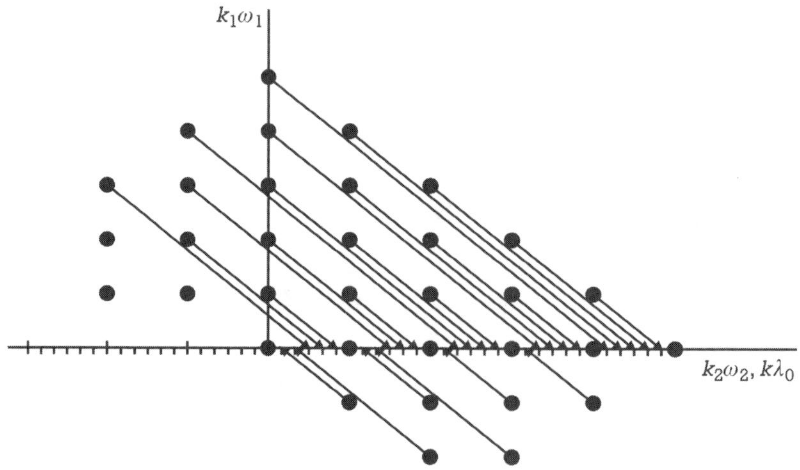

Figure 3.34 Two-dimensional to one-dimensional transformation performed by a two-tone diamond truncation AFM technique.

Another generalization of this AFM, with important practical interest, uses the fact that a multitone signal composed of Q equally spaced tones is not characterized by Q independent base frequencies (as in the case of truly uncommensurated tones) but only by two. Indeed, note that if the Q tones share an exact common separation $\Delta\omega$, then all Q tones can be uniquely identified by the frequency of one of them—for example, the one of lower frequency, ω_0, and a multiple of the separation $\Delta\omega$: $\omega_q = \omega_0 + (q-1)\Delta\omega$; $q = 1, \ldots, Q$.

An example of such type of spectrum can be seen in Figure 3.35, where the number of tones is Q and truncation order is $K = 3$.

Again, we can obtain the desired dense and harmonically related spectrum by simply eliminating the null amplitude positions.

The total number of (positive and negative) mixing products, obtained after diamond truncation up to order K, can be calculated by summing up the number of mixing products in the even and odd-order clusters:

If K is odd:	No. of odd order clusters	$= K + 1$
	No. of products per odd order cluster	$= QK - (K-1)$
	No. of even order clusters	$= K$
	No. of products per even order cluster	$= Q(K-1) - (K-2)$
	Total number of products	$= 2QK^2 - 2K^2 + 2K + 1$
If K is even:	No. of odd order clusters	$= K$
	No. of products per odd order cluster	$= Q(K-1) - (K-2)$
	No. of even order clusters	$= K + 1$
	No. of products per even order cluster	$= QK - (K-1)$
	Total number of products	$= 2QK^2 - 2K^2 + 2K + 1$

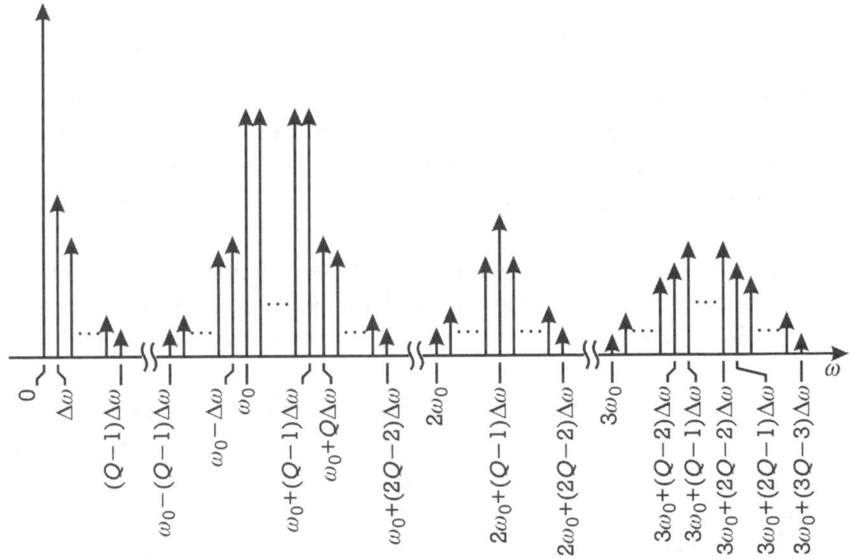

Figure 3.35 Original uniformly distributed multitone spectrum with diamond truncation.

Thus, the total number of mixing products is

$$2QK^2 - 2K^2 + 2K + 1 \tag{3.246}$$

for either odd or even K. The mapping functions can be given by

$$\begin{aligned} \omega_1 &\rightarrow [K(Q-1) - Q + 2]\lambda_0 \\ \omega_2 &\rightarrow [K(Q-1) - Q + 3]\lambda_0 \\ &\vdots \\ \omega_Q &\rightarrow [K(Q-1) + 1]\lambda_0 \end{aligned} \tag{3.247a}$$

or

$$\omega_k = \sum_{q=1}^{Q} k_q \omega_q \rightarrow \lambda_k = \left\{ \sum_{q=1}^{Q} k_q [K(Q-1) - Q + 1 + q] \right\} \lambda_0 \tag{3.247b}$$

Artificial Frequency Mapping for Multitone Signals with Combined Box-Diamond Truncation

Comparing box truncation and diamond truncation, we conclude that although the multitone box truncation scheme is applicable to any number of uncorrelated signals, it can be computational expensive in presence of a large number of input tones. Diamond truncation produces clearly more efficient index vectors, but it is restricted to equally spaced signals. So, none of these AFMs is appropriate for a very important case: large-signal multitone mixer analysis. There, we can have a large number of equally spaced tones composing the radio-frequency signal input, RF excitation, plus one very strong local oscillator signal of uncorrelated frequency, LO pump. The two AFM arrangements presented next were exactly conceived to satisfy this need.

The first case to be considered is an AFM that handles a mixed truncation scheme: diamond truncation for the RF equally spaced multitone signal, and box truncation for the LO. The algorithm necessary for its mapped frequency index vector generation can be divided into two steps.

The first step consists of applying the AFM for diamond truncation spectra, described above, to our equally spaced multitone RF input. This mapped spectrum, composed of a certain number of adjacent clustered mixing products, is shown in Figure 3.36.

In the second step, this diamond truncated mapped spectrum is considered as a new composite tone to be mixed with the LO spectral line. The resultant mixing products are then box truncated as seen in Figure 3.37, and another AFM is considered. The obtained spectrum corresponds to the dense and periodic frequency index vector used for simulating the mixer.

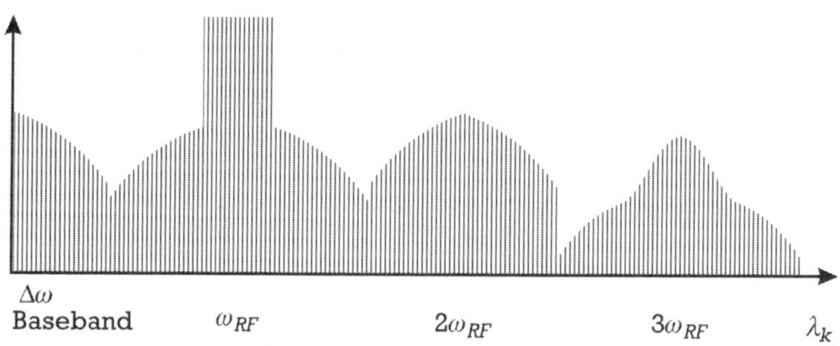

Figure 3.36 Artificial frequency mapped output of the diamond truncated RF spectrum.

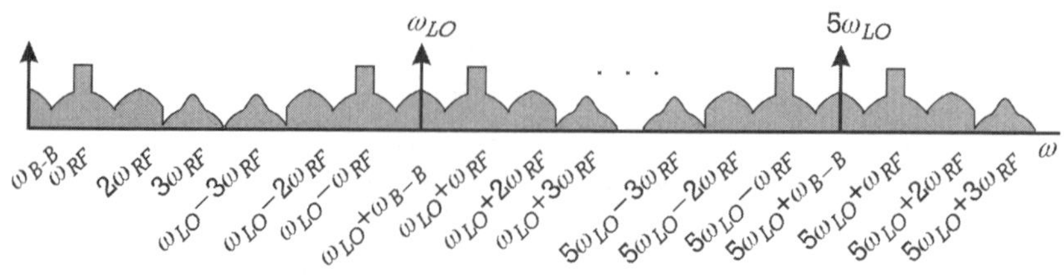

Figure 3.37 Final mapped spectrum obtained from a diamond truncation AFM to the RF, followed by a box truncation AFM to combine the RF and LO excitations.

Expression (3.248) presents the total number of resulting clusters that must be considered:

$$N \approx 2\left(\frac{O_{RF}+1}{2}O_{RF}+\frac{O_{RF}}{2}(O_{RF}-1)\right)(1+2O_{LO}) \tag{3.248}$$

where O_{RF} and O_{LO} are the RF nonlinear order and the LO nonlinear order considered, respectively.

The alternative AFM is based on a diamond truncation scheme specially designed for uniformly discretized spectra mixed with another uncorrelated LO tone ($\omega_{RF} = \omega_{RF_0} + k_1\Delta\omega$, and $\omega_{LO} \neq \omega_{RF_0} + k_2\Delta\omega$, $k_1, k_2 \in Z$).

To build the AFM index vector, we begin by first viewing the complete excitation as a two-tone signal. One of these imaginary tones is the local oscillator, while the other is located at the center of the RF signal. The mixing process thus generates new frequency components at positions given by Table 3.1, for the example of a third-order nonlinearity. Each of these mixing terms creates a clustered spectrum.

Knowing the number of RF terms produced in each cluster, the second step consists of generating the corresponding spectral regrowth, using the formulas already developed for the diamond truncation.

Because the direct application of the general rules of diamond truncation AFM would not generate a periodic mapped spectrum, the third step enforces that periodicity by inserting a certain number of zeros between spectrum clusters. This number of zeros can be determined by first considering the LO and RF as a two-tone signal, and (as before) simply ignoring the zeros present between each of these clusters. Then, the various numbers of zeros between the output RF (spectral regrowth included) and the LO are determined for each cluster. The number of zeros required for guaranteeing a harmonically related mapped spectrum is equal

Table 3.1 Two-Tone Cluster Generation for a Third-Order Nonlinearity

Number	*Cluster*	*Mixing Terms*
First	Second-order distortion difference frequencies	dc
		$\omega_{RF} - \omega_{LO}$
Second	Fundamental third-order inband distortion	$2\omega_{LO} - \omega_{RF}$
		ω_{LO}
		ω_{RF}
		$2\omega_{RF} - \omega_{LO}$
Third	Second-order distortion sum frequencies	$2\omega_{LO}$
		$\omega_{LO} + \omega_{RF}$
		$2\omega_{RF}$
Fourth	Third-order out-of-band distortion	$3\omega_{LO}$
		$2\omega_{LO} + \omega_{RF}$
		$2\omega_{RF} + \omega_{LO}$
		$3\omega_{RF}$

to the minimum of those. This way, it is possible to generate a new artificially mapped spectrum that is periodic and compact. Although the mapped spectrum is not completely dense, the resulting final spectrum is much more efficient to handle than the original one.

Figure 3.38 is an illustration of the resulting mapped spectrum.

3.3.5 Harmonic Balance Applied to Network Analysis

There are two different basic approaches for analyzing a nonlinear network with HB, depending on the way the linear subnetwork is treated. The one known as *piecewise harmonic balance* [18] breaks the circuit into a multiport linear subnetwork to whose ports the excitation sources and nonlinear elements are connected. Then, a harmonic balance equation is written for every port shared by a nonlinear element. The other was named *nodal harmonic balance* [6], as it treats any node of the circuit in an uniform way, whether it has nonlinear element connections or not. Therefore, nodal HB writes down an HB equation for each circuit node, and this leads to much larger systems of equations and unknowns than piecewise HB. This is particularly true in RF and microwave circuits, where they are usually composed of much larger number of linear elements than nonlinear ones. Because of that, in the following we will address the piecewise HB approach.

Let us begin by considering the circuit of Figure 3.39, which includes a certain number of nonlinear elements and excitation sources connected to a linear subnetwork. As was explained in Section 3.2.2.1 for the nonlinear currents method of Volterra series, this linear subnetwork must have as many ports as the sum of the number of nonlinear elements in the circuit ($m = 1, \ldots, M$), plus their controlling voltages (if not defined in any other created port) ($l = 1, \ldots, L$), excitation sources ($f = 1, \ldots, F$) and, finally, the output. It is assumed that the nonlinear elements are represented as voltage-controlled current sources, for which the general constitutive relation applies:

$$i_{NL}^{(m)}(t) = f_{NL}^{(m)}[v^{(1)}(t), \ldots, v^{(L)}(t)] \tag{3.249}$$

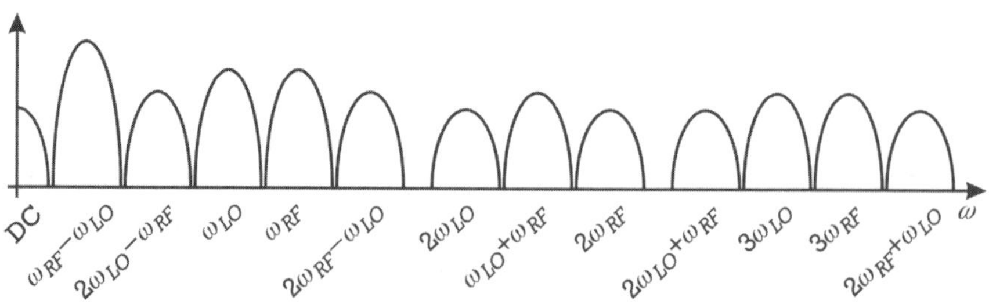

Figure 3.38 Output mapped spectrum for the diamond/diamond truncation AFM.

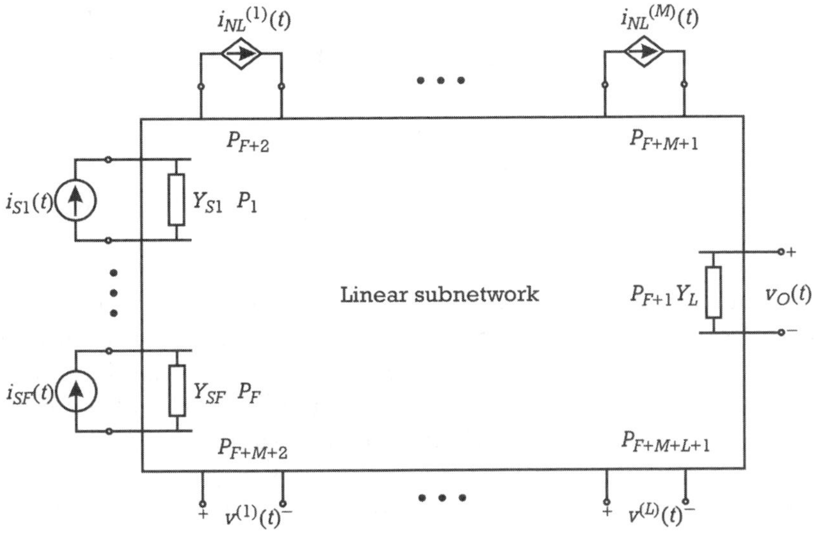

Figure 3.39 General nonlinear network description used for piecewise HB analysis.

and the linear subnetwork is described by its $(F + M + L + 1) \times (F + M + L + 1)$ admittance parameter matrix $\mathbf{Y}(\omega)$.

Since any circuit branch current or node voltage can be derived from the $(F + M + L + 1)$ port voltages, the problem is reduced to finding $v_1(t), \ldots, v_F(t)$, $[v_{(F+1)}(t) = v_O(t)]$, $v_{(F+2)}(t), \ldots, v_{(F+M+1)}(t)$, $[v_{(F+M+2)}(t) = v^{(1)}(t)], \ldots$, $[v_{(F+M+L+1)}(t) = v^{(L)}(t)]$ which are represented, in frequency-domain after spectrum truncation, by $\mathbf{V}(\omega) = [V_1(\omega) \ \ldots \ V_F(\omega) V_o(\omega) V_{(F+2)}(\omega) \ \ldots \ V_{(F+M+1)}(\omega) V_{(F+M+2)}(\omega) \ldots V_{(F+M+L+1)}(\omega)]^{\mathrm{T}}$.

To do that, we begin by imposing F plus $L + 1$ boundary conditions for the appropriate port currents:

$$i_1(t) = i_{S_1}(t), \ldots, i_f(t) = i_{S_f}(t), \ldots, i_F(t) = i_{S_F}(t) \tag{3.250a}$$

and

$$i_{(F+1)}(t) = i_{(F+M+2)}(t) = \ldots = i_{(F+M+l+1)}(t) = \ldots = i_{(F+M+L+1)}(t) = 0 \tag{3.250b}$$

Then, the remaining set of M HB nodal equations are written:

$$1 \rightarrow \sum_{p=1}^{F+M+L+1} Y_{1,p}(\omega) V_p(\omega) + I_{nl}^{(1)}[\mathbf{V}(\omega)] = 0$$

$$\vdots$$

$$m \rightarrow \sum_{p=1}^{F+M+L+1} Y_{m,p}(\omega)V_p(\omega) + I_{nl}^{(m)}[\mathbf{V}(\omega)] = 0$$

$$\vdots$$

$$M \rightarrow \sum_{p=1}^{F+M+L+1} Y_{M,p}(\omega)V_p(\omega) + I_{nl}^{(M)}[\mathbf{V}(\omega)] = 0 \qquad (3.250c)$$

For calculating the currents and voltages at the M nonlinear ports, these $F + M + L + 1$ equations can be reduced to a minimum set of $M + L$ size, if the current sources are embedded in the linear network. That diminishes Jacobian size and thus allows important savings in memory storage and computing time. However, it demands for a redefinition of the admittance matrix, $Y(\omega)$, a reduced voltage vector, $\mathbf{V}(\omega) = [V_1(\omega) \ \ldots \ V_M(\omega)V_{M+1}(\omega) \ \ldots \ V_{(M+L)}(\omega)]^T$, and the calculation of the current gains between each source $I_{s_f}(\omega)$ and current at port p, $I_p(\omega)$, $A_{p,f}(\omega)$. So, the final system of HB equations comes as

$$1 \rightarrow \sum_{p=1}^{M+L} Y_{1,p}(\omega)V_p(\omega) + \sum_{f=1}^{F} A_{1,f}(\omega)I_{s_f}(\omega) + I_{nl}^{(1)}[\mathbf{V}(\omega)] = 0$$

$$\vdots$$

$$M \rightarrow \sum_{p=1}^{M+L} Y_{M,p}(\omega)V_p(\omega) + \sum_{f=1}^{F} A_{M,f}(\omega)I_{s_f}(\omega) + I_{nl}^{(M)}[\mathbf{V}(\omega)] = 0$$

$$M+1 \rightarrow \sum_{p=1}^{M+L} Y_{(M+1),p}(\omega)V_p(\omega) + \sum_{f=1}^{F} A_{(M+1),f}(\omega)I_{s_f}(\omega) = 0$$

$$\vdots$$

$$M+L \rightarrow \sum_{p=1}^{M+L} Y_{(M+L),p}(\omega)V_p(\omega) + \sum_{f=1}^{F} A_{(M+L),f}(\omega)I_{s_f}(\omega) = 0 \qquad (3.251a)$$

or, in matrix form,

$$\mathbf{Y}(\omega)\mathbf{V}(\omega) + \mathbf{A}(\omega)\mathbf{I}_s(\omega) + \mathbf{I}_{nl}[\mathbf{V}(\omega)] = \mathbf{I}_L[\mathbf{V}(\omega)] + \mathbf{I}_{NL}[\mathbf{V}(\omega)] = 0 \qquad (3.251b)$$

where $I_{nl}^{(p)}[\mathbf{V}(\omega)]$ is zero at all ports p greater than M.

Equation (3.251) [along with (3.250)] is now solved for $\mathbf{V}(\omega)$ using the harmonic-Newton method, for which it is rewritten as

$$\mathbf{F}[{}^{i+1}\mathbf{V}(\omega)] \approx \mathbf{F}[{}^{i}\mathbf{V}(\omega)] + \mathbf{J}[{}^{i}\mathbf{V}(\omega)][{}^{i+1}\mathbf{V}(\omega) - {}^{i}\mathbf{V}(\omega)] \qquad (3.252)$$

In the usual way, the voltage vector update, ${}^{i+1}\mathbf{V}(\omega)$, can be obtained from the old one, ${}^{i}\mathbf{V}(\omega)$, by

$$ {}^{i+1}\mathbf{V}(\omega) = {}^{i}\mathbf{V}(\omega) - \{\mathbf{J}[{}^{i}\mathbf{V}(\omega)]\}^{-1}\mathbf{F}[{}^{i}\mathbf{V}(\omega)] \tag{3.253} $$

If the frequency components and nodes are organized such that (3.251) reads as

$$ \begin{bmatrix} I_{L_{1,-K}} \\ \vdots \\ I_{L_{1,K}} \\ \vdots \\ I_{L_{M,-K}} \\ \vdots \\ I_{L_{M,K}} \\ I_{L_{M+1,-K}} \\ \vdots \\ I_{L_{M+1,K}} \\ \vdots \\ I_{L_{M+L,-K}} \\ \vdots \\ I_{L_{M+L,K}} \end{bmatrix} + \begin{bmatrix} I_{NL_{1,-K}} \\ \vdots \\ I_{NL_{1,K}} \\ \vdots \\ I_{NL_{M,-K}} \\ \vdots \\ I_{NL_{M,K}} \\ 0 \\ \vdots \\ 0 \\ \vdots \\ 0 \\ \vdots \\ 0 \end{bmatrix} = \begin{bmatrix} 0 \\ \vdots \\ 0 \\ \vdots \\ 0 \\ \vdots \\ 0 \\ 0 \\ \vdots \\ 0 \\ \vdots \\ 0 \\ \vdots \\ 0 \end{bmatrix} \tag{3.254} $$

the Jacobian is an $(M+L)(2K+1) \times (M+L)(2K+1)$ matrix and has the following structure:

$$ \mathbf{J}[\mathbf{V}(\omega)] = \begin{bmatrix} \frac{\partial F_{1,-K}}{\partial V_{1,-K}} & \cdots & \frac{\partial F_{1,-K}}{\partial V_{1,K}} & \cdots\cdots & \frac{\partial F_{1,-K}}{\partial V_{(M+L),-K}} & \cdots & \frac{\partial F_{1,-K}}{\partial V_{(M+L),K}} \\ \vdots & & \vdots & & \vdots & & \vdots \\ \frac{\partial F_{1,K}}{\partial V_{1,-K}} & \cdots & \frac{\partial F_{1,K}}{\partial V_{1,K}} & \cdots\cdots & \frac{\partial F_{1,K}}{\partial V_{(M+L),-K}} & \cdots & \frac{\partial F_{1,K}}{\partial V_{(M+L),K}} \\ \vdots & & \vdots & & \vdots & & \vdots \\ \vdots & & \vdots & & \vdots & & \vdots \\ \frac{\partial F_{(M+L),-K}}{\partial V_{1,-K}} & \cdots & \frac{\partial F_{(M+L),-K}}{\partial V_{1,K}} & \cdots\cdots & \frac{\partial F_{(M+L),-K}}{\partial V_{(M+L),-K}} & \cdots & \frac{\partial F_{(M+L),-K}}{\partial V_{(M+L),K}} \\ \vdots & & \vdots & & \vdots & & \vdots \\ \frac{\partial F_{(M+L),K}}{\partial V_{1,-K}} & \cdots & \frac{\partial F_{(M+L),K}}{\partial V_{1,K}} & \cdots\cdots & \frac{\partial F_{(M+L),K}}{\partial V_{(M+L),-K}} & \cdots & \frac{\partial F_{(M+L),K}}{\partial V_{(M+L),K}} \end{bmatrix} \tag{3.255} $$

3.4 Time-Domain Techniques for Distortion Analysis

Let us consider again the circuit example of Figure 3.1 herein represented by its nonlinear ODE:

$$Gv_O(t) + \frac{dq_{NL}[v_O(t)]}{dt} + i_{NL}[v_O(t)] = i_S(t) \tag{3.256}$$

The most natural way of finding $v_O(t)$ is to solve (3.256) directly in time-domain. For that, the nonlinear differential equation is converted into a nonlinear difference equation, in which the time derivatives are approximated by appropriate incremental ratios. For example, using the scheme known as Backward Euler Rule, we could have

$$\left.\frac{dq_{NL}[v_O(t)]}{dt}\right|_{t=t_k} \approx \frac{q_{NL}[v_O(t_k)] - q_{NL}[v_O(t_k - h)]}{h} \tag{3.257}$$

where h is the time-step, or sampling period. Since the derivative is defined as the limit of this incremental ratio when h tends to zero, it is obvious that the two sides of (3.257) are progressively approximated when the time discretization is denser. This, however, has a direct impact on the simulation time, even in zones of solution's quietness. A good compromise between accuracy and simulation time is achieved if h is dynamically selected according to the solution's rate of change. In fact, the gain in computation time is usually so high that all time-domain simulators use this technique. And, as a consequence, the solution is obtained in a nonuniform grid of samples. If this does not constitute any problem in general, it can have a catastrophic effect on the quality of simulations of circuits presenting low distortion levels, as we shall explain later.

3.4.1 Time-Step Integration Basics

Substituting (3.257) into (3.256) and assuming dynamic time-step selection, we have

$$Gv_O(t_k) + \frac{q_{NL}[v_O(t_k)] - q_{NL}[v_O(t_{k-1})]}{h_k} + i_{NL}[v_O(t_k)] = i_S(t_k) \tag{3.258a}$$

or

$$h_k G v_O(t_k) + q_{NL}[v_O(t_k)] + h_k i_{NL}[v_O(t_k)] = h_k i_S(t_k) + q_{NL}[v_O(t_{k-1})] \tag{3.258b}$$

which shows we can determine $v_O(t)$ for any time-sample, t_k, from the knowledge of the forcing function at that point, $i_S(t_k)$, and the past circuit solution, or system state, $v_O(t_{k-1})$.[11]

So, $v_O(t)$ is calculated for all time $t_0 < t < t_K$ beginning with the knowledge of the initial condition $v_O(t_0)$, and solving the nonlinear algebraic equation (3.258) for each time-step:

$$\begin{aligned}
&v_O(t_0)\\
&h_1 G v_O(t_1) + q_{NL}[v_O(t_1)] + h_1 i_{NL}[v_O(t_1)] = h_1 i_S(t_1) + q_{NL}[v_O(t_0)]\\
&\quad\vdots\\
&h_k G v_O(t_k) + q_{NL}[v_O(t_k)] + h_k i_{NL}[v_O(t_k)] = h_k i_S(t_k) + q_{NL}[v_O(t_{k-1})]\\
&\quad\vdots\\
&h_K G v_O(t_K) + q_{NL}[v_O(t_K)] + h_K i_{NL}[v_O(t_K)] = h_K i_S(t_K) + q_{NL}[v_O(t_{K-1})]
\end{aligned} \tag{3.259}$$

which results in a set of K nonlinear algebraic equations that can be successively solved for $v_O(t_1), \ldots, v_O(t_k), \ldots, v_O(t_K)$ using the Newton-Raphson scheme.

This formulation follows directly from our intuitive knowledge of dynamic systems' operation. So, it should be of no surprise that it was used in the first digital computer programs of circuit analysis and is still nowadays the most widely used numerical method for that purpose. It is present in all SPICE or SPICE-like computer programs.

Time-step integration is a particular implementation of what is usually known in differential equations as an *initial value problem,* because it solves $v_O(t)$ for all t_k from the knowledge of the initial condition (or state) $v_O(t_0)$. Therefore, it is clearly tailored for finding the circuit's transient responses. However, if the objective is the determination of the steady-state, there is no other way than to pass through the painful process of integrating all transients, and expecting them to vanish. In circuits having extremely different time constants, or high Q resonances, time-step integration can be very inefficient. And, unfortunately, this is typical of RF and microwave circuits, because, not only they are narrowband tuned, as their bias networks present time constants that are various orders of magnitude higher than the excitation period, or the time constants of the signal networks.

Since distortion problems demand for very high numerical dynamic ranges, and the transition from transient to steady-state behavior is gradual, the response periodicity must be guaranteed with high precision, which contributes to also

11. Because our circuit is of first-order—only first-order time-derivatives are involved—the system state is characterized by only $v_O(t_{k-1})$. In a circuit of nth order, the state would require knowledge of $v_O(t_{k-1})$, $\ldots$, $v_O(t_{k-n})$.

exacerbate simulation time. But, this is not the only problem time-step integration poses to distortion simulation. For example, a two-tone excitation involves two different time-scales, which correspond to the tones' period and the tones' separation, or envelope period. If the tones are commensurated, but close in frequency, the number of time-steps can be very large. If they are uncommensurated, the period tends to infinity, and the circuit will never reach a periodic steady-state. Obviously, postprocessing these simulation results by some kind of windowing and FFT computation will always involve a certain amount of error [19]. This error, appearing as a spectrum "noise floor," can easily mask small amplitude distortion components. Another common origin of this numerical noise floor comes from the dynamic step usually adopted to reduce simulation time. Because FFT algorithms require a uniform time-sampling, this type of postprocessing is always preceded by interpolation. And this introduces an amount of noise which often cannot be tolerated. In such cases, there is no alternative way than to impose a fixed time-step (e.g., by declaring a very conservative time-step ceiling) and cope with the resultant huge amount of data points and simulation time.

Beyond these drawbacks, time-step integration shares with any other time-domain methods a disadvantage that is particularly significant to microwave circuits. It cannot handle directly circuit elements having a frequency-domain representation. Examples of these are dispersive transmission lines, transmission line discontinuities (as microstrip cross-junctions, bends, impedance-steps, coupled-lines, etc.), or, in general, any admittance or scattering matrix coming from laboratorial network analysis. To circumvent that, some time-domain simulators suggest the use of approximated lumped equivalent circuits [20]. Nevertheless, these lumped equivalents are so difficult to extract and involve such a large number of elements that in those cases a frequency-domain simulator, as the ones using HB, is usually preferable.

Time-domain methods present, however, an important advantage over HB for lumped circuits: they are capable of handling much stronger nonlinearities. Facing strong nonlinear regimes, HB requires Fourier expansions with a large number of coefficients and its harmonic-Newton Jacobian matrix loses its characteristic diagonal dominance. As a consequence, harmonic-Newton becomes very inefficient, both in memory storage and simulation time. Time-domain methods do not suffer from these problems as the time-variable can be used as a natural continuation parameter: circuit solution at the previous time-step is always used as the initial estimate for the next Newton-Raphson iteration. In this respect, time-step integration is so good, compared to HB, that there has been a continuous push of that simulation method into the RF domain and a steadily proposal of new time-domain methods that circumvent some of the above-mentioned drawbacks.

The following sections will briefly review some of these alternatives for time-domain simulation. Steady-state sinusoidal excitation is addressed first, and then guidelines for its generalization to multitone will be given.

3.4.2 Steady-State Response Using Shooting-Newton

As was stated above, time-step integration is tailored to transient simulation, but inadequate for calculating steady-state responses. This problem comes from the fact that there is only one precise initial condition, or state, $v_O(t_0)$, for the input $i_S(t_0)$, that will lead to the steady-state in the period $T[v_O(t_0 + T) = v_O(t_0)$ for $i_S(t_0 + T) = i_S(t_0)]$, but it is unknown a priori. So, the simulator starts from another initial condition (generally determined from a previous dc analysis) and must integrate the resulting transients.

What happens is that we are trying to solve a *boundary value problem* using an initial value solution technique. In fact, the steady-state solution of our ODE can be formulated in the following way: What initial condition, or left boundary, $v_O(t_0)$, should be selected to our ODE [forced by a periodic function $i_S(t)$ such that $i_S(t_0 + T) = i_S(t_0)$] that leads to a periodic solution also obeying the final condition, or right boundary, $v_O(t_0 + T) = v_O(t_0)$? One possible way to transform our initial value solution procedure into a boundary value problem solver consists of guessing an initial estimate of $v_O(t_0)$, or *shooting* for $v_O(t_0)$, solving the ODE using the normal initial value solver, comparing the resulting $v_O(t_0 + T)$ with $v_O(t_0)$, and then wisely update the initial condition guess. For example, the operator could try first a simulation for one period of the excitation using a certain initial condition, ${}^0v_O(t_0)$, and then observe the resulting ${}^0v_O(t_0 + T) - {}^0v_O(t_0)$. He could then estimate the sensitivity of $v_O(t_0 + T)$ to the adopted $v_O(t_0)$ by slightly perturbing ${}^0v_O(t_0)$, creating a new ${}^1v_O(t_0) = {}^0v_O(t_0) + \Delta v_O(t_0)$. Another simulation using that ${}^1v_O(t_0)$ would lead to a ${}^1v_O(t_0 + T)$, which can be expressed as ${}^1v_O(t_0 + T) = {}^0v_O(t_0 + T) + \Delta v_O(t_0 + T)$. A closer look on this $\Delta v_O(t_0 + T)$ could then be used to build a better estimate of $v_O(t_0)$, ${}^2v_O(t_0)$, and thus accelerate the route to steady-state.

The first nice property of this procedure is that it converges to the steady-state solution much faster than the normal time-step integration. Really, our first studied method uses the natural initial condition update algorithm of starting from a predetermined ${}^0v_O(t_0)$, and making ${}^1v_O(t_0) = {}^0v_O(t_0 + T)$, ${}^2v_O(t_0) = {}^1v_O(t_0 + T)$, ..., ${}^nv_O(t_0) = {}^{n-1}v_O(t_0 + T)$, until ${}^Nv_O(t_0) = {}^{N-1}v_O(t_0)$.

The second nice property of the proposed algorithm is that it can be easily automated. One way to do that constitutes the so-called *shooting-Newton* technique [21].

As any other shooting method, shooting-Newton relies on guessed initial conditions. But, contrary to other methods, it takes advantage of the observed fact that, although electrical circuits can be very nonlinear, their state-transition functions,[12] $\phi[v_O(t_0), t]$, are usually quite linear. That is, small perturbations on the initial condition, or starting state, produce almost proportional perturbations in the

12. State transition function, $\phi[v_O(t_0), t]$, is a mapping that describes the evolution of the system, or its trajectory along time, in the sate-space, such that $v_O(t) = \phi[v_O(t_0), t]$.

subsequent time states. Therefore, the state transition function can be approximated by a first-order Taylor series, and thus our steady-state boundary condition,

$$v_O(t_0 + T) = v_O(t_0) \tag{3.260a}$$

or

$$v_O(t_0 + T) - v_O(t_0) = \phi[v_O(t_0), T] - v_O(t_0) = 0 \tag{3.260b}$$

can be iteratively solved for $v_O(t_0)$ by

$$\phi[{}^0v_O(t_0), T] - {}^0v_O(t_0) + \left[\left. \frac{\partial \phi[v_O(t_0), T]}{\partial v_O(t_0)} \right|_{v_O(t_0) = {}^0v_O(t_0)} - 1 \right] \cdot [{}^1v_O(t_0) - {}^0v_O(t_0)] = 0 \tag{3.261a}$$

or, generally,

$${}^{i+1}v_O(t_0) = {}^iv_O(t_0) - \left[\left. \frac{\partial \phi[v_O(t_0), T]}{\partial v_O(t_0)} \right|_{v_O(t_0) = {}^iv_O(t_0)} - 1 \right]^{-1} \cdot [\phi[{}^iv_O(t_0), T] - {}^iv_O(t_0)] \tag{3.261b}$$

Since $\phi[{}^iv_O(t_0), T]$ is simply the $v_O(t_0 + T)$ resulting from the ODE integration with the initial condition ${}^iv_O(t_0)$, the only entity of (3.261b) that is difficult to be computed is the state transition function's sensitivity.

To calculate this sensitivity, we should first realize that the chain differentiation rule imposes that, since $\phi[v_O(t_0), T] \equiv \phi[v_O(t_0), t_K]$ is a function of $\phi[v_O(t_0), t_{K-1}]$, which, itself also depends on $\phi[v_O(t_0), t_{K-2}]$, and so forth; $\partial\phi[v_O(t_0), T]/\partial v_O(t_0)$ can be given by

$$\frac{\partial \phi[v_O(t_0), T]}{\partial v_O(t_0)} = \frac{\partial \phi[v_O(t_0), t_K]}{\partial \phi[v_O(t_0), t_{K-1}]} \cdot \frac{\partial \phi[v_O(t_0), t_{K-1}]}{\partial \phi[v_O(t_0), t_{K-2}]} \cdot \ldots \cdot \frac{\partial \phi[v_O(t_0), t_k]}{\partial \phi[v_O(t_0), t_{k-1}]} \cdot \ldots \cdot \frac{\partial \phi[v_O(t_0), t_1]}{\partial v_O(t_0)} \tag{3.262a}$$

or

$$\frac{\partial v_O(t_K)}{\partial v_O(t_0)} = \frac{\partial v_O(t_K)}{\partial v_O(t_{K-1})} \cdot \frac{\partial v_O(t_{K-1})}{\partial v_O(t_{K-2})} \cdot \ldots \cdot \frac{\partial v_O(t_k)}{\partial v_O(t_{k-1})} \cdot \ldots \cdot \frac{\partial v_O(t_1)}{\partial v_O(t_0)} \tag{3.262b}$$

Now, looking onto the time-step iteration scheme given by (3.259), which states that any state $v_O(t_k)$ can be computed from the previous one $v_O(t_{k-1})$ by

$$h_k G v_O(t_k) + q_{NL}[v_O(t_k)] + h_k i_{NL}[v_O(t_k)] = h_k i_S(t_k) + q_{NL}[v_O(t_{k-1})] \tag{3.263}$$

it is easily concluded that all derivatives of (3.262) can be computed along the time-step integration process, because $\partial\phi[v_O(t_0), t_k]/\partial\phi[v_O(t_0), t_{k-1}]$ can be obtained by simply deriving (3.263) with respect to $v_O(t_{k-1})$:

$$h_k G \frac{\partial v_O(t_k)}{\partial v_O(t_{k-1})} + \frac{\partial q_{NL}[v_O(t_k)]}{\partial v_O(t_k)} \frac{\partial v_O(t_k)}{\partial v_O(t_{k-1})} + h_k \frac{\partial i_{NL}[v_O(t_k)]}{\partial v_O(t_k)} \frac{\partial v_O(t_k)}{\partial v_O(t_{k-1})}$$
$$= \frac{\partial q_{NL}[v_O(t_{k-1})]}{\partial v_O(t_{k-1})} \tag{3.264a}$$

or

$$\frac{\partial v_O(t_k)}{\partial v_O(t_{k-1})} = \{h_k G + J_q[v_O(t_k)] + h_k J_i[v_O(t_k)]\}^{-1} J_q[v_O(t_{k-1})] \tag{3.264b}$$

where $J_q[.]$ and $J_i[.]$ are the charge and current entries of the Jacobian, computed when solving (3.259) using the Newton-Raphson iteration.

3.4.3 Finite-Differences in Time-Domain

A different approach for determining the periodic steady-state response of our circuit consists of solving the ODE directly as a boundary value problem. For that, we first define a certain time grid, $t_0, t_1, \ldots, t_{K-1}, t_K = t_0 + T$, and impose the finite-differences discretization of our ODE, as in (3.259), to all internal time-points plus the final one:

$$h_k G v_O(t_k) + q_{NL}[v_O(t_k)] + h_k i_{NL}[v_O(t_k)] - h_k i_S(t_k) - q_{NL}[v_O(t_{k-1})] = 0 \tag{3.265a}$$

or

$$f_{NL}[v_O(t_k), v_O(t_{k-1}), i_S(t_k)] = 0 \tag{3.265b}$$

This leads to a system of K equations in $(K + 1)$ unknowns, $v_O(t_0), \ldots, v_O(t_K)$, which can be solved using another equation describing the periodic regime we seek:

$$v_O(t_0 + T) = v_O(t_K) = v_O(t_0) \tag{3.266}$$

So, a final system of K equations in K unknowns can be written as

$$\begin{gathered} f_{NL}[v_O(t_1), v_O(t_0), i_S(t_1)] = 0 \\ \vdots \\ f_{NL}[v_O(t_k), v_O(t_{k-1}), i_S(t_k)] = 0 \\ \vdots \\ f_{NL}[v_O(t_0), v_O(t_{K-1}), i_S(t_0)] = 0 \end{gathered} \tag{3.267}$$

Again, this is a nonlinear system that may be solved using a K-dimensional Newton-Raphson iteration. In that case, we start by assuming a certain estimate for the vector $[{}^0v_O(t_0), \ldots, {}^0v_O(t_{k-1})]$, that, in principle, does not verify our system, and then hope it iteratively relaxes to the steady-state solution.

Note that, contrary to the initial value strategy of shooting, in which all intermediate iterations verify the system's ODE but not the boundary condition $v_O(t_0 + T) = v_O(t_0)$, in this FDTD approach the boundary condition is always verified, but the ODE is not.

Another important difference between these two methods is that while shooting-Newton solves for only $v_O(t_k)$ at instant t_k, FDTD must solve for all $[v_O(t_0), \ldots, v_O(t_{K-1})]$ simultaneously. So, shooting-Newton relies on K one-dimensional Newton-Raphson iteration schemes, while FDTD requires one K-dimensional Newton-Raphson iterative solver. The amount of storage needed for FDTD is clearly much larger than the one required for shooting-Newton, which has prevented its use in general-purpose commercial simulators. Furthermore, because the state-transition function is almost linear, shooting-Newton generally needs less Newton-Raphson iterations to converge than FDTD, and thus it can be more robust and fast.

Conversely, because FDTD has to handle all time-points simultaneously, it can deal with convolutive relations. And this may be very useful for treating RF or microwave elements with frequency-domain representations, as their responses can always be computed as the convolution of their impulse responses (inverse Fourier transform of the frequency-domain transfer functions) with the time-domain inputs [1].

3.4.4 Quasiperiodic Steady-State Solutions in Time-Domain

All the methods until now presented in the literature to extend the traditional sinusoidal steady-state solvers to multitone excitations assume that those multiple tones are uncommensurated [22]. So, for example, in the two-tone case, ω_1 and

ω_2, $k_1\omega_1 + k_2\omega_1 \neq 0$ for any nonsimultaneous null integers k_1 and k_2. In this sense, the phase variables $\theta_1 \equiv \omega_1 t$, $\theta_2 \equiv \omega_2 t$ are independent, which allows the statements $\theta_1 = \omega_1 t_1$, $\theta_2 = \omega_2 t_2$, and that the circuit is no longer dependent on a single time-scale, but on as many time-scales as the number of uncommensurated excitations. Therefore, our previous nonlinear dynamic system described by the ODE,

$$Gv_O(t) + \frac{dq_{NL}[v_O(t)]}{dt} + i_{NL}[v_O(t)] = i_S(t) \tag{3.268}$$

will now be mathematically represented by the following nonlinear *multirate partial differential equation* (MPDE) [23]:

$$Gv_O(t_1, t_2) + \frac{\partial q_{NL}[v_O(t_1, t_2)]}{\partial t_1} + \frac{\partial q_{NL}[v_O(t_1, t_2)]}{\partial t_2} + i_{NL}[v_O(t_1, t_2)] = i_S(t_1, t_2) \tag{3.269}$$

This MPDE can be discretized over a rectangular time-grid of $(K_1 + 1) \times (K_2 + 1)$ points, using again the backward Euler rule, which leads to a general multirate difference equation for the two-dimensional point (t_{k_1}, t_{k_2}) as

$$\begin{aligned} &Gv_O(t_{k_1}, t_{k_2}) + \frac{q_{NL}[v_O(t_{k_1}, t_{k_2})] - q_{NL}[v_O(t_{k_1-1}, t_{k_2})]}{h_{k_1}} \\ &+ \frac{q_{NL}[v_O(t_{k_1}, t_{k_2})] - q_{NL}[v_O(t_{k_1}, t_{k_2-1})]}{h_{k_2}} + i_{NL}[v_O(t_{k_1}, t_{k_2})] \\ &= i_S(t_{k_1}, t_{k_2}) \end{aligned} \tag{3.270a}$$

or

$$\begin{aligned} &h_{k_1}h_{k_2}Gv_O(t_{k_1}, t_{k_2}) + (h_{k_1} + h_{k_2})q_{NL}[v_O(t_{k_1}, t_{k_2})] \\ &+ h_{k_1}h_{k_2}i_{NL}[v_O(t_{k_1}, t_{k_2})] \\ &= h_{k_1}h_{k_2}i_S(t_{k_1}, t_{k_2}) - h_{k_2}q_{NL}[v_O(t_{k_1-1}, t_{k_2})] - h_{k_1}q_{NL}[v_O(t_{k_1}, t_{k_2-1})] \end{aligned} \tag{3.270b}$$

This equation, associated with the initial conditions imposed on $v_O(t_{k_1}, t_0)$ and $v_O(t_0, t_{k_2})$ [determined by time-step integration along the axes $(t_k, 0)$ and $(0, t_k)$], allows the calculation of any $v_O(t_{k_1}, t_{k_2})$. Note that, since $v_O(t)$ was mapped into $v_O(t, t)$, the sought solution of our original ODE under multirate excitation is the one-dimensional subset $v_O(t_{k_1}, t_{k_2})$ in which $t_{k_1} = t_{k_2}$.

With this MPDE formulation in mind, the generalization of shooting-Newton or FDTD to multitone excitations is, in concept, straightforward, although it may become extremely laborious. In fact, as the number of excitation tones increases, problem complexity also increases, and the number of grid points to be determined rises exponentially. Actually, technical publications [22, 23] only report two-tone MPDE implementations.

3.4.5 Mixed-Mode Simulation Techniques

Until now we have reviewed some simulation techniques that operate either in frequency- or in time-domain. However, since we have seen that any circuit facing a multitone excitation can be modeled as a MPDE, it seems possible to use a mixed-mode simulation technique, handling the solution dependence on some of its time variables in time-domain, and the course of the solution to other time variables in frequency-domain. This idea is particularly attractive when the excitation is dependent on two or more time-scales that differ by many orders of magnitude.

An example of practical interest refers to modulated signals in which the information signal is typically aperiodic and has a spectral content of much lower frequency than the periodic carrier. In that case, it can be useful to apply HB for simulating the circuit behavior to the carrier, and time-step integration for determining the response to the slowly varying envelope. A special implementation of this concept is known as *envelope transient harmonic balance* (ETHB) [24, 25], and constitutes the subject of the remainder of this section.

The basic thought behind the application of ETHB to the distortion simulation problem consists of considering a certain class of RF signals as high-frequency carriers modulated by low-frequency complex envelopes. For example, the equally spaced multitone excitation spectrum of Figure 3.40(a),

$$i_S(t) = I_{S_0} + \sum_{\substack{q=-Q \\ (q \neq 0)}}^{Q} I_{S_q} e^{j\omega_q t} \tag{3.271}$$

can be rewritten as

$$i_S(t) = I_{S_0} + \left[\sum_{m=-(Q-1)/2}^{(Q-1)/2} I_{S_m} e^{j\Omega_m t} \right] (e^{-j\omega_c t} + e^{j\omega_c t}) \tag{3.272}$$

if Q is odd. Equation (3.272) can be interpreted as if the RF carrier,

$$i_{S_c}(t) = 2 \cos \omega_c t; \; \omega_c = \omega_q : q = \frac{Q+1}{2} \tag{3.273}$$

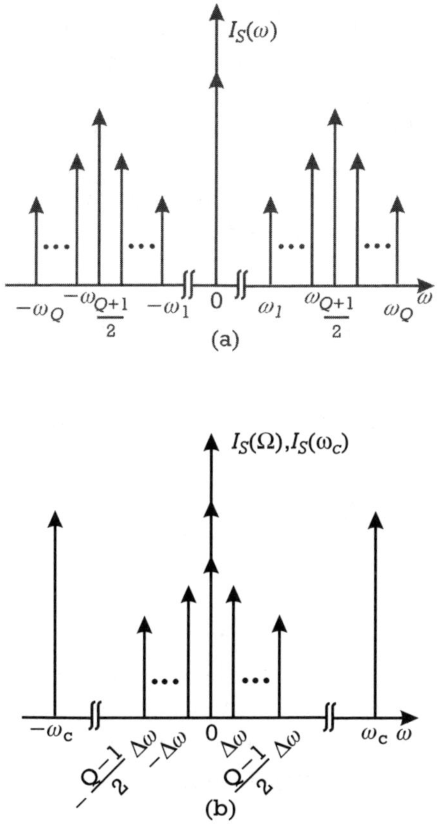

Figure 3.40 Interpretation of the (a) original excitation as (b) an RF carrier modulated by a baseband complex envelope.

were amplitude modulated by the following baseband envelope:

$$i_{S_e}(t) = I_{se_0} + 2 \sum_{m=1}^{(Q-1)/2} |I_{s_m}| \cos(\Omega_m t + \phi_m);\ \Omega_m = m\Delta\omega \tag{3.274}$$

as seen in Figure 3.40(b).

Now, imagine we would like to compute the transient response of a linear circuit (e.g., a linearized version of our test circuit),

$$Gv_O(t) + c_1 \frac{dv_O(t)}{dt} + g_1 v_O(t) = i_S(t) \tag{3.275}$$

to the envelope of an excitation $i_S(t)$ of this type.

The direct integration of (3.275) would be very inefficient since the time-step ceiling would be imposed by the high-frequency RF carrier, $h < \pi/\omega_c$, while the total integration time would be determined by the period of the slowly varying envelope, $T = 2\pi/\Delta\omega$. The transient envelope method circumvents this problem by assuming that $v_O(t)$ is dependent on two time variables—t_c for the carrier time-scale, and t_e for the envelope time-scale—which correspond to also two distinct frequency-domain scales, ω_c and Ω.

$$v_O(t_c, t_e) = V_O + v_{O_e}(t_e)v_{O_c}(t_c) \leftrightarrow V_o(\omega_c, \Omega) \tag{3.276}$$

In this sense, (3.275) can then be rewritten to read as

$$Gv_O(t_c, t_e) + c_1 \frac{\partial v_O(t_c, t_e)}{\partial t_c} + c_1 \frac{\partial v_O(t_c, t_e)}{\partial t_e} + g_1 v_O(t_c, t_e) = i_S(t_c, t_e) \tag{3.277}$$

The assumed periodicity of the carrier and the possible aperiodic behavior of the envelope indicates that we should compute the envelope transient in the time-domain, t_e, and the carrier behavior in the frequency-domain, ω_c.

For that, we begin by expressing the output voltage $v_O(t_c, t_e)$ *as a modulated carrier:*

$$v_O(t_c, t_e) = V_O + v_{O_e}(t_e)v_{O_c}(t_c) = \sum_{k=-1}^{1} V_{O_k}(t_e)e^{jk\omega_c t_c} \tag{3.278}$$

Then, substituting (3.278) into a discretized version of (3.277) we obtain

$$\begin{aligned} &G\sum_k V_{O_k}(t_{e_s})e^{jk\omega_c t_c} + \sum_k jk\omega_c c_1 V_{O_k}(t_{e_s})e^{jk\omega_c t_c} \\ &+ c_1 \sum_k \frac{V_{O_k}(t_{e_s}) - V_{O_k}(t_{e_s-1})}{h_s} e^{jk\omega_c t_c} + g_1 \sum_k V_{O_k}(t_{e_s})e^{jk\omega_c t_c} \\ &= \sum_k I_{S_k}(t_{e_s})e^{jk\omega_c t_c} \end{aligned} \tag{3.279a}$$

or

$$h_s \sum_k (G + g_1 + jk\omega_c c_1) V_{O_k}(t_{e_s}) e^{jk\omega_c t_c} + c_1 \sum_k V_{O_k}(t_{e_s}) e^{jk\omega_c t_c}$$

$$= h_s \sum_k I_{s_k}(t_{e_s}) e^{jk\omega_c t_c} + c_1 \sum_k V_{O_k}(t_{e_s-1}) e^{jk\omega_c t_c} \quad (3.279b)$$

or even

$$\mathbf{Y}(\omega_c)\mathbf{V}_O(t_{e_s}) = h_s \mathbf{I}_s(\omega_c) + c_1 \mathbf{V}_O(t_{e_s-1}) \quad (3.279c)$$

which allows the desired time-step integration of the envelope.

At this time, it should be clear that the extension of this process to nonlinear circuits would be straight forward if the $\mathbf{V}_O(t_{e_s})$ were interpreted as t_e-varying (i.e., following the envelope dynamics) Fourier coefficients of $v_O(t_c)$ and the resulting equation—similar to (3.279c)—as a t_e-varying harmonic balance, or the so-called ETHB equation.

Nevertheless, it should be also recognized that the generalization of this ETHB formulation to a general nonlinear network would lead to a nodal HB scheme, which may be inefficient if the actual circuit involves a large and dynamic linear subnetwork (see discussion in Section 3.3.5).

One proposed way to enable the application of these mixed-mode methods to the piecewise harmonic balance [24, 25] restricts the envelope dynamics to be very slow when compared with the carrier. By doing that, we would be implicitly limiting the envelope bandwidth, Bw, to be much smaller than the carrier; or, in other words, saying that the envelope frequency, Ω, constitutes a small deviation from a fixed carrier frequency, ω_c. The gain in the analysis is that now any linear subnetwork compacted into a $\mathbf{Y}(\omega)$ admittance matrix can be represented by only its dominant dynamic behavior if its $\mathbf{Y}(\omega)$ is approximated by a Taylor series around ω_c with a relatively small number of terms:

$$\mathbf{Y}(\omega) - \mathbf{Y}(k\omega_c) \approx \left.\frac{d\mathbf{Y}(\omega)}{d\omega}\right|_{\omega = k\omega_c} \Omega + \frac{1}{2!}\left.\frac{d^2\mathbf{Y}(\omega)}{d\omega^2}\right|_{\omega = k\omega_c} \Omega^2 + \ldots$$
$$(\omega = k\omega_c + \Omega) \quad (3.280a)$$

Actually, rewriting this Taylor expansion as

$$\mathbf{Y}(\omega) - \mathbf{Y}(k\omega_c) \approx \frac{1}{j}\left.\frac{d\mathbf{Y}(\omega)}{d\omega}\right|_{\omega = k\omega_c} j\Omega + \frac{1}{2j^2}\left.\frac{d^2\mathbf{Y}(\omega)}{d\omega^2}\right|_{\omega = k\omega_c} (j\Omega)^2 + \ldots$$
$$(\omega = k\omega_c + \Omega) \quad (3.280b)$$

and recognizing that any $d^n\mathbf{Y}(\omega)/d\omega^n$ evaluated at $\omega = k\omega_c$ is a constant, and that the frequency-domain multiplication by $j\Omega$ corresponds to the time-domain derivative in respect to t_e, we may use (3.280b) to represent the dominant behavior of the linear subnetwork with only a few t_e-derivatives:

$$\frac{1}{2\pi}\int_{-Bw/2}^{Bw/2} \mathbf{Y}(\omega)\mathbf{V}_o(\Omega)e^{j\Omega t_e}\,d\Omega$$

$$\approx \mathbf{Y}(k\omega_c)\mathbf{V}_o(k\omega_c, t_e) + \frac{1}{j}\left.\frac{d\mathbf{Y}(\omega)}{d\omega}\right|_{\omega = k\omega_c} \frac{\partial \mathbf{V}_o(k\omega_c, t_e)}{\partial t_e}$$

$$+ \frac{1}{2j^2}\left.\frac{d^2\mathbf{Y}(\omega)}{d\omega^2}\right|_{\omega = k\omega_c} \frac{\partial^2 \mathbf{V}_o(k\omega_c, t_e)}{\partial t_e^2} + \ldots \quad (3.281)$$

Using this concept in the piecewise HB equation [actually, a generalized version of (3.251)],

$$\mathbf{Y}(\omega)\mathbf{V}_o(k\omega_c, \Omega) + \mathbf{A}(\omega)\mathbf{I}_s(k\omega_c, \Omega) + \mathbf{I}_{nl}[\mathbf{V}_o(k\omega_c, \Omega)] = 0 \quad (3.282)$$

would lead to

$$\mathbf{Y}(k\omega_c)\mathbf{V}_o(k\omega_c, \Omega) + \frac{1}{j}\left.\frac{d\mathbf{Y}(\omega)}{d\omega}\right|_{\omega = k\omega_c} (j\Omega)\mathbf{V}_o(k\omega_c, \Omega)$$

$$+ \frac{1}{2j^2}\left.\frac{d^2\mathbf{Y}(\omega)}{d\omega^2}\right|_{\omega = k\omega_c} (j\Omega)^2\mathbf{V}_o(k\omega_c, \Omega) + \ldots$$

$$+ \mathbf{A}(k\omega_c)\mathbf{I}_s(k\omega_c, \Omega) + \frac{1}{j}\left.\frac{d\mathbf{A}(\omega)}{d\omega}\right|_{\omega = k\omega_c} (j\Omega)\mathbf{I}_s(k\omega_c, \Omega)$$

$$+ \frac{1}{2j^2}\left.\frac{d^2\mathbf{A}(\omega)}{d\omega^2}\right|_{\omega = k\omega_c} (j\Omega)^2\mathbf{I}_s(k\omega_c, \Omega) + \ldots$$

$$+ \mathbf{I}_{nl}[\mathbf{V}_o(k\omega_c, \Omega)] = 0 \quad (3.283)$$

and calculating the inverse Fourier transform in $\Omega \rightarrow t_e$ gives, finally, the desired $(2K + 1)$ ETHB equations:

$$
\begin{aligned}
&\mathbf{Y}(k\omega_c)\mathbf{V}_o(k\omega_c, t_e) + \frac{1}{j}\left.\frac{d\mathbf{Y}(\omega)}{d\omega}\right|_{\omega = k\omega_c} \frac{\partial \mathbf{V}_o(k\omega_c, t_e)}{\partial t_e} \\
&+ \frac{1}{2j^2}\left.\frac{d^2\mathbf{Y}(\omega)}{d\omega^2}\right|_{\omega = k\omega_c} \frac{\partial^2 \mathbf{V}_o(k\omega_c, t_e)}{\partial t_e^2} + \ldots \\
&+ \mathbf{A}(k\omega_c)\mathbf{I}_s(k\omega_c, t_e) + \frac{1}{j}\left.\frac{d\mathbf{A}(\omega)}{d\omega}\right|_{\omega = k\omega_c} \frac{\partial \mathbf{I}_s(k\omega_c, t_e)}{\partial t_e} \\
&+ \frac{1}{2j^2}\left.\frac{d^2\mathbf{A}(\omega)}{d\omega^2}\right|_{\omega = k\omega_c} \frac{\partial^2 \mathbf{I}_s(k\omega_c, t_e)}{\partial t_e^2} + \ldots \\
&+ \mathbf{I}_{nl}[\mathbf{V}_o(k\omega_c, t_e)] = 0
\end{aligned}
\qquad (3.284)
$$

By using an appropriate discretization for t_e, the complete set of $(2K + 1)$ differential equations as (3.284) is converted into a conventional HB equation for each time-sample. So, solving (3.284) in t_e corresponds to the determination of the varying envelope of $\mathbf{V}_o(\omega_c)$, or all its Fourier coefficients. These t_e-varying coefficients can be converted again into the desired spectrum representation if a Fourier transformation in $t_e \rightarrow \Omega$ is now performed and it is again considered that $\mathbf{V}_o(\omega) = \mathbf{V}_o(k\omega_c + \Omega)$.

This procedure is graphically illustrated in Figure 3.41.

3.5 Summary of Nonlinear Analysis Techniques for Distortion Evaluation

To close this chapter on nonlinear analysis techniques for distortion evaluation, let us go over its most important ideas. The summary thus obtained will also help with getting an overview of the currently available methods, and to rapidly choose the one mostly amenable to the particular problem in question. But, beforehand, it must be said that not only the judging statements are the personal view of the authors, as they were drawn from a distortion evaluation perspective. For example, although Volterra series plays a major role on distortion analysis, it is almost unknown out of this field. And even though time-step integration is probably the method suffering from the larger set of disadvantages for analyzing distortion impairments, it is still one of the most utilized methods (if not the most, indeed) in circuit analysis, even among RF engineers.

Volterra Series

In the authors' opinion, Volterra series is, by far, the best nonlinear analysis method for distortion evaluation, whenever applicable.

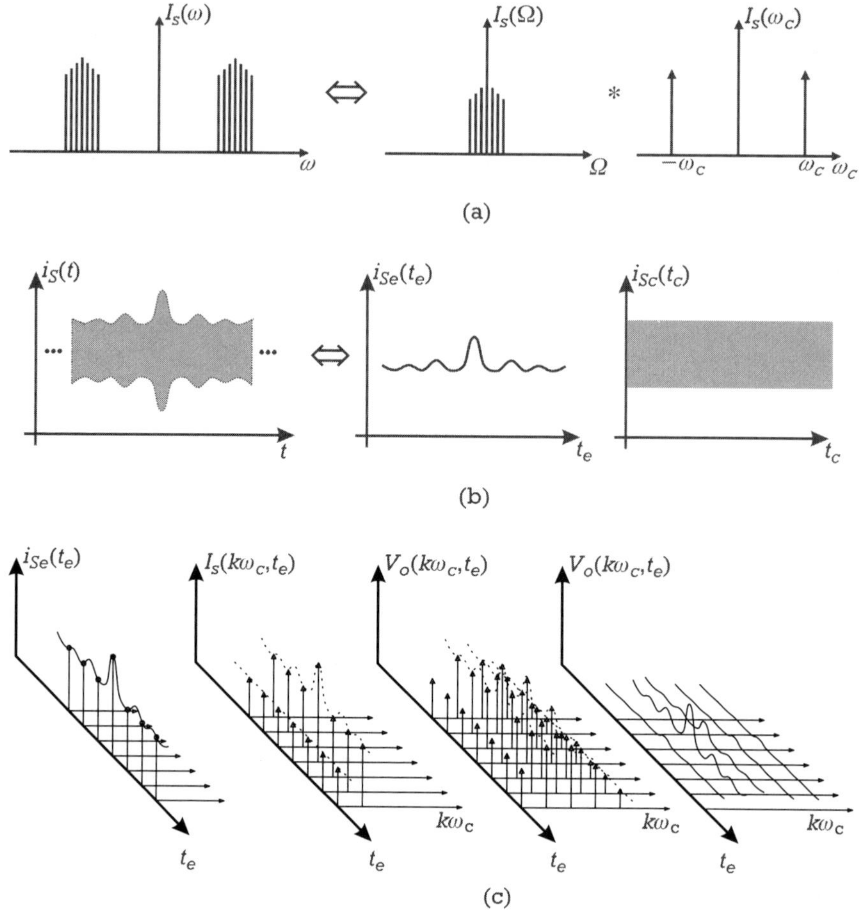

Figure 3.41 Envelope transient harmonic balance procedure. (a) Decomposition of the original spectrum in a RF carrier modulated by a base-band envelope. (b) Time-domain representation of the composite signal, RF carrier and baseband envelope. (c) Time-step integration of the slowly varying Fourier coefficients. (d) Frequency-domain reconstitution of the desired composite circuit response.

First of all, it does not rely on any iterative procedure, being, in contrast, a recursive method. In fact, by substituting the circuit's nonlinearities by Taylor series approximations of nth degree, it gives a solution to that nth-order problem by solving n times the same linear circuit. Relying in a linear solver, it allows analysis in both time and frequency-domains. Therefore, and contrary to any other nonlinear method, it enables handy calculations and provides closed-form expressions for the sought nonlinear solutions. Behaving this way as a true nonlinear modeling tool for devices, circuits, or complex systems, it offers detailed qualitative information on the system's properties, enabling analysis, performance optimization, and design tasks.

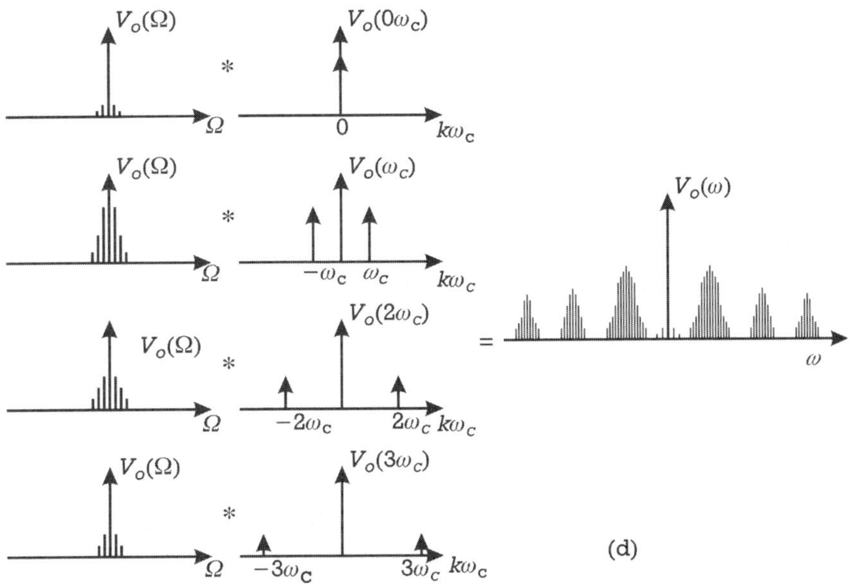

Figure 3.41 (continued).

Since it can operate entirely in the frequency-domain, it imposes no restrictions on the excitation signal spectrum, making it the ideal method for multitone distortion analysis.

Despite these outstanding benefits, Volterra series suffers from an important disadvantage: it cannot be applied to strongly nonlinear problems. Actually, either because the series simply does not converge or requires an intractable number of terms for required accuracy, Volterra series is usually limited to quite smooth nonlinearities subject to small amplitude signals. These are the so-called mildly nonlinear problems. In practice, the series' range of applicability becomes restricted to signal levels comfortably behind the 1-dB compression point, leaving outside all class AB, B, or class C amplifiers, saturated mixers, etc. Moreover, even in the case of a mildly nonlinear circuit, although Volterra series can handle the nonlinear effects pressed onto the signal, it cannot cope with the strong nonlinearities usually associated with the quiescent point calculation. For example, despite Volterra series being the best method to predict distortion behavior of small-signal amplifiers or mixers, it can only be applied after the bias point calculation or the local oscillator pumping have been determined by some other nonlinear analysis technique. Finally, it should be also stressed that only the engineer's intuition and experience can tell him when the series' results are no longer useful, since there is no such indication available from the method. That is, one may perform a power sweep simulation up to a stimulus level where the series no longer produce any useful results without the slightest error or warning.

From the authors' point of view, these restrictions are the ones mostly responsible for the limited range of acceptance of the Volterra series in commercial nonlinear simulators. In fact, and despite one recent exception, Volterra series seams to be still confined to some academic or industrial scientific laboratories.

Harmonic Balance

From the contents of previous sections it is easily concluded that harmonic balance should be the adopted method for general nonlinear analysis of RF and microwave circuits.

By handling the full nonlinear expressions, it does not share the signal level limitations with the Volterra series. In this sense, it is no longer a method for quasi-linear regimes, but a true nonlinear analysis technique. This is obtained at the cost of being a numerical iterative method that does not allow handy calculations, nor provides symbolic expressions for the solutions. Therefore, although it may be used in an algorithm with feedback to produce performance optimization, it is not truly a technique amenable for circuit design. Its most common form, the harmonic-Newton, works by initially estimating a vector of Fourier coefficients for the solution, and then successively refining that estimate using a Newton-Raphson nonlinear iterative solver.

Although it is classified as a frequency-domain technique, because it solves the circuit for the Fourier coefficients of the voltages and currents, it still requires time-domain calculations. Actually, a harmonic balance engine relies on balancing the harmonic levels of node currents arising from the circuit's linear-dynamic and nonlinear-memoryless elements. Having an estimate of the node voltage, it must find a way to determine those levels in the linear and nonlinear elements. If that is immediate in a linear element when it is described in admittance form, it is not so obvious for the nonlinear elements. In this case, the HB machine converts the voltage into time-domain, using the inverse discrete Fourier transform (usually its fast algorithm the IFFT), computes the nonlinear algebraic currents in a time-sample per time-sample basis, and then converts again this time-domain current back into the frequency-domain using the DFT. The harmonic balance method is, therefore, mainly constrained by that Fourier transform. The handled signals must be periodic and their spectra truncated up to a convenient number of harmonics. If those two conditions are not met, the results' accuracy becomes severely compromised by spectral leakage or aliasing errors. Moreover, if the number of harmonics is too small, then convergence problems may be faced and the HB routine may never reach a solution.

Nevertheless, a great research effort has been continuously put into the HB method for the last 20 years, which permitted to overcome some of these limitations. In what distortion simulation is concerned, the most important is the required signals periodicity. That was partially solved for quasiperiodic signals, with the MDFT or the AFM techniques. Recognizing that n incommensurate frequencies produce n phase uncorrelated arguments, in the same way as if they were created by n independent time variables, the MDFT technique samples that n-dimensional waveform along these new n axes, thus obviating common DFT spectral leakage. The algorithm becomes, however, rather complex and time-consuming, which obviates its use whenever n is greater then two or three. An alternative way to this process takes profit of the fact that, because the nonlinearities are memoryless, their results are independent on the absolute frequency value. Therefore, the AFM technique transforms the original time-domain quasiperiodic waveform into a new artificial domain in which the mapped version is periodic. Since the waveform is now periodic, and the spectrum is harmonically related, the DFT can again be applied in those artificially mapped domains, as usual.

With these modifications, the harmonic balance becomes indeed a very powerful tool even for distortion analysis. There are various commercially available nonlinear simulators using this method.

Time-Domain and Mixed-Mode Techniques

Although time-step integration, used in the SPICE-like programs, is still the nonlinear analysis method of wider acceptance, it suffers from several disadvantages in RF and microwave distortion simulation. First of all, it was initially conceived to simulate the circuit's transient response, while our interest normally resides on the steady-state. So, because it has to wait until all transients have vanished, it is quite inefficient for that purpose. Also, since it operates entirely in time-domain, it cannot handle linear elements having a frequency-domain description, like dispersive distributed transmission media. Finally, even if that drawback could be circumvented (for example, by approximating these elements by lumped networks) the necessity of operating in the time-domain, while the input and resulting signals needed to be presented in spectra, would end up in all difficulties associated with the DFT, which were already referred for the HB technique.

Fortunately, some time-domain alternatives to the initial time-step integration method, like the shooting-Newton or the ETHB may circumvent some of these difficulties in the future. Shooting-Newton bypasses the transient response, therefore obviating the waste of time needed to let it vanish. The ETHB, and all the other methods based in multirate partial differential equation descriptions, seem to be promising alternatives for multitone simulation in the time-domain.

In the meantime, time-domain methods benefit from two important advantages. First, since they rely on the SPICE simulator engine, they are well known and are available from many simulator vendors. And second, as they use time as a natural continuation parameter, they are likely to be the ones supporting strongest nonlinear regimes.

Summarizing Table

Table 3.2 concludes this section by summarizing the determining characteristics of each of the analysis methods, in the context of distortion evaluation.

Table 3.2 Summary of the Determining Characteristics of Nonlinear Analysis Techniques for Distortion Evaluation

Nonlinear Analysis Method	*Method's Variant*	*Main Features for Distortion Evaluation*
Volterra series	Power series	—Weakly nonlinear memoryless systems —Very easy to use: handy calculations —Provides first insight on system's behavior —Approximate circuit design
	Time-invariant Volterra series	—Analytical expressions of weakly nonlinear system's responses —Easy to use: handy calculations —Analysis and design of small-signal or class A power amplifiers
	Time-varying Volterra series	—Analysis and design of small-signal mixers, electronic attenuators or switches —Requires another large-signal method for local oscillator, or control signal pumping
Harmonic balance	Quasiperiodic harmonic balance	—Iterative procedure: only useful for analysis and optimization —Large-signal distortion analysis of amplifiers and mixers —Somewhat restricted in allowed excitation types: only a small number of tones —Mature technology with widespread and flexible commercially available software packages
Time-domain techniques	Time-step integration using shooting-Newton and mixed-mode techniques	—Iterative procedure: used only for analysis —Especially amenable for multirate excitations as complex modulated signals —Appropriate for the analysis of transients —Allows very strong nonlinear regimes

References

[1] Kundert, K., J. White, and A. Sangiovanni-Vicentelli, *Steady-State Methods for Simulating Analog and Microwave Circuits,* Norwell, MA: Kluwer Academic Publishers, 1990.

[2] Maas, S. A., *Nonlinear Microwave Circuits,* Norwood, MA: Artech House, 1988.

[3] Rodrigues, P. J., *Computer-Aided Analysis of Nonlinear Microwave Circuits,* Norwood, MA: Artech House, 1998.

[4] Schetzen, M., *The Volterra and Wiener Theories of Nonlinear Systems,* New York: John Wiley & Sons, 1980.

[5] Maas, S. A., *Microwave Mixers,* Norwood, MA: Artech House, 1986.

[6] Rizzoli, V., and A. Neri, "State of the Art and Present Trends in Nonlinear Microwave CAD Techniques," *IEEE Transactions on Microwave Theory and Techniques,* Vol. 36, No. 2, 1988, pp. 343–365.

[7] Steer, M. B., C. Ren-Chang, and G. W. Rhyne, "Computer Aided Analysis of Nonlinear Microwave Circuits Using Frequency Domain Nonlinear Analysis Techniques: The State-of-the Art," *International Journal of Microwave and Millimeter Wave Computer Aided Engineering,* Vol. 1, No. 2, 1991, pp. 181–200.

[8] Kroezer, V., and H. L. Hartnagel, "Large-Signal Analysis of Nonlinear Microwave Circuits Using Modified Volterra Series," *Proc. IEEE Microwave Theory and Techniques-Integrated Nonlinear Microwave and Millimetre Wave Circuits, First International Workshop of the West Germany IEEE MTT/AP Digest,* Duisburg, FRG, Oct. 1990, pp. 197–211.

[9] Eijnde, E., Van den, *Steady-State Analysis of Strongly Nonlinear Circuits,* Vrije Universiteit Brussel, Fac. Toegespaste Wetenschappen, Department ELEC, Brussels, Belgium, Ph.D. dissertation, 1989.

[10] Chang, C. R., and M. B. Steer, "Frequency-Domain Nonlinear Microwave Circuit simulation Using the Arithmetic Operator Method," *IEEE Transactions on Microwave Theory and Techniques,* Vol. 38, No. 8, 1990, pp. 1139–1143.

[11] Carvalho, N. B., and J. C. Pedro, "Multi-Tone Frequency Domain Simulation of Nonlinear Circuits in Large and Small Signal Regimes," *IEEE Transactions on Microwave Theory and Techniques,* Vol. 46, No. 12, 1999, pp. 2016–2024.

[12] Kundert, K. S., G. B. Sorkin, and A. Sangiovanni-Vicentelli, "Applying Harmonic Balance to Almost-Periodic Circuits," *IEEE Transactions on Microwave Theory and Techniques,* Vol. 36, No. 2, 1988, pp. 366–378.

[13] Rizzoli, V., C. Cecchetti, and A. Lipparni, "A General-Purpose Program for the Analysis of Nonlinear Microwave Circuits Under Multitone Excitation by Multidimensional Fourier Transform," *Proc. 17th European Microwave Conference,* Rome, 1987, pp. 635–340.

[14] Brachtendorf, H., et al., "Numerical Steady-State Analysis of Electronic Circuits Driven By Multi-Tone Signals," *Electrical Engineering,* Vol. 79, No. 2, 1996, pp. 103–112.

[15] Hente, D., and R. H. Jansen, "Frequency Domain Continuation Method for the Analysis and Stability Investigation of Nonlinear Microwave Circuits," *IEE Proceedings-H Microwaves Antennas and Propagation,* Vol. 133, No. 5, 1986, pp. 351–362.

[16] Rodrigues, P. J., "An Orthogonal Almost Periodic Fourier Transform for Use in Nonlinear Circuit Simulation," *IEEE Microwave and Guided Wave Letters,* Vol. 4, No. 3, 1994, pp. 74–76.

[17] Pedro, J. C., and N. B. Carvalho, "Efficient Harmonic Balance Computation of Microwave Circuits' Response to Multi-Tone Spectra," *Proc. 29th European Microwave Conference,* Munchen, Oct. 1999, pp. 103–106.

[18] Naklıla, M. S., and J. Vlach, "A Piecewise Harmonic Balance Technique for Determination of Periodic Response of Nonlinear Systems," *IEEE Transactions on Circuits and Systems,* Vol. 23, No. 2, 1976, pp. 85–91.

[19] Ushida, A., and L. Chua, "Frequency-Domain Analysis of Nonlinear Circuits Driven by Multi-Tone Signals," *IEEE Transactions on Circuits and Systems,* Vol. 31, No. 9, 1984, pp. 766–779.

[20] Nagel, L. W., *Spice2: A Computer Program to Simulate Semiconductor Circuits,* Electronic Research Laboratory, University of California-Berkeley, Memo ERL-M520, 1975.

[21] Aprille, T. J., and T. N. Trick, "Steady-State Analysis of Nonlinear Circuits with Periodic Inputs," *Proceedings of the IEEE,* Vol. 60, No. 1, 1972, pp. 108–114.

[22] Mayaram, K., et al., "Computer-Aided Circuit Analysis Tools for RFIC Simulation: Algorithms, Features, and Limitations," *IEEE Transactions on Circuits and Systems—II,* Vol. 47, No. 4, 2000, pp. 274–286.

[23] Roychowdhury, J., "Analysing Circuits with Widely Separated Time Scales Using Numerical PDE Methods," *IEEE Transactions on Circuits and Systems—I,* Vol. 48, No. 5, 2001, pp. 578–594.

[24] Ngoya, E., and R. Larchevèque, "Envelop Transient Analysis: A New Method for the Transient and Steady State Analysis of Microwave Communication Circuits and Systems," *Proc. IEEE Microwave Theory and Techniques Symposium Digest,* California, 1996, pp. 1365–1368.

[25] Rizzoli, V., A. Neri, and F. Mastri, "A Modulation-Oriented Piecewise Harmonic Balance Technique Suitable for Transient Analysis and Digitally Modulated Analysis," *Proc. 26th European Microwave Conference,* Prague, Czech Republic, 1996, pp. 546–550.

CHAPTER 4

Nonlinear Device Modeling

4.1 Introduction

This chapter is dedicated to nonlinear device modeling. Here, the reader can find general information concerning the models of the devices responsible for the generation of intermodulation distortion in typical microwave and wireless circuits. Because there is a wide variety of different nonlinear electron devices, and also a reasonably large number of nonlinear model descriptions for some families, it would be impossible to address all of them in detail. Even if that were possible, it probably would not be essential, as one can easily find entire books dedicated to some of these devices. So, the objective pursued in this text is primarily to present a general view of the various model formulation and parameter extraction strategies, and to discuss them under the purpose of nonlinear circuit analysis or simulation for distortion prediction. Instead of describing in detail the physics associated to the current-voltage, I/V, or charge-voltage, Q/V, characteristics of each nonlinear element, the emphasis will be put on a criteria set necessary to compare various model formats and predict, or simply understand, the distortion characteristics of the circuits based on these devices. This will be complemented by discussing a few number of electron device models selected for their importance in predicting nonlinear distortion of microwave and wireless circuits.

Beginning with a brief reference to model classification, there are two basic groups in which mathematical representations of real devices can be organized: *physical modeling* and *empirical modeling*.

In the context of nonlinear electron device modeling, a certain representation is classified as being physical, or of having a physical basis, if it is drawn from the knowledge of the device's geometrical and physical structure, and from the application of a certain set of physics laws. Examples of purely physical models require the solution of a set of coupled nonlinear partial differential equations describing the internal fields of the device and electrical charge transport [1]. However, some analytical representations common in electrical engineering, as the I/V characteristics of a linear resistor ($v = Ri$) and of a Schottky junction ($i = I_S$ $[\exp(v/\eta V_T) - 1]$), can also be considered as semiphysical models. Unfortunately, most practical devices are extremely more complex than resistors or Schottky junctions, impeding the derivation of physical models.

Actually, since the application of basic physics laws to the structure always requires approximations and measurement of certain physical quantities, there are no ideally pure physical models. So, even the so-called physical models result from some compromise between representation accuracy and model involvedness.

On one of these extremes we have comprehensive models providing very accurate descriptions of the devices' external characteristics, for a comparably small amount of measurement data. Unfortunately, they consist of a set of time-space nonlinear partial differential equations that cannot be solved in analytical form. So, they not only require an extremely high computation cost for evaluation (it is not unusual that the evaluation of the model in a single I/V point requires much more time and memory storage than what is necessary to make all the calculations corresponding to the external embedding circuit), as, incapable of giving qualitative information on the device behavior, they are useless for first step hand circuit analysis or design. In summary, there is no such pure physical model (a model that, in theory, would only need the information provided by the foundry process). And even if there were, it probably would not be very useful for electrical engineers interested in designing circuits with prescribed specifications.

The solution to this problem seems to stand, therefore, in alleviating the model mathematical complexity, substituting physical information by empirical descriptions gathered from laboratory observations (i.e., measurement data). Moving that way, we reach the other extreme of model format: empirical or *behavioral modeling*. In fact, a model (or a part of it) is said to be of empirical nature if its format has not been derived from any basic physical knowledge of device operation, but simply from the necessity to represent, in a behavioral sense, measured characteristics of that device. Examples of empirical device models are the various matrix descriptions of linear networks or devices, like the S-parameter matrix, and the quadratic law of a RF diode detector. But, as we will discuss later in Section 4.4, behavioral modeling can even be found to describe higher levels of system operation.

In comparison to physical models, an empirical representation is much more compact, amenable for fast computer calculations, or even hand evaluation, and provides direct qualitative information of the device performance. On the contrary, its description capabilities are almost restricted to the type of measurements from which it was derived, and its accuracy is immediately determined by the adopted fitting process. As a matter of fact, while physical models tend to only require the measurement of some physical quantities (like geometrical dimensions of semiconductor layers, doping profiles, etc.), purely empirical models tend to be only capable of describing the device characteristics from which they were extracted. Their accuracy is a priori determined by the error accepted in the model extraction process, and their predictive capabilities are almost inexistent.

Beyond that classification in physical and empirical nature, we should also subdivide empirical representations in global and local models. A *local model* is one empirical representation capable of accurately representing the device behavior only in a very restricted domain of its input variables, whereas a *global model* seeks for a much wider applicability.

This way defined, we should immediately question ourselves why there is such a classification if a global model is clearly better. The reason for this is that there is a trade-off between representation range and accuracy. In fact, a local model is a low order Taylor series (or Volterra series) recognized to produce very fine representations near the fixed point around which it is expanded, but then rapidly diverges if pushed off of its limited approximation range. So, it is optimum for small-signal distortion analysis, but it is neither applicable in bias point calculations, nor in large-signal ac predictions.

An empirical global model, on the other hand, is a mathematical function conceived for general-purpose analysis. It is thus capable of representing the device in a much wider zone of input stimuli, but with poorer approximation accuracy. For example, it can be simply a fitting function optimized for an average low error in the whole of its domain, or a detailed piecewise approximant. In the first case, it will only give a rough estimate of the general device behavior, while, in the second, it will produce nonphysical responses in the break points, due to the discontinuity of its local derivatives.

In summary, there is a long pathway between pure physical descriptions and pure empirical, local or global, ones. Thus, it is the wise choice of a convenient point in between, for a certain application, that makes what we could loosely classify as the "optimum model."

4.2 Device Models Based on Equivalent Circuits

Although system analysis usually relies on pure behavioral models (i.e., abstract mathematical representations of observed input-output characteristics), nonlinear device models used in circuit analysis are always based on equivalent circuits. Those are characterized by a specific topology, I/V or Q/V relations for the elements and their set of fitting parameters.

For example, for a junction field effect transistor, like a microwave MESFET, whose idealized geometrical structure can be seen in Figure 4.1, a convenient equivalent circuit topology could be the one depicted in Figure 4.2.

Note that, as shown in Figure 4.3, the equivalent circuit topology of Figure 4.2 started from an approximate physical description of the device structure. This means that the topology is not only driven by the necessity to reproduce the measurement data, but also to represent many electromagnetic effects caused by the particular device structure. So, it should not be strange to recognize that any equivalent circuit comes always divided in an inner and outer part.

The inner part, called the *intrinsic model,* represents the elements that are specific to the operation of the device. For example, in a MESFET, it models the

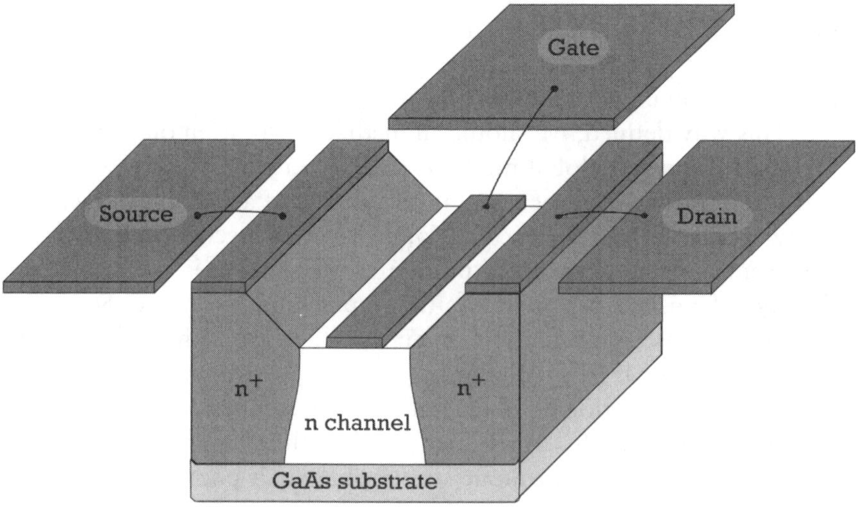

Figure 4.1 Geometrical structure of a microwave FET.

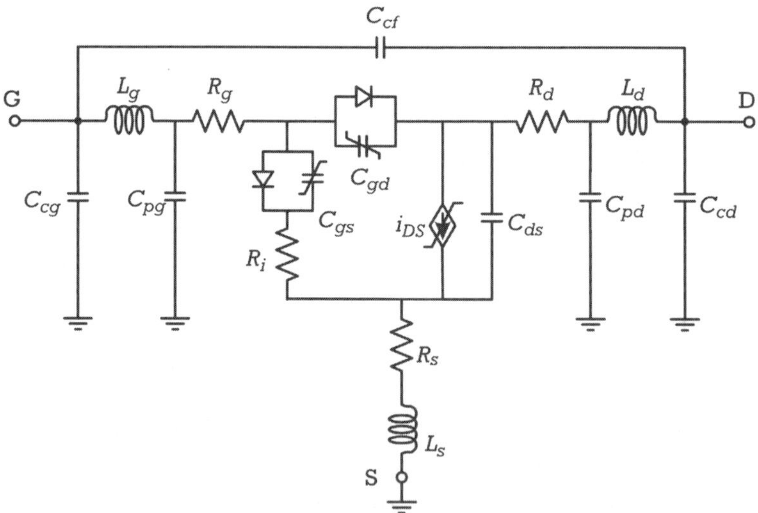

Figure 4.2 Equivalent circuit model topology of the microwave FET shown in Figure 4.1, when mounted in common-source.

distributed channel charge controlled by the gate-channel potential with two lumped nonlinear capacitances, gate-source capacitance, $C_{gs}(v_{GS})$, and gate-drain capacitance, $C_{gd}(v_{GD})$, and a linear resistor, R_i. And it also models channel current by a bidimensional voltage-dependent current source, $i_{DS}(v_{GS}, v_{DS})$.

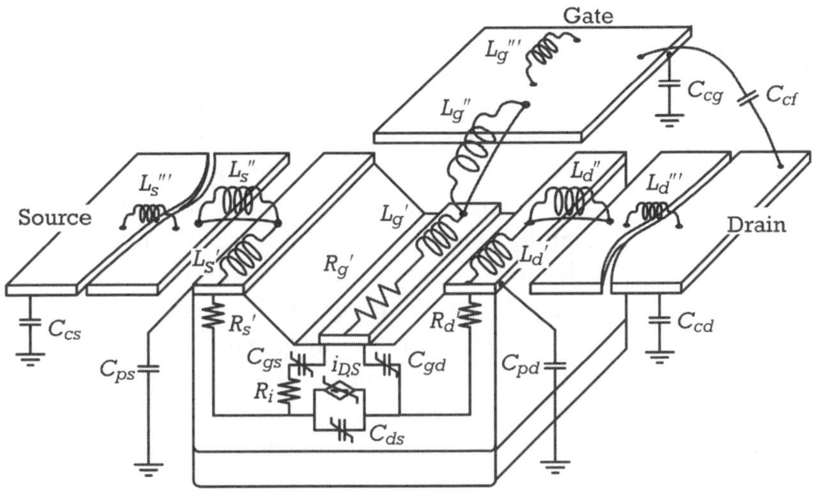

Figure 4.3 Illustration of the physical nature of the equivalent circuit topology presented in Figure 4.2.

The outer elements, surrounding the intrinsic model, are parasitic to the device's ideal behavior, and are thus called the *extrinsic model*. They represent contact and semiconductor bulk resistances as R_g, R_s, and R_d, contact and bond-wire inductances like L_g, L_s, and L_d, beyond several other elements used to model the package. Like many of the intrinsic elements, the extrinsic model is usually composed of lumped elements trying to emulate actual distributed effects. For example, in Figure 4.2, C_{pd}, L_d, and C_{cd} constitute, in fact, a π lumped network approximation of the distributed effects caused by the drain chip pad, the chip-package bondwire, and the drain lead.

Contrary to the intrinsic subcircuit, which is dependent on the device, the extrinsic part is mainly dependent on the environment embedding the device. So, while the value and the topology of those parasitic elements can vary significantly for different packages, or from these to chip devices, the topology and values of the intrinsic elements depend mainly on the device structure. For example, a packaged MESFET and a packaged BJT may have a similar extrinsic model, if their packages are equal, but their intrinsic models are necessarily different; whereas a chip and a packaged version of the same device family will have equal intrinsic models but completely distinct extrinsic subcircuits.

Having selected the equivalent circuit topology, it is then necessary to determine values for each of its elements. This procedure assumes that, first, those elements have been separated in linear and nonlinear elements. Linear elements maintain their values constant with applied signal and bias, and so are perfectly identified by a single value. For example, it is almost certain that the drain pad capacitance does not depend on the applied voltage, and so we can assume that this capacitance is constant and equal to a fixed value. This is the typical situation for all parasitic elements, and gives very good results for the generality of RF analyses.

On the contrary, the intrinsic elements are usually dependent on the applied signals. So, the problem is no longer to determine a convenient value, but an appropriate functional description. And, for this task, once again the physical and empirical modeling philosophies can be followed. That is, on one hand, we can study the physics of the device operation and then derive a mathematical description based on that knowledge. Or, on the other hand, measure I/V or Q/V characteristics, propose a functional format capable of approximately reproducing those curves, and then simply select a set of parameters that guarantees a best fit between the predicted and measured data. Although, in theory, this problem appears trivial, it is so delicate in modeling for distortion prediction that it deserves special attention.

4.2.1 Selecting an Appropriate Nonlinear Functional Description

In order to clarify the modeling problems involved, let us particularize for the case of the MESFET $i_{DS}(v_{GS}, v_{DS})$ channel current.

The physical approach of deriving that mathematical representation consists of solving a system of nonlinear differential equations that relate the potential (or electric field), the accumulated channel charge, and carrier concentration inside the various semiconductor layers, under the boundary conditions imposed by the geometric limits of the structure and by the external voltage applied to the contacts. Due to the vertical nature of the electric field imposed by the gate-channel voltage, and the horizontal nature of that same field determined by the drain-source voltage, the FET is, in essence, a bidimensional device. The coupled nonlinear differential equations are, actually, partial differential equations in space and time, and an analytical solution is inevitably out of sight. Keeping our will of having a closed-form description for $i_{DS}(v_{GS}, v_{DS})$ pushes us to accept an empirical model.

Since an empirical model is nothing but an interpolating function to a set of measured data, our imagination is the ultimate limit for possible functional descriptions. Also, keeping in mind that a model is something we use to reproduce a certain set of practical observations, we can restrict our empirical model validity to the domain in which it is intended to be used. For example, if we were thinking in modeling outputs to excitations of infinitesimal amplitude, we could perfectly rely on a simple local linear model. Or, if we expect to use the model for small-signal distortion predictions, we could include other higher order coefficients of the Taylor series expansion, and thus augment its validity domain. In any case, those mathematical representations are only locally valid, and so limited to a restricted range of possible control voltages around the quiescent point.

Contrary to physical models, which are global representations expected to automatically reproduce local behavior, empirical models usually present a compromise between the capability of accurately reproducing the global general characteristics of the device and of preserving the local details.

To illustrate this trade-off with a practical case, let us admit an $i_{DS}(v_{GS}, v_{DS})$ model of separated variables (i.e., given by the product of two functions), one only dependent on v_{GS}, $f_g(v_{GS})$, and the other on v_{DS}, $f_d(v_{DS})$:

$$i_{DS}(v_{GS}, v_{DS}) = f_g(v_{GS})\, f_d(v_{DS}) \tag{4.1}$$

So, assuming we have selected a V_{DS} bias in the saturation zone, and have kept it fixed, we have to find an appropriate interpolating function to reproduce $i_{DS}(v_{GS})$ whose sample measured data is shown in Figure 4.4.

If that model is also intended for small-signal applications, it should be also capable of reproducing the $f_g(v_{GS})$ Taylor series coefficients around any quiescent point (V_{GS}, V_{DS})

$$i_{DS}(v_{GS}, V_{DS}) = I_{DS} + G_m v_{gs} + G_{m2} v_{gs}^2 + G_{m3} v_{gs}^3 \tag{4.2}$$

$(v_{gs} \equiv v_{GS} - V_{GS})$ whose corresponding measured data is depicted in Figure 4.5. That is, the interpolating function must not only approximate $f_g(v_{GS})$ but osculate some of its first-order derivatives. That guarantees accurate large and small-signal ac predictions, but also consistency between them. In practical terms this implies

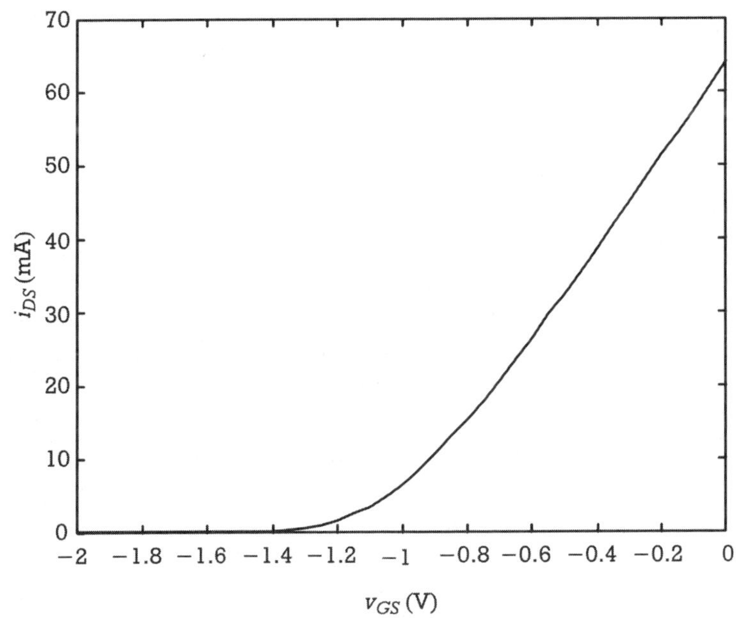

Figure 4.4 Sample measured data of $i_{DS}(v_{GS}, V_{DS})$ when V_{DS} is kept fixed in the saturation zone.

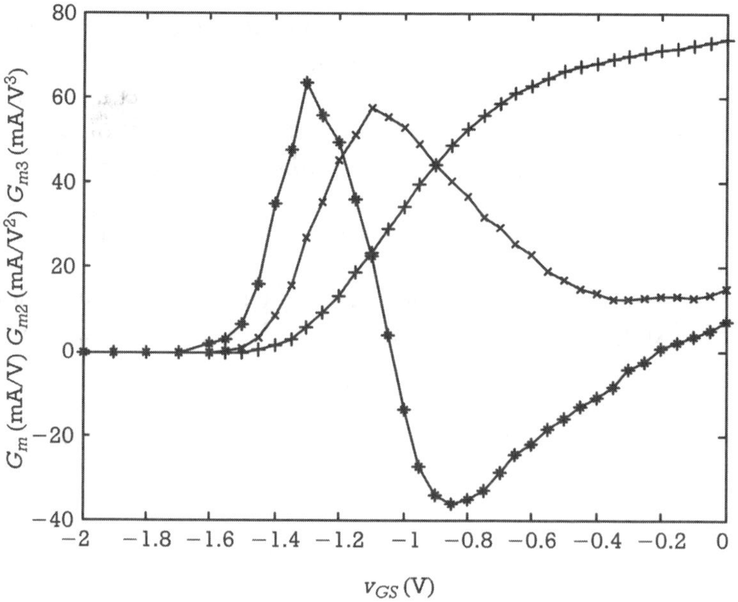

Figure 4.5 G_m (-+-), G_{m2} (-x-), and G_{m3} (-*-) Taylor series coefficients of $i_{DS}(v_{GS}, V_{DS})$ shown in Figure 4.4.

that, for example, predicted large-signal power gain of an amplifier smoothly tends to the linear forward gain parameter, $|S_{21}|^2$, when the amplitude excitation is sufficiently decreased.

The first and most important impact of these considerations is that selecting an appropriate model is a much more involved task than simply trying to find an adequate interpolating function under a least squares fit.

For example, any discontinuous function, or even continuous but presenting a discontinuous nth order derivative, G_{mn}, is totally unacceptable as it implies an infinite G_{mn+1} coefficient, and thus a nonphysical prediction of small-signal $(n + 1)$th order distortion. And this refers to all piecewise linear, quadratic, etc. approximations, as the ones seen in most textbooks and simulators for describing the so-called quadratic behavior of FET current [2, 3].

Another example must refer some fitting functions that guarantee a very small approximating error (i.e., the difference between the predicted values and the actual observed ones) by wandering around the data—although very close to it—or even passing exactly over the data points.

Such a situation is illustrated in Figure 4.6 and Figure 4.7, where two different $f_g(v_{GS})$ fitting functions were considered for an imaginary $i_{DS}(v_{GS})$ data, corresponding also to two different levels of fitting error. It is clear that a $f_g(v_{GS})$ that does not follow the measured $i_{DS}(v_{GS})$ extremely well, but even so is capable of

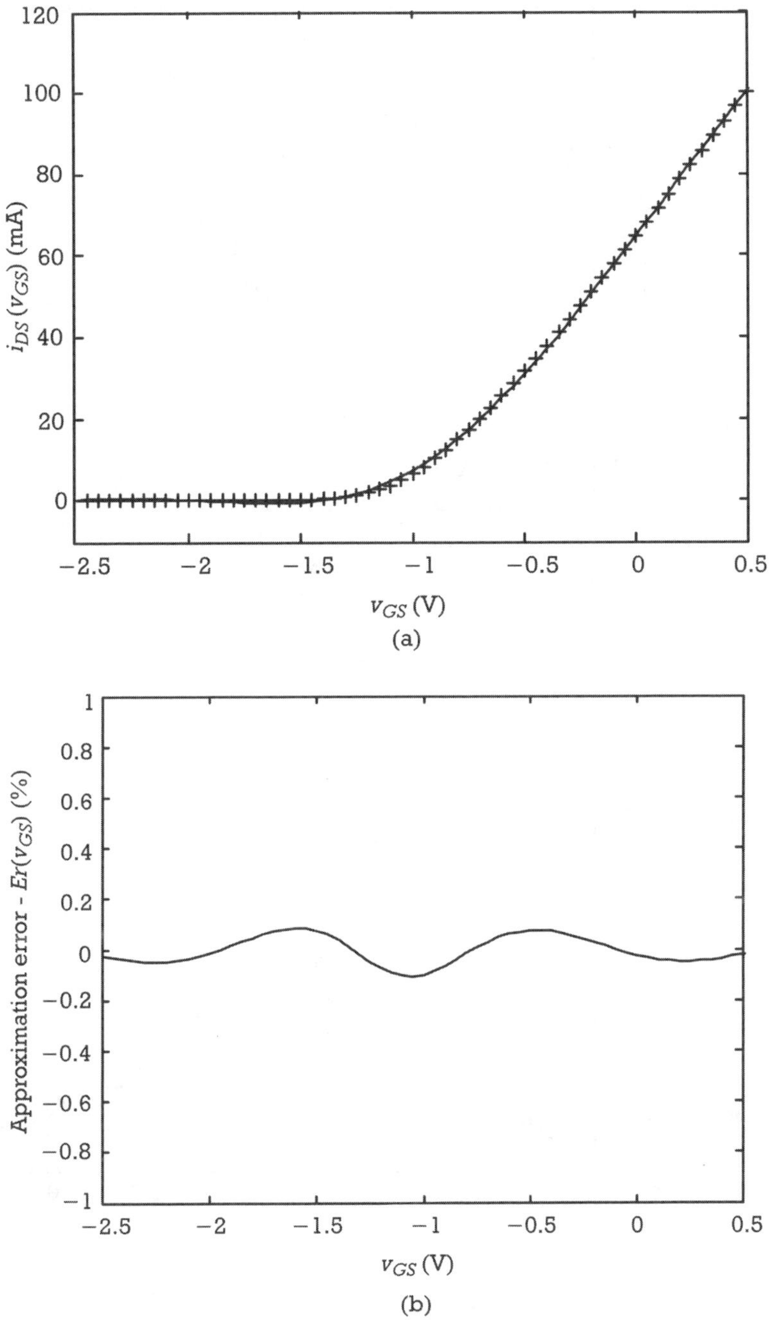

Figure 4.6 Illustration of why fitting data with a small error does not necessarily lead to a good model intended for small-signal distortion predictions: (a) tenth-degree polynomial fit: $i_{DS}(v_{GS})$ original data (+) and model (–); (b) fitting error; (c) first-order derivative, G_m; (d) second-order derivative, G_{m2}; and (e) third-order derivative, G_{m3}.

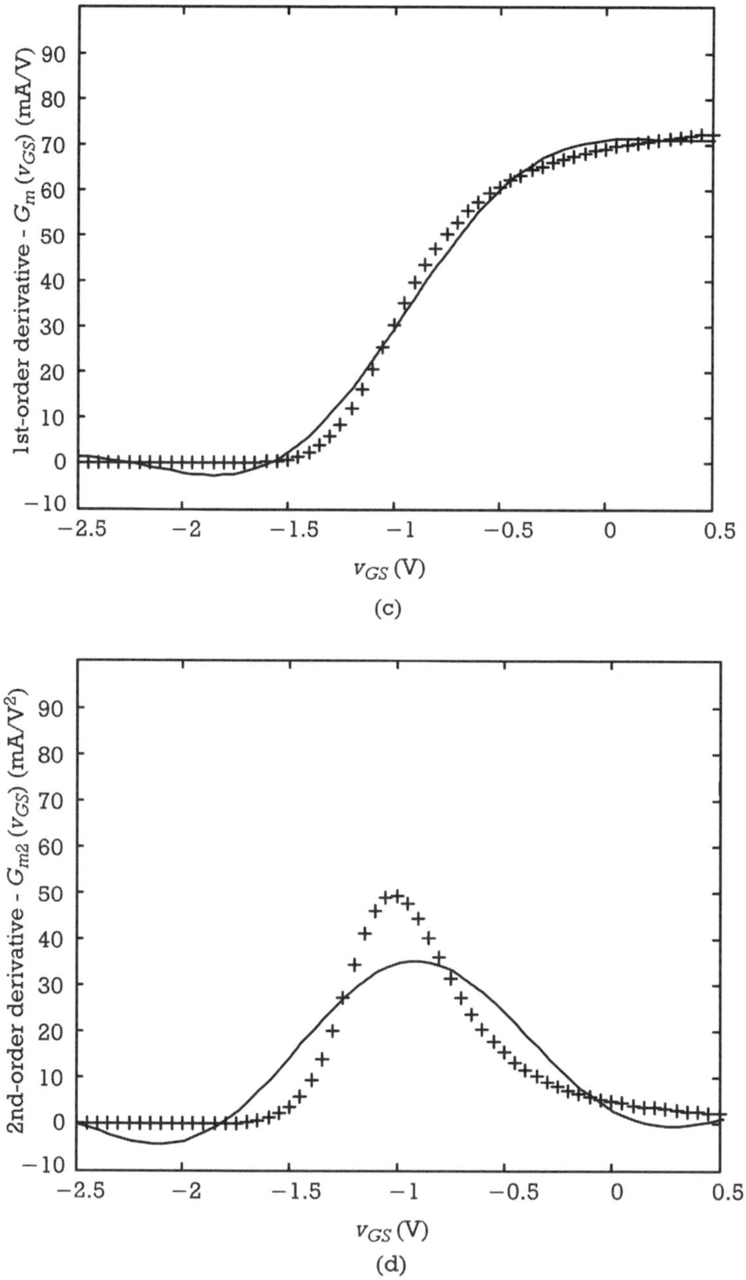

Figure 4.6 (continued).

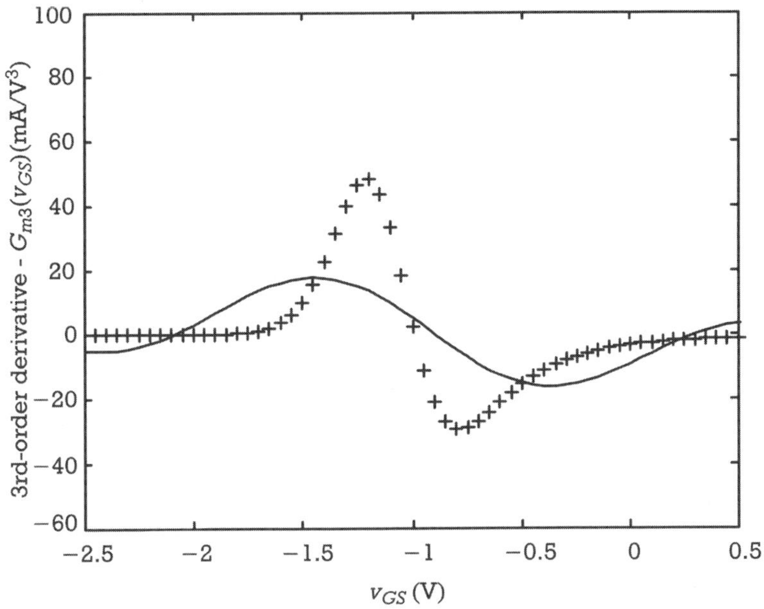

(e)

Figure 4.6 (continued).

reproducing the general trend (Figure 4.7), is preferable to another one involving a lower error but at the expense of completely failing the derivatives (Figure 4.6). Note that such approximation even produces nonphysical results as negative current and transconductance values.

But, an even more astonishing result can be obtained from data perturbed by measurement error. In those cases, it is exactly the random behavior of measurement noise that produces data wandering around the true device's characteristics. And so a blind fitting process, willing to approximate that data with minimized error, would simply be disastrous. For example, since approximation theory states that every data set of dimension N can be exactly fitted by a $(N-1)$th degree polynomial, one could be tempted to simply follow that result (known as the Lagrange interpolation formula) to create an approximating function that passes through all the measured points. The problem is that, in cases where measurements are corrupted by noise, we would not be modeling the device's characteristics but the actual measurement errors.

In summary, there are two different types of empirical models that can be considered.

Local models have a restricted range of validity but, as we have seen from the Volterra series analysis of Chapter 3, can produce very accurate small-signal

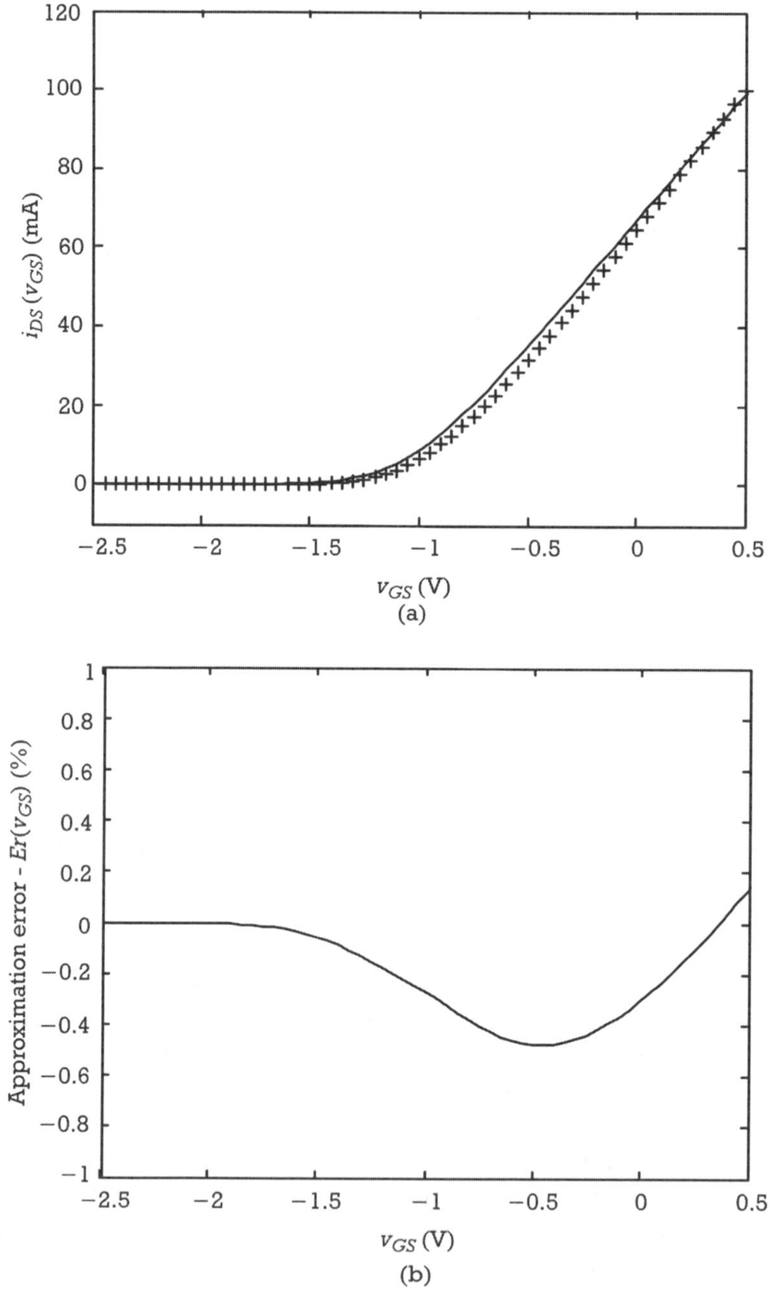

Figure 4.7 Illustration of why fitting data with a larger error may lead to a good model intended for small-signal distortion predictions, provided the derivatives are also approximated: (a) fit with a $x + \ln[2.\cosh(x)]$ function: $i_{DS}(v_{GS})$ original data (+) and model (–); (b) fitting error; (c) first-order derivative, G_m; (d) second-order derivative, G_{m2}; and (e) third-order derivative, G_{m3}.

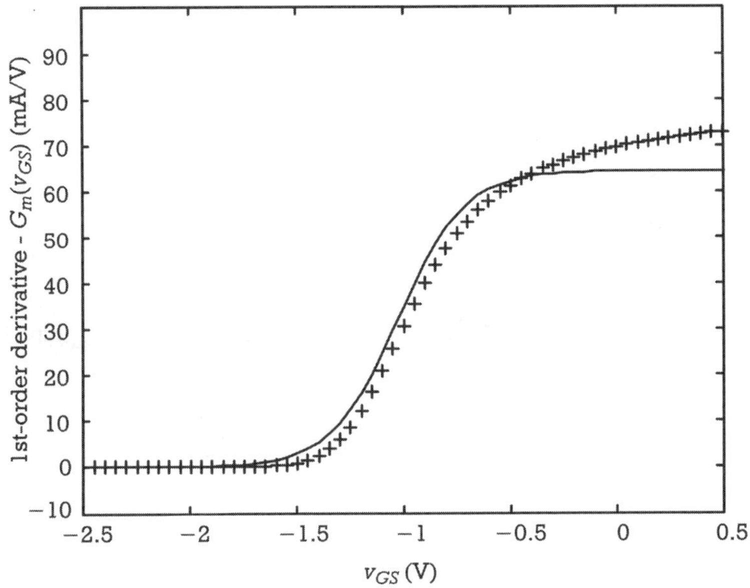

(c)

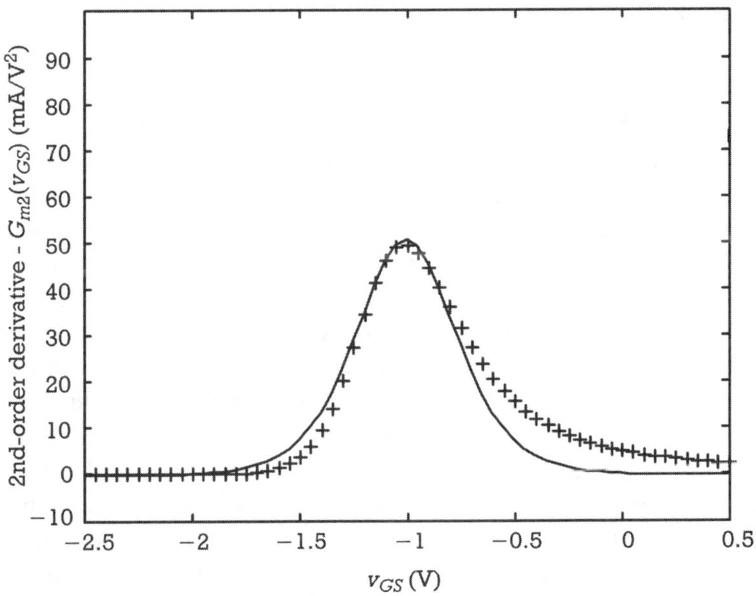

(d)

Figure 4.7 (continued).

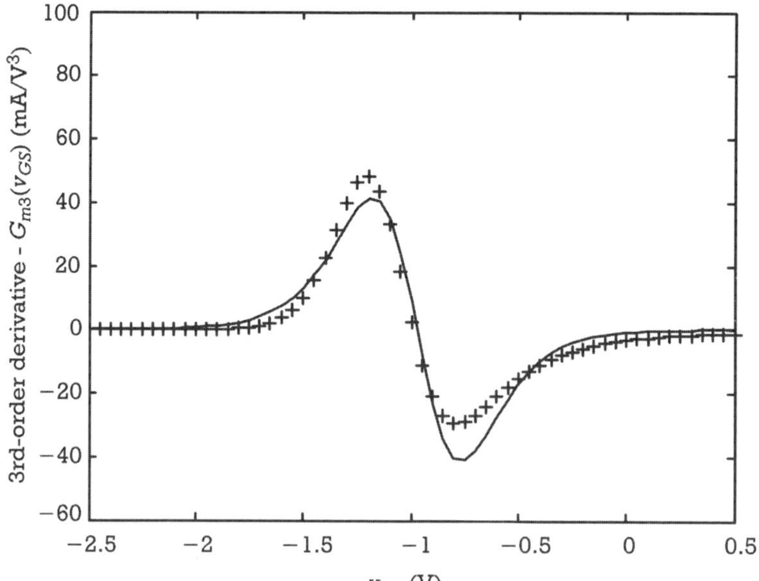

(e)

Figure 4.7 (continued).

predictions. Since they are simply Taylor series expansions around the quiescent point, their format is a priori fixed, and we only have to determine their coefficients.

On the other hand, global models are valid in a much wider stimulus range, and they are useful for both bias point calculations and large-signal analysis. However, to be adequate to small-signal distortion calculations they must also accurately fit the derivatives, at least up to the highest order of the desired mixing products. So, because they also have to represent the general data trend, they are far from being usual interpolating functions. From the authors' own experience, it seems that the best approximants are the ones that would result from the mathematical integration of a function used to fit one of the higher order derivatives (whenever analytically possible). Or, since that is generally impossible, model formats that are derived from the knowledge of the device's physical behavior, as they are forced to represent, from their genesis, the device's characteristics. In fact, we are again advising a middle term between the ideal, but inexistent, physical model in analytical form, and the totally blind empirical model.

4.2.2 Equivalent Circuit Model Extraction

The process with which an empirical model is particularized for a certain device is called model extraction. Strictly speaking, we are never extracting the model format but only its parameter set. In what the device's equivalent circuit is concerned, we are not extracting the topology, but only its element values. When referring to the functional description of the I/V or Q/V nonlinearities, we start by a predefined expression, and then try to find the values of a set of tuning parameters.

Starting by the extraction of the equivalent circuit elements' values, there are basically two different approaches to fit the model to a set of linear data, typically bias-dependent S-parameters measured in a very wide frequency range.

The first approach consists of an optimization process, in which simulated results of linear analysis, obtained for each bias point, are compared to the measured ones. The error function generated in this way is then minimized by changing the values of the elements under extraction. Although general and conceptually simple, this method suffers from several drawbacks. First, we never know if the reached error minimum really corresponds to the optimum values of the elements for that particular topology. Actually, the existence of local (or secondary) minima in error functions of several variables is common, and there is no way to distinguish them from the desired global minimum, except for the associated error magnitude. And, convergence to local minima, instead of the desired global minimum, is normal in optimization engines based on the error gradient. The second disadvantage, related to the use of optimization processes, refers to the sensitivity of the solution to the starting condition. Since we are trying to minimize an error function, we must begin with some, more or less arbitrary, starting values. It is often observed that the solution obtained by the process changes appreciably when we select different starting values. Although one might object that any solution that matches the measurements should be acceptable from the empirical modeling philosophy standpoint, this argument may not be true for our particular case of distortion analysis. In fact, remember that we are extracting the model from linear measurements and then want to extrapolate its results to nonlinear predictions. For example, in the model of Figure 4.2 it is sometimes possible to find different R_g, R_i, and R_s combinations that produce similar linear S-parameter results but, because R_s is a feedback resistor, lead to substantially dissimilar nonlinear predictions.

The second approach for obtaining the values of the equivalent circuit elements is known as direct extraction. It consists of deriving a system of equations whose left sides are the analytical description of the linear equivalent circuit behavior, and the right sides the corresponding measured data. The solution of this system for the unknowns is the sought parameter set. To be useful, this system must have a number of equations at least equal to the number of unknowns, and those equations have to be independent. Having measured S-matrices for each bias point, the maximum number of equations one can obtain is four (corresponding to the measured four S_{ij}) times the number of frequencies. This would indicate, in principle, a favorable scenario.

Unfortunately, the situation is much more difficult than it appears. First of all, note that, despite the analysis being linear (for the excitation), the system of equations will be, in general, nonlinear for the unknown values. Second, the sensitivity of the S-parameters measured or calculated at the device terminals to some of the internal elements' values may be so low that the system will be badly conditioned for those unknowns. And third, even if we could measure S-parameters for a fairly large number of frequencies, we would reach a point where a further rise in the dimension of the data set would not bring any new essential information (at least clearly above the associated measurement errors). That is, measuring more data points may not imply any increase in the number of clearly independent equations. Therefore, though theoretically possible, this method is never directly applied to extract equivalent circuits whose complexity level is close to the ones found for the majority of practical devices, as illustrated in Figure 4.2.

The difficulty of the problem needs to be reduced by first extracting the extrinsic elements' values, deembedding the intrinsic model from these already known parasitics, and then finally proceeding to the extraction of the intrinsic elements. This extraction of the extrinsic model without being significantly perturbed by the intrinsic elements is usually made possible by putting the device in abnormal states of operation. For example, for the extraction of a FET equivalent circuit, the device is biased below cut-off and for $V_{DS} = 0$V, which is the reason why it is known as the "cold-FET" model extraction procedure [4]. In the first case, $i_{DS}(v_{GS}, v_{DS})$ is zero and the C_{gs} and C_{gd} depletion capacitances tend to their minimum value. The intrinsic FET tends to an open-circuit, as shown in Figure 4.8(a), which enables the extraction of the parallel parasitic elements. In the second situation, the FET is biased in its linear region and with its gate-channel junction forward biased, determining very high input and output conductances. So, as shown in Figure 4.8(b), the intrinsic FET tends to be short-circuited, which allows the extraction of the parasitic series elements.

In conclusion, even if that direct equivalent circuit extraction methodology may not always be possible (and when it is, it certainly is not easy) and may require a slight fine tune at the end, it provides much more reliable, and physically meaningful, results than the blind optimization. This is the reason why direct extraction is now often preferred in detriment of optimization procedures more common in the past.

4.2.3 Parameter Set Extraction of the Model's Nonlinearities

After having extracted the equivalent circuit topology, it is now time to extract the parameter set of the nonlinearities' functional descriptions. Again, this can be done comparing predictions of the model with measured data, either using some kind of multidimensional nonlinear optimization, or by solving a system of equations following a direct extraction scheme.

Because an empirical model is nothing but a functional description that approximately reproduces a set of measurement data, the type, accuracy, and amount of those observations are essential to the success of the model extraction process. So,

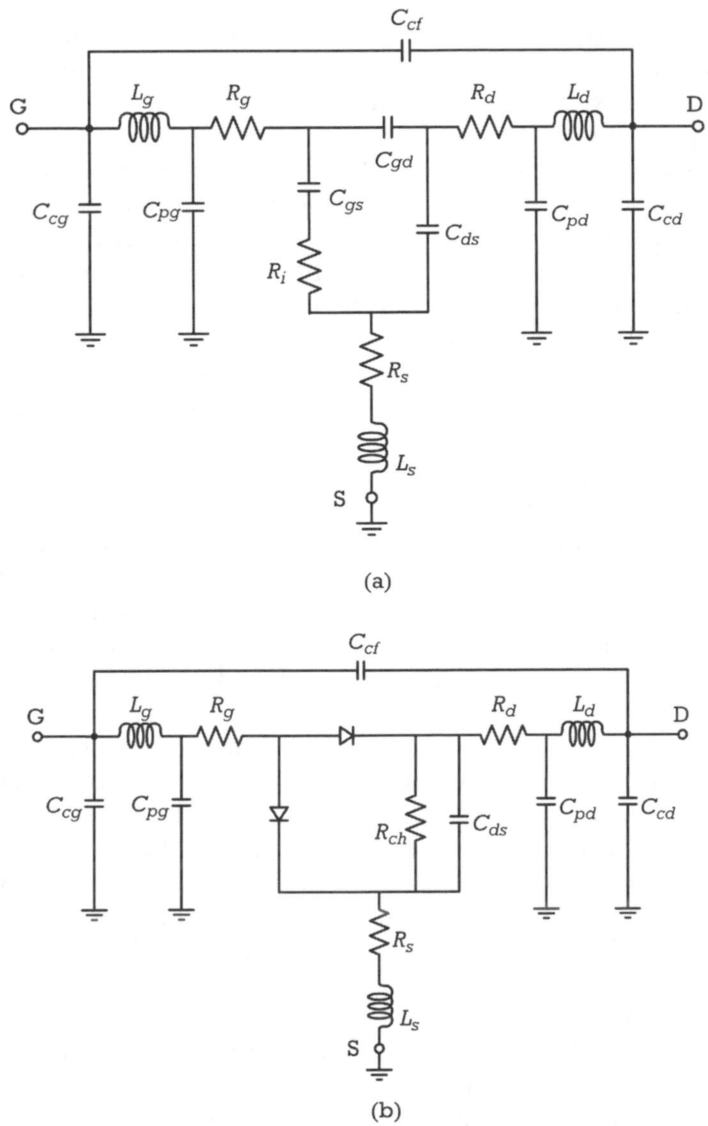

Figure 4.8 Parasitic extraction using the cold-FET procedure: (a) MESFET equivalent circuit when the device is biased below cut-off; and (b) MESFET equivalent circuit when the device is biased in the linear region.

recovering our previous discussion on channel current modeling, the first question that needs an answer is if an $i_{DS}(v_{GS}, v_{DS})$ description derived from dc is still valid for ac.

Although it is known that semiconductor charges cannot be instantaneously rearranged every time control voltages are changed, and so that $i_{DS}(v_{GS}, v_{DS})$, $Q_{gs}(v_{GS})$, and $Q_{gd}(v_{GD})$ are dynamic nonlinearities of their control voltages, it is generally considered that, for the typical frequencies in which the RF and microwave devices are operated, they can be represented as memoryless functions. This is the quasistatic assumption underlying all nonlinear device models accepted by the usual nonlinear circuit simulators using time-step integration or harmonic-balance. Under this quasistatic approximation, the response of each of these nonlinearities at a certain time point does not depend on past time, and ac behavior can be viewed as a succession of static dc excitations. That is, under the quasistatic assumption, dc behavior can indeed be used to predict ac performance. Note that, by this, we are not saying that the device is memoryless—as it indeed includes capacitances—we are only expressing the idea that its charges are static functions of control voltages. Memory is actually due to the derivative over time required to transform those storage charges into currents across those capacitances.

Being acceptable in theory, the use of dc data to predict ac performance is generally not adequate in practice, for two main reasons.

First of all, despite RF and microwave devices are expected to show significant memory effects only when the period of excitation becomes close to the time-constants determining their maximum frequency of operation, they sometimes also present long-time memory, or, in other words, comparably large time-constants. These are many orders of magnitude higher then the ones imposed by the intrinsic or extrinsic capacitances above referred, and have their origins in semiconductor trapping effects [5], thermal time-constants, or even in bias networks. For example, for a device intended to operate at 10 GHz or more, those time-constants manifest themselves at frequencies up to a few MHz. Trapping effects are usually observed in GaAs FET devices, but not on Si FETs or BJTs. Thermal time-constants are of concern in medium and high power devices. And, finally, time-constants associated to the bias circuitry are obviously only noticed whenever complete RF circuits are being modeled. So, from now on, and unless otherwise stated, dc data means the extrapolated to 0-Hz ac behavior measured in a region where neither short nor long-memory effects are noticed, or, alternatively, dc data obtained from pulsed I/V measurement setups [6].

The second order of reasons that refrains us from using dc data to extrapolate ac performance deals with the aggravation of errors in numerical differentiation. To understand this fundamental problem in modeling for nonlinear distortion calculations, let us remember that, in the framework of Volterra series, small-signal nonlinear distortion can be modeled by a Taylor series expansion of the nonlinearity around the bias point. And, all coefficients up to the nth degree contribute to determine the nth-order response of the device. For example, in the bidimensional expansion of $i_{DS}(v_{GS}, v_{DS})$,

$$i_{DS}(v_{GS}, v_{DS}) = I_{DS} + G_m v_{gs} + G_{ds} v_{ds} + G_{m2} v_{gs}^2 + G_{md} v_{gs} v_{ds} + G_{d2} v_{ds}^2 + G_{m3} v_{gs}^3 + G_{m2d} v_{gs}^2 v_{ds} + G_{md2} v_{gs} v_{ds}^2 + G_{d3} v_{ds}^3 + \ldots \quad (4.3)$$

all these nine coefficients determine the small-signal nonlinear distortion. Due to the fact that each of these nth-degree coefficients is defined as the ith partial derivative of $i_{DS}(v_{GS}, v_{DS})$ in respect to v_{GS}, and the jth partial derivative of $i_{DS}(v_{GS}, v_{DS})$ in respect to $v_{DS}(n = i + j)$, evaluated in the bias point $I_{DS}(V_{GS}, V_{DS})$, we might have thought that similar results would be obtained if dc I_{DS} (or zero order) data were measured and then numerically differentiated n times, or, conversely, the nth degree coefficients were measured, and then numerically integrated n times. But, unfortunately, that is not true. And the reason is not the unknown constant lost in any differentiation (since, in this case, it can be made zero), but refers to the way measurement noise is dealt with.

Everyone knows that after eliminating systematic errors by an appropriate calibration, random errors can be also reduced by repeated measurement followed by averaging. Since this measurement noise has an energy distributed in a frequency band much wider than the one of the signal, averaging—which is an integration, or, in other words, lowpass filtering—increases signal-to-noise ratio (i.e., measurement accuracy). So, for the same order of reasons, differentiation acts exactly on the opposite direction, aggravating measurement noise. In practice, if one measures I_{DS} dc data and then differentiates it to obtain G_m, that G_m will be quite noisy. As illustrated in Figure 4.9, another differentiation of G_m to obtain G_{m2} would produce meaningless results.

Even if measurement noise aggravation could be minimized by smoothing the data (careful digital lowpass filtering, so that no important information is lost) before differentiating it, experience shows that, in general, numerically differentiating G_m and G_{ds} data obtained from the linear equivalent circuit extraction is dangerous, and numerically differentiating dc I/V is completely useless.

In conclusion, since any measurement is always affected by additive noise, (i.e., instrumentation random errors and quantization noise due to finite resolution), any empirical model intended to predict very high signal-to-distortion ratios must be extracted from, or at least adjusted to, measured higher order ac data. Otherwise, there is no guarantee that the model will be capable of predicting accurate nonlinear distortion performance.

In the same way the linear equivalent circuit model (e.g., first-degree coefficients of G_m and G_{ds}) was extracted from measured linear S-parameter data, higher degree Taylor series coefficients must be extracted from higher order ac behavior. And this means either CW harmonic distortion, or multitone intermodulation.

Since the equivalent circuit model was previously extracted, all of its inner elements can now be deembedded. The quasistatic nonlinear elements can then be directly measured, and, as those are memoryless, they can be extracted using only CW tests. According to what we have already discussed for extracting the equivalent circuit model's first-order elements, we can again choose between a direct extraction and a nonlinear optimization process. And, using the same arguments as before, it is the direct extraction that will be chosen. So, the idea is, once more, to predict

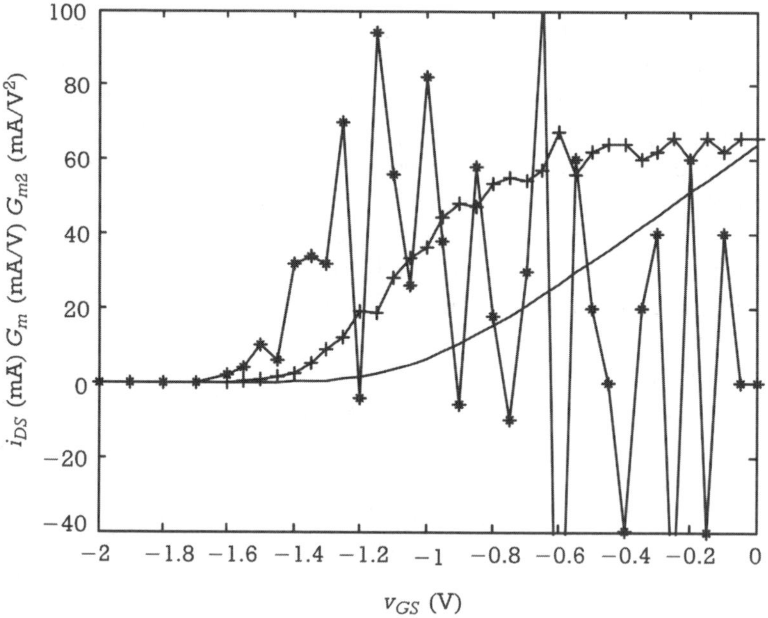

Figure 4.9 Noisy G_m and G_{m2} obtained by direct and successive differentiation of dc $I_{DS}(V_{GS})$ values.

an appropriate set of nth-order output distortion voltages or currents (in this case using Volterra series analysis since it is the only technique capable of giving solutions in analytical form), which will be dependent on the unknown coefficients, and then compare them to the ones actually observed in practice.

For example, if we were interested in extracting the $i_{DS}(v_{GS}, v_{DS})$ Taylor series coefficients, we would need three independent second-order observations for extracting the three G_{m2}, G_{md}, and G_{d2} coefficients, plus four independent third-order measurements for extracting the four G_{m3}, G_{m2d}, G_{md2}, and G_{d3} coefficients. As a single tone excitation would lead to only one independent measurement of second order (at 2ω), and another one of third order (at 3ω), it is clear that for extracting the coefficients of a bidimensional series we really need to simultaneously excite the device at the input and output with two tones (ω_1 and ω_2), and then observe the output distortion at $\omega_1 - \omega_2$, $2\omega_1$, $\omega_1 + \omega_2$, $2\omega_2$ and at $2\omega_1 - \omega_2$, ω_1, ω_2, $2\omega_2 - \omega_1$, $3\omega_1$, $2\omega_1 + \omega_2$, $2\omega_2 + \omega_1$, and $3\omega_2$.[1]

Actually, we can select a set of any three second-order products from the four available, and any four third-order products from the eight generated, and neglect

1. As a matter of fact, some alternative schemes using a single tone excitation have also been published. They basically substitute the degree of freedom lost when eliminating the source driving the output by changing the load impedance [7]. However, practical limitations in the possible range of this output termination lead to reduced extraction accuracy [8].

the others. Because we are dealing with small-signal distortion, very high signal-to distortion-ratios are expected. So, with the objective of guaranteeing that the output first-order signal will not generate perturbing distortion in the measurement instruments' nonlinearities, it is convenient that those components are filtered out. That can be easily done with a highpass filter, if the two tones ω_1 and ω_2 are close in frequency and the selected distortion products are $2\omega_1$, $\omega_1 + \omega_2$, $2\omega_2$ and $3\omega_1$, $2\omega_1 + \omega_2$, $2\omega_2 + \omega_1$, $3\omega_2$. This is illustrated in the coefficients' extraction setup shown in Figure 4.10. There, the LPF and ATTN blocks connected to each of the signal sources are lowpass filters and attenuator pads intended to filter out any spurious harmonic distortion, while guaranteeing the necessary broadband matching at these ports. The block tagged Diplexer is composed of two branches, one passive, of lowpass nature, and another one, active, that includes the above referred highpass filters.

A Volterra series analysis of the circuit shown in Figure 4.11 results in the following system of coupled equations:

$$\begin{bmatrix} K_{GS}^2 & K_{GS}K_{DS} & K_{DS}^2 \\ 2K_{GS}K_{GL} & (K_{GS}K_{DL}+K_{GL}K_{DS}) & 2K_{DS}K_{DL} \\ K_{GL}^2 & K_{GL}K_{DL} & K_{DL}^2 \end{bmatrix} \begin{bmatrix} G_{m2} \\ G_{md} \\ G_{d2} \end{bmatrix} = \begin{bmatrix} \dfrac{2N_{Ids}(2\omega_1)}{|V_S|^2} \\ \dfrac{2N_{Ids}(\omega_1+\omega_2)}{|V_S||V_L|} \\ \dfrac{2N_{Ids}(2\omega_2)}{|V_L|^2} \end{bmatrix} \tag{4.4a}$$

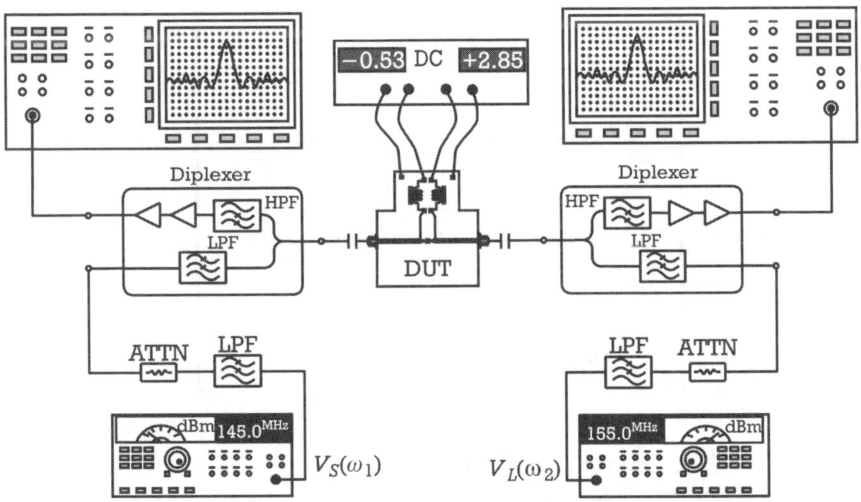

Figure 4.10 Experimental setup for the extraction of the $i_{DS}(v_{GS}, v_{DS})$ coefficients from harmonic and intermodulation power measurements.

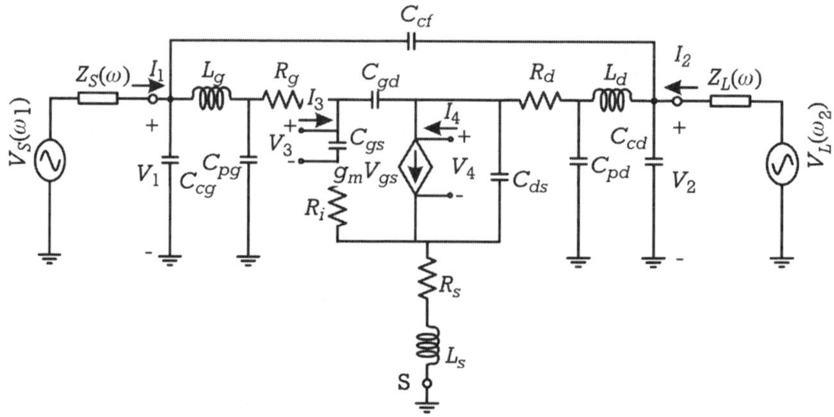

Figure 4.11 Equivalent circuit used for calculating output second and third-order distortion voltages by Volterra series analysis.

and

$$\begin{bmatrix} K_{GS}^3 & K_{GS}^3 K_{DS} & K_{GS} K_{DS}^2 & K_{DS}^3 \\ 3K_{GS}^2 K_{GL} & (K_{GS}^2 K_{DL} + 2K_{GS} K_{GL} K_{DS}) & (K_{GL} K_{DS}^2 + 2K_{GS} K_{DL} K_{DS}) & 3K_{DS}^2 K_{DL} \\ 3K_{GL}^2 K_{GS} & (K_{GL}^2 K_{DS} + 2K_{GL} K_{GS} K_{DL}) & (K_{GS} K_{DL}^2 + 2K_{GL} K_{DL} K_{DS}) & 3K_{DL}^2 K_{DS} \\ K_{GL}^3 & K_{GL}^2 K_{DL} & K_{GL} K_{DL}^2 & K_{DL}^3 \end{bmatrix}$$

$$\cdot \begin{bmatrix} G_{m3} \\ G_{m2d} \\ G_{md2} \\ G_{d3} \end{bmatrix} = \begin{bmatrix} 4\dfrac{N_{Ids}(3\omega_1) - N_{Ids_{23}}(3\omega_1)}{|V_S|^3} \\ 4\dfrac{N_{Ids}(2\omega_1 + \omega_2) - N_{Ids_{23}}(2\omega_1 + \omega_2)}{|V_S|^2 |V_L|} \\ 4\dfrac{N_{Ids}(\omega_1 + 2\omega_2) - N_{Ids_{23}}(\omega_1 + 2\omega_2)}{|V_S||V_L|^2} \\ 4\dfrac{N_{Ids}(3\omega_2) - N_{Ids_{23}}(3\omega_2)}{|V_L|^3} \end{bmatrix} \tag{4.4b}$$

in which $N_{Ids_{23}}(\omega)$ represents the third-order $N_{Ids}(\omega)$ current component generated in the second-degree coefficients, and the gain constants are derived from the analysis of the first-order equivalent circuit of Figure 4.11, in accordance to the application of the method of nonlinear currents:

$$\left\{\begin{array}{l}\begin{bmatrix} I_1 \\ I_2 \\ I_3 \\ I_4 \end{bmatrix} = \begin{bmatrix} Y_{11} & Y_{12} & Y_{13} & Y_{14} \\ Y_{21} & Y_{22} & Y_{23} & Y_{24} \\ Y_{31} & Y_{32} & Y_{33} & Y_{34} \\ Y_{41} & Y_{42} & Y_{43} & Y_{44} \end{bmatrix} \cdot \begin{bmatrix} V_1 \\ V_2 \\ V_3 \\ V_4 \end{bmatrix} \\ V_1 = V_S - Z_S I_1 \\ V_2 = V_L - Z_L I_2 \\ V_3 = V_{gs} = K_{GS} V_S + K_{GL} V_L + K_{RG} N_{Ids} \\ V_4 = V_{ds} = K_{DS} V_S + K_{DL} V_L + K_{RD} N_{Ids} \\ I_3 = 0 \\ I_4 = -NI_{ds} \end{array}\right. \tag{4.5}$$

Using (4.4) and (4.5) it is then possible to predict, for example, the terminal voltages, $V_1(\omega)$ and $V_2(\omega)$, at the seven mixing products of interest. Then, comparing them to the correspondent measured voltages we can obtain the desired direct extraction of the seven coefficients. There are, however, some practical details that must be clarified.

First of all, we must note that, since the $i_{DS}(v_{GS}, v_{DS})$ source drives the FET's output mesh, it suffices to predict and measure $V_2(\omega)$. $V_1(\omega)$ should be also considered in case we were extracting any other nonlinear element driving the input, as is the case of $C_{gs}(v_{GS})$ [9]. (There, I_3 would no longer be defined by $I_3 = 0$, but by $I_3 = -NI_{gs}$.)

Second, because we have to deembed the nonlinear element from the other linear elements pertaining to the equivalent circuit topology, and it is intuitive that the error associated to this process increases with the impact of those linear elements, ω_1 and ω_2 should be chosen such that the reactive elements have a negligible effect, but high enough to avoid low frequency dispersion as self-heating and trapping effects. This means using sufficiently low test frequencies, an attitude that has the favorable side effect of also preventing the generation of nonlinear distortion in $C_{gs}(v_{GS})$, which, otherwise, could perturb the extraction of the i_{DS} model.

And third, unless we decide to use sophisticated test equipment, capable of measuring the phase of the distortion products, we will have to complete the task relying only on amplitude information. Actually, that is the solely data we can get from a conventional spectrum analyzer. Nevertheless, it is also true that the referred use of low frequencies for V_S and V_L determines that all coefficients of expressions (4.4) and (4.5) become purely real, implying that the infinite possible range of unknown phases is, in fact, restricted to only two values of 0 or 180 degrees. That is, there is only one undefinition in the sign of the measured terminal voltages or nonlinear currents, which can be easily resolved using simple physical reasoning. For example, since below cut-off any device has null current, and thus also null

nth-degree coefficients of (4.3), and then starts to conduct when this threshold is passed, every nth-degree coefficient must start with a positive value. And, it will keep this sign until the $(n-1)$th-degree coefficient passes through a maximum and then begins to decrease. Since the nth-degree coefficient is simply the derivative of the $(n-1)$th coefficient, it is obvious that, at this point, the nth coefficient must pass through a zero and then become negative [10, 11]. This is illustrated in Figure 4.12, where the observation of extracted $G_{m2}(v_{GS})$ allowed the selection of the appropriate signs for the various regions of $G_{m3}(v_{GS})$.

After completing the extraction of the $i_{DS}(v_{GS}, v_{DS})$ model, the nonlinear channel current effects can be subtracted from the device's output distortion at substantially higher frequencies, and a similar procedure can then be pursued to extract the $C_{gs}(v_{GS})$ coefficients. Such a procedure is explained in detail in [9].

4.3 Electron Device Models for Nonlinear Distortion Prediction

This section presents an illustration of some nonlinear device models (sometimes referred to in the literature as compact models) that can be used to predict nonlinear

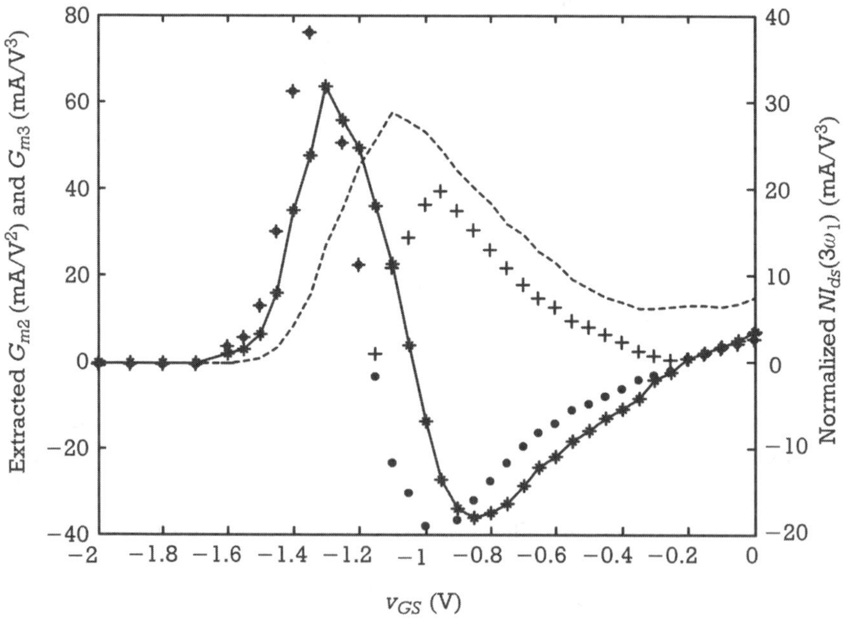

Figure 4.12 Example of measured values of normalized nonlinear current at $3\omega_1$, $NI_{ds}(3\omega_1)$ (+), signed correspondents after correct signal estimation (•), and extracted G_{m3} (-*-). The shape of previously extracted G_{m2} (--) was used for correct $NI_{ds}(3\omega_1)$ signal estimation.

distortion in time-domain or frequency-domain simulators with reasonable accuracy. Because these models are required to represent both dc, small and large-signal ac behavior, they are inherently global models. Computational efficiency reasons determined that they must be fully analytical, and so, they are either approximate semiphysical models or completely empirical.

Since all the nonlinear device models presented are to be included in some kind of circuit level simulator, they are equivalent circuit based. Nevertheless, because the extrinsic equivalent circuit model strongly depends on the particular device package, we will only concentrate on their common intrinsic parts. Using the quasi-static approximation, all nonlinear elements will be described as voltage-dependent current sources or voltage-dependent charge sources.

Obviously, a full treatment, even of only a few nonlinear device model families, is far beyond the scope of this book. For that, the interested reader can consult the abundant literature on the subject [12–15]. Here, we will solely point out some particular features of the most important models for nonlinear distortion simulation of microwave and wireless circuits, giving special attention to the ones that we think better match the representation criteria explained in the previous section.

The section opens by briefly addressing the nonlinear current and charge characteristics of the p-n or Schottky junctions, due to their role as stand-alone P-N and metal-semiconductor junction diodes, but also to their presence in many other nonlinear devices.

Because of its importance in analog circuit design, and, in particular, in the highest frequency ranges, the first transistor family to be addressed is the field effect transistor. We will start by one example of the junction FET (JFET), the metal-semiconductor FET (MESFET), and its heterojunction counterpart, the high electron mobility FET (HEMT) [sometimes also referred to as the modulation doped FET (MODFET), or heterojunction FET (HFET)], to then pass to the metal-oxide FET (MOSFET), and one of its most relevant examples for high-power RF applications, the lateral diffused MOSFET (LDMOS).

Finally, the bipolar junction transistor is addressed, either in its homojunction, BJT, or heterojunction, HBT, structures.

4.3.1 Diodes and Other Semiconductor Junctions

Nonlinearity is present in all metal or semiconductor junctions, whether these junctions are accidental or intentionally created.

There is evidence that any time two different metals are brought into contact, their conduction properties do not become ideally ohmic, but present a certain degree of nonlinearity. That depends on many distinct variables as the type of metals, contact pressure, contact surface roughness, possible oxidation, and current density. As metal junctions are commonplace in electronic equipment, these nonlinearities appear even from the most unexpected sources as RF connectors, metallic shields, and antennas. They play such an important role in higher power telecommunication systems' design that they have received the attention of one of the more difficult, and still misunderstood, nonlinear phenomena fields of study: passive intermodulation (PIM) [16].

On the side of deliberately made nonlinear junctions we find both semiconductor-semiconductor and metal-semiconductor junctions. The former is usually made from the physical union of two doped semiconductors, of type p and n, made of similar (homojunctions) or different (heterojunctions) materials. It is thus named a P-N junction or diode. The latter arises, for example, whenever a metal-semiconductor junction is built, and this is known as a Schottky junction.

The diffusion of electrons, or holes, from one junction side (where they are majority carriers) to the other (where they become minority carriers) restores thermodynamic equilibrium, but also creates a zone of charged ions, the space-charge, depleted of carriers, which builds up a certain potential barrier. External applied voltage with a sign that overcomes that potential barrier creates an energy unbalance that facilitates current conduction (forward biasing), while applied voltage with an opposite sign (reverse biasing) reinforces the potential barrier impeding current conduction. Therefore, a P-N junction presents a distinct nonlinearity that has been for a long time used as a rectifier, voltage-controlled current switch or resistor, amplitude demodulator, or mixing device.

Actually, the diode I/V characteristic, $i_D(v_D)$, is one of the strongest nonlinearities found in nature, which can be represented by the well-known semiphysical exponential model:

$$i_D(v_D) = I_S(e^{qv_D/\eta kT} - 1) \tag{4.6}$$

where I_S is a scaling parameter, k is the Boltzmann constant, T is the junction temperature, q is the electronic charge, and η is an empirical ideality factor, used for modeling imperfections in the junction.

Unfortunately, the variation of junction accumulated charge with the external applied voltage results in a dynamic nonlinearity, which is modeled as a nonlinear capacitance. Actually, there are two possible sources of accumulated charge within a P-N junction [1, 15].

The first one is the space-charge associated with the depletion region, which is usually modeled by

$$q_d(v_D) = Q_{j0}\left(1 - \frac{v_D}{\phi}\right)^{1-m} \tag{4.7}$$

and thus leads to a nonlinear depletion capacitance of

$$C_d(v_D) \equiv \frac{\partial q_d(v_D)}{\partial v_D} = C_{j0}\left(1 - \frac{v_D}{\phi}\right)^{-m} \tag{4.8}$$

where, ϕ is the built-in potential, Q_{j0} and C_{j0} are two scaling parameters that represent the zero-voltage charge and capacitance, respectively, and m is a coefficient that depends on the doping profile. For the most common case of uniform doping profiles, m is close to 0.5. However, other P-N or Schottky junctions are found where m can assume clearly distinct values. That is the case, for example, of abrupt or hyper-abrupt junctions specially conceived to present predetermined $C_d(v_D)$ characteristics, and thus to be used as voltage-controlled capacitors named as varicaps or varactors.

Note that the depletion capacitance model has a discontinuity at $v_D = \phi$, becoming imaginary beyond that. This nonphysical characteristic is due to the so-called abrupt depletion region approximation adopted for the semiphysical charge model derivation, and must be circumvented in actual model implementations because it can generate convergence difficulties and nonphysical simulation results.

The second source of accumulated charge within a P-N junction is due to the finite minority carrier lifetime. When the junction is forward biased, electrons are diffused from the N to the P side and holes from the P to the N side. Since these minority carriers cannot be immediately recombined with the corresponding majority holes or electrons, there will be an accumulated diffusion charge. This charge can be approximately modeled as being proportional to the minority carrier lifetime, τ, and forward current, so that

$$q_D = \tau i_D \tag{4.9}$$

and

$$C_D(v_D) \equiv \frac{\partial q_D(v_D)}{\partial v_D} = \tau \frac{qI_S}{\eta kT} e^{qv_D/\eta kT} \tag{4.10}$$

Obviously, since a Schottky diode is a majority carrier junction it is only affected by the depletion capacitance.

As the diode was modeled by an exponential voltage-dependent current source, plus depletion and diffusion accumulated charges, it conforms to the equivalent circuit model depicted in Figure 4.13 [13]. There, the series access resistor, R_s, represents the distributed ohmic behavior of the device, as seen from its external terminals, and models the semiconductor bulk and ohmic contact losses, but also the parasitic resistances of the bond pads, bond wires, and package terminals. Despite it being known that bulk resistance can change with applied voltage (due to undepleted semiconductor width variations and current crowding effects [15]),

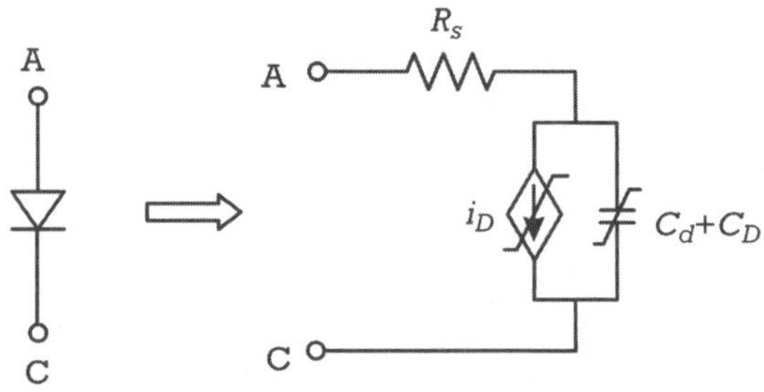

Figure 4.13 Junction diode equivalent circuit model.

R_s is usually still considered as a constant (and thus, sometimes, extrinsic) element, since its parasitic terms generally dominate the others.

4.3.2 Field Effect Transistors

Due to its geometrical structure, the FET is, inherently, a bidimensional device. The applied drain-source voltage, v_{DS}, determines a strong longitudinal channel electrical field component, but the gate-channel control voltage (or gate-source voltage, v_{GS}, when it is referred to the source) also imposes a nonnegligible transversal channel component. Therefore, any physical modeling approach leads to the solution of a set of nonlinear partial deferential equations in space and time, which is highly inefficient. FET models thus use extensive empirical knowledge [12].

4.3.2.1 The MESFET Model

A MESFET is a junction FET whose gate-channel contact is a Schottky junction. At the equivalent circuit level, this distributed gate-channel Schottky junction is actually represented by two lumped gate-source and gate-drain diodes, as shown in Figure 4.14 [4]. They are primarily used to simulate gate-channel conduction and breakdown—two fundamental effects for the description of the device's RF power saturation. But, with the addition of two lumped depletion capacitances from gate to source and to drain, C_{gs} and C_{gd}, [9] they can also describe nonlinear dynamic effects.

Contrary to C_{gd} and, especially, C_{gs}, which are distinctly nonlinear, the drain-source capacitance, C_{ds}, is normally accepted to be linear, as it mainly describes the geometric capacitance effects between the drain and source terminals.

The input resistance, R_i, is sometimes interpreted as an equivalent lumped representation of the channel distributed losses below the gate. And, although this physical foundation is still object of controversy, its need for correctly representing measured small-signal S_{11} data seems to be consensual.

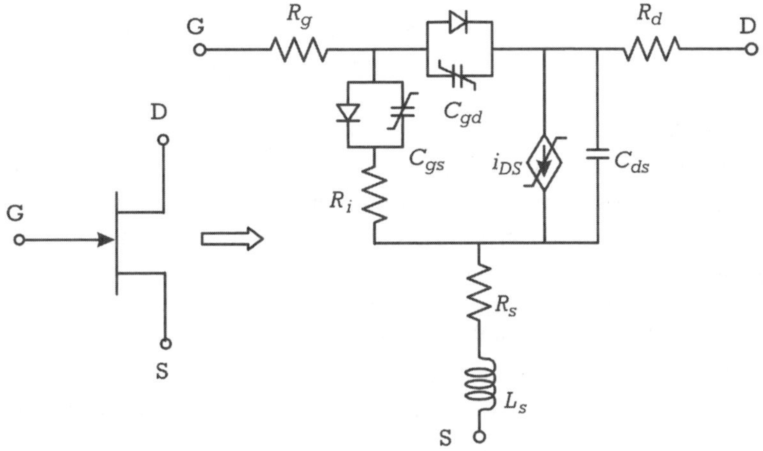

Figure 4.14 MESFET equivalent circuit topology.

The three access resistors, R_g, R_s, and R_d represent the distributed ohmic behavior of the device, as already discussed for the diode. Similarly, they are also sometimes considered as extrinsic elements.

Finally, the voltage-controlled current source, i_{DS}, is the core of the equivalent circuit model and constitutes the FET's main nonlinear distortion generator. Describing the basic FET operation, it is simultaneously dependent on the longitudinal channel and gate-channel transversal potentials, herein represented by two control variables: drain-source voltage, v_{DS}, and gate-source voltage, v_{GS}.

One of the possible continuous functions used for $i_{DS}(v_{GS}, v_{DS})$ is the one sometimes referred to as the Pedro's model [17]:

$$v_p(v_{DS}) = V_{p0} + \gamma v_{DS} \tag{4.11}$$

$$h(v_{GS}, v_{DS}) = A\left[1 - \sqrt{\frac{v_{bi} - v_{GS}}{v_p(v_{DS})}}\right] \tag{4.12}$$

$$u(v_{GS}, v_{DS}) = \frac{h(v_{GS}, v_{DS}) - C}{2} \tag{4.13}$$

$$I_{dssat}(v_{GS}, v_{DS}) = u(v_{GS}, v_{DS}) + \ln\left[e^{u(v_{GS}, v_{DS})} + e^{-u(v_{GS}, v_{DS})}\right] \tag{4.14}$$

$$i_{DS}(v_{GS}, v_{DS}) = \beta I_{dssat}(v_{GS}, v_{DS}) \tanh(\alpha v_{DS}) \tag{4.15}$$

In this model, the i_{DS} dependence on v_{DS} is described at two different levels: the hyperbolic tangent and the pinch-off voltage, v_P, formulation. The former describes the linear to saturation zone transition (whose abruptness is controlled by α) while the latter models (with the empirical parameter γ) the threshold dependence on v_{DS}, but also the device's output conductance in saturation.

The dominant behavior of $i_{DS}(v_{GS})$ is represented with the Shockley model [18] under the depletion approximation. Assuming a whole current-saturated channel, $h(v_{GS}, v_{DS})$ is the effective channel height. The FET's soft turn-on is represented by $u + \ln[\exp(u) + \exp(-u)]$, which was found to be very useful in describing typical MESFET transconductance $G_m(v_{GS})$ shapes. Finally, A and C are two empirical coefficients that control, respectively, $G_m(v_{GS})$ abruptness and threshold position, while β is a scaling parameter.

As a first illustration of its intermodulation modeling capabilities, Figures 4.15 and 4.16 show measured and modeled values of i_{DS}, and its first three derivatives in order to v_{GS}: G_m, G_{m2}, and G_{m3}.

4.3.2.2 The HEMT Model

A HEMT is a device similar to the MESFET, despite its heterojunction channel. So, they share the equivalent circuit topology (see Figure 4.14) and extraction procedure. Their main difference, from the nonlinear distortion point of view, is the HEMT's lack of transconductance expansion for high values of v_{GS}, which prevents the existence of the G_{m3} null (a bias point of very good small-signal inband IMD) verified in some MESFETs biased at reasonably high gate voltages. Indeed, the HEMT's so-called parasitic MESFET effect [12] causes a decrease of transconductance for higher gate voltages, which is often observed as a tendency to $i_{DS}(v_{GS})$ saturation.

This transconductance collapse is, actually, the most important effect described by the Angelov-Zirath HEMT drain-source current model [19]:

$$\psi(v_{GS}) = P_1(v_{GS} - V_{pk}) + P_2(v_{GS} - V_{pk})^2 + P_3(v_{GS} - V_{pk})^3 \tag{4.16}$$

$$i_{DS}(v_{GS}, v_{DS}) = I_{pk}\{1 + \tanh[\psi(v_{GS})]\}(1 + \lambda v_{DS}) \tanh(\alpha v_{DS}) \tag{4.17}$$

In these expressions, the various P_i $(i = 1, 2, 3)$ are empirical polynomial fitting parameters describing the dependence of an effective gate potential, ψ, on v_{GS}, and V_{pk} is the gate voltage where transconductance shows its peak. λ is, like the MESFET's γ, an empirical output conductance parameter modeling the behavior of the saturation current with v_{DS}, and α was inherited from the MESFET model.

Beyond this $i_{DS}(v_{GS}, v_{DS})$ representation, the Angelov-Zirath HEMT model [19] also presents empirical functions for the HEMT's $C_{gs}(v_{GS}, v_{DS})$ and $C_{gd}(v_{GS}, v_{DS})$ capacitances.

As done for the MESFET model, Figure 4.17 presents a comparison between measured and modeled i_{DS} current and its derivatives.

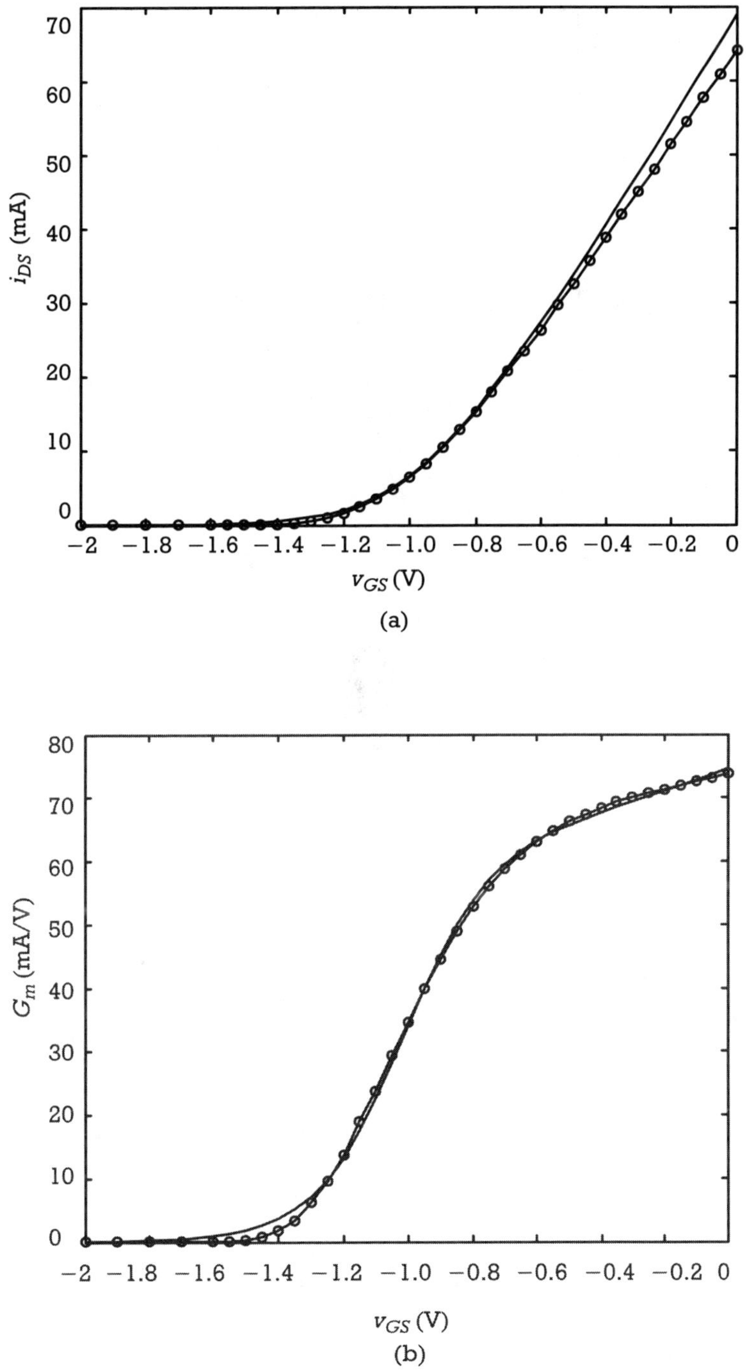

Figure 4.15 Measured (-o-) and modeled (-) (a) i_{DS} and (b) G_m versus v_{GS} when the FET is biased in the saturation region with $V_{DS} = 3V$.

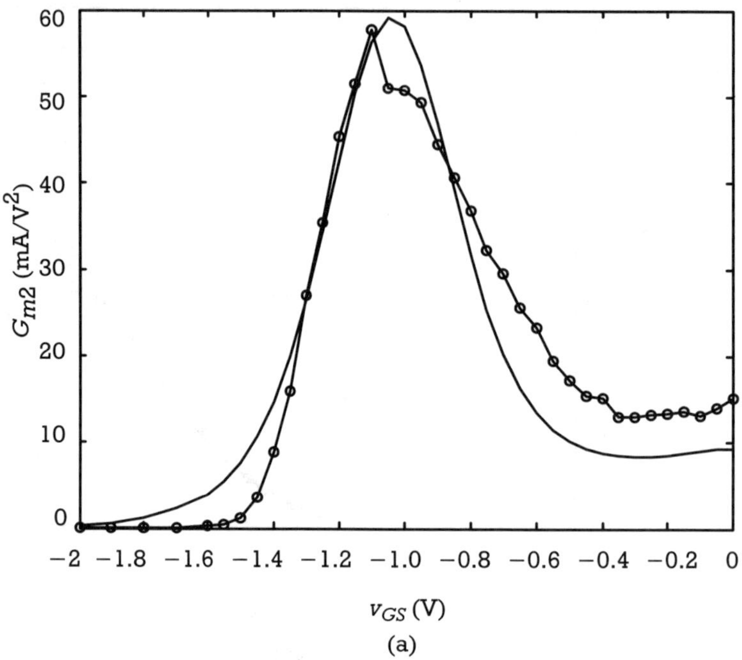

(a)

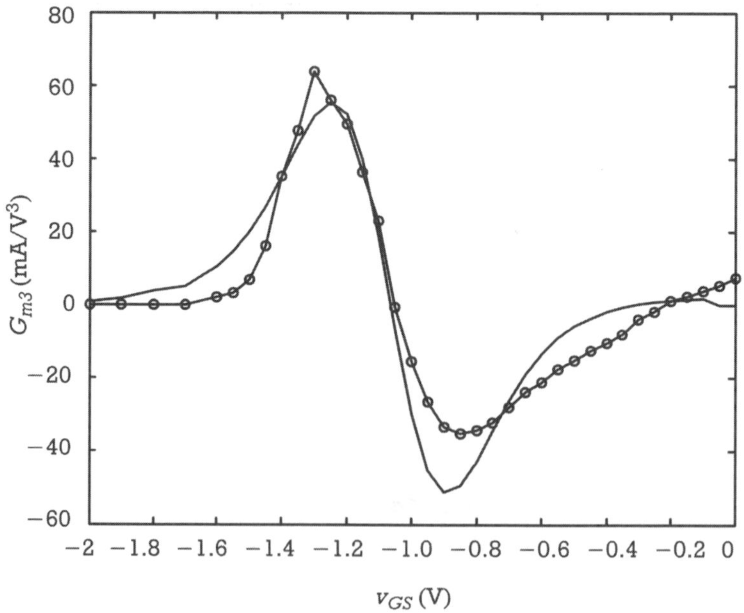

(b)

Figure 4.16 Measured (-o-) and modeled (-) (a) G_{m2} and (b) G_{m3} versus v_{GS} when the FET is biased in the saturation region with $V_{DS} = 3V$.

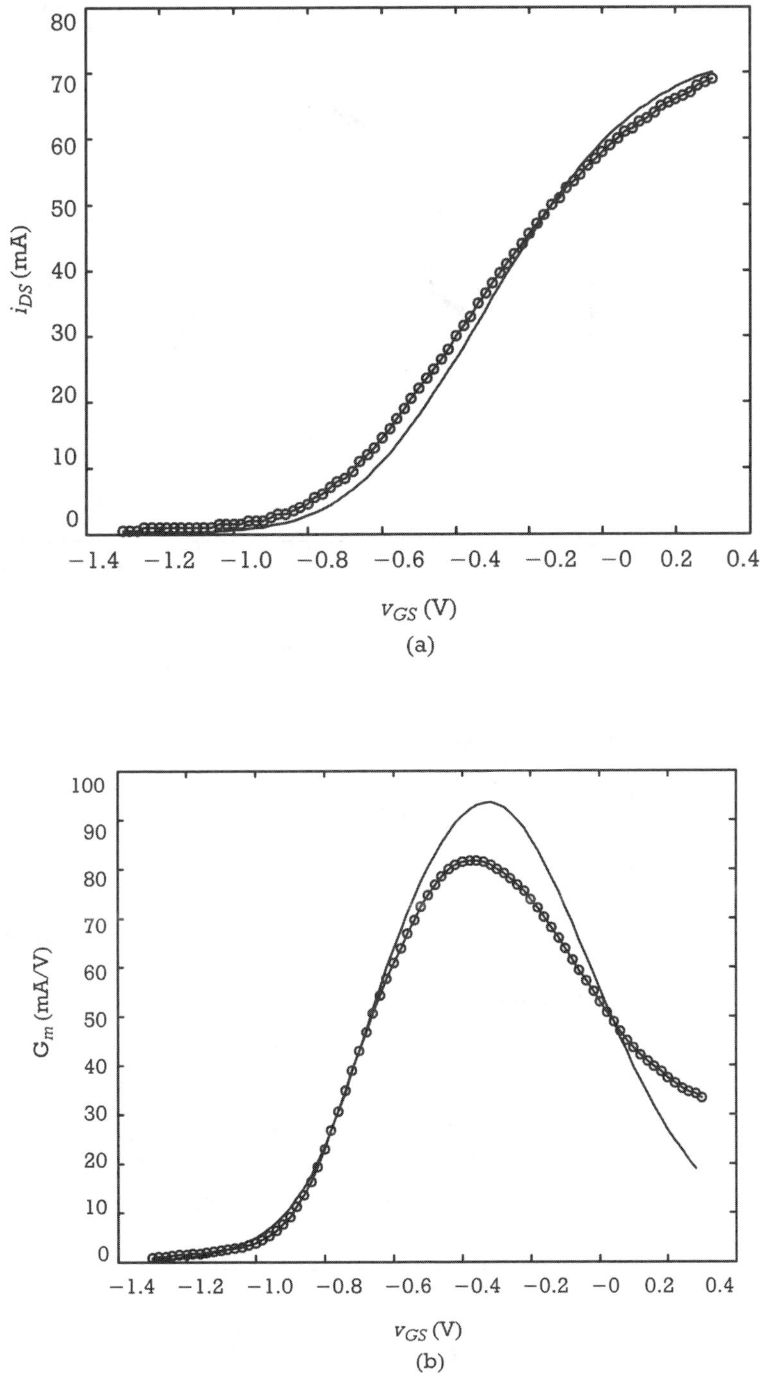

Figure 4.17 (a) Measured (-o-) and modeled (-) (a) i_{DS}, (b) G_m, (c) G_{m2}, and (d) G_{m3} versus v_{GS} when the HEMT is biased in the saturation region with $V_{DS} = 3V$.

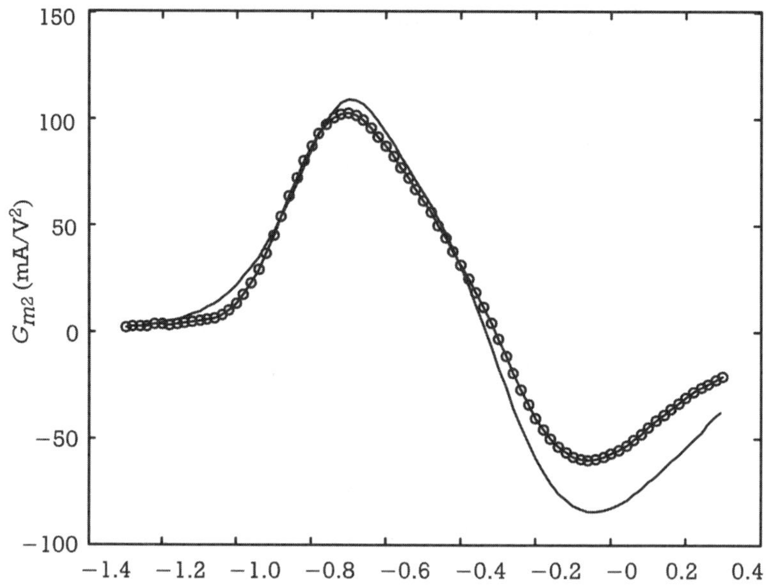

(c)

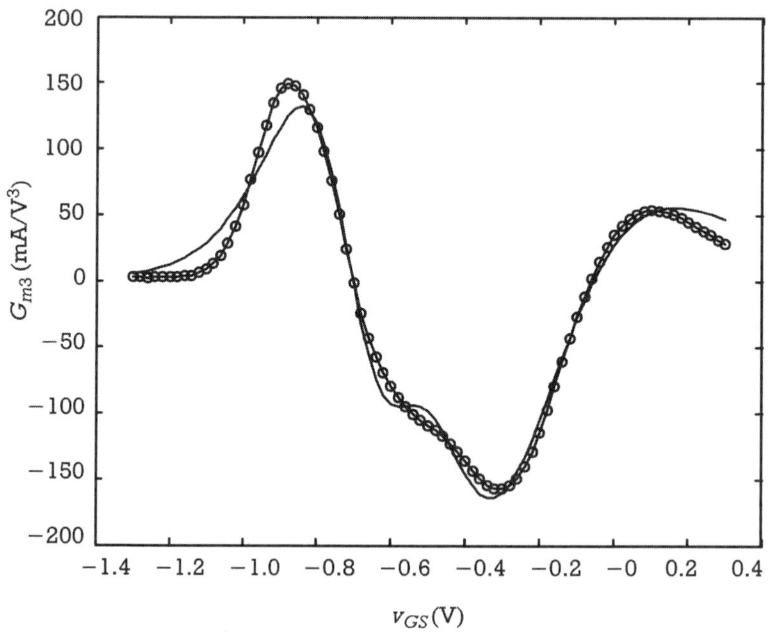

(d)

Figure 4.17 (continued).

4.3.2.3 The MOSFET and LDMOS Models

There are two MOSFET devices of significant value for microwave and wireless applications: the traditional MOS, as used in CMOS and BiCMOS IC technologies [14, 20, 21], and the high-power LDMOS [22–25]. Despite their significantly distinct drain-source current characteristics, they share the same equivalent circuit topology shown in Figure 4.18 [22].

Similarly to the MESFET, R_g, R_d, and R_s are access resistances. C_{gs} and C_{gd} are gate-source and gate-drain nonlinear lumped capacitances representing the distributed metal-oxide-channel accumulated charge.

C_{ds} is, again, a drain-source capacitance. But, contrary to the MESFET, it now manifests a nonnegligible v_{DS} dependence, and thus, nonlinearity.

Although there is a very large number of available MOS drain-source current empirical descriptions, the Berkeley BSIM3 MOSFET model [14] seems to be the most amenable for analog circuits' design, and, in particular, for the prediction of their nonlinear distortion.

This model has its roots in a very simplified and empirical description of the MOS nonlinear behavior. But, after being many times reformulated, it is currently a very detailed, semiphysical, representation of the device's channel current. For this reason, the MOSFET model became so complex (and difficult to extract) that complete books were written to address it. So, in the present text, there is not much we can do than to direct the interested reader to some of those titles, for example, [14, 20, 21].

The LDMOS is a transistor specially conceived for withstanding large amounts of power in a small active device, a goal that was fulfilled by the addition of a specific lateral diffused channel region near the drain [23]. However, that lightly doped region acts as a grounded gate JFET in series with the traditional MOS

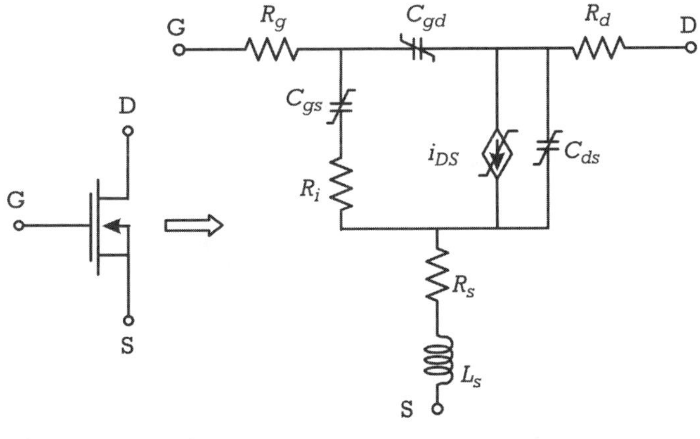

Figure 4.18 The MOSFET equivalent circuit topology.

transistor (see Figure 4.19), which has a dramatic impact on the observed $i_{DS}(v_{GS})$ characteristics. For high v_{GS} voltages the i_{DS} current is so large that the voltage drop in the highly resistive low-doped region is enough to draw the JFET into saturation. The MOSFET v_{DS} voltage is suddenly reduced, and this device is pushed into its linear region. When this happens, v_{GS} looses control on i_{DS}, and the LDMOS shows an evident $i_{DS}(v_{GS})$ current saturation, and thus, transconductance collapse.

Because of the referred high power applications, this LDMOS transconductance collapse is even aggravated by self-heating effects. The overall impact on the device performance can be so high that, not only completely new $i_{DS}(v_{GS}, v_{DS})$ descriptions (compared to its low power counterpart, the MOSFET) had to be proposed,

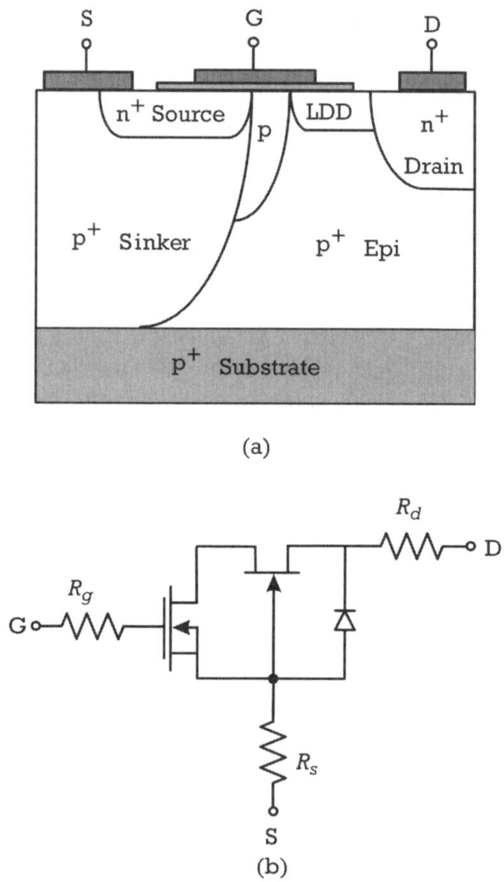

Figure 4.19 (a) A detailed view of a typical LDMOS physical structure; and (b) its correspondent equivalent circuit, showing the MOSFET device in series with the grounded JFET caused by the low-doped drift (LDD) region.

as they had to be coupled to some thermal model. One of these electrothermal current descriptions is the LDMOS MET Model [24], whose $i_{DS}(v_{GS}, v_{DS})$ electrical part is given by

$$v_{GS1}(v_{GS}) = v_{GS} - V_T \tag{4.18}$$

$$v_{GS2}(v_{GS1}) = v_{GS1} - \frac{1}{2}\left(v_{GS1} + \sqrt{(v_{GS1} - VK)^2 + \Delta^2} - \sqrt{VK^2 + \Delta^2}\right) \tag{4.19}$$

$$v_{GSt}(v_{GS2}) = VST \ln\left(1 + e^{v_{GS2}/VST}\right) \tag{4.20}$$

$$i_{DS}(v_{GS}, v_{DS}) = \beta v_{GSt}^{V_{GEXP}}(1 + \lambda v_{DS}) \tanh\left(\frac{\alpha v_{DS}}{v_{GSt}}\right) \tag{4.21}$$

$$V_T(v_{DS}) = V_{T0} + \gamma v_{DS} \tag{4.22}$$

and whose thermal model uses also an equivalent circuit representation as shown in Figure 4.20. In the thermal equivalent subcircuit, absolute temperature is modeled by voltage, heat energy by charge—and thus thermal power by current—thermal resistance (whose dimensions are $[R_{Th}] = \mathrm{KW}^{-1}$) by an electrical resistance, and, finally, thermal capacitance (whose dimensions are $[C_{Th}] = \mathrm{JK}^{-1}$) is represented with an electrical capacitance. The forcing current source, I_{Th} is given by the device's total power dissipation, while the forcing voltage source, V_{Tsink}, is the

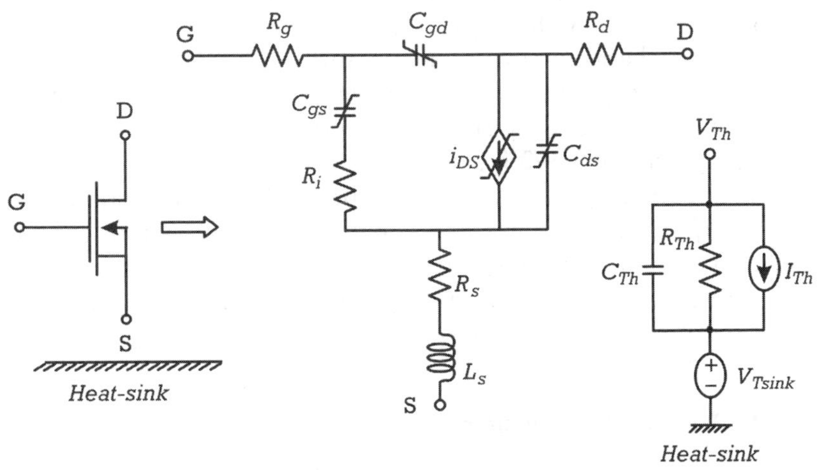

Figure 4.20 Electrothermal equivalent circuit model for an RF power LDMOS device.

environment absolute temperature, usually the transistor's mount heat-sink temperature. The model parameters dependence on temperature is described by

$$R_g = R_{g\_0} + R_{g\_1}(T - Tnom) \tag{4.23}$$

$$R_d = R_{d\_0} + R_{d\_1}(T - Tnom) \tag{4.24}$$

$$R_s = R_{s\_0} + R_{s\_1}(T - Tnom) \tag{4.25}$$

$$V_{T0} = V_{T0\_0} + V_{T0\_1}(T - Tnom) \tag{4.26}$$

$$\beta = \beta_0 + \beta_1(T - Tnom) \tag{4.27}$$

where $Tnom$ is the nominal temperature at which R_{g_0}, R_{d_0}, R_{s_0}, V_{T0_0}, and $\beta_{\_0}$ were measured, and R_{g_1}, R_{d_1}, R_{s_1}, V_{T0_1}, and $\beta_{\_1}$ are first-order approximation thermal sensitivity parameters.

As shown in (4.21), the LDMOS shares with the MESFET and HEMT the hyperbolic tangent for describing the $i_{DS}(v_{DS})$ dependence. Nonetheless, a new hyperbolic tangent argument variation with v_{GS} was included to describe the recognized dependence of the linear-to-saturation knee voltage with this input control voltage. By modifying the effective value of the v_{DS} scaling parameter α, that $i_{DS}(v_{DS})$ functional description also carries the benefit of better representing the abruptness variation with v_{GS} manifested by the linear-to-saturation transition.

The description of the i_{DS} dependence on v_{GS} is a certain power (of V_{GEXP} exponent) of a function similar to the one already described for the MESFET, the $x + \ln[\exp(x) + \exp(-x)]$, except that now its argument is no longer a scaled version of the undepleted channel height, but an effective gate-source voltage, $v_{GS2}(v_{GS} - V_T)$. By that, $i_{DS}(v_{GS})$ presents the desired subthreshold conduction exponential characteristic, a smooth turn-on (whose abruptness is controlled by V_{ST}), a nearly quadratic v_{GS} dependence (whose rate is controlled by V_{GEXP}), and finally, the necessary current compression when $v_{GS2}(v_{GS} - V_T)$ tends to its saturated value of $V_K/2$ [25].

V_{T0}, β, and λ are standard model parameters used to control threshold voltage, to scale $i_{DS}(v_{GS}, v_{DS})$ and to account for the finite output conductance when the device is in saturation, respectively. γ and Δ are simply fitting parameters.

Figure 4.21 compares measured and modeled results of i_{DS} current and its first three derivatives in order to v_{GS}.

4.3.3 The Bipolar Transistor Family

The structure and mode of operation of a bipolar junction transistor (BJT) is completely distinct from the FET. Being composed of two back-to-back P-N

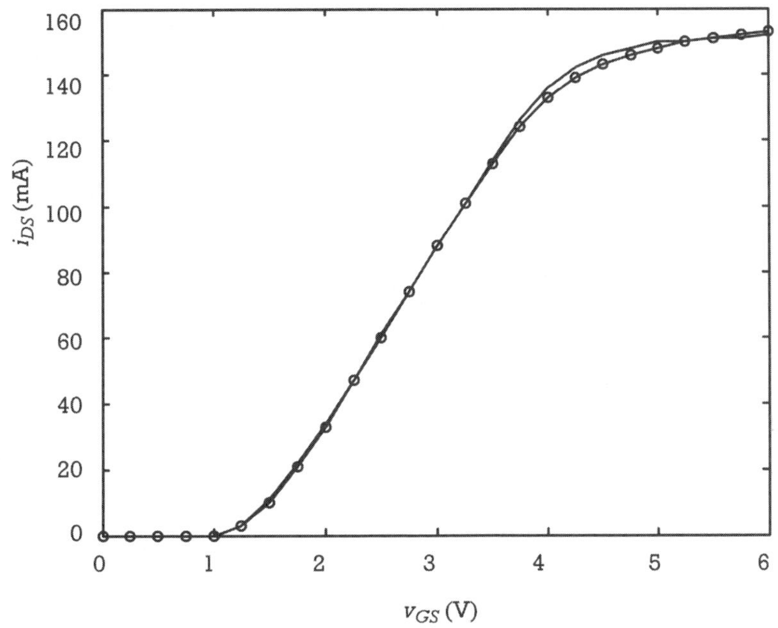

(a)

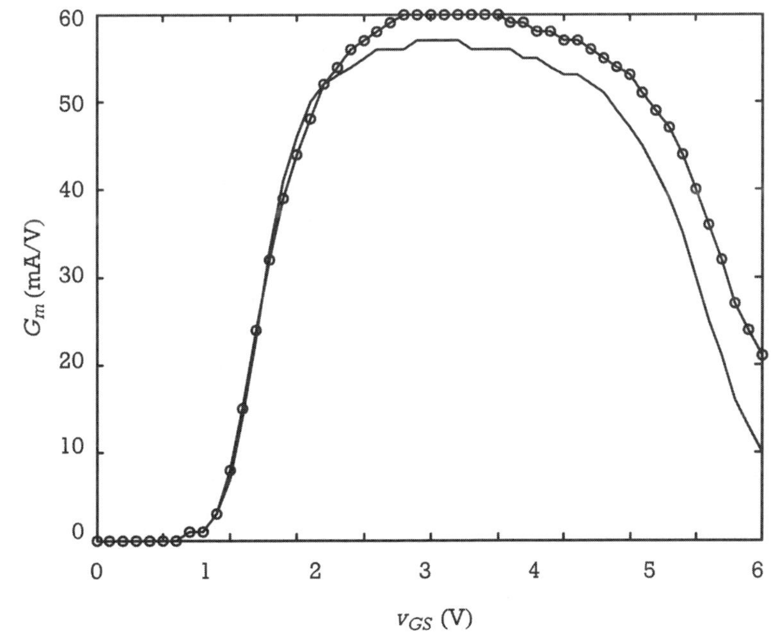

(b)

Figure 4.21 Measured (-o-) and modeled (-) (a) i_{DS}, (b) G_m, (c) G_{m2}, and (d) G_{m3} versus v_{GS} when the LDMOS is biased in the saturation region with $V_{DS} = 10\text{V}$.

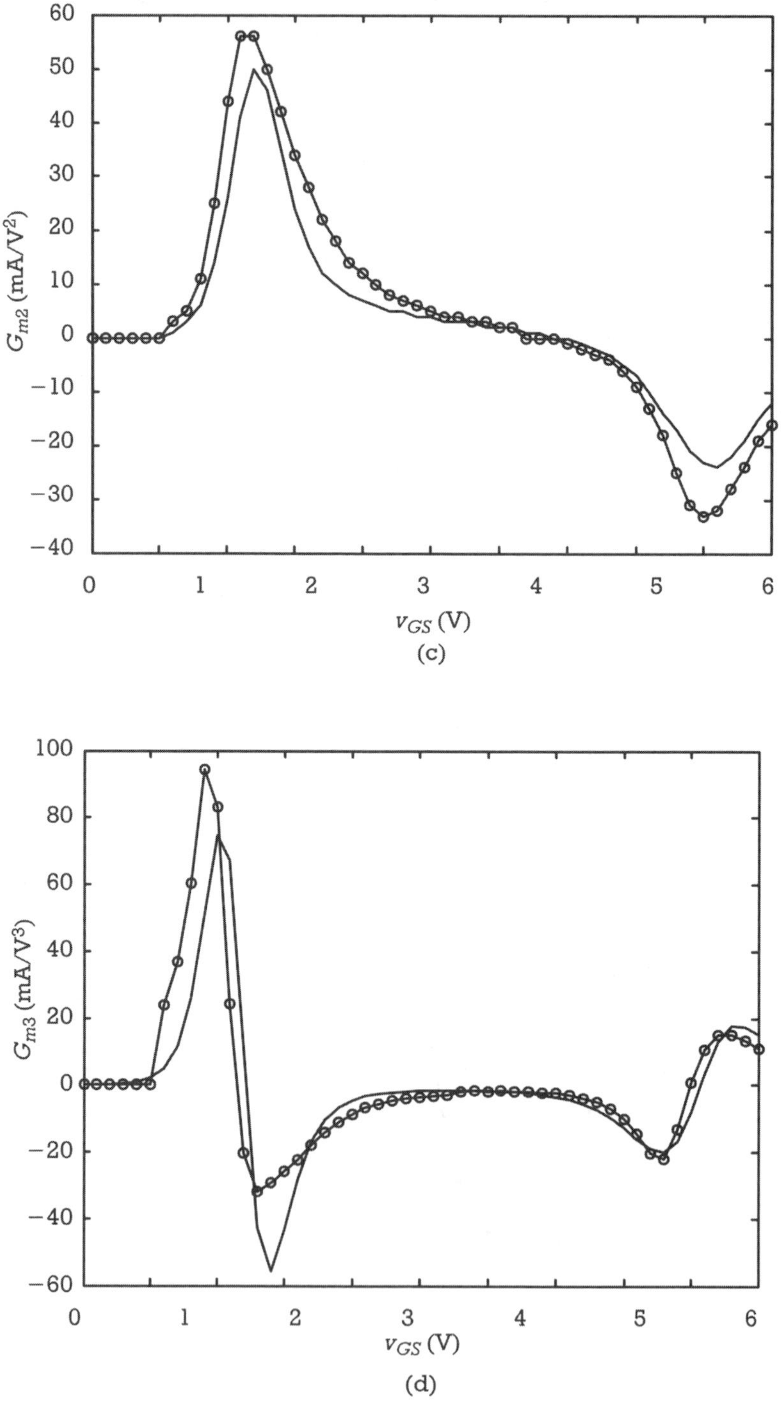

Figure 4.21 (continued).

junctions, in a one-dimensional way, allows a semiphysical modeling approach. Actually, this inherently T structure can be represented with reasonable accuracy by either a T or π equivalent circuit topology, whose voltage-controlled current sources and charges can be derived from the P-N junction model described above [15].

Due to its importance in BJT and, more recently, also in HBT representation [26], in this section we will briefly address the Gummel-Poon model (actually its SPICE implementation)[2] [15, 20, 29, 30], which uses the π equivalent circuit topology of Figure 4.22.

The Gummel-Poon model is based on an integral charge control relation, which connects the total charge of majority carriers in the base to the device's terminal characteristics. This way, it describes the two main sources of BJT nonlinearity (i.e., the base and the collector currents), but sometimes also the base resistance nonlinear behavior is accounted for.

As previously mentioned, the functional descriptions of these modeled currents are based on the P-N junction exponential characteristic, for which the base-emitter and base-collector voltages were adopted as the controlling variables. So, the collector current is given by

$$i_C(v_{BE}, v_{BC}) = \frac{I_S}{q_b(v_{BE}, v_{BC})}\left[\left(e^{qv_{BE}/kT} - 1\right) - \left(e^{qv_{BC}/kT} - 1\right)\right] \quad (4.28)$$

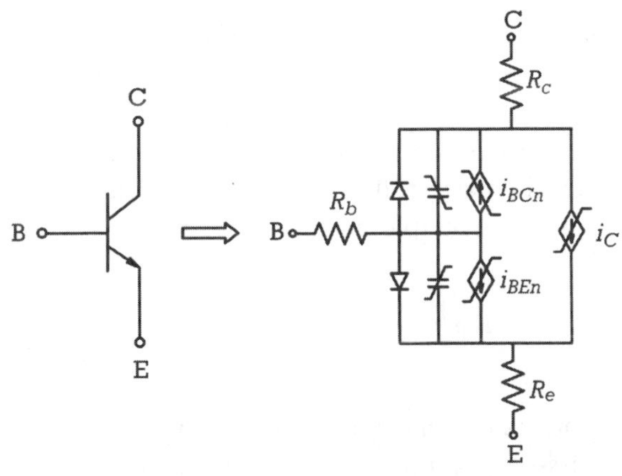

Figure 4.22 π equivalent circuit topology adopted for the Gummel-Poon BJT model.

2. Even though the SPICE Gummel-Poon model still constitutes the standard bipolar transistor representation, some other alternatives have recently been proposed (like the MEXTRAM [27] and the VBIC [28] models) to overcome some of its deficiencies. From these stand, for their impact in high-speed bipolar devices, quasisaturation effects, avalanche multiplication, bias dependent transit time, and self-heating.

where I_S, q, k, and T are directly mirrored in the P-N junction I/V characteristic of (4.6), and q_b is the normalized majority base charge, which can be expressed in implicit form as

$$q_b(v_{BE}, v_{BC}) = 1 + \frac{v_{BE}}{V_B} + \frac{v_{BC}}{V_A} + \frac{\tau_{BF}}{Q_{B0}} I_{SS} \frac{e^{qv_{BE}/kT} - 1}{q_b} + \frac{\tau_{BR}}{Q_{B0}} I_{SS} \frac{e^{qv_{BC}/kT} - 1}{q_b} \tag{4.29}$$

V_A and V_B are, respectively, the forward and reverse Early voltages, while τ_{BF} and τ_{BR} are the forward and reverse base transit times. Finally, Q_{B0} is the zero bias majority base charge.

The base current reflects the emitter and collector junction currents seen at the collector terminal, although scaled by the corresponding transistor's forward and reverse current gains, β_{FM} and β_{RM}. It also includes two other nonideal current sources, i_{BEn} and i_{BCn} which account for carrier recombination in the semiconductor surfaces and the emitter-base and collector-base space-charge layers [15]. So, in the Gummel-Poon model, $i_B(v_{BE}, v_{BC})$ is described by

$$\begin{aligned} i_B(v_{BE}, v_{BC}) = {} & \frac{I_S}{\beta_{FM}}\left(e^{qv_{BE}/kT} - 1\right) + C_2 I_S\left(e^{qv_{BE}/n_{EL}kT} - 1\right) \\ & + \frac{I_S}{\beta_{RM}}\left(e^{qv_{BC}/kT} - 1\right) + C_4 I_S\left(e^{qv_{BC}/n_{CL}kT} - 1\right) \end{aligned} \tag{4.30}$$

where the coefficients n_{EL} and n_{CL} are the base-emitter and base-collector leakage emission coefficients, respectively, and C_2 and C_4 are simply empirical scaling parameters.

Finally, the resistors R_b, R_c, and R_e, seen in the equivalent circuit model of Figure 4.22, are the base, collector, and emitter access resistances. From these, only R_b manifests an appreciable nonlinearity, which is controlled by base current, as

$$R_B(z) = r_{BM} + 3(r_B - r_{BM})\left(\frac{\tan(z) - z}{z\tan^2(z)}\right) \tag{4.31}$$

The parameters on this expression are the minimum base resistance—occurring at high currents—r_{BM}, and the base resistance at zero bias, r_B. z is a variable of base resistivity, thermal voltage, and intrinsic base length, which can be approximately given by [15]

$$z(i_B) = \frac{-1 + \sqrt{1 + 144 i_B/\pi^2 I_{rB}}}{24/\pi^2 \sqrt{i_B/I_{rB}}} \tag{4.32}$$

where I_{rB} is the current where the base resistance falls halfway to its minimum value.

4.4 Behavioral Models for System Level Simulation

As previously stated in the classification of models, there are various levels of abstraction in the way we can represent real-world nonlinear systems. Or, in other words, there is a gradual pathway between the pure physically conceived models to black box models. These can be applied to nonlinear elements, devices, circuits, or even complete systems.

Many practical situations are found where a pure behavioral approach is preferable. That is the case, for example, when there is not enough physical information of the modeling object for extracting an equivalent circuit representation, or it seems impossible to describe that object by any equivalent circuit with enough accuracy, or even when the entity to be modeled is so complex that such a low level representation becomes extremely inefficient. One of those examples is a traveling-wave tube amplifier (TWTA), for which it has been difficult to propose any useful equivalent circuit or physical description.

Pure behavioral models play such an important role in system simulation that they deserve to be mentioned even in a text focused on circuit analysis. So, the next few pages are intended to provide a first step overview of that system modeling strategy.

In the context of telecommunication systems, the fidelity with which the information signal is processed constitutes the ultimate goal of any nonlinear distortion analysis. So, the highest level of abstraction corresponds to behavioral models specially conceived for that information signal. That is, a system representation is sought where the input/output function no longer handles a time-domain or frequency-domain representation of the modulated RF signal, but a conceptual information signal which is known as the modulation envelope [31].

Actually, such an envelope-oriented system description is only meaningful when the information signal bandwidth is a small fraction of the RF carrier frequency (i.e., when we are dealing with systems of a distinct bandpass characteristic). In communication systems theory, these are known as bandpass nonlinearities [32].

For any other system, it seems that only the usual RF signal driven behavioral representation is applicable.

To understand the underlying ideas behind the envelope-oriented black box modeling approach, let us imagine a simple memoryless mildly nonlinear system described by the following third-degree power series:

$$y(t) = a_1 x(t) + a_2 x(t)^2 + a_3 x(t)^3 \tag{4.33}$$

which we assume is excited by an equal-amplitude two-tone RF stimulus:

$$x(t) = A \cos \omega_1 t + A \cos \omega_2 t \tag{4.34}$$

Computing the response of (4.34) to (4.33) results in various clusters of frequency components appearing near dc, ω_1 and ω_2, and their second and third-harmonics. But, since for bandpass systems the information spectrum is located around the carrier, in what is called the first, principal or fundamental zone of the output, the components of interest will be only

$$\begin{aligned} y(t) = {} & \left(a_1 A + \frac{9}{4} a_3 A^3\right) \cos \omega_1 t + \left(a_1 A + \frac{9}{4} a_3 A^3\right) \cos \omega_2 t \\ & + \frac{3}{4} a_3 A^3 \cos [(2\omega_1 - \omega_2)t] + \frac{3}{4} a_3 A^3 \cos [(2\omega_2 - \omega_1)t] \end{aligned} \tag{4.35}$$

Looking at (4.34) from a telecommunication systems point of view, it can be rewritten as

$$x(t) = 2A \cos \left(\frac{\omega_1 - \omega_2}{2} t\right) \cos \left(\frac{\omega_1 + \omega_2}{2} t\right) \tag{4.36}$$

which indicates that our stimulus can in fact be perceived as a sinusoidal RF carrier of frequency $\omega_c = (\omega_1 + \omega_2)/2$ amplitude modulated by a sinusoidal envelope of frequency $\omega_m = (\omega_1 - \omega_2)/2$:

$$x(t) = A_m \cos \omega_m t \cos \omega_c t \tag{4.37}$$

And, following the same reasoning, (4.35) can be rewritten as

$$\begin{aligned} y(t) = {} & 2\left(a_1 A + \frac{9}{4} a_3 A^3\right) \cos \left(\frac{\omega_1 - \omega_2}{2} t\right) \cos \left(\frac{\omega_1 + \omega_2}{2} t\right) \\ & + \frac{6}{4} a_3 A^3 \cos \left(3 \frac{\omega_1 - \omega_2}{2} t\right) \cos \left(\frac{\omega_1 + \omega_2}{2} t\right) \\ = {} & \left[\left(a_1 A_m + \frac{9}{16} a_3 A_m^3\right) \cos \omega_m t + \frac{3}{16} a_3 A_m^3 \cos 3\omega_m t\right] \cos \omega_c t \end{aligned} \tag{4.38}$$

A natural step in system identification would then be to conceive a new system that is directly excited by the envelope,

$$\tilde{x}(t) = A_m \cos \omega_m t \tag{4.39}$$

and whose output would be

$$\tilde{y}(t) = \left(a_1 A_m + \frac{9}{16} a_3 A_m^3\right) \cos \omega_m t + \frac{3}{16} a_3 A_m^3 \cos 3\omega_m t \qquad (4.40)$$

A polynomial behavioral model for our new envelope driven system, $\tilde{y}(t) = \tilde{S}[\tilde{x}(t)]$ would then be

$$\tilde{y}(t) = \tilde{a}_1 \tilde{x}(t) + \tilde{a}_2 \tilde{x}(t)^2 + \tilde{a}_3 \tilde{x}(t)^3 \qquad (4.41)$$

whose coefficients can be readily obtained from the ones of (4.33) recognizing that the response of $\tilde{S}[\tilde{x}(t)]$,

$$\tilde{y}(t) = \tilde{S}[A_m \cos \omega_m t] = \frac{1}{2} \tilde{a}_2 A_m^2 + \left(\tilde{a}_1 A_m + \frac{3}{4} \tilde{a}_3 A_m^3\right) \cos \omega_m t$$
$$+ \frac{1}{2} \tilde{a}_2 A_m^2 \cos 2\omega_m t + \frac{1}{4} \tilde{a}_3 A_m^3 \cos 3\omega_m t \qquad (4.42)$$

must equal (4.40).[3] So, $\tilde{a}_1 = a_1$, $\tilde{a}_2 = 0$, $\tilde{a}_3 = (3/4) a_3$.

This is the basis of envelope-driven behavioral modeling.

Naturally, our introductory example was oversimplified, and it would have to be more involved if we would like to use this black box modeling concept in practical situations.

The first thing to do in that respect is to allow for more complicated envelopes. That is, we must eliminate the restriction of having two tones of equal amplitude. Doing this corresponds to considering an asymmetric envelope spectrum, in which $\tilde{X}(\Omega) \neq \tilde{X}(-\Omega)^*$, and so supposing that $\tilde{x}(t)$ can be a complex (i.e., with real and imaginary parts) information signal. In that case, the general amplitude and phase modulated excitation signal would be

$$x(t) = A(t) \cos[\omega_c t + \theta(t)] = \mathrm{Re}\left[A(t) e^{j\theta(t)} e^{j\omega_c t}\right] \qquad (4.43)$$

whose complex envelope is [31]

$$\tilde{x}(t) = A(t) e^{j\theta(t)} \qquad (4.44)$$

The next step would be to eliminate the memoryless restriction, allowing for the description of dynamic systems. Unlike memoryless systems, whose outputs

3. It can be shown that the outcome $\tilde{a}_2 = 0$ is not accidental but results from the general rule saying that only odd order products contribute to the fundamental zone.

are instantaneous functions of the input, dynamic systems can present memory effects that can be short or long compared to the time periods of their excitations, $T_c = 2\pi/\omega_c$ or $T_m = 2\pi/\omega_m$.

Obviously, when the system impulse response tail has a much smaller duration than the carrier period [i.e., $\tau << T_c$ (where τ: $h(t) \approx 0$ for $t \geq \tau$)], the system can be approximated as being memoryless. In that case, both $y(t)$ and $\tilde{y}(t)$ will be almost instantaneous functions of their inputs $x(t)$ and $\tilde{x}(t)$, and we end up in the simplified situation already discussed.

When τ is comparable to T_c, it is likely that it will be negligible if compared to T_m as $T_m >> T_c$. So, the system will present memory effects to the carrier, but not to the envelope. This means $y(t) = S[x(t)]$ must be represented by some dynamic model, whereas $\tilde{y}(t) = \tilde{S}[\tilde{x}(t)]$ can still be considered memoryless.

When τ is comparable to T_m, then both $y(t)$ and $\tilde{y}(t)$ have to be described by dynamic models of $x(t)$ and $\tilde{x}(t)$, unless the impulse response $h(t)$ varies so little up to T_c that the system behaves as being almost memoryless to the carrier, but dynamic to the envelope.

An illustration of those different situations can be gathered from the hypothetical system of Figure 4.23.

This system is basically composed of a memoryless forward path $G[e(t)]$ to which a linear dynamic feedback path, $F(\omega)$, was applied.

Similarly to $F(\omega)$, $O(\omega)$ is also a linear filter—typically a bandpass filter intended to preserve only the fundamental zone output—while $G[e(t)]$ is a memoryless nonlinearity, whose output is given by (4.33). Assuming (4.33) is a third-degree Taylor series and the various subsystems do not interact with each other, our system can be analyzed with the harmonic input method of Volterra series giving the following first three NLTFs:

$$S_1(\omega) = \frac{O(\omega)}{D(\omega)} a_1 \tag{4.45}$$

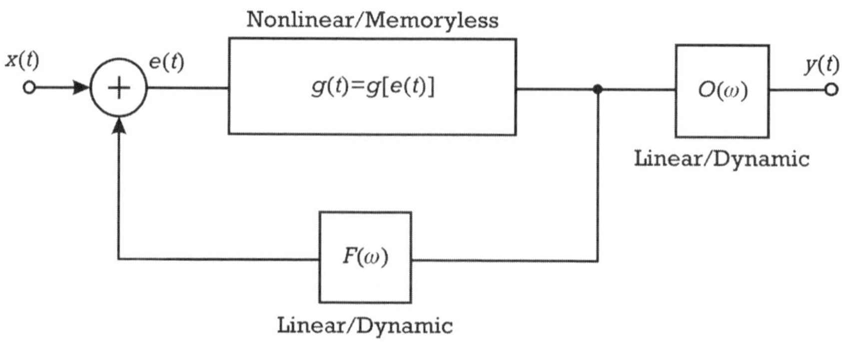

Figure 4.23 General dynamic nonlinear system.

$$S_2(\omega_1, \omega_2) = \frac{O(\omega_1 + \omega_2)}{D(\omega_1 + \omega_2)} \frac{a_2}{D(\omega_1)D(\omega_2)} \tag{4.46}$$

and

$$S_3(\omega_1, \omega_2, \omega_3) = \frac{O(\omega_1 + \omega_2 + \omega_3)}{D(\omega_1 + \omega_2 + \omega_3)} \frac{1}{D(\omega_1)D(\omega_2)D(\omega_3)}$$
$$\left\{a_3 + \frac{2}{3}a_2^2\left[\frac{F(\omega_1 + \omega_2)}{D(\omega_1 + \omega_2)} + \frac{F(\omega_1 + \omega_3)}{D(\omega_1 + \omega_3)} + \frac{F(\omega_2 + \omega_3)}{D(\omega_2 + \omega_3)}\right]\right\} \tag{4.47}$$

where $D(\omega)$ is the feedback desensitization factor: $D(\omega) = 1 - a_1F(\omega)$.

Even though our example system shows a feedback path, the actual represented system may not present any explicit feedback. It suffices that the memoryless nonlinearity interacts with a linear dynamic element. And, that happens anytime a certain mixing product of the controlled variable produces a corresponding component of the controlling variable in the dynamic element, which is then remixed to generate higher order outputs.

The first situation we will study is the one of absent feedback. That corresponds to a simple model in which a memoryless nonlinearity is followed by a zonal filter [32]. If $F(\omega)$ is made zero in (4.45) to (4.47), $D(\omega)$ is unity and those odd-order NLTF's become

$$S_1(\omega) = O(\omega)a_1 \tag{4.48}$$

and

$$S_3(\omega_1, \omega_2, \omega_3) = O(\omega_1 + \omega_2 + \omega_3)a_3 \tag{4.49}$$

Even though the dependence on frequency of these NLTFs is an indication of memory, this memory effect can, actually, be separated from the nonlinear system behavior, as it does not really interact with the memoryless nonlinearity. Moreover, the dependence of $S_n(\omega_1, \ldots, \omega_n)$ $(n = 1, 3)$ on n different frequencies is now illusive as both first and third-order transfer functions only depend on one degree of freedom, the frequency sum, which, in our case of a two-tone test, is ω_1, ω_2, $\omega_1 + \omega_1 - \omega_2 = 2\omega_1 - \omega_2$, $\omega_1 + \omega_1 - \omega_1 = \omega_1 + \omega_2 - \omega_2 = \omega_1$, $\omega_2 + \omega_1 - \omega_1 = \omega_2 + \omega_2 - \omega_2 = \omega_2$ and $\omega_2 + \omega_2 - \omega_1 = 2\omega_2 - \omega_1$. Since the system still behaves in a memoryless fashion, any odd order $S_n(\omega_1, \ldots, \omega_n)$ can be extracted from the response to the two-tone stimulus, or even to a sinusoid at ω. In this latter case, the instantaneous response amplitude should be measured against the instantaneous excitation amplitude as in a conventional AM-AM test. Furthermore, note that, because of the similarity of $S_1(\omega)$ to $S_3(\omega, \omega, -\omega)$, the first-order and third-order responses will always be in-phase, (when $a_1 a_3 > 0$) or always in opposite-phase (when $a_1 a_3 < 0$) and the system cannot show any AM-PM conversion.

In this case, the stimulus of (4.43) will produce a response of

$$y(t) = \tilde{g}[A(t)] \cos[\omega_c t + \theta(t)] \tag{4.50}$$

whose lowpass equivalent [31], or complex envelope, will be

$$\tilde{y}(t) = \tilde{g}[A(t)] e^{j\theta(t)} \tag{4.51}$$

which can be modeled by the lowpass equivalent behavioral model of Figure 4.24, using a measured or simulated AM-AM, $\tilde{g}[A(t)]$, characteristic.

The situation changes dramatically if a non zero $F(\omega)$ is included.

If now $F(\omega)$ is a bandpass filter to the fundamental zone, for which Imag $[F(\omega)] \neq 0$ and $F(\omega_i + \omega_j) \approx 0$ ($i, j = 1, 2, 3$), the system's odd-order NLTFs will be

$$S_1(\omega) = \frac{O(\omega)}{D(\omega)} a_1 \tag{4.52}$$

and

$$S_3(\omega_1, \omega_2, \omega_3) = \frac{O(\omega_1 + \omega_2 + \omega_3)}{D(\omega_1 + \omega_2 + \omega_3)} \frac{a_3}{D(\omega_1) D(\omega_2) D(\omega_3)} \tag{4.53}$$

Now, memory already interacts with the nonlinearity, although the NLTFs still do not depend on the separation (or envelope) frequency. This guarantees that our system does not present any memory effects to the envelope, and thus can be

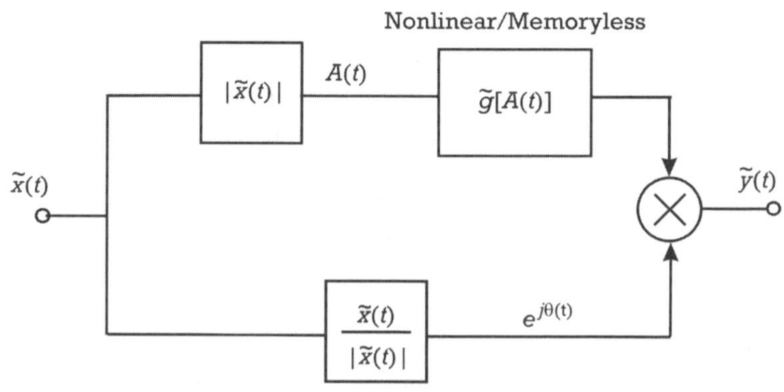

Figure 4.24 Lowpass equivalent behavioral model of a memoryless bandpass nonlinearity with amplitude distortion, AM-AM, only.

extracted from a one-tone test. However, $S_1(\omega)$ and $S_3(\omega, \omega, -\omega)$ will no longer be in-phase and the system will show AM-PM conversion beyond its previously noticed AM-AM characteristic.

Now, the stimulus of (4.43) will produce a response of

$$y(t) = \tilde{g}[A(t)] \cos\{\omega_c t + \theta(t) + \tilde{\phi}[A(t)]\} \tag{4.54}$$

whose lowpass equivalent [31], or complex envelope, will be

$$\tilde{y}(t) = \tilde{g}[A(t)] e^{j\{\theta(t)+\tilde{\phi}[A(t)]\}} \tag{4.55}$$

and which can be modeled by the lowpass equivalent quadrature behavioral model of Figure 4.25, using measured or simulated AM-AM, $\tilde{g}[A(t)]$, and AM-PM characteristics, $\tilde{\phi}[A(t)]$.

This model indicates that the lowpass equivalent behavioral system output is still a memoryless function of the complex envelope, despite of its original dynamic bandpass behavior. This is the reason why such systems are usually known as memoryless bandpass nonlinearities [32].[4]

If, finally, $F(\omega)$ varies with the envelope frequency, then the full odd-order NLTFs of (4.45) and (4.47) must be accounted for. $S_3(\omega_1, \omega_2, \omega_3)$ depends on the frequency sum, on every input frequency, but also on all second-order mixing

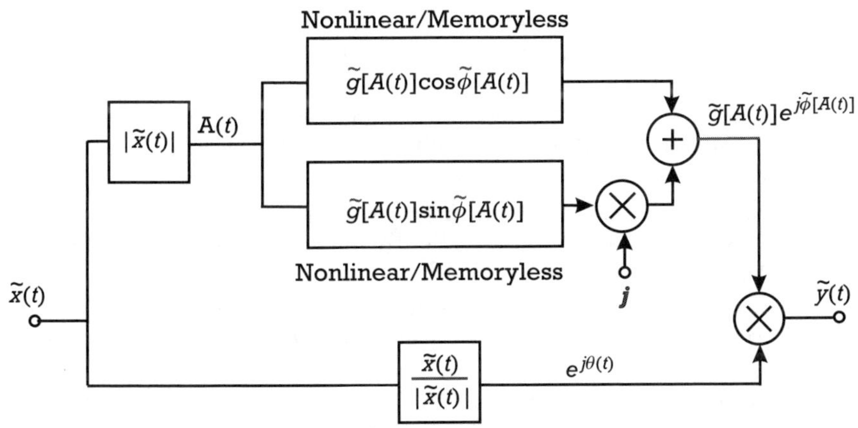

Figure 4.25 Lowpass equivalent behavioral model of a memoryless bandpass nonlinearity with amplitude, AM-AM, and phase, AM-PM, distortion.

4. At this time the reader may think that naming a bandpass system as "memoryless" is a countersense. But, it even sounds more absurd admitting that a memoryless nonlinearity can present AM/PM conversion [33]. Actually, both of these statements result from the zero memory characteristic of the lowpass equivalents of some bandpass nonlinear systems.

products $\omega_j + \omega_k$ (j, k = 1, 2, 3). Any complete system identification based on one-tone measurements (AM-AM and AM-PM characteristics) would fail as the output envelope no longer varies instantaneously with the envelope of the input. But, the extraction would also be impossible with a two-tone excitation as we then would be only capturing those memory effects on one sinusoid, which means at a single envelope frequency.

Behavioral modeling of those dynamic nonlinear systems, either in its original bandpass, or lowpass equivalent, forms is a very challenging subject. And, even if the Volterra series could be accepted as a behavioral modeling tool in their mildly nonlinear operation regime, a full behavioral model has been an object of strong research [34–38].

References

[1] Selberherr, S., *Analysis and Simulation of Semiconductor Devices,* New York, Wien: Springer-Verlag, 1984.

[2] Curtice, W. R., "A MESFET Model for Use in the Design of GaAs Integrated Circuits," *IEEE Transactions on Microwave Theory and Techniques,* Vol. 28, No. 7, 1980, pp. 448–456.

[3] Curtice, W. R., and M. Ettenberg, "A Nonlinear GaAs FET Model for Use in the Design of Output Circuits for Power Amplifiers," *IEEE Transactions on Microwave Theory and Techniques,* Vol. 33, No. 12, 1985, pp. 1383–1394.

[4] Dambrine, G., et al., " A New Method for Determining the FET Small-Signal Equivalent Circuit," *IEEE Transactions on Microwave Theory and Techniques,* Vol. 36, No. 7, 1988, pp. 1151–1159.

[5] Camacho-Peñalosa, C., and C. Aitchinson, "Modelling Frequency Dependencies of Output Impedance of a Microwave MESFET at Low Frequencies," *Electronic Letters,* Vol. 21, No. 6, 1985, pp. 528–529.

[6] Fernandez, T., et al., "Extracting a Bias-dependent Large Signal MESFET Model from Pulsed I/V Measurements," *IEEE Transactions on Microwave Theory and Techniques,* Vol. 44, No. 3, 1996, pp. 372–378.

[7] Qu, G., and A. Parker, "New Model Extraction for Predicting Distortion in HEMT and MESFET Circuits," *IEEE Microwave & Guided Wave Letters,* Vol. 9, No. 9, 1999, pp. 363–365.

[8] Pedro, J., J. C. Madaleno, and J. A. Garcia, "Theoretical Basis for the Extraction of Mildly Nonlinear Behavioral Models," *International Journal of RF and Microwave Computer-Aided Engineering,* Vol. 13, No. 1, 2003, pp. 40–53.

[9] Garcia, J., et al., "Characterizing the Gate to Source Nonlinear Capacitor Role on GaAs FET IMD Performance," *IEEE Transactions on Microwave Theory and Techniques,* Vol. 46, No. 12, 1998, pp. 2344–2355.

[10] Maas, S. A., and A. Crosmun, "Modeling the Gate I/V Characteristic of a GaAs MESFET for Volterra-Series Analysis," *IEEE Transactions on Microwave Theory and Techniques,* Vol. 37, No. 7, 1989, pp. 1134–1136.

[11] Pedro, J., and J. Perez, "Accurate Simulation of GaAs MESFET's Intermodulation Distortion Using a New Drain-Source Current Model," *IEEE Transactions on Microwave Theory and Techniques,* Vol. 42, No. 1, 1994, pp. 25–33.

[12] Golio, M., *Microwave MESFETs & HEMTs,* Norwood, MA: Artech House, 1991.

[13] Maas, S., *Nonlinear Microwave Circuits,* Norwood, MA: Artech House, 1988.

[14] Cheng, Y., and C. Hu, *Mosfet Modeling & BSIM3 Users's Guide,* Boston: Kluwer Academic Publishers, 1999.

[15] Antognetti, P., and G. Massobrio, *Semiconductor Device Modeling with SPICE*, London: McGraw-Hill International Editions, 1988.

[16] Liu, P., "Passive Intermodulation Interference in Communication Systems," *Electronics & Communication Engineering Journal,* Vol. 2, No. 3, 1990, pp. 109–118.

[17] Pedro, J. C., "Physics Based MESFET Empirical Model," *Proc. 1994 IEEE MTT-S International Microwave Symposium Digest,* San Diego, 1994, pp. 973–976.

[18] Shockley, W., "A Unipolar Field Effect Transistor," *Proceedings IRE*, Vol. 40, Nov. 1952, pp. 1365–1376.

[19] Angelov, I., H. Zirath, and N. Rorsman, "A New Empirical Nonlinear Model for HEMT and MESFET Devices," *IEEE Transactions on Microwave Theory and Techniques,* Vol. 40, No. 12, 1992, pp. 2258–2266.

[20] Wambacq, P., and W. Sansen, W., *Distortion Analysis of Analog Integrated Circuits,* Boston: Kluwer Academic Publishers, 1998.

[21] Tsividis, Y., *Operation and Modeling of the MOS Transistor,* Second Edition, Boston: WCB/McGraw-Hill, 1999.

[22] Lovelace, D., J. Costa, and N. Camilleri, "Extrating Small Signal Model Parameters of Silicon MOSFET Transistors," *Proc. 1994 IEEE MTT-S International Microwave Symposium Digest,* San Diego, June 1994, pp. 865–868.

[23] Perugupalli, P., et al., "Modeling and Characterization of an 80 V Silicon LDMOSFET for Emerging RFIC Applications," *IEEE Transactions on Electron Devices,* Vol. 45, No. 7, 1998, pp. 1468–1478.

[24] Curtice, W. R., et al., "A New Dynamic Electro-Thermal Nonlinear Model for Silicon RF LDMOS FETs," *Proc. 1999 IEEE MTT-S International Microwave Symposium Digest,* Anaheim, CA, 1999, pp. 419–422.

[25] Fager, C., et al., "Prediction of IMD in LDMOS Transistor Amplifiers Using a New Large-Signal Model," *IEEE Transactions on Microwave Theory and Techniques,* Vol. 50, No. 12, 2002, pp. 2834–2842.

[26] Maas, S. A., B. L. Nelson, and D. L. Tait, "Intermodulation in Heterojunction Bipolar Transistors," *IEEE Transactions on Microwave Theory and Techniques,* Vol. 40, No. 3, 1992, pp. 442–448.

[27] Graaff, H. C., and F. M. Klaassen, *Compact Transistor Modeling for Circuit Design,* New York: Springer-Verlag, 1990.

[28] McAndrew, C. C., et al. "VBIC95, The Vertical Bipolar Inter-Company Model," *IEEE Journal of Solid-State Circuits,* Vol. 31, No. 10, 1996, pp. 1476–1483.

[29] Gummel, H. K., and H. C. Poon, "An Integral Charge Control Model of Bipolar Transistors," *Bell System Technical Journal,* Vol. 49, No. 5, 1970, p. 827.

[30] Getreu, I., *Modeling the Bipolar Transistor,* New York: Elsevier, 1976.

[31] Jeruchim, M., P. Balaban, and K. Shanmugan, *Simulation of Communication Systems,* Second Edition, Boston: Kluwer Academic Publishers, 2000.

[32] Blachman, N. M., "Bandpass Nonlinearities," *IEEE Transactions on Information Theory,* Vol. 10, No. 1, 1964, pp. 162–164.

[33] Minkoff, J., "The Role of AM-to-PM Conversion in Memoryless Nonlinear Systems," *IEEE Transactions on Communications,* Vol. 33, No. 2, 1985, pp. 139–144.

[34] Schetzen, M., "Nonlinear System Modeling Based on the Wiener Theory," *Proceedings IEEE,* Vol. 69, No. 12, 1981, pp. 1557–1573.

[35] Ku, H., M. Mckinley, and J. S. Kenney, "Extraction of Accurate Behavioral Models for Power Amplifiers with Memory Effects Using Two-Tone Measurements," *Proc. 2002 IEEE MTT-S International Microwave Symposium Digest,* Seattle, WA, 2002, pp. 139–142.

[36] Soury, A., E. Ngoya, and J. M. Nebus, "A New Behavioral Model Taking into Account Nonlinear Memory Effects and Transient Behaviors in Wideband SSPAs," *Proc. 2002 IEEE MTT-S International Microwave Symposium Digest,* Seattle, WA, 2002, pp. 853–856.

[37] Clark, C. J., et al., "Time-Domain Envelope Measurement Technique with Application to Wideband Power Amplifier Modeling," *IEEE Transactions on Microwave Theory and Techniques,* Vol. 46, No. 12, 1998, pp. 2531–2540.

[38] Silva, C. P., et al., "Application of Polyspectral Techniques to Nonlinear Modeling and Compensation," *Proc. 2001 IEEE MTT-S International Microwave Symposium Digest,* Phoenix, AZ, 2001, pp. 13–16.

CHAPTER 5

Highly Linear Circuit Design

5.1 Introduction

The present chapter deals with the distortion performance of typical microwave and wireless circuits as small-signal amplifiers, power amplifiers, and mixers.

It begins with a general glance on the intermodulation problem, as seen from the system point of view, to define the important concept of dynamic range. Since this figure of merit is determined by distortion but also by system noise, an outline of this noteworthy subject is first given. This is followed by a review of the traditional linear amplifier design methodology, as an introduction to the analysis of nonlinear distortion in small-signal amplifiers.

Field effect and bipolar transistor-based weakly nonlinear amplifiers are addressed. For that, we present detailed Volterra series analyses of two typical circuits, which enable the extraction of broad results on their distortion behavior, and thus some general design rules for highly linear small-signal amplifiers.

Large-signal or power amplifiers are treated next. Again, we introduce this subject with a review of the traditional power amplifier concepts and design methodology. Because of the inherent complexity of distortion generation mechanisms in those circuits, we first present a very simple memoryless model of the amplifier. This provides a fruitful insight into the power amplifier inband distortion performance, particularly its dependence on the bias point load impedance and driving level. These results are extended to more involved circuit models, which can serve as linear power amplifier design rules, or provide the qualitative knowledge required for a first design approach amenable to be complemented by some automated design process.

Next comes an analysis of distortion generation mechanisms in frequency conversion circuits. Because of the enormous amount of possible mixer topologies, we concentrated on two common circuits: the FET gate active mixer and the singly balanced diode mixer. Although necessarily simplified, the time-varying Volterra analysis that was used led to interesting qualitative conclusions on the linearity optimization of those circuits. Again, we believe this to be valuable for a starting mixer design, which should be then complemented by some sort of automated design aid.

Finally, the chapter closes by analyzing the distortion performance of balanced arrangements of multiple amplifier and mixer devices.

Except for those balanced configurations, no other form of external linearization is addressed. That should, by no means, be interpreted as an indication that linearizers have no significant role to play in microwave and wireless system design. On the contrary! But, since they are usually treated at the system level—whereas the following sections are focused on circuits—and they have already been thoroughly covered in the literature, we believe the interested reader will face no difficulty in finding a wide range of publications (even whole books) dedicated to that broad subject [1–3].

5.2 High Dynamic Range Amplifier Design

5.2.1 Concepts and Systemic Considerations

High dynamic range amplifiers are those that are expected to behave, simultaneously, in a very linear way—in order to preserve the integrity of the highest-level signals—and to add small amounts of noise—to prevent degradation of the signal-to-noise ratio of weak signals. Contrary to other closely related circuits located at the system's output, the power amplifiers or their drivers, these preamplifiers handle very low power levels, and thus are not designed for maximized efficiency or output power capability. Instead, they should generate as little noise as possible while providing fairly large amounts of gain. This helps in preserving the signal from noise originated in subsequent stages.

Because of the very low signal amplitudes involved, preamplifiers are usually taken as nonlinear distortion-free devices. This erroneous assumption comes from the fact that everyone thinks on the desired signal level but generally forgets possible high power inband interferers. In telecommunication systems, these can arise from other services sharing the same physical transmission channel, unintentional (or intentional) perturbations (like man-made impulsive noise or jamming in electronics warfare environments), or even from the same transceiver when full-duplex transmission is used. From those, the last is of particular interest, as is clear from the numbers involved.

Consider, for example, a full-duplex transceiver whose previewed receiver input power is –80 dBm at 1,800 MHz, and required radiated power is 1W (30 dBm) at 1,830 MHz. If a single antenna were to be used, and its associated diplexer filters could provide a marvelous figure of 70-dB isolation, the receiver input stage would be driven by the natural excitation of –80 dBm along with an unexpected interferer 10,000 times (40 dB) stronger.

5.2.1.1 System Sensitivity and Dynamic Range

The situation referred to above is illustrated in Figure 5.1, which also motivates the following definition of *spurious free dynamic range,* or simply dynamic range (DR): The dynamic range of a system is the ratio between the maximum and the minimum detectable signals for a prescribed performance quality.

In this context, a system's quality is measured in terms of the *signal-to-noise ratio* (SNR). Actually, here we are admitting a broad significance for noise, as it includes all forms of perturbation, either additive noise or "distortion noise." Therefore, this SNR is, in fact, an abbreviated term for what is usually called *signal-to-noise-and-distortion ratio* (SINAD).

The *minimum detectable signal,* $P_{i_{min}}$, *or input sensitivity,* S_i, is thus equal to the total input referred system's noise power, N_i (integrated within the system's ultimate equivalent noise bandwidth, *Bw,* [4]), plus the SNR (in logarithmic units):

$$S_{i_{dBm}} = N_{i_{dBm/Hz}} + B_{w_{dBHz}} + SNR_{dB} \tag{5.1}$$

The *maximum detectable signal,* $P_{i_{Max}}$ is taken as the interferer power that generates an amount of distortion exactly SNR_{dB} below the minimum signal, S_i.[1] In a small-signal system, in which nth-order distortion is dominant, and thus can be described by the output nth-order intercept point, IP_n, this $P_{i_{Max}}$ is such that

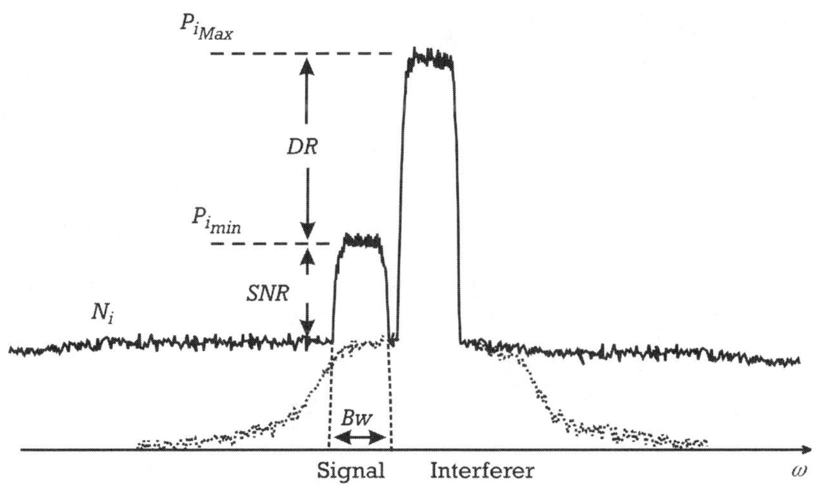

Figure 5.1 Graphical illustration of dynamic range definition.

1. Although this is the general way maximum detectable signal is defined, we would like to point out that the combined effects of additive noise and nonlinear distortion lead to an overall SNR value 3 dB below the specifications!

$$P_{IMD,n_i} = P_{i_{Max\,dBm}} - (n-1)\left(IP_{n_{dBm}} - G_{dB} - P_{i_{Max\,dBm}}\right) = S_{i_{dBm}} - SNR_{dB} \tag{5.2a}$$

or, referring IP_n to the input, IP_{n_i}

$$P_{i_{Max\,dBm}} = \frac{n-1}{n} IP_{n_{i_{dBm}}} + \frac{1}{n}\left(S_{i_{dBm}} - SNR_{dB}\right) \tag{5.2b}$$

which leads to a DR of

$$DR_{dB} = P_{i_{Max\,dBm}} - S_{i_{dBm}} = \frac{n-1}{n}\left(IP_{n_{i_{dBm}}} - S_{i_{dBm}} - \frac{1}{n-1} SNR_{dB}\right) \tag{5.3}$$

As is easily seen from the above, dynamic range maximization can be obtained either by increasing input IP_n, decreasing system's sensitivity, or both. The former requires highly linear designs, while the latter is demanding for reduced noise floor. As this noise floor results from the combination of source noise (external to the system under consideration) and system's added noise, S_i improvement must be achieved by low-noise designs. So, we will begin by introducing low-noise design concepts.

5.2.1.2 System Noise Figure

Let us begin by recalling some basic concepts of electronic noise modeling.

The most important of those concepts is *noise figure* (NF). NF is defined in a spot frequency as the ratio of (1) total available output noise power spectral density function (PSD) at that frequency, when the input source is at the reference temperature ($T_0 = 290$K), to (2) available output noise PSD (in the same conditions as above) only due to the excitation source.

If the available source PSD seen at the output is N_{So}, while the one added by the system is N_{ao}, then, according to the definition, NF is given by

$$NF = \frac{N_{So} + N_{ao}}{N_{So}} \tag{5.4}$$

To get a more useful significance of NF, let us assume the system has a power gain (at the spot frequency) of G,[2] such that the input signal available power, S_S, will appear at the output as $S_o = GS_S$, and its associated noise, N_S, as $N_{So} = GN_S$.

2. Because *NF* is defined in terms of available output *PSD*s this gain, G, herein loosely called "power gain" is, in fact, the available power gain of the network, G_A, as will be defined in Section 5.2.2.

The input available SNR will be $SNR_S = S_S/N_S$ and the output available SNR, $SNR_o = S_o/N_o$. Therefore (5.4) can be rewritten as

$$NF = \frac{N_o}{N_{So}} = \frac{N_o}{GN_S}\frac{GS_S}{S_o} = \frac{S_S/N_S}{S_o/N_o} = \frac{SNR_S}{SNR_o} \tag{5.5}$$

which shows that NF is really a measure of the available SNR degradation imposed by the network. Since no physical network can reduce the amount of noise power spectral density already present at the input, NF is always greater or equal than one ($NF_{dB} \geq 0$), being the equality reserved for the ideal noiseless network. Therefore, the introduction of any physical device always degrades SNR.[3]

The noise figure of a cascade of M noisy blocks, like the one depicted in Figure 5.2, can be directly derived from the NF definition. Assuming the noise level is so low that every block m behaves linearly with a gain G_m and a noise figure of NF_m, the overall output available noise PSD is then

$$N_o = G_1 \ldots G_M N_S + G_2 \ldots G_M N_{a_1} + \ldots + G_{m+1} \ldots G_M N_{a_m} + \ldots + N_{a_M} \tag{5.6}$$

while the output available noise PSD due to only the input is

$$N_{So} = G_1 \ldots G_M N_S \tag{5.7}$$

From the NF definition, the system NF is given by

$$NF = \frac{N_o}{N_{So}} = \frac{G_1 N_S + N_{a_1}}{G_1 N_S} + \ldots + \frac{N_{a_m}}{G_1 \ldots G_m N_S} + \ldots + \frac{N_{a_M}}{G_1 \ldots G_M N_S} \tag{5.8}$$

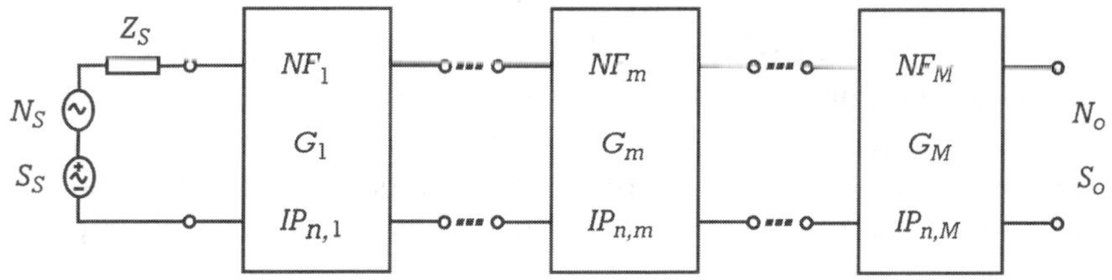

Figure 5.2 Noisy block chain for overall system dynamic range calculations.

3. This is actually a consequence of the adopted NF definition, which refers to power spectral densities and not to integrated power. Therefore, a filter block designed to restrict noise bandwidth just to the signal's bandwidth really improves the system's quality or SNR.

which, using (5.4), can be put in the form

$$NF = NF_1 + \ldots + \frac{NF_m - 1}{G_1 \ldots G_{m-1}} + \ldots + \frac{NF_M - 1}{G_1 \ldots G_{M-1}} \tag{5.9}$$

Equation (5.9) is nothing more than the rigorous statement of the intuitive idea that the noise performance of a block chain is greatly determined by the noise figure of the first stage, provided its gain is sufficiently high to downgrade the impact of noise added by subsequent blocks. Or, in other words, it tells us that a good sensitivity chain requires a high-gain and low-noise preamplifier.

5.2.1.3 System Linearity and Automatic Gain Control

Now, turning our attention to the upper end of signal level, high dynamic ranges require highly linear devices. Since we are dealing with small-signal operation regimes, linearity is a synonym of large IP_n. So, we will begin the discussion of linearity issues by deriving an expression for the overall system IP_n, similar to (5.9).

Starting from the definition of nth-order output intercept point of a general block m of gain G_m, excited by an input signal power of $P_{i,m}$, it follows that its output nth-order distortion power can be given by

$$P_{IMDn,m} = G_m P_{i,m} \left(\frac{IP_{n,m}}{G_m P_{i,m}} \right)^{-(n-1)} = G_m^n IP_{n,m}^{(1-n)} P_{i,m}^n \tag{5.10}$$

So, the nth-order output P_{IMD} of first block is simply

$$P_{IMDn,1} = G_1^n IP_{n,1}^{(1-n)} P_i^n \tag{5.11}$$

in which P_i is the chain's input power.

To determine $P_{IMDn,2}$ we now assume that the various blocks do not interact, and that the mixing of lower order distortion products generating nth-order components is negligible. Actually, if this condition cannot be guaranteed, and the number of blocks is greater than two or three, the calculation of P_{IMDn} would be nearly impossible by hand. (Just look at the results obtained in Section 3.2.4.2, for the cascade of two blocks.) Under this simplifying assumption, $P_{IMDn,2}$ is the combination of two terms:

$$P_{IMDn,2} = G_1^n G_2 IP_{n,1}^{(1-n)} P_i^n + G_1^n G_2^n IP_{n,2}^{(1-n)} P_i^n \tag{5.12}$$

The first term is the amplified $P_{IMDn,1}$, and the other is the P_{IMD} generated in block 2. Again, since these two parts are correlated in phase, a rigorous analysis would demand for a vector addition. The resulting amplitude of that vector addition would range from an extremely rare case of total cancellation, and so an nth-order distortion-free system (only partially verified in the so-called pre- or postdistortion linearizer arrangements), to a worst-case of in-phase combination, in which (5.12) would read as

$$P_{IMDn,2} = \left[G_1^{n/2} G_2^{1/2} IP_{n,1}^{(1-n)/2} + G_1^{n/2} G_2^{n/2} IP_{n,2}^{(1-n)/2}\right]^2 P_i^n \tag{5.13}$$

The power addition adopted for (5.12) is thus an "averaged" indicative value that can be taken as a conservative lower end of system distortion, while the voltagewise addition of (5.13) should be considered as its higher end [5].

The generalization of (5.12) to block m gives

$$\begin{aligned} P_{IMDn,m} = {} & \left(G_1^n G_2 \ldots G_m IP_{n,1}^{(1-n)} + G_1^n G_2^n G_3 \ldots G_m IP_{n,2}^{(1-n)} + \ldots \right. \\ & \left. + G_1^n \ldots G_m^n IP_{n,m}^{(1-n)}\right) P_i^n \end{aligned} \tag{5.14}$$

Now, noticing that, from (5.10), the overall chain P_{IMDn} must be

$$P_{IMDn} = G_1^n \ldots G_M^n IP_n^{(1-n)} P_i^n \tag{5.15}$$

it follows that the chain IP_n under the powerwise addition is given by

$$\begin{aligned} IP_n^{(1-n)} = {} & G_2^{(1-n)} \ldots G_M^{(1-n)} IP_{n,1}^{(1-n)} + \ldots \\ & + G_{m+1}^{(1-n)} \ldots G_M^{(1-n)} IP_{n,m}^{(1-n)} + \ldots + IP_{n,M}^{(1-n)} \end{aligned} \tag{5.16a}$$

or

$$\begin{aligned} IP_n = {} & \left[(G_2 \ldots G_M IP_{n,1})^{(1-n)} + \ldots \right. \\ & \left. + (G_{m+1} \ldots G_M IP_{n,m})^{(1-n)} + \ldots + IP_{n,M}^{(1-n)}\right]^{1/(1-n)} \end{aligned} \tag{5.16b}$$

while its worst case voltagewise counterpart would be

$$\begin{aligned} IP_n^{(1-n)/2} = {} & (G_2 \ldots G_M IP_{n,1})^{(1-n)/2} + \ldots \\ & + (G_{m+1} \ldots G_M IP_{n,m})^{(1-n)/2} + \ldots + IP_{n,M}^{(1-n)/2} \end{aligned} \tag{5.17a}$$

or

$$IP_n = \left[(G_2 \ldots G_M IP_{n,1})^{(1-n)/2} + \ldots \right. \tag{5.17b}$$
$$\left. + (G_{m+1} \ldots G_M IP_{n,m})^{(1-n)/2} + \ldots + IP_{n,M}^{(1-n)/2} \right]^{2/(1-n)}$$

If nth-order input intercept point, IP_{n_i}, is desired as a function of the input intercept points of each block, $IP_{n_i,m}$, we simply need to divide (5.16) and (5.17) by the overall cascaded gain, $G_1 \ldots G_m \ldots G_M$, to obtain

$$IP_{n_i} = \tag{5.18}$$
$$\left[IP_{n_i,1}^{(1-n)} + \ldots + \left(\frac{IP_{n_i,m}}{G_1 \ldots G_{m-1}} \right)^{(1-n)} + \ldots + \left(\frac{IP_{n_i,M}}{G_1 \ldots G_{M-1}} \right)^{(1-n)} \right]^{1/(1-n)}$$

and

$$IP_{n_i} = \tag{5.19}$$
$$\left[IP_{n_i,1}^{(1-n)/2} + \ldots + \left(\frac{IP_{n_i,m}}{G_1 \ldots G_{m-1}} \right)^{(1-n)/2} + \ldots + \left(\frac{IP_{n_i,M}}{G_1 \ldots G_{M-1}} \right)^{(1-n)/2} \right]^{2/(1-n)}$$

Expressions (5.18) and (5.19) state that the weight with which general block m contributes to the overall *IMD* is proportional to the net gain from the system's input up to that block. Therefore, and in the opposite direction to what we have learned from additive noise considerations, it is the stages located near the system's output that determine the linearity performance of an amplifier chain. Once again, this should be of no surprise, since the signal level increases as we proceed from the system's input to its output.

Another interesting conclusion that may be drawn from (5.16) through (5.19) is that, unless a fortunate and unpredicted phase opposition exists between distortion components generated in different blocks (the pre- or postdistortion linearization scheme above referred), the addition of another block to any system always degrades distortion. Only if the new block is ideally linear (infinite IP_n) is the overall distortion maintained. This is similar to what we have previously seen from the cascade noise figure expression, which tells us that the distribution of gain blocks within an amplifier chain should be carefully conducted just to shift the various signal levels into the block's dynamic ranges.

Alternatively, a gain variation, dynamically controlled by the signal's level, could be tried. This is the essence of *automatic gain control* (AGC). Although put in these terms, AGC seems to be a magical solution, it cannot be applied if the chain is expected to simultaneously handle signals of significantly different amplitudes. Also, even in single-channel systems, AGC can only be used to control chains where it is guaranteed that the interesting signal is already isolated from possible interferers. Typical AGC loops in super heterodyne receivers actually derive its control from the signal demodulator, and vary the gain of the intermediate frequency or RF bandpass stages. If it were not done this way, the AGC feedback loop could be controlled and/or actuate over a strong interferer. In any case, this would result in a dramatic system desensitization.

A final conclusion that can be gathered from the expressions of IP_n and IP_{n_i} is that input intercept point does not depend on the gain of last stage, while, if it were referred to the output, it would not depend on the gain of first stage. Therefore, any time linearity performance is to be evaluated in a mildly nonlinear block chain, it seems we should use IP_n whenever changes are made at the input, and IP_{n_i} when they are performed at the output, so that we can compensate for the intercept point variations induced by mere gain changes. This is the case, for example, of amplifier distortion performance optimization tasks, in which IP_n should be the preferred choice, whenever input source match impact is being studied, and IP_{n_i}, whenever load match is under test.

5.2.2 Small-Signal Amplifier Design—General Remarks

Although some low-frequency circuits can be analyzed by hand—using simplified equivalent circuit models for the active devices—this is in general not practical in RF and microwave electronics because of the additional complexity imposed by the parasitics. Therefore, nowadays engineers rely on either CAD techniques or two-port behavioral representations. Contrary to the former, a network analysis tool permits analytic treatment of the circuit, leading to closed-form solutions. It is, thus, appropriated for (at least a first step) design, while CAD is usually reserved for analysis and performance optimization.

There are various two-port network representations available for active devices. Nevertheless, only two have gained wide acceptance in amplifier design: the admittance parameter matrix, [Y], mostly used for RF frequencies, and the scattering parameter matrix [S], of unquestionable utilization at microwave and millimeter-wave bands. Despite their obvious differences, in terms of considered inputs (port voltages versus normalized incident power waves) and outputs (port currents versus normalized reflected power waves) they are exactly equivalent, since they do nothing more than to express the two-ports linearity: each of the outputs is given as a linear combination of the inputs. In fact, they are simply expressed in different, but equivalent, domains, the one of admittances, Y, and of reflection coefficients, Γ. If the [S] matrix is defined using a purely real reference impedance (admittance), Z_0 (Y_0), (by far the most frequently encountered case), then the two domains can be easily converted from one into the other using the transformations [6]

$$\Gamma = \frac{Y_0 - Y}{Y_0 + Y} \tag{5.20a}$$

and

$$Y = Y_0 \frac{1 - \Gamma}{1 + \Gamma} \tag{5.20b}$$

These are known as bilinear transformations, which have the important feature of transforming straight lines or circles into circles. The right-half admittance plane (positive conductances, or passive admittances) is transformed into a circle of unity radius in the reflection coefficient plane (the Smith chart), and every construction made in one domain can be easily mapped into the other. Also, there are simple formulas to convert an [S] matrix into a [Y] matrix and vice versa [6]:

$$\mathbf{S} = (\mathbf{I} - \mathbf{y})(\mathbf{I} + \mathbf{y})^{-1} \tag{5.21a}$$

and

$$\mathbf{Y} = \mathbf{Y}_0(\mathbf{I} - \mathbf{S})(\mathbf{I} + \mathbf{S})^{-1} \tag{5.21b}$$

where $\mathbf{I}$ is the identity matrix, $\mathbf{y}$ is the normalized (to Y_0) admittance matrix, and $\mathbf{Y}_0$ is a diagonal matrix of all nonzero elements equal to Y_0. Knowing (5.20) and (5.21) it becomes theoretically irrelevant which type of parameter set to use, and so that choice is almost always conditioned by the information format present in the active devices data sheet, or the type of CAD package used. Since nowadays almost all RF CAD packages use S-parameter formulation, and a lot of RF devices are specified in terms of frequency and/or bias-dependent [S] matrices, we will adopt this type of formulation in the major part of this chapter.

5.2.2.1 Amplifier Transducer Power Gain

Let us consider the amplifier block diagram of Figure 5.3.

It is assumed that the active device (or devices) has been biased in a predetermined quiescent point, where the frequency-dependent [S] matrix was measured.

Various definitions of gain can be adopted to describe the signal energy relations between the amplifier's input and output, but there is one of special physical significance called *transducer power gain,* G_T. It is defined as the ratio between power actually delivered to the load, P_L, and the maximum power that could be obtained from the source (source available power), P_{avS}. Since P_L depends on the mismatch between the load and the network output (relative to the condition of

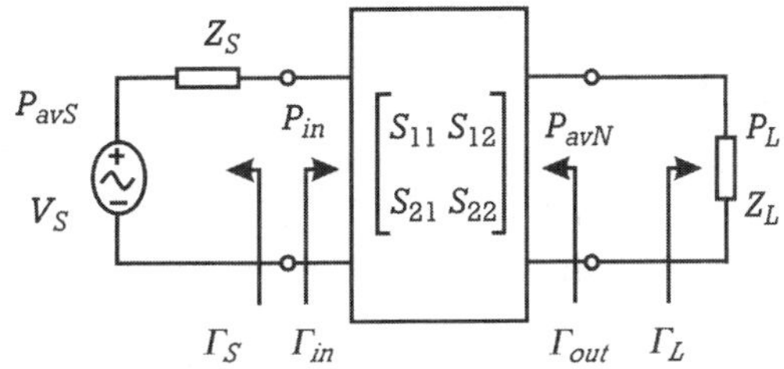

Figure 5.3 Basic amplifier block diagram.

maximum power transfer, or conjugate matching), it will not be equal, in general, to the available power at that port, P_{avN}. Also, and for similar reasons, the power delivered to the network, P_{in}, is not necessarily equal to P_{avS}. Therefore, G_T accounts for the two mismatches, being dependent on Γ_S and Γ_L, plus the network S-parameters. It can be expressed by

$$G_T \equiv \frac{P_L}{P_{avS}} = \frac{1 - |\Gamma_S|^2}{|1 - \Gamma_{in}\Gamma_S|^2} |S_{21}|^2 \frac{1 - |\Gamma_L|^2}{|1 - S_{22}\Gamma_L|^2} \tag{5.22a}$$

or even by

$$G_T \equiv \frac{P_L}{P_{avS}} = \frac{1 - |\Gamma_S|^2}{|1 - S_{11}\Gamma_S|^2} |S_{21}|^2 \frac{1 - |\Gamma_L|^2}{|1 - \Gamma_{out}\Gamma_L|^2} \tag{5.22b}$$

Because both Γ_S and Γ_L are complex quantities, a design for gain leads to an undetermined mathematical problem, where only one equality of real numbers, (5.22), must be used to determine four unknowns. Therefore, an amplifier design procedure for gain leaves room for other constraints like linearity, noise figure, or stability.

5.2.2.2 Stability Considerations

Stability issues definitely assume a central role in amplifier design as even any bad, but stable, amplifier is preferable to an unwanted oscillator. Indeed, a steady-state oscillating amplifier is a circuit capable of developing a new signal of exponentially rising amplitude, which is then limited to a constant value when it "touches" the active devices' nonlinearities. At this point, the forward gain has been reduced to a value just enough to compensate the oscillator's feedback path loss. So, an oscillator can be viewed as one circuit, which, by its nature, creates its own interfering signal (the oscillation) of so strong amplitude that it automatically produces amplifier desensitization. Therefore, every oscillation is dangerous and should be eliminated, even if it appears at a frequency so much different from the desired signal that it is not supposed to produce any linear perturbation. And this is true even for the ones at very low or high frequencies, which are observed to have very small amplitudes. Actually, their impact is essentially the same as the ones having stronger levels located within the pass band. They only look negligible, at the amplifier's terminals, because they were attenuated by the filtering characteristics of the amplifier's input and output reactive networks. They still have strong levels inside the nonlinear active device, exactly where they cause harm.

Because transistor feedback (either due to C_{bc} or L_e in bipolars, or to C_{gd} and L_s in FETs) and path delays (in lumped or distributed elements) increase with frequency, parasitic oscillations are much more common in RF and microwave bands than in low-frequency designs. Therefore, against the usual procedure followed in low-frequency amplifiers, where stability conditions are tested only after an almost completed design, at high frequencies, stability must be a priori guaranteed. Also, care in this respect is taken to the point that RF engineers do not test if the design is stable, but if it has any possibility of oscillating. Knowing that any passive one-port is stable if an also passive termination is connected across its terminals, testing for potential instability is equivalent to checking for negative resistances seen into the active device's ports [i.e., $|\Gamma_{in}| > 1$ and $|\Gamma_{out}| > 1$, for any passive Γ_S and Γ_L ($|\Gamma_S|$, $|\Gamma_L| < 1$]. By expressing Γ_{in} and Γ_{out} as

$$\Gamma_{in} = S_{11} + \frac{S_{12}S_{21}\Gamma_L}{1 - S_{22}\Gamma_L} \tag{5.23}$$

and

$$\Gamma_{out} = S_{22} + \frac{S_{12}S_{21}\Gamma_S}{1 - S_{11}\Gamma_S} \tag{5.24}$$

it could be shown [6] that the necessary and sufficient conditions for unconditional stability are that

$$K \equiv \frac{1 - |S_{11}|^2 - |S_{22}|^2 + \Delta^2}{2|S_{12}S_{21}|} > 1 \tag{5.25}$$

and

$$|\Delta| = |S_{11}S_{22} - S_{12}S_{21}| < 1 \tag{5.26}$$

If either (5.25) or (5.26) are not met, then the device is said to be potentially unstable, in the sense that there are certain passive terminations Γ_S and Γ_L that lead to an oscillating design. To find out the location of these dangerous zones of Γ_S and Γ_L, $|\Gamma_{in}|$ in (5.23) is set to one, and $|\Gamma_{out}|$ in (5.24) is also set to one, respectively. On the Smith chart of Γ_S, this corresponds to the *input stability circle* whose center is

$$C_S = \frac{(S_{11} - \Delta S_{22}^*)^*}{|S_{11}|^2 - |\Delta|^2} \tag{5.27}$$

and has a radius of

$$r_S = \left| \frac{S_{12}S_{21}}{|S_{11}|^2 - |\Delta|^2} \right| \tag{5.28}$$

while on the Smith chart of Γ_L this defines the *output stability circle* of center at

$$C_L = \frac{(S_{22} - \Delta S_{11}^*)^*}{|S_{22}|^2 - |\Delta|^2} \tag{5.29}$$

and radius of

$$r_L = \left| \frac{S_{12}S_{21}}{|S_{22}|^2 - |\Delta|^2} \right| \tag{5.30}$$

Since these stability circles only define the border of the desired Γ_S and Γ_L, another condition is necessary to determine if those stable Γ_S and Γ_L are located inside or outside their stability circle. One practical way to do that is to check one point where $|\Gamma_{in}| < 1$ or $|\Gamma_{out}| < 1$ conditions are easily verified. This point is $\Gamma_S = 0$ or $\Gamma_L = 0$ ($Z_S = Z_0$ or $Z_L = Z_0$) since these conditions impose $\Gamma_{in} = S_{11}$ and $\Gamma_{out} = S_{22}$, respectively. Therefore, if $|S_{22}| < 1$ ($|S_{11}| < 1$)—by far the most usual situation encountered in practice—we know that the center of the Smith chart of Γ_S (Γ_L), along with all Γ_S (Γ_L) in the same side of the border, lead to stable designs, while the ones located on the other side lead to potentially unstable circuits.

5.2.2.3 Conditions of Simultaneous Conjugate Match

In the case of unconditionally stable devices (i.e., when $K > 1$ and $|\Delta| < 1$), the whole Γ_S and Γ_L can be used since they all produce stable designs. Therefore, a design for a maximum gain of

$$G_{T_{Max}} = \frac{1}{1 - |\Gamma_{MS}|^2} |S_{21}|^2 \frac{1 - |\Gamma_{ML}|^2}{|1 - S_{22}\Gamma_{ML}|^2} \tag{5.31}$$

can be tried, which is reached in the following conditions of simultaneous conjugate match:

$$\Gamma_{MS} = \Gamma_{in}^* = \frac{B_1 \pm \sqrt{B_1^2 - 4|C_1|^2}}{2C_1} \tag{5.32a}$$

where

$$B_1 = 1 + |S_{11}|^2 - |S_{22}|^2 - |\Delta|^2 \tag{5.32b}$$

and

$$C_1 = S_{11} - \Delta S_{22}^* \tag{5.32c}$$

and also

$$\Gamma_{ML} = \Gamma_{out}^* = \frac{B_2 \pm \sqrt{B_2^2 - 4|C_2|^2}}{2C_2} \tag{5.33a}$$

in which

$$B_2 = 1 + |S_{22}|^2 - |S_{11}|^2 - |\Delta|^2 \tag{5.33b}$$

and

$$C_2 = S_{22} - \Delta S_{11}^* \tag{5.33c}$$

5.2.2.4 Available Power Gain and Operative Power Gain Circles

For potentially unstable devices there are no maximum gain conditions. In fact, the gain of a potentially unstable device can rise without limit, until oscillatory behavior occurs (i.e., nonzero output for zero input) or infinite gain. In these cases, or whenever $G_{T_{Max}}$ is greater than the specified G_T, there are no unique solutions for Γ_S and Γ_L, but an infinite number of possible combinations. This is when the amplifier design process really begins.

To help the task of Γ_S and Γ_L selection, two other auxiliary gains have been proposed. One, the *available power gain,* G_A, is defined as the ratio between output available power, P_{avN}, and source available power, P_{avS}. Since it involves only the power available from the network, not the real power delivered to the load, it cannot account for the output mismatch, being only dependent on Γ_S.

$$G_A \equiv \frac{P_{avN}}{P_{avS}} = \frac{1 - |\Gamma_S|^2}{|1 - S_{11}\Gamma_S|^2} |S_{21}|^2 \frac{1}{1 - |\Gamma_{out}|^2} \tag{5.34}$$

Therefore, G_A is particularly useful for designs driven by specifications (other than gain) relative to the network's input. Examples of these are low-noise amplifiers, as we shall detail in the next section.

On the input Smith chart, the locus of Γ_S that leads to a constant G_A is known as a *constant available gain circle.* It has center at

$$C_a = \frac{g_a\left(S_{11} - \Delta S_{22}^*\right)^*}{1 + g_a\left(|S_{11}|^2 - |\Delta|^2\right)} \tag{5.35}$$

and radius

$$r_a = \frac{\left(1 - 2Kg_a|S_{12}S_{21}| + g_a^2|S_{12}S_{21}|^2\right)^{1/2}}{\left|1 + g_a\left(|S_{11}|^2 - |\Delta|^2\right)\right|} \tag{5.36}$$

in which K is given by (5.25) and g_a is the normalized available power gain: $g_a = G_A/|S_{21}|^2$.

Obviously, the actual amplifier gain, G_T, will only be equal to the designed G_A if the output is matched (i.e., if the load termination is chosen so that $\Gamma_L = \Gamma_{out}^*$). Thus, constant available power gain circles can only be viewed as real G_T circles whenever the network output is matched.

The other auxiliary gain widely used in amplifier design is the *operative power gain,* G_P, defined as the ratio of the power delivered to the load, P_L, to the power actually delivered to the network input port, P_{in}. Therefore, as G_P involves the power actually delivered to the network, not source available power, it cannot account for any possible input mismatch, being only dependent on Γ_L.

$$G_P \equiv \frac{P_L}{P_{in}} = \frac{1}{1 - |\Gamma_{in}|^2} |S_{21}|^2 \frac{1 - |\Gamma_L|^2}{|1 - S_{22}\Gamma_L|^2} \tag{5.37}$$

G_P is used when the design is driven by specifications involving the network output port. For example, as we shall see in Section 5.3.2, amplifiers designed for maximum output power are critically dependent on Γ_L, while Γ_S is generally selected for input conjugate matching.

Comparing the form of the expressions for G_A (5.34) and G_P (5.37), it should be of no surprise that the locus of Γ_L for constant G_P, in the Smith chart, defines the so-called *constant operative power gain circle,* of center at

$$C_p = \frac{g_p\left(S_{22} - \Delta S_{11}^*\right)^*}{1 + g_p\left(|S_{22}|^2 - |\Delta|^2\right)} \tag{5.38}$$

and radius

$$r_p = \frac{\left(1 - 2Kg_p|S_{12}S_{21}| + g_p^2|S_{12}S_{21}|^2\right)^{1/2}}{|1 + g_p\left(|S_{22}|^2 - |\Delta|^2\right)|} \tag{5.39}$$

in which K is again given by (5.25) and g_p is the normalized operative power gain: $g_p = G_P/|S_{21}|^2$. Similarly to what we have already said for G_A, this G_P only becomes real transducer gain, G_T, when the input is matched: $\Gamma_S = \Gamma_{in}^*$. In this case the operative power gain circles are constant G_T circles.

5.2.2.5 Matching and Bias Networks

Since only by a strange coincidence the determined Γ_S and Γ_L equal the terminations imposed by the preceding and subsequent stages where our amplifier will operate (typically 50Ω), the next design step consists of synthesizing two impedance transformer networks. These, usually called input and output matching networks, can be made of reactive, resistive, or even active elements.

The first case is the one of much wider acceptance in RF and microwave fields, and can include lumped or distributed elements, whether the design is intended for RF or microwave frequencies, respectively. However, microwave monolithic implementations (MMIC) still use lumped elements because of the chip area required by distributed elements.

Resistive and active matching are used for general purpose or broadband designs, especially at low frequencies. They have been extensively applied in low-frequency integrated circuits, although they do not behave as well as reactive matching. The problem is that both resistive and active elements add more noise, and active elements contribute with additional sources of nonlinearity. Resistive matching is also sometimes used as a means to help amplifier stability.

Finally, the amplifier design process is completed with the synthesis of the necessary bias networks. Although they are usually built in a way that does not perturb the amplifier passband characteristics, they can be central to dc coupled designs and determine the device terminations at very low frequencies. So, they are utilized to prevent low-frequency oscillations, and to control the impedance seen by the signal envelopes, a very important issue in amplifiers intended for modulated or multitone signals. Active biasing is also an alternative to obtain tightly controlled or stabilized quiescent points. Fortunately, these added active elements do not add noise or nonlinearity if they are kept outside the signal path by convenient decoupling of the RF and the envelope frequency components.

5.2.3 Low-Noise Amplifier Design

5.2.3.1 Equivalent Two-Port Noise Model

For the purpose of circuit analysis and design, linear noisy networks are usually represented by their equivalent input referred noise voltage source $\langle u^2 \rangle$, noise current source $\langle i^2 \rangle$, and correlation admittance Y_c [7]. This allows the substitution of the original noisy network by one that is exactly equal to the former except that it is now noiseless and is preceded by a noise equivalent circuit network as depicted in Figure 5.4.

As most of the cases $\langle u^2 \rangle$ and $\langle i^2 \rangle$ can be modeled as white noise sources (constant power spectral densities), they are normally represented by equivalent noise resistance and conductance (assumed at reference temperature $T_0 = 290$K), such that

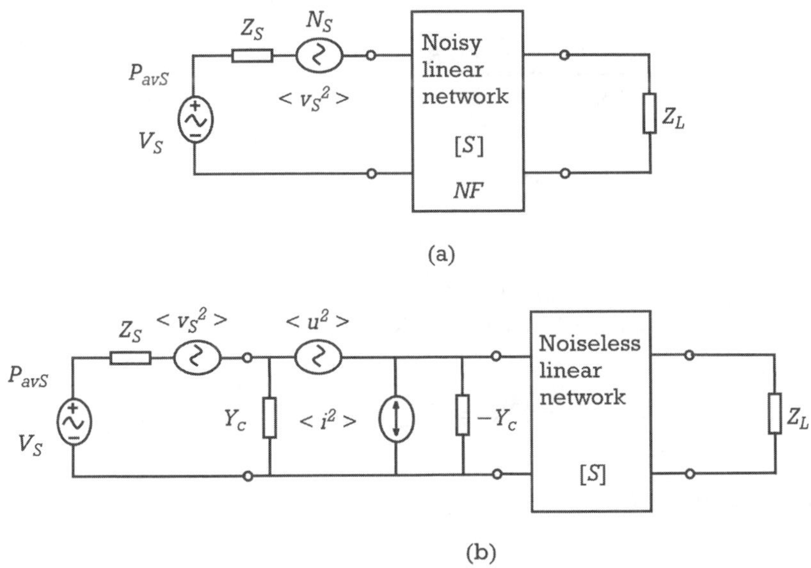

Figure 5.4 Equivalent circuit model representation of a general linear noisy two-port network. (a) Original circuit, and (b) noise equivalent plus noiseless network decomposition.

$$\langle u^2 \rangle = 4kT_0 R_n \Delta f \tag{5.40}$$

and

$$\langle i^2 \rangle = 4kT_0 G_n \Delta f \tag{5.41}$$

in which k is the Boltzman constant ($k = 1.38 \times 10^{-23}$ J.K^{-1}) and Δf is the noise bandwidth in hertz. Similarly, the source available noise power, $N_S \Delta f$, is represented by its Thévenin equivalent noise voltage,

$$\langle v_S^2 \rangle = 4kT_0 R_S \Delta f \tag{5.42}$$

and noise resistance,

$$R_S = \mathrm{Re}[Z_S] \tag{5.43}$$

Since the circuit of Figure 5.4(b) is composed by the cascade of a noisy two-port with another noiseless one, its noise figure is simply the noise figure of the noisy two-port. For determining this NF, we begin by realizing that the available noise power density at the output of the noisy two-port is the power density that would be delivered to a load of Z_S^*. So,

$$NF = \frac{N_{S_0} + N_{a_0}}{N_{S_0}} = \frac{\frac{1}{4\Delta f}\frac{\langle v_S^2 \rangle}{R_S} + \frac{1}{4\Delta f} R_S \langle i^2 \rangle + \frac{1}{4\Delta f} R_S \langle u^2 \rangle |Y_S + Y_c|^2}{\frac{1}{4\Delta f}\frac{\langle v_S^2 \rangle}{R_S}}$$

$$= 1 + \frac{G_n}{G_S} + \frac{R_n}{G_S}|Y_S + Y_c|^2 \tag{5.44}$$

in which $Y_S = Z_S^{-1} = G_S + jB_S$.

It can be shown [7] that this NF is minimized when

$$Y_S = Y_{opt} = G_{opt} + jB_{opt} = \sqrt{\frac{G_n}{R_n} + G_c^2} - jB_c \tag{5.45}$$

leading to a NF_{min} of

$$NF_{min} = 1 + 2R_n(G_c + G_{opt}) \tag{5.46}$$

Using this result in (5.44), it is possible to rewrite NF as

$$NF = NF_{min} + \frac{R_n}{G_S}|Y_S - Y_{opt}|^2 \tag{5.47}$$

which, in the reflection coefficients' domain ($Y_S \rightarrow \Gamma_S$, $Y_{opt} \rightarrow \Gamma_{opt}$) becomes

$$NF = NF_{min} + 4r_n \frac{|\Gamma_S - \Gamma_{opt}|^2}{(1 - |\Gamma_S|^2)|1 + \Gamma_{opt}|^2} \tag{5.48}$$

where r_n is R_n normalized to Z_0.

This is the central expression in low-noise amplifier design. Equations (5.47) and (5.48) are so widely used by RF engineers that nowadays all device manufacturers characterize noise performance of their transistors presenting values for the noise parameters: NF_{min}, r_n (or R_n), and Γ_{opt} (or Y_{opt}).

Before moving on with the design procedure, let us clarify some important aspects of (5.48).

The first thing that can cause some surprise is an *NF* independent on the device's linear behavior (there are no *S*-parameters involved). This is only apparent, as the equivalent noise sources $\langle u^2 \rangle$, $\langle i^2 \rangle$ and the correlation admittance actually vary according to the device's linear equivalent circuit model, and so also NF_{min}, r_n, and Γ_{opt} depend on that model. Furthermore, since these equivalent noise representations were derived from a bias-dependent equivalent circuit model, with also bias dependent intrinsic noise sources, NF_{min}, r_n, and Γ_{opt} are going to show a certain variation with bias. In particular, NF_{min} of FETs and bipolars typically shows a pattern versus drain or collector current that decreases with the rapid gain rise associated to turn-on, reaches its lowest value for a fairly low quiescent current, and then increases, accompanying the growth of its intrinsic noise sources.

The second thing to be noticed on these *NF* expressions is that they only depend on the amplifier input termination, Γ_S, and not on the output, Γ_L. This is a consequence of the adopted noise figure definition. Since it relates output available power, not actual power delivered to the load, it had to be independent on Γ_L. So, (5.47) and (5.48) determine that a low-noise design should concentrate on selecting the appropriate Γ_S.

5.2.3.2 Constant Noise Figure Circles and Low-Noise Design

A detailed analysis of (5.48) [6] showed that the locus of Γ_S corresponding to a constant *NF* value is a circle centered at

$$C_n = \frac{\Gamma_{opt}}{1 + N_i} \tag{5.49}$$

and having a radius of

$$r_{NF} = \frac{1}{1 + N_i} \sqrt{N_i^2 + N_i\left(1 - |\Gamma_{opt}|^2\right)} \quad (5.50)$$

where the parameter N_i is defined as

$$N_i = \frac{NF - NF_{min}}{4r_n} |1 + \Gamma_{opt}|^2 \quad (5.51)$$

These *constant noise figure circles* play a major role on low-noise amplifier design. The procedure normally takes the following steps:

1. Select an appropriate device and bias point appropriate to meet the desired specifications of gain, *NF*, and possibly input/output mismatch (usually specified as standing wave ratio or return-loss). Note that since best noise performance frequently shows up at low bias currents, while small-signal gain demands for fairly high currents, you face a first design compromise. Also, and unless you have detailed bias-dependent equivalent circuit models and intrinsic noise source models—which are necessary to get bias-dependent *S*-parameters and NF_{min}, Γ_{opt}, and r_n—you will have to rely on the information present in the device's data sheet. For common low-noise transistors, this typically provides two bias points, one for optimized *NF*, and another one for optimized gain. Figure 5.5 depicts typical [S] matrices and noise parameters for these two bias points.
2. On the Smith chart of Γ_S, draw a family of noise figure circles for various *NF* close to the desired value. On the same chart, include also the input stability circle (unless the transistor is unconditionally stable) and another family of available gain circles. An example of such plots is shown in Figure 5.6, where the chosen device was the one of Figure 5.5(a). It is this chart that will guide the selection of Γ_S. As is seen from Figure 5.6, there is a clear compromise between noise, gain, and stability. Although, in this case, G_A of up to 16 dB or *NF* on the order of 1.2 dB can be easily obtained without compromising stability, they cannot be met simultaneously. Before making the decision of *NF* minimization, assuming that gain limitations can always be compensated afterwards by subsequent amplifier stages, remember that this solution may not lead to an overall optimized noise performance, especially if those subsequent stages have poor noise figures.

   Fortunately, there is a technique that relaxes this trade-off between gain and *NF*. It consists of adding a small amount of current-series inductive feedback to the active device, which has the effect of approximating *NF*

	BJT common emitter S-parameters @V_{CE} = 3V I_C = 5mA # GHz S MA R 50Ω							
Freq (GHz)	S_{11} MAG	S_{11} ANG	S_{21} MAG	S_{21} ANG	S_{12} MAG	S_{12} ANG	S_{22} MAG	S_{22} ANG
0.010	0.7903	−1.0	15.143	179.2	0.0012	83.4	0.9881	−0.7
0.150	0.7858	−17.7	14.919	167.4	0.0133	80.6	0.9761	−9.7
0.400	0.7344	−46.0	13.689	147.3	0.0332	67.5	0.8999	−24.7
0.600	0.6788	−65.5	12.302	133.5	0.0448	58.9	0.8123	−34.2
0.800	0.6286	−83.6	10.863	122.0	0.0539	51.3	0.7282	−42.5
1.200	0.5603	−113	8.454	104.0	0.0656	42.6	0.5841	−54.8
1.400	0.5340	−125	7.508	97.1	0.0695	39.7	0.5228	−58.6
1.800	0.5105	−146	6.132	84.7	0.0765	36.1	0.4305	−65.6
2.000	0.5065	−156	5.603	79.4	0.0798	34.6	0.4006	−70.3
2.400	0.5015	−172	4.738	70.1	0.0859	32.8	0.3450	−76.8
2.800	0.5063	174	4.104	61.5	0.0925	31.0	0.2999	−83.3
3.000	0.5106	168	3.840	57.1	0.0957	29.8	0.2674	−84.2
3.500	0.5296	153	3.298	47.6	0.1040	27.2	0.2375	−94.0
4.000	0.5487	142	2.868	38.5	0.1121	25.1	0.2068	−110
4.500	0.5805	131	2.520	30.0	0.1210	22.3	0.1725	−127
5.000	0.6036	123	2.260	22.1	0.1285	19.4	0.1502	−137
5.500	0.6249	116	2.047	14.4	0.1368	16.7	0.1637	−148
6.000	0.6329	110	1.861	6.7	0.1442	13.1	0.1726	169

Freq (GHz)	NF_{min}	Γ_{opt} Mag	Γ_{opt} Phase	r_n
0.450	0.87	0.19	13	0.16
0.900	0.88	0.19	31	0.15
1.800	1.05	0.10	92	0.12
2.400	1.13	0.14	121	0.12
3.000	1.26	0.16	155	0.12
4.000	1.42	0.31	−145	0.13

(a)

Figure 5.5 Typical [S] matrices and noise parameters for two different bias points. Proposed bias point (a) for low-noise operation, and (b) for high gain.

and Γ_A circles [8]. It is known as source or emitter degeneration (when applied to FETs or bipolars, respectively) and should be used carefully as the added feedback may lead to instability.

3. After having chosen Γ_S, determine Γ_{out} with (5.24) and then select Γ_L for a prescribed gain. Note that G_A values previously considered will only be converted into specified transducer power gain if a matched output is adopted. If not, a gain reduction of

BJT common emitter S-parameters @V_{CE} = 3V I_C = 30mA
GHz S MA R 50Ω

Freq (GHz)	S_{11} MAG	S_{11} ANG	S_{21} MAG	S_{21} ANG	S_{12} MAG	S_{12} ANG	S_{22} MAG	S_{22} ANG
0.010	0.3297	−3.2	44.240	178.2	0.0011	85.5	0.9356	−0.2
0.150	0.3312	−49.0	40.707	154.5	0.0095	75.6	0.8899	−20.9
0.400	0.3427	−104	29.060	124.7	0.0199	65.9	0.6555	−44.3
0.600	0.3402	−128	22.124	110.3	0.0253	62.7	0.5129	−53.7
0.800	0.3477	−146	17.517	100.8	0.0313	62.4	0.4208	−61.0
1.200	0.3692	−167	12.092	87.2	0.0425	61.3	0.3078	−71.0
1.400	0.3742	−175	10.421	82.4	0.0482	60.7	0.2661	−73.2
1.800	0.3900	172	8.126	73.3	0.0593	59.0	0.2109	−81.3
2.000	0.4015	167	7.313	69.4	0.0654	57.8	0.1981	−88.3
2.400	0.4140	158	6.088	62.3	0.0777	55.1	0.1661	−98.2
2.800	0.4359	149	5.212	55.4	0.0888	51.6	0.1430	−110
3.000	0.4447	145	4.850	52.0	0.0950	49.6	0.1174	−114
3.500	0.4757	136	4.129	44.0	0.1082	44.6	0.1134	−135
4.000	0.5011	127	3.573	36.4	0.1215	40.2	0.1273	−163
4.500	0.5398	120	3.130	29.1	0.1347	35.1	0.1428	169
5.000	0.5618	114	2.804	22.3	0.1459	29.9	0.1432	154
5.500	0.5863	109	2.539	15.2	0.1567	25.4	0.1623	150
6.000	0.5925	104	2.310	8.2	0.1667	20.0	0.2001	138

Freq (GHz)	NF_{min}	Γ_{opt} Mag	Γ_{opt} Phase	r_n
0.450	1.65	0.28	−165	0.14
0.900	1.67	0.28	−155	0.12
1.800	1.77	0.34	−142	0.13
2.400	1.94	0.29	−136	0.19
3.000	2.00	0.39	−126	0.19
4.000	2.12	0.48	−111	0.31

(b)

Figure 5.5 (continued).

$$\frac{G_T}{G_A} = \frac{\left(1 - |\Gamma_L|^2\right)\left(1 - |\Gamma_{out}|^2\right)}{|1 - \Gamma_{out}\Gamma_L|^2} \tag{5.52}$$

should be accounted for.

4. Design the necessary input and output matching networks, along with bias circuitry. After the whole amplifier circuit is set, verify out-of-band stability from dc up to the maximum frequency where the transistor is active.

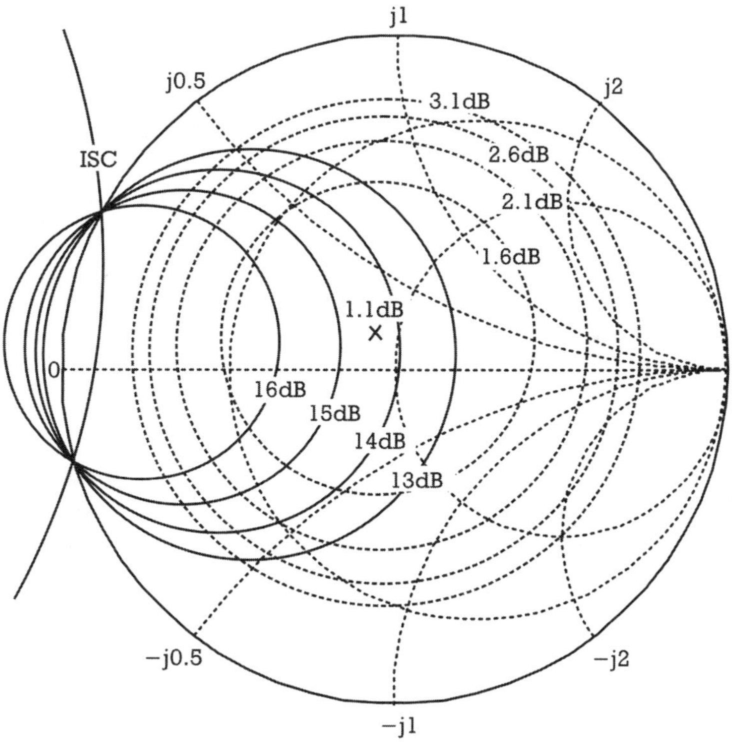

Figure 5.6 Example of noise figure, *NF*; available gain, G_A; and input stability circles, ISC, for the device of Figure 5.5 biased for low *NF*.

Remember that resistive loading applied to the bias circuits can be a small effort remedy for many low-frequency oscillations.

5. Especially at high frequencies, substitute any ideal passive element with its "real" equivalent circuit model and optimize your design. Expect for a detuning, caused by the reactive parasitic elements, and also a reduction in gain and *NF* degradation, imposed by the element's losses.

5.2.4 Nonlinear Distortion in Small-Signal Amplifiers

The two amplifier design criteria addressed up until now were gain and noise figure. Since they are exclusively determined by linear relations, device-independent formulations could be used. This led to some well-established general conclusions and design rules. Unfortunately, small-signal intermodulation distortion is controlled by the specific device's nonlinearities, which impedes any attempts of drawing device-independent conclusions. Also, there are no black-box models available for IMD, and the analysis or design requires handling the full nonlinear equivalent circuit model of the actual amplifier active device.

Because MESFETs and BJTs are widely accepted devices for low noise and high linearity at RF and microwave bands, we will use these transistors in the following two design studies. As a matter of fact, we would like to note that, even though the conclusions drawn cannot be directly extrapolated for HEMTs and MOSFETs or HBTs, the design procedure is essentially the same. Furthermore, because of the similarity of I/V and Q/V characteristics and of equivalent circuit model topologies, the first design study dealing with the MESFET typifies IMD predictions of small-signal amplifiers based on field effect devices, while the second one using the BJT instance is a good illustration of the bipolar-based small-signal amplifier group.

5.2.4.1 Selecting an Appropriate FET Device Model

As small-signal IMD is determined by the devices' mild nonlinearities, good models of those characteristics are instrumental for the success of highly linear amplifier designs. Also, the guarantee that the distortion indeed comes from the device's weak nonlinearities, and not from its hard nonlinear characteristics, is of primary importance if a reasonably low level of distortion is sought. So, the first thing to do is to identify the physical origins of weak and strong nonlinear behavior.

In an FET device, linear amplification demands for operation in the saturation zone. As seen in Figure 5.7, this area is limited, in drain-source voltage, by knee voltage, V_K, and drain-gate breakdown, V_{BR}; and, in current, by drain current cut-off and maximum channel opening or gate-channel junction conduction.

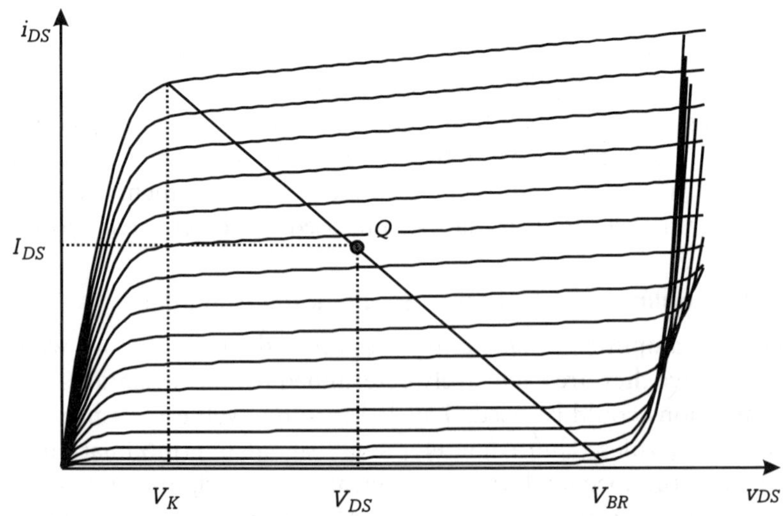

Figure 5.7 Typical MESFET output I/V curves, depicting strong nonlinear distortion-free operation. Q is the quiescent point and the diagonal line is the dynamic load line.

Knee voltage refers to the point where the transistor leaves current saturation, entering the linear or triode zone. So, the transistor output ceases behaving as an almost ideal current source and becomes a quite low channel resistance. Drain-gate breakdown is similar in its effects to the transition between saturation and linear zones, as it also transfers substantial i_{DS} control from gate-source voltage to drain-source voltage. Thus, it must also be avoided to prevent important strong nonlinear distortion components. In terms of current, it is obvious that both current cut-off and maximum channel current or gate-channel junction conduction produce signal clipping, also being strong nonlinear contributors.

In summary, the FET should be biased for class A operation and maintained comfortably inside the rectangle limited by the referred hard nonlinearity borders, in all possible ranges of signal excursion. In that respect, the best quiescent point is the one exactly located at the geometrical center of the saturation zone rectangle (point Q in Figure 5.7), since it will allow maximized output linear power, when associated to the diagonal load-line depicted in the same figure.

Such a strategy leads to the widely accepted rule of thumb that states that a linear amplifier has its active device biased for class A, at a current close to $I_{DS} = I_{ds_s}/2$ [$I_{ds_s} \equiv I_{DS}(V_{GS} = 0\text{V})$] and $V_{DS} = (V_K + V_{BR})/2$, and is loaded by the output termination that allows maximized signal excursion without clipping [i.e., an intrinsic R_L not far from $R_L = (V_{DS} - V_K)/I_{DS}$]. Although this first step design cannot be considered as a dramatically bad solution, it gives room for exploring a fairly wide margin optimization. To do this, we must rely on a more detailed view of device operation and go through a nonlinear analysis of the amplifier circuit.

The amplifier models adopted for the following nonlinear study are shown in Figures 5.8 and 5.9.

Figure 5.8 shows a general amplifier model, and Figure 5.9 is another representation where the active device was particularized for the π equivalent circuit of an

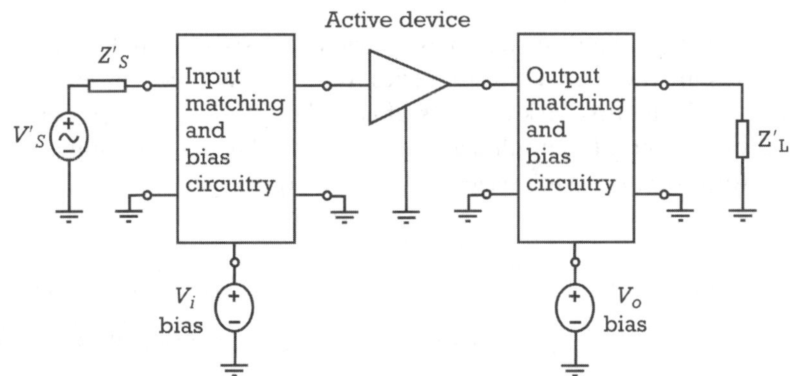

Figure 5.8 Functional diagram of nonlinear amplifier model used for distortion analysis.

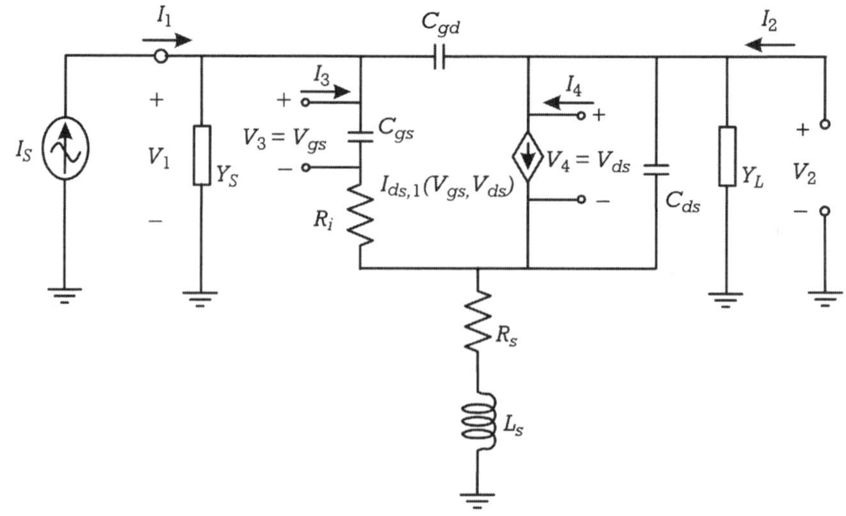

Figure 5.9 Detailed equivalent circuit model of the amplifier depicted in Figure 5.8.

FET (our GaAs MESFET). (I_S, Y_S) stands for the Norton equivalent of the excitation source, as seen from the input matching network and bias circuitry, plus any parasitic elements connected between the extrinsic and intrinsic gate terminals. Y_L is the corresponding Norton equivalent seen into the output node.

According to the MESFET model description given in Chapter 4, since the FET is expected to remain in saturation, only $i_{DS}(v_{GS}, v_{DS})$ and $C_{gs}(v_{GS})$ nonlinearities have to be considered, while C_{gd} can be assumed as a linear capacitance.

5.2.4.2 Small-Signal FET Amplifier Distortion Prediction

As our present analysis target is restricted to small-signal behavior, the nonlinear currents method of Volterra series will be used. So, as explained in detail in Chapter 3, the first task consists of calculating the quiescent control voltages, V_{GS} and V_{DS}, using a full nonlinear analysis at dc. Then, the nonlinearities are expanded in a third-order Taylor series around this quiescent point as

$$i_{ds}(v_{gs}, v_{ds}) = G_m v_{gs} + G_{ds} v_{ds} + G_{m2} v_{gs}^2 + G_{md} v_{gs} v_{ds} + G_{d2} v_{ds}^2 + G_{m3} v_{gs}^3 + G_{m2d} v_{gs}^2 v_{ds} + G_{md2} v_{gs} v_{ds}^2 + G_{d3} v_{ds}^3 \quad (5.53)$$

and

$$
\begin{aligned}
i_{gs}(v_{gs}) &= \frac{\mathrm{d}}{\mathrm{d}t}\left[q_{gs}(v_{gs})\right] \\
&= \frac{\mathrm{d}}{\mathrm{d}v_{gs}}\left[C_{gs}v_{gs} + C_{gs2}v_{gs}^2 + C_{gs3}v_{gs}^3\right]\frac{\mathrm{d}v_{gs}}{\mathrm{d}t} \\
&= \left(C_{gs} + 2C_{gs2}v_{gs} + 3C_{gs3}v_{gs}^2\right)\frac{\mathrm{d}v_{gs}}{\mathrm{d}t}
\end{aligned}
\tag{5.54}
$$

According to the expansion of Section 3.2.2.1, and because we have two nonlinear elements ($M = 2$), depending on two control voltages $v^{(l_1)} = v_3 = v_{gs}$ and $v^{(l_2)} = v_4 = v_{ds}$, defined at the terminals of the two nonlinear ports already considered, plus the excitation and output ports, we only need four ports to describe the linear subnetwork. Therefore, in the frequency-domain we will have

$$
\begin{bmatrix} V_1(\omega) \\ V_2(\omega) \\ V_3(\omega) \\ V_4(\omega) \end{bmatrix} = \begin{bmatrix} Z_{11} & Z_{12} & Z_{13} & Z_{14} \\ Z_{21} & Z_{22} & Z_{23} & Z_{24} \\ Z_{31} & Z_{32} & Z_{33} & Z_{34} \\ Z_{41} & Z_{42} & Z_{43} & Z_{44} \end{bmatrix} \cdot \begin{bmatrix} I_1(\omega) \\ I_2(\omega) \\ I_3(\omega) \\ I_4(\omega) \end{bmatrix}
\tag{5.55}
$$

and

$$
I_1(\omega) = I_s(\omega)\text{: } i_S(t) = \frac{1}{2}\sum_{\substack{q=-Q \\ q\neq 0}}^{Q} I_{s_q} e^{j\omega_q t}
\tag{5.56}
$$

$$
I_2(\omega) = 0
\tag{5.57}
$$

$$
\begin{aligned}
I_3(\omega) &= -[I_{c2}(\omega) + I_{c3}(\omega)] \\
&= -2C_{gs_2}V_{3,1}(\omega_{q_1}) * \left[j(\omega_{q_2})V_{3,1}(\omega_{q_2})\right] \\
&\quad -2C_{gs_2}V_{3,1}(\omega_{q_1}) * \left[j(\omega_{q_2}+\omega_{q_3})V_{3,2}(\omega_{q_2}+\omega_{q_3})\right] \\
&\quad -2C_{gs_2}V_{3,2}(\omega_{q_1}+\omega_{q_2}) * \left[j(\omega_{q_3})V_{3,1}(\omega_{q_3})\right] \\
&\quad -3C_{gs_3}\left[V_{3,1}(\omega_{q_1}) * V_{3,1}(\omega_{q_2})\right] * \left[j(\omega_{q_3})V_{3,1}(\omega_{q_3})\right]
\end{aligned}
\tag{5.58}
$$

and

$$
\begin{aligned}
I_4(\omega) &= -[I_{ds_2}(\omega) + I_{ds_3}(\omega)] \\
&= -G_{m2} V_{3,1}(\omega_{q_1}) * V_{3,1}(\omega_{q_2}) \\
&\quad -G_{md} V_{3,1}(\omega_{q_1}) * V_{4,1}(\omega_{q_2}) - G_{d2} V_{4,1}(\omega_{q_1}) * V_{4,1}(\omega_{q_2}) \\
&\quad -2G_{m2} V_{3,1}(\omega_{q_1}) * V_{3,2}(\omega_{q_2} + \omega_{q_3}) - G_{md} V_{3,1}(\omega_{q_1}) * V_{4,2}(\omega_{q_2} + \omega_{q_3}) \\
&\quad -G_{md} V_{3,2}(\omega_{q_1} + \omega_{q_2}) * V_{4,1}(\omega_{q_3}) - 2G_{d2} V_{4,1}(\omega_{q_1}) * V_{4,2}(\omega_{q_2} + \omega_{q_3}) \\
&\quad -G_{m3} V_{3,1}(\omega_{q_1}) * V_{3,1}(\omega_{q_2}) * V_{3,1}(\omega_{q_3}) - G_{m2d} V_{3,1}(\omega_{q_1}) * V_{3,1}(\omega_{q_2}) \\
&\qquad * V_{4,1}(\omega_{q_3}) \\
&\quad -G_{md2} V_{3,1}(\omega_{q_1}) * V_{4,1}(\omega_{q_2}) * V_{4,1}(\omega_{q_3}) - G_{d3} V_{4,1}(\omega_{q_1}) * V_{4,1}(\omega_{q_2}) \\
&\qquad * V_{4,1}(\omega_{q_3})
\end{aligned}
\tag{5.59}
$$

in which the operator "*" in (5.58) and (5.59) represents pseudoconvolution as given by the product summations of (3.97), (3.101) and (3.98), (3.102), respectively.

These are general relations applicable to any bias point, excitation signal, and input/output terminations. However, they are not of much use for analytical design, unless some simplifications are made.

For example, if we were interested in the assessment of second and third-order nonlinear distortions evaluating the intercept points that correspond to the difference or the sum frequencies, and the inband distortion components, we would restrict $I_S(\omega)$ to be an equal amplitude two-tone excitation [$Q = 2$ in (5.56)] of frequencies ω_1 and ω_2.

In this context we may expect that $\Delta\omega \equiv \omega_1 - \omega_2 << \omega_1, \omega_2$; $\Sigma\omega \equiv \omega_1 + \omega_2 \approx 2\omega_1 \approx 2\omega_2 \equiv 2\omega$, and that the linear circuit behavior, represented by $\mathbf{Z}(\omega)$ of (5.55) obeys $Z_{ij}(\omega_1) \approx Z_{ij}(\omega_2) \approx Z_{ij}(2\omega_1 - \omega_2) \approx Z_{ij}(2\omega_2 - \omega_1) \equiv Z_{ij}(\omega_0)$. Furthermore, we may also consider some other common simplifying assumptions in the FET's equivalent circuit: $G_m|R_s + j\omega L_s| << 1$, $\omega^2 C_{gs}^2 R_i^2 << 1$, $\omega C_{gd} << G_m$ and also $\omega C_{gd} << |G_{ds} + j\omega C_{ds} + Y_L(\omega)|$.

In this situation, the relevant $\mathbf{Z}$ parameters can be expressed by these simple relations:

$$
Z_{21}(\omega) \approx -\frac{G_m}{Y_D(\omega) Y_G(\omega)} \approx Z_{23}(\omega) \approx Z_{43}(\omega) \approx Z_{41}(\omega) \tag{5.60}
$$

$$
Z_{24}(\omega) \approx Z_D(\omega) \frac{Y_S(\omega) + j\omega C_T}{Y_G(\omega)} \approx Z_{44}(\omega) \tag{5.61}
$$

$$Z_{31}(\omega) \approx \frac{1}{Y_G(\omega)} \approx Z_{33}(\omega) \tag{5.62}$$

$$Z_{34}(\omega) \approx Z_D(\omega)\frac{j\omega C_{gd}}{Y_G(\omega)} \tag{5.63}$$

where $C_T = C_{gs} + C_{gd}$, $Y_D(\omega) = Z_D(\omega)^{-1} = G_{ds} + j\omega C_{ds} + Y_L(\omega)$, and

$$Y_G(\omega) = Y_S(\omega) + j\omega C_T + \frac{j\omega C_{gd}\, G_m}{G_{ds} + j\omega C_{ds} + Y_L(\omega)} \tag{5.64}$$

The first step in the nonlinear currents method application consists of deriving the fundamental output voltage, $V_{2,1}(\omega_0)$, and the first-order control voltages, $V_{3,1}(\omega_0)$ and $V_{4,1}(\omega_0)$.

$$V_{2,1}(\omega_0) = Z_{21}(\omega_0)\frac{I_S(\omega_0)}{2} \tag{5.65}$$

$$V_{3,1}(\omega_0) = Z_{31}(\omega_0)\frac{I_S(\omega_0)}{2} \tag{5.66}$$

$$V_{4,1}(\omega_0) = Z_{41}(\omega_0)\frac{I_S(\omega_0)}{2} \tag{5.67}$$

So, second-order nonlinear current components at $\Delta\omega = \omega_1 - \omega_2$, $\Sigma\omega = \omega_1 + \omega_2$ and $2\omega \approx 2\omega_1 \approx 2\omega_2$, produced by C_{gs} are

$$I_{3,2}(\Delta\omega) = -2j\Delta\omega C_{gs2}\,|Z_{31}(\omega_0)|^2\,\frac{|I_S|^2}{4} \tag{5.68}$$

$$I_{3,2}(\Sigma\omega) = -2j\Sigma\omega C_{gs2} Z_{31}(\omega_0)^2\,\frac{I_S^2}{4} \tag{5.69}$$

$$I_{3,2}(2\omega) = -2j\omega_0 C_{gs2} Z_{31}(\omega_0)^2\,\frac{I_S^2}{4} \tag{5.70}$$

where I_S is a simplified, and in this context equivalent, notation for $I_S(\omega_0)$. Since we have assumed $\Delta\omega << \omega_0$, it might be expected that the input capacitance does not contribute with any appreciable current at $\Delta\omega$. Thus, $I_{3,2}(\Delta\omega) \approx 0$.

Similarly, second-order I_{ds} current components at $\Delta\omega$, $\Sigma\omega$, and 2ω are given by

$$I_{4,2}(\Delta\omega) = -\left[2G_{m2}|Z_{31}(\omega_0)|^2 + G_{md}Z_{31}(\omega_0)Z_{41}(\omega_0)^* + G_{md}Z_{31}(\omega_0)^*Z_{41}(\omega_0) + 2G_{d2}|Z_{41}(\omega_0)|^2\right]\frac{|I_S|^2}{4} \quad (5.71)$$

$$I_{4,2}(\Sigma\omega) = -2\left[G_{m2}Z_{31}(\omega_0)^2 + G_{md}Z_{31}(\omega_0)Z_{41}(\omega_0) + G_{d2}Z_{41}(\omega_0)^2\right]\frac{I_S^2}{4} \quad (5.72)$$

$$I_{4,2}(2\omega) = -\left[G_{m2}Z_{31}(\omega_0)^2 + G_{md}Z_{31}(\omega_0)Z_{41}(\omega_0) + G_{d2}Z_{41}(\omega_0)^2\right]\frac{I_S^2}{4} \quad (5.73)$$

Then, the second-order output and control voltages become

$$V_{2,2}(\Delta\omega) = Z_{24}(\Delta\omega)I_{4,2}(\Delta\omega) \quad (5.74)$$

$$V_{3,2}(\Delta\omega) = Z_{34}(\Delta\omega)I_{4,2}(\Delta\omega) \quad (5.75)$$

$$V_{4,2}(\Delta\omega) = Z_{44}(\Delta\omega)I_{4,2}(\Delta\omega) \quad (5.76)$$

$$V_{2,2}(\Sigma\omega) = Z_{23}(\Sigma\omega)I_{3,2}(\Sigma\omega) + Z_{24}(\Sigma\omega)I_{4,2}(\Sigma\omega) \quad (5.77)$$

$$V_{3,2}(\Sigma\omega) = Z_{33}(\Sigma\omega)I_{3,2}(\Sigma\omega) + Z_{34}(\Sigma\omega)I_{4,2}(\Sigma\omega) \quad (5.78)$$

$$V_{4,2}(\Sigma\omega) = Z_{43}(\Sigma\omega)I_{3,2}(\Sigma\omega) + Z_{44}(\Sigma\omega)I_{4,2}(\Sigma\omega) \quad (5.79)$$

$$V_{2,2}(2\omega) = Z_{23}(2\omega)I_{3,2}(2\omega) + Z_{24}(2\omega)I_{4,2}(2\omega) \quad (5.80)$$

$$V_{3,2}(2\omega) = Z_{33}(2\omega)I_{3,2}(2\omega) + Z_{34}(2\omega)I_{4,2}(2\omega) \quad (5.81)$$

$$V_{4,2}(2\omega) = Z_{43}(2\omega)I_{3,2}(2\omega) + Z_{44}(2\omega)I_{4,2}(2\omega) \quad (5.82)$$

Given first and second-order control voltages, third-order nonlinear currents at $2\omega_1 - \omega_2$ or $2\omega_2 - \omega_1$ will be approximately equal in amplitude and given by

$$I_{3,3}(2\omega_1 - \omega_2) \approx -2j\omega_0 C_{gs2}Z_{31}(\omega_0)^*\left[Z_{33}(2\omega)I_{3,2}(2\omega) + Z_{34}(2\omega)I_{4,2}(2\omega)\right]\frac{I_S^*}{2} - 2j\omega_0 C_{gs2}Z_{31}(\omega_0)Z_{34}(\Delta\omega)I_{4,2}(\Delta\omega)\frac{I_S}{2} - 3j\omega_0 C_{gs3}Z_{31}(\omega_0)|Z_{31}(\omega_0)|^2\frac{I_S|I_S|^2}{8} \quad (5.83)$$

and

$$
\begin{aligned}
I_{4,3}(2\omega_1 - \omega_2) \approx & -2G_{m2} Z_{31}(\omega_0)^* \left[Z_{33}(2\omega) I_{3,2}(2\omega) + Z_{34}(2\omega) I_{4,2}(2\omega) \right] \frac{I_S^*}{2} \\
& - G_{md} Z_{31}(\omega_0)^* \left[Z_{43}(2\omega) I_{3,2}(2\omega) + Z_{44}(2\omega) I_{4,2}(2\omega) \right] \frac{I_S^*}{2} \\
& - G_{md} Z_{41}(\omega_0)^* \left[Z_{33}(2\omega) I_{3,2}(2\omega) + Z_{34}(2\omega) I_{4,2}(2\omega) \right] \frac{I_S^*}{2} \\
& - 2G_{d2} Z_{41}(\omega_0)^* \left[Z_{43}(2\omega) I_{3,2}(2\omega) + Z_{44}(2\omega) I_{4,2}(2\omega) \right] \frac{I_S^*}{2} \\
& - 2G_{m2} Z_{31}(\omega_0) Z_{34}(\Delta\omega) I_{4,2}(\Delta\omega) \frac{I_S}{2} \\
& - G_{md} Z_{31}(\omega_0) Z_{44}(\Delta\omega) I_{4,2}(\Delta\omega) \frac{I_S}{2} \\
& - G_{md} Z_{41}(\omega_0) Z_{34}(\Delta\omega) I_{4,2}(\Delta\omega) \frac{I_S}{2} \\
& - 2G_{d2} Z_{41}(\omega_0) Z_{44}(\Delta\omega) I_{4,2}(\Delta\omega) \frac{I_S}{2} \\
& - \left[3G_{m3} Z_{31}(\omega_0) |Z_{31}(\omega_0)|^2 \right. \\
& + G_{m2d} Z_{31}(\omega_0)^2 Z_{41}(\omega_0)^* + 2G_{m2d} |Z_{31}(\omega_0)|^2 Z_{41}(\omega_0) \\
& + G_{md2} Z_{31}(\omega_0)^* Z_{41}(\omega_0)^2 + 2G_{md2} Z_{31}(\omega_0) |Z_{41}(\omega_0)|^2 \\
& \left. + 3G_{d3} Z_{41}(\omega_0) |Z_{41}(\omega_0)|^2 \right] \frac{I_S |I_S|^2}{8}
\end{aligned}
\tag{5.84}
$$

Finally, the inband distortion output voltage component comes:

$$
V_{2,3}(2\omega_1 - \omega_2) \approx Z_{23}(\omega_0) I_{3,3}(2\omega_1 - \omega_2) + Z_{24}(\omega_0) I_{4,3}(2\omega_1 - \omega_2) \tag{5.85}
$$

The desired signal second and third-order output powers and intercept point can now be calculated from $V_{2,1}(\omega_0)$ – (5.65), $V_{2,2}(\omega_1 - \omega_2)$ – (5.74), $V_{2,2}(\omega_1 + \omega_2)$ – (5.77), and $V_{2,3}(2\omega_1 - \omega_2)$ – (5.85) as follows.

First, the source available power per tone, P_{av_S}, and the power actually delivered to the amplifier input port, P_{in}, are calculated. Since the excitation given by (5.56) corresponds to a peak amplitude current of $|I_S|$,

$$P_{av_S} = \frac{|I_S|^2}{8G_S(\omega_0)} \tag{5.86}$$

where $G_S(\omega) = \text{Re}[Y_S(\omega)]$.

The amplifier input admittance, $Y_{in}(\omega)$, can be obtained from $Y_{in}(\omega) = G_{in}(\omega) + jB_{in}(\omega) = Z_{11}(\omega)^{-1} - Y_S(\omega)$, and so,

$$P_{in} = \frac{1}{2} G_{in}(\omega_0)|Z_{11}(\omega_0)|^2|I_S|^2 = 4G_{in}(\omega_0)G_S(\omega_0)|Z_{11}(\omega_0)|^2 P_{av_S} \tag{5.87}$$

Now, we calculate the output voltage amplitude at the fundamental tones,

$$|V_2(\omega_0)| = |Z_{21}(\omega_0)||I_S| \tag{5.88}$$

which leads to an output power per tone of

$$P_L = \frac{1}{2} G_L(\omega_0)|V_L(\omega_0)|^2 = \frac{1}{2} G_L(\omega_0)|Z_{21}(\omega_0)|^2|I_S|^2 \tag{5.89}$$

and thus operative and transducer power gains of

$$G_P = \frac{P_L}{P_{in}} = \frac{G_L(\omega_0)}{G_{in}(\omega_0)}\left|\frac{Z_{21}(\omega_0)}{Z_{11}(\omega_0)}\right|^2 \tag{5.90}$$

and

$$G_T = \frac{P_L}{P_{av_S}} = 4G_L(\omega_0)G_S(\omega_0)|Z_{21}(\omega_0)|^2 \tag{5.91}$$

Given the calculated distortion output voltages [whose amplitude is actually the double of the calculated positive spectrum frequency components, $V_{2,n}(\omega)$], the corresponding desired output nonlinear distortion powers are

$$P_{L_{2\Delta}} = 2G_L(\Delta\omega)|V_{2,2}(\Delta\omega)|^2 \tag{5.92}$$

$$P_{L_{2\Sigma}} = 2G_L(\Sigma\omega)|V_{2,2}(\Sigma\omega)|^2 \tag{5.93}$$

and

$$P_{L_3} = 2G_L(\omega_0)|V_{2,3}(2\omega_1 - \omega_2)|^2 \tag{5.94}$$

Except for the narrowband excitation assumption [$\Delta\omega << \omega_1$, ω_2 and $\mathbf{Z}(\omega_1) \approx \mathbf{Z}(\omega_2) \approx \mathbf{Z}(2\omega_1 - \omega_2) \approx \mathbf{Z}(2\omega_2 - \omega_1)$] and approximations made for the device's linear equivalent circuit, the achieved results are general and may be used to investigate the FET's distortion dependence on many amplifier parameters like bias, frequency, and input/output terminations. Also, since they are fully analytical, they allow qualitative small-signal amplifier distortion analysis, optimization, and design. Unfortunately, this wide range of applications has a big price in the expressions' complexity, obviating any handy calculations. So, we will begin by restricting the analysis to a simplified case, and then extrapolate its results to more involved situations.

As was expected from the nature of C_{gs} nonlinearity, all distortion currents generated by this element are proportional to frequency, and so they diminish in relevance as operating frequency goes down. Because these current components must be converted into input control voltages when they circulate in the gate mesh, C_{gs} distortion is a strong function of source impedance [via $Z_{23}(\omega)$, $Z_{33}(\omega)$, and $Z_{43}(\omega)$]. Therefore, C_{gs} will also have a negligible impact on the overall amplifier distortion whenever it is short-circuited by small input impedance terminations.

Furthermore, as these C_{gs} contributions are added to the distortion generated in I_{ds} nonlinearity, a decrease in frequency or source impedance, or even a rise in the ratio between the FET's resistive and reactive nonlinearities, they all produce same results. So, they will be treated in one single case: $I_{3,n}(\omega) \approx 0$ ($n = 2, 3$). In this situation, applicable to a broad range of low-frequency or RF designs, the distortion relations become quite simplified, being possible to express second and third-order intercept points by meaningful formulae. For example, the second-order intercept point for the distortion power arising at the difference frequency would be ideally met when the excitation current, $|I_S|$, had an amplitude such that

$$P_{L_{2\Delta}} = 2G_L(\Delta\omega)|V_{2,2}(\Delta\omega)|^2 = P_L = \frac{1}{2}G_L(\omega_0)|Z_{21}(\omega_0)|^2|I_S|^2 \tag{5.95}$$

Solving (5.95) for $|I_S|$, and substituting this value into P_L gives

$$IP_{2\Delta} \approx 2\frac{G_L(\omega_0)^2|Z_{21}(\omega_0)|^4}{G_L(\Delta\omega)|Z_{24}(\Delta\omega)|^2}\Big|2G_{m2}|Z_{31}(\omega_0)|^2 + G_{md}Z_{31}(\omega_0)Z_{41}(\omega_0)^*$$
$$+ G_{md}Z_{31}(\omega_0)^*Z_{41}(\omega_0) + 2G_{d2}|Z_{41}(\omega_0)|^2\Big|^{-2} \tag{5.96}$$

Similarly, the second-order intercept point at the sum frequency can be derived from (5.72) and (5.77),

$$IP_{2\Sigma} \approx \frac{1}{2}\frac{G_L(\omega_0)^2|Z_{21}(\omega_0)|^4}{G_L(\Sigma\omega)|Z_{24}(\Sigma\omega)|^2} \tag{5.97}$$
$$\left|G_{m2}Z_{31}(\omega_0)^2 + G_{md}Z_{31}(\omega_0)Z_{41}(\omega_0) + G_{d2}Z_{41}(\omega_0)^2\right|^{-2}$$

and the third-order intercept point for the inband distortion, IP_3, will be from (5.85),

$$\begin{aligned}
IP_3 \approx\ & 2G_L(\omega_0)\frac{|Z_{21}(\omega_0)|^3}{|Z_{24}(\omega_0)|} \\
&\cdot \Big|\big[2G_{m2}Z_{31}(\omega_0)^*Z_{34}(2\omega) + G_{md}Z_{31}(\omega_0)^*Z_{44}(2\omega) \\
&\quad + G_{md}Z_{41}(\omega_0)^*Z_{34}(2\omega) + 2G_{d2}Z_{41}(\omega_0)^*Z_{44}(2\omega)\big] \\
&\cdot \big[G_{m2}Z_{31}(\omega_0)^2 + G_{md}Z_{31}(\omega_0)Z_{41}(\omega_0) + G_{d2}Z_{41}(\omega_0)^2\big] \\
&+ \big[2G_{m2}Z_{31}(\omega_0)Z_{34}(\Delta\omega) + G_{md}Z_{31}(\omega_0)Z_{44}(\Delta\omega) \\
&\quad + G_{md}Z_{41}(\omega_0)Z_{34}(\Delta\omega) + 2G_{d2}Z_{41}(\omega_0)Z_{44}(\Delta\omega)\big] \\
&\cdot \big[2G_{m2}|Z_{31}(\omega_0)|^2 + G_{md}Z_{31}(\omega_0)Z_{41}(\omega_0)^* \\
&\quad + G_{md}Z_{31}(\omega_0)^*Z_{41}(\omega_0) + 2G_{d2}|Z_{41}(\omega_0)|^2\big] \\
&- \big[3G_{m3}Z_{31}(\omega_0)|Z_{31}(\omega_0)|^2 \\
&\quad + G_{m2d}Z_{31}(\omega_0)^2Z_{41}(\omega_0)^* + 2G_{m2d}|Z_{31}(\omega_0)|^2Z_{41}(\omega_0) \\
&\quad + G_{md2}Z_{31}(\omega_0)^*Z_{41}(\omega_0)^2 + 2G_{md2}Z_{31}(\omega_0)|Z_{41}(\omega_0)|^2 \\
&\quad + 3G_{d3}Z_{41}(\omega_0)|Z_{41}(\omega_0)|^2\big]\Big|^{-1}
\end{aligned} \tag{5.98}$$

Although these output intercept points are widely accepted figures of merit for comparing different amplifiers with similar characteristics, they can produce misleading conclusions when used for linearity optimization. As we have already seen in Section 5.2.1, the reason is that one can detect a certain IP_n increase, which may correspond to only a gain change imposed by an output gain factor (e.g., load impedance), and not to the desired dynamic range optimization. So, and according to the dynamic range expression (5.3), we will use input intercept points, $IP_{n_i} = IP_n/G_P$, instead, in the following discussions.

5.2.4.3 Second-Order Distortion Optimization in FET-Based Small-Signal Amplifiers

Comparing $IP_{2\Delta}$ and $IP_{2\Sigma}$, there is a clear resemblance between them. Therefore, we will address only one of these (e.g., $IP_{2\Delta}$) as the conclusions drawn for it can be directly extrapolated to the other.

Noticing that $Z_{21}(\omega) \approx Z_{41}(\omega) \approx -G_m Z_D(\omega) Z_S(\omega) \approx A_v(\omega) Z_S(\omega)$ [where $A_v(\omega)$ is the intrinsic device's voltage gain $A_v(\omega) \equiv V_4(\omega)/V_3(\omega) = V_{ds}(\omega)/V_{gs}(\omega)$], $Z_{11}(\omega) \approx [Y_{in}(\omega) + Y_S(\omega)]^{-1}$, $Z_{31}(\omega) \approx Z_S(\omega)$ and $Z_{24}(\omega) \approx Z_D(\omega)$ we may write:

$$IP_{2\Delta_i} = \frac{IP_{2\Delta}}{G_P} \approx 2G_{in}(\omega_0) \left| \frac{Z_{in}(\omega_0)}{Z_{in}(\omega_0) + Z_S(\omega_0)} \right|^2 \frac{G_L(\omega_0)}{G_L(\Delta\omega)} \left| \frac{A_v(\omega_0)}{Z_D(\Delta\omega)} \right|^2 \cdot \left| 2G_{m2} + G_{md} A_v(\omega_0) + G_{md} A_v(\omega_0)^* + 2G_{d2} |A_v(\omega_0)|^2 \right|^{-2} \tag{5.99}$$

The first conclusion we may extract from this expression is that second-order distortion is only weakly dependent on the input termination, especially at very low frequencies where the FET input tends to an open-circuit. In general, as long as i_{DS} nonlinearity is dominant, and there is no strong internal feedback (via C_{gd} or R_s and L_s), distortion control is restricted to the output port.

For discussing the $IP_{2\Delta_i}$ dependence on $Z_L(\omega)$ we must begin by realizing that if second-order distortion is important, it is because it still falls inside the operation bandwidth. Therefore, we may expect $Z_L(\Delta\omega) \approx Z_L(\omega_0)$, which implies that the only noticeable dependence of $IP_{2\Delta_i}$ on the output termination must come from the varying relative contributions of G_{m2}, $G_{md}A_v$, and $G_{d2}A_v^2$. Since the effect of G_{m2} usually dominates over the ones produced by the other two nonlinear coefficients, $A_v(\omega) \approx -G_m Z_D(\omega)$, and we know that an FET presents a strong transconductance variation with bias, it is likely that $IP_{2\Delta_i}$ will not change too much with $Z_L(\omega)$, at least when low quiescent currents are used.

Figure 5.10 confirms this hypothesis by showing the relative contributions of these three terms for a typical small-signal MESFET, loaded with a reference termination of 50Ω. Since the effect of G_{d2} is negligible, and G_{md} falls to a very low value when G_{m2} reaches its minimum, G_{m2} indeed dominates over $G_m A_v$ and $G_{d2}A_v^2$ in the whole range of V_{GS} bias, limiting any possible control of second-order distortion by load impedance tuning.

Figure 5.10 also shows how the quiescent point affects second-order distortion. It is clear that, after the FET's turn-on, second-order distortion power monotonically decreases with increasing bias, until it reaches a point (sometimes not far from $I_{ds_s}/2$) beyond which it remains nearly constant.

In conclusion, we can expect second-order distortion to be almost independent on the input and output terminations, but show a significant variation with bias

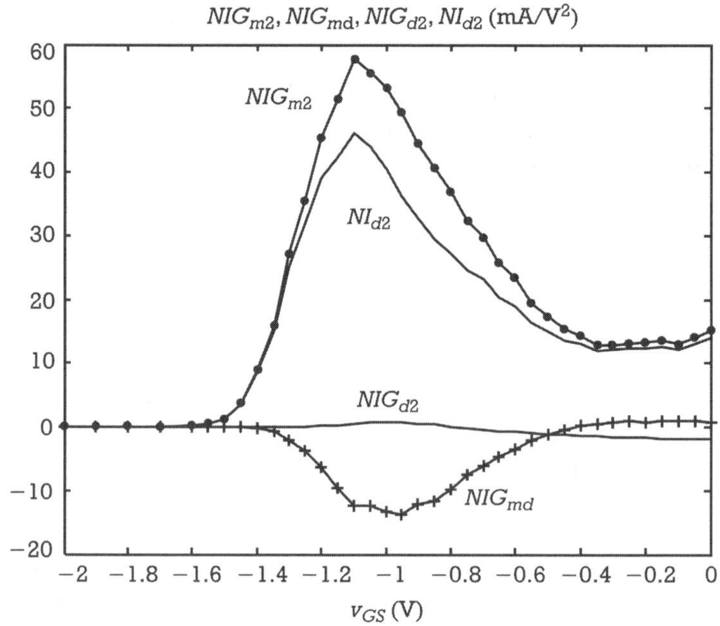

Figure 5.10 Bias point variation of second-order distortion contributions in a small-signal MESFET loaded with 50Ω. NIG_{m2} is the normalized current contribution of the term involving G_{m2} in (5.99), NIG_{md} is the one depending on G_{md}, NIG_{d2} is that of G_{d2}, and NI_{d2} refers to their overall addition.

current. Best linearity is achieved for quiescent currents greater then $I_{ds_s}/2$, which is compatible with gain requirements but presents a significant trade-off with noise figure.

5.2.4.4 Third-Order Distortion Optimization in FET-Based Small-Signal Amplifiers

Now we turn our attention to third-order distortion. Following the above discussion, the third-order input intercept point can be expressed as

$$IP_{3_i} = \frac{IP_3}{G_P} \approx 2G_{in}(\omega_0)\left|\frac{Z_{in}(\omega_0)}{Z_{in}(\omega_0)+Z_S(\omega_0)}\right|^2 \left|\frac{A_v(\omega_0)}{Z_D(\omega_0)}\right|$$
$$\cdot \Big|\big[2G_{m2}Z_{34}(2\omega) + G_{md}Z_D(2\omega)$$
$$+ G_{md}A_v(\omega_0)^* Z_{34}(2\omega) + 2G_{d2}A_v(\omega_0)^* Z_D(2\omega)\big]$$
$$\cdot \left[G_{m2} + G_{md}A_v(\omega_0) + G_{d2}A_v(\omega_0)^2\right]$$

$$
\begin{aligned}
&+ \big[2G_{m2} Z_{34}(\Delta\omega) + G_{md} Z_D(\Delta\omega) \\
&\quad + G_{md} A_v(\omega_0) Z_{34}(\Delta\omega) + 2G_{d2} A_v(\omega_0) Z_D(\Delta\omega)\big] \\
&\cdot \big[2G_{m2} + G_{md} A_v(\omega_0)^* + G_{md} A_v(\omega_0) + 2G_{d2}|A_v(\omega_0)|^2\big] \\
&+ \big[3G_{m3} + G_{m2d} A_v(\omega_0)^* + 2G_{m2d} A_v(\omega_0) + G_{md2} A_v(\omega_0)^2 \\
&\quad + 2G_{md2}|A_v(\omega_0)|^2 + 3G_{d3} A_v(\omega_0)|A_v(\omega_0)|^2\big]\Big|^{-1}
\end{aligned} \tag{5.100}
$$

As stated in (5.84), (5.98), and (5.100), third-order distortion at $2\omega_1 - \omega_2$ can be attributed to two different origins. One is due to remixing of second-order control voltages at $\omega_1 - \omega_2$ with first-order ones at ω_1, or at $2\omega_1$ with $-\omega_2$, while the other arises directly from third-order mixing of ω_1, ω_1, and $-\omega_2$. The first group is recognized from being dependent on the product of two second-degree I_{ds} coefficients, while the other is controlled by terms involving only third-degree coefficients. In common designs, $Z_L(\Delta\omega)$ and $Z_L(2\omega)$ are kept low in order to prevent out-of-band oscillations, which leads to a dominance of the third-degree coefficients' group over the other. However, even if this is not verified, it is almost sure that the internal feedback dependent contributions [the ones involving $Z_{34}(\omega) \approx Z_S(\omega) Z_D(\omega) j\omega C_{gd}$] will be negligible when compared to all the others. This will be especially true for both 2ω and $\Delta\omega$ in low-frequency designs, and almost always obeyed at the difference frequency.

With these considerations in mind, and remembering that G_{m2} usually dominates over $G_{md} A_v$ or $G_{d2} A_v^2$, (5.100) can be further simplified to read

$$
\begin{aligned}
IP_{3_i} &\approx 2G_{in}(\omega_0) \left|\frac{Z_{in}(\omega_0)}{Z_{in}(\omega_0) + Z_S(\omega_0)}\right|^2 \left|\frac{A_v(\omega_0)}{Z_D(\omega_0)}\right| \\
&\cdot \Big|G_{m2}\big[G_{md} Z_D(2\omega) + 2G_{d2} A_v(\omega_0)^* Z_D(2\omega)\big] \\
&+ 2G_{m2}\big[G_{md} Z_D(\Delta\omega) + 2G_{d2} A_v(\omega_0) Z_D(\Delta\omega)\big] \\
&+ \big[3G_{m3} + G_{m2d} A_v(\omega_0)^* + 2G_{m2d} A_v(\omega_0) + G_{md2} A_v(\omega_0)^2 \\
&\quad + 2G_{md2}|A_v(\omega_0)|^2 + 3G_{d3} A_v(\omega_0)|A_v(\omega_0)|^2\big]\Big|^{-1}
\end{aligned} \tag{5.101}
$$

Similarly to what we had already concluded for IP_{2_i}, third-order distortion is again nearly independent of the input termination. This conclusion can now be also extended to the source terminations, at either the difference frequency or the second-harmonic. This is, once more, a consequence of the devices' low internal feedback, and could be confirmed to be true even in presence of C_{gs} nonlinearity and at microwave frequencies [9].

Where load impedance is concerned, it appears at first sight that both inband and out-of-band terminations can be used to control $IP3_i$. Unfortunately, since we are dealing with small-signal amplifiers conceived to operate in highly linear modes and thus producing very low levels at $\Delta\omega$ and $2\omega_0$, in practice, third-order distortion only barely depends on these out-of-band terminations. Actually, a close look into (5.100) and (5.101) shows that these remixing components have a magnitude proportional to the product of two second-degree coefficients, and thus a level (up to 10 times lower) easily absorbed by the direct third-order mixing products. Reference [9] shows that this is verified even in the microwave range, and in presence of a nonlinear C_{gs}. The only way it seems possible that out-of-band effects may show up, is to dramatically reduce G_{m3} contribution. The work of [10] proved that such a situation is indeed possible if the FET is biased in a small-signal IMD sweet-spot (a bias point close to $G_{m3} = 0$), and showed that $Z_L(\Delta\omega)$ may be responsible for long-time constant memory effects manifested as asymmetries in the amplitude levels of distortion falling at $2\omega_1 - \omega_2$ and $2\omega_2 - \omega_1$. As we will see later on, many small-signal amplifiers are biased near one of these IMD sweet-spots. In those cases, precise prediction of third-order IMD is a very difficult task because it becomes dependent on G_{m2d}, G_{md2}, and G_{d3}, the coefficients of mostly inaccurate extraction. Nevertheless, second-order remixing becomes approximately proportional to $G_{m2}G_{md}[Z_D(2\omega_0) + Z_D(\Delta\omega)]$, thus demanding short-circuits for the output terminations at the difference frequency and second harmonics.

IP_3 variation with $Z_L(\omega_0)$ is a widely known effect in MESFET small-signal amplifiers, confirmed by several authors [11–13]. Equations (5.100) and (5.101) show that this behavior is due to the varying relations between the third-degree I_{ds} coefficient terms, as they change with $A_v(\omega)$.

Contrary to what was previously seen for second-degree coefficients, G_{m2d} and G_{md2} can now have an important impact on the overall distortion, as shown in Figure 5.11. Furthermore, since the gate coefficient G_{m3} is still dominant, but has the same sign of G_{m2d}, these two contributions can be combined to produce a certain amount of distortion compensation. According to (5.101), that compensation is optimized when $G_{m2d}[A_v(\omega_0)^* + 2A_v(\omega_0)]$ is the highest positive real number. Since this intrinsic voltage gain was calculated to be $A_v = -G_m Z_D(\omega) = G_m[G_{ds} + j\omega C_{ds} + Y_L(\omega)]^{-1}$, it is clear that the best distortion performance for each $G_L(\omega)$ is obtained when the imaginary part of Y_L, $B_{L_{opt}}$, exactly cancels the imaginary part of the device's output admittance, $B_{out}(\omega)$. This means that constant carrier-to-intermodulation performance, IMR, load-pull contours will be tangent to the Smith chart g_L circles exactly on the line $b_{L_{opt}} = -b_{out}$, and the best value of G_L will be the one that approximately minimizes $[G_{m3} - G_{m2d}G_m(G_L + G_{ds})^{-1} + G_{md2}G_m^2(G_L + G_{ds})^{-2}]$.

Differentiating this expression with respect to G_L and equating to zero, gives a result of

$$G_{L_{opt}} \approx 2G_m \frac{G_{md2}}{G_{m2d}} - G_{ds} \tag{5.102}$$

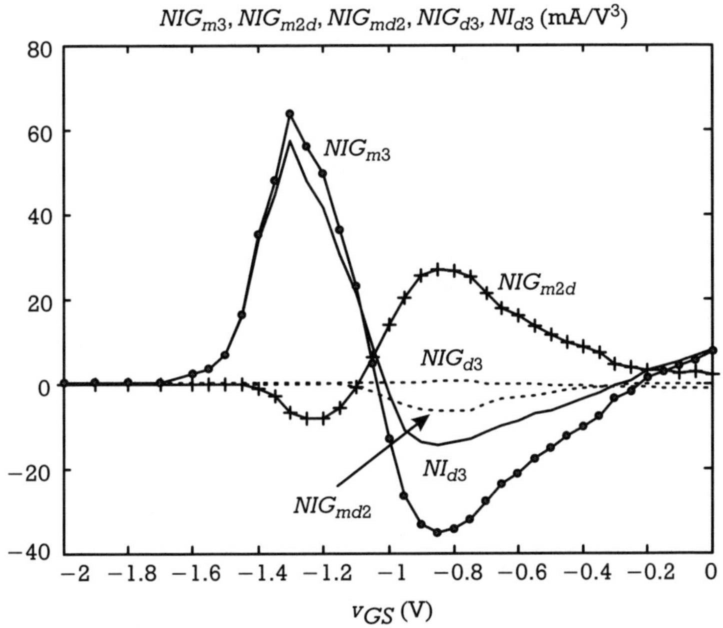

Figure 5.11 Bias point variation of third-order distortion contributions in a small-signal MESFET loaded with 50Ω. NIG_{m3} is the normalized current contribution of the term involving G_{m3} in (5.101), NIG_{m2d} is the one depending on G_{m2d}, NIG_{md2} is that of G_{md2}, NIG_{d3} is that of G_{d3}, and NI_{d3} refers to their overall addition.

which, associated to $B_{L_{opt}} = -B_{out}$, identifies an optimum load impedance for best third-order distortion performance. Obviously, these results should be used with care as they were derived from a very simplified condition. However, as long as the FET's internal feedback is not too high, these theoretical load-pull conclusions can be used, at least, as first-order estimates of optimum output termination.

Those ideas are illustrated in Figure 5.12, in which measured load-pull contours of constant carrier-to-intermodulation ratio, IMR, at 2 GHz are plotted in a Smith chart of $\Gamma_L(\omega_0)$. The line of $b_{L_{opt}} = -b_{out}$ was also drawn passing through the points of predicted tangency of constant g_L circles with the IMR contours.

Figure 5.13 shows simulated IMR load-pull contours corresponding to the ones presented in Figure 5.12, along with G_T circles. It also shows that, for this type of devices, $\Gamma_{L_{opt}}(\omega_0)$ is almost coincident with the zone of optimum gain, a hypothesis also experimentally confirmed in the work reported in [11].

Results depicted in Figure 5.11 show how the various distortion contributions produced by third-degree I_{ds} coefficients change with V_{GS} bias, when the FET is again loaded with a 50Ω reference impedance. The most important information it gives is that the FET presents two third-order small-signal IMD sweet-spots, which are roughly coincident with the nulls of G_{m3}.

The first one corresponds to the G_{m2} maximum, and is a consequence of the sudden rise of G_m determined by the device's turn on. Therefore, a FET biased at

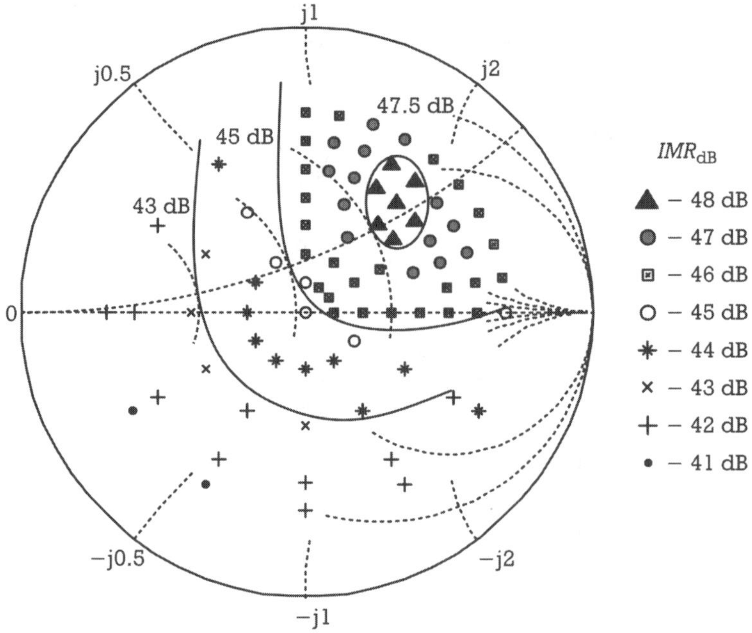

Figure 5.12 Measured constant *IMR* load-pull contours of a small-signal MESFET at 2 GHz.

this sweet-spot would present a poor gain and the worst possible second-order distortion. Also, a strong second-order behavior would inevitably generate third-order distortion by remixing out-of-band products with the fundamentals, unless the load impedances for $\Delta\omega$ and $2\omega_0$ were carefully selected as short-circuits.

The second of these third-order IMD sweet-spots is located in the zone of saturated G_m, and corresponds to the minimum G_{m2}. So, it constitutes an optimum point in terms of both gain, second-order and third-order distortion. Its only drawbacks are that, first, and contrary to the turn-on sweet-spot that is present in all modern FET's, this one is visible in some MESFET's, eventually by some Si JFET's, but never by MOSFET's or HEMT's. Second, it usually appears for quite high quiescent currents, sometimes close to, or even higher than I_{ds_s}, reducing available signal excursion. And third, its associated high currents usually lead to poor noise performance.

In summary, we have seen that third-order distortion is almost insensitive to input terminations, either at inband or out-of-band, leaving room for available power gain or noise figure optimization.

With respect to the output, it could be concluded that the contributions coming from remixing second-order products are usually negligible, unless third-degree coefficients are located near a small-signal IMD sweet-spot. In that case, all out-of-band terminations should be selected as short-circuits. A very low termination at

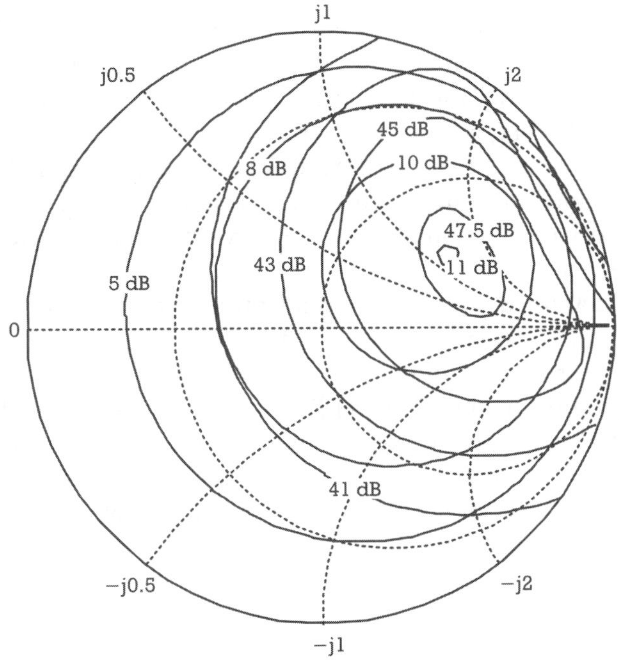

Figure 5.13 Simulated constant transducer power gain circles and *IMR* load-pull contours corresponding to the ones depicted in Figure 5.12.

the difference frequency is normally obtained by simply controlling the decoupling capacitors and RF chokes included in the drain bias network. At $2\omega_0$, this may not be so easy as the required short-circuit condition may conflict with the desired $Z_L(\omega_0)$. Nevertheless, trying to keep low $Z_L(\Delta\omega)$ and $Z_L(2\omega_0)$ is always a good policy, since it helps achieving out-of-band stability and prevents undesirable long-term memory effects (noticed by the RF signal's envelope).

For selecting the appropriate inband output termination, we saw that there is an optimum load admittance, $Y_{L_{opt}}$, which is almost coincident with the one that maximizes gain.

Finally, bias point considerations indicated that third-order distortion is a strong function of the quiescent point. Higher currents generally lead to better IP_{3_i} performance, unless the active device presents a useful small-signal IMD sweet-spot, where it should be then biased. Unfortunately, only a few FET devices present these IMD sweet-spots, and these can be so close to I_{ds_s} that the amplifier becomes limited in output signal excursion.

5.2.4.5 Selecting an Appropriate BJT Device Model

Now we turn our attention to the nonlinear distortion performance of bipolar transistor-based amplifiers.

Since we are interested in small-signal amplifier behavior, it is assumed that our bipolar device—an RF Si NPN BJT, for example—is biased in the forward active region. There, $v_{BE} >> 0$ and $v_{BC} << 0$, forward biasing base-emitter junction and reverse biasing base-collector junction. Under small-signal operation, it is also assumed that base-collector breakdown effects are negligible. For the purpose of enabling hand calculations, and the extraction of simple qualitative conclusions, distributed base phenomena and collector-base capacitance nonlinearity are also neglected.

The simplified equivalent circuit model of our small-signal BJT amplifier, in common-emitter configuration, is drawn in Figure 5.14. A T equivalent circuit topology is herein adopted in detriment of the π network seen in the SPICE implementation of the Gummel-Poon large-signal model shown is Section 4.3.3, because it allows a more explicit and intuitive treatment of most important small-signal bipolar transistor nonlinearities: exponential emitter junction current, exponential emitter junction diffusion charge, and forward current-gain nonlinearity.

In Figure 5.14 (I_S, Y_S) is, again, the Norton equivalent circuit of an equal amplitude two-tone signal source

$$i_S(t) = \frac{1}{2} \sum_{\substack{q=-2 \\ q \neq 0}}^{2} I_{s_q} e^{j\omega_q t} \tag{5.103}$$

plus the input matching network, base bias circuitry and base parasitics. Y_L stands for the equivalent load terminating impedance, lumping the load impedance, output matching circuit, collector bias circuitry, and collector parasitics. This small-signal

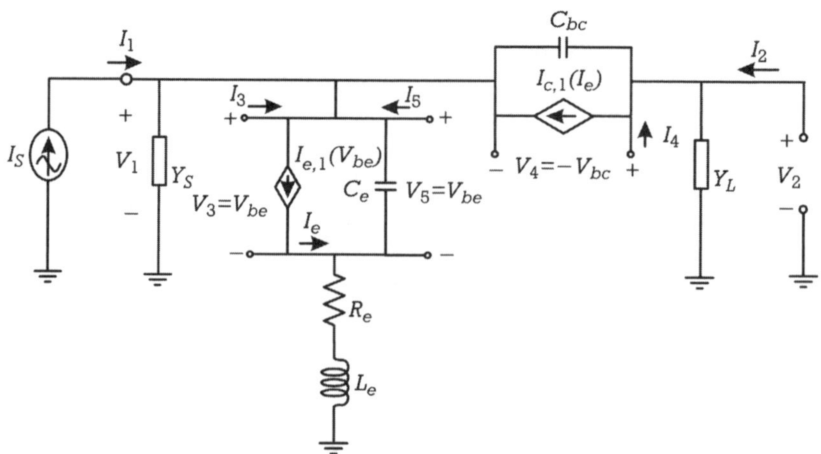

Figure 5.14 Detailed equivalent circuit model of a BJT-based amplifier.

model assumes that $i_e(v_{be}, v_{bc})$ must be given by a bidimensional Taylor series of the signal components v_{be} and v_{bc}:

$$i_e(v_{be}, v_{bc}) = G_e v_{be} + G_c v_{bc} + G_{e2} v_{be}^2 + G_{ec} v_{be} v_{bc} + G_{c2} v_{bc}^2 + G_{e3} v_{be}^3 + G_{e2c} v_{be}^2 v_{bc} + G_{ec2} v_{be} v_{bc}^2 + G_{c3} v_{bc}^3 \quad (5.104)$$

accounting for the BJT base-emitter junction and Early effect nonlinearities. Then, $i_c(v_{be}, v_{bc})$ is expressed as a nonlinear function of $i_e(v_{be}, v_{bc})$ or $i_c[i_e(v_{be}, v_{bc})]$:

$$i_c(i_e) = \alpha(i_e) i_e = \alpha_1 i_e + \alpha_2 i_e^2 + \alpha_3 i_e^3 \quad (5.105)$$

this way modeling forward current gain nonlinearity, since the transistor's β can be given by $\beta = \alpha/(1 - \alpha)$.

Because collector-to-emitter resistance of a BJT is usually negligible in comparison to the much lower load resistances normally used in RF circuits, and because it is known that exponential emitter junction nonlinearity usually dominates over the Early effect, it can be anticipated that the benefit of including v_{bc} in (5.104) will certainly not pay for the excess labor required to handle a bidimensional series. So, even though that complete i_e form should be used in a general purpose computer program, in the following simplified hand analysis we will assume that any term involving v_{bc} is zero.

Finally, the current through the input diffusion charge, $q_{be}(v_{be})$, is modeled by

$$i_q = \frac{d[q_{be}(v_{be})]}{dt} = \frac{d}{dv_{be}}\left(C_e v_{be} + C_{e2} v_{be}^2 + C_{e3} v_{be}^3\right)\frac{dv_{be}}{dt} = \left(C_e + 2C_{e2} v_{be} + 3C_{e3} v_{be}^2\right)\frac{dv_{be}}{dt} \quad (5.106)$$

5.2.4.6 Small-Signal BJT Amplifier Distortion Prediction

Following the procedure already used for FET-based small-signal amplifiers, the frequency-domain equations describing the linear relations between port currents and voltages are

$$\begin{bmatrix} V_1(\omega) \\ V_2(\omega) \\ V_3(\omega) \\ V_4(\omega) \\ V_5(\omega) \end{bmatrix} = \begin{bmatrix} Z_{11} & Z_{12} & Z_{13} & Z_{14} & Z_{15} \\ Z_{21} & Z_{22} & Z_{23} & Z_{24} & Z_{25} \\ Z_{31} & Z_{32} & Z_{33} & Z_{34} & Z_{35} \\ Z_{41} & Z_{42} & Z_{43} & Z_{44} & Z_{45} \\ Z_{51} & Z_{52} & Z_{53} & Z_{54} & Z_{55} \end{bmatrix} \cdot \begin{bmatrix} I_1(\omega) \\ I_2(\omega) \\ I_3(\omega) \\ I_4(\omega) \\ I_5(\omega) \end{bmatrix} \quad (5.107)$$

and

$$I_1(\omega) = I_S(\omega) \tag{5.108}$$

$$I_2(\omega) = 0 \tag{5.109}$$

$$\begin{aligned} I_3(\omega) &= -[I_{e2}(\omega) + I_{e3}(\omega)] \\ &= -G_{e2} V_{3,1}(\omega_{q_1}) * V_{3,1}(\omega_{q_2}) \\ &\quad -2G_{e2} V_{3,1}(\omega_{q_1}) * V_{3,2}(\omega_{q_2} + \omega_{q_3}) \\ &\quad - G_{e3} V_{3,1}(\omega_{q_1}) * V_{3,1}(\omega_{q_2}) * V_{3,1}(\omega_{q_3}) \end{aligned} \tag{5.110}$$

$$\begin{aligned} I_4(\omega) &= -[I_{c2}(\omega) + I_{c3}(\omega)] \\ &= -\alpha_2 I_{e,1}(\omega_{q_1}) * I_{e,1}(\omega_{q_2}) \\ &\quad -2\alpha_2 I_{e,1}(\omega_{q_1}) * I_{e,2}(\omega_{q_2} + \omega_{q_3}) \\ &\quad - \alpha_3 I_{e,1}(\omega_{q_1}) * I_{e,1}(\omega_{q_2}) * I_{e,1}(\omega_{q_3}) \end{aligned} \tag{5.111}$$

$$\begin{aligned} I_5(\omega) &= -[I_{q2}(\omega) + I_{q3}(\omega)] \\ &= -2C_{e2} V_{5,1}(\omega_{q_1}) * \left[j\omega_{q_2} V_{5,1}(\omega_{q_2}) \right] \\ &\quad -2C_{e2} V_{5,1}(\omega_{q_1}) * \left[j(\omega_{q_2} + \omega_{q_3}) V_{5,2}(\omega_{q_2} + \omega_{q_3}) \right] \\ &\quad -2C_{e2} V_{5,2}(\omega_{q_1} + \omega_{q_2}) * \left[j\omega_{q_3} V_{5,1}(\omega_{q_3}) \right] \\ &\quad -3C_{e3} \left[V_{5,1}(\omega_{q_1}) * V_{5,1}(\omega_{q_2}) \right] * \left[j\omega_{q_3} V_{5,1}(\omega_{q_3}) \right] \end{aligned} \tag{5.112}$$

where the spectra operator "*" represents again the pseudoconvolutions expressed in Section 3.2.2.1.

Finally,

$$V_3(\omega) = V_5(\omega) = V_{be}(\omega) \tag{5.113}$$

$$V_4(\omega) = -V_{bc}(\omega) \tag{5.114}$$

Taking into account that our two-tone excitation is again composed of closely spaced frequencies $\omega_1 \approx \omega_2 \approx \omega_0$, we have $\Delta\omega \equiv \omega_1 - \omega_2 << \omega_1, \omega_2$; $\Sigma\omega \equiv \omega_1 + \omega_2 \approx 2\omega_1 \approx 2\omega_2 \equiv 2\omega$ and thus $Z_{ij}(\omega_1) \approx Z_{ij}(\omega_2) \approx Z_{ij}(2\omega_1 - \omega_2) \approx Z_{ij}(2\omega_2 - \omega_1) \approx Z_{ij}(\omega_0)$. Under this situation, the relevant Z-parameters are

$$Z_{21}(\omega) = \frac{\alpha_1 G_e - j\omega C_{bc}(1 + Z_e Y_{be})}{\Delta(\omega)} \tag{5.115}$$

$$Z_{23}(\omega) = -\frac{\alpha_1 [Y_S + j\omega C_e(1 + Z_e Y_S)] + j\omega C_{bc}}{\Delta(\omega)} \tag{5.116}$$

$$Z_{24}(\omega) = -\frac{Y_{be} + Y_S(1 + Z_e Y_{be})}{\Delta(\omega)} \tag{5.117}$$

$$Z_{25}(\omega) = \frac{\alpha_1 G_e(1 + Z_e Y_S) - j\omega C_{bc}}{\Delta(\omega)} \tag{5.118}$$

$$Z_{31}(\omega) = Z_{51}(\omega) = -\frac{Y_L + j\omega C_{bc}}{\Delta(\omega)} \tag{5.119}$$

$$Z_{33}(\omega) = Z_{53}(\omega) = -\frac{Z_e Y_L(Y_S + j\omega C_{bc}) + j\omega C_{bc}(1 + Z_e Y_S) + (1 - \alpha_1) Y_L}{\Delta(\omega)} \tag{5.120}$$

$$Z_{34}(\omega) = Z_{54}(\omega) = \frac{Y_L}{\Delta(\omega)} \tag{5.121}$$

$$Z_{35}(\omega) = Z_{55}(\omega) = -\frac{[Z_e(Y_S + j\omega C_{bc}) + 1] Y_L + j\omega C_{bc}(1 + Z_e Y_S)}{\Delta(\omega)} \tag{5.122}$$

in which $Z_e \equiv Z_e(\omega) = R_e + j\omega L_e$, $Y_{be} \equiv Y_{be}(\omega) = G_e + j\omega C_e$, $Y_S \equiv Y_S(\omega) = 1/Z_S(\omega)$ and $Y_L \equiv Y_L(\omega) = 1/Z_L(\omega)$ and $\Delta(\omega) \equiv [\alpha_1 G_e - (Y_S + j\omega C_{bc})(1 + Z_e Y_{be}) - Y_{be}] Y_L - j\omega C_{bc}[Y_{be} + Y_S(1 + Z_e Y_{be})]$.

Contrary to the more usual nonlinear currents' method implementation, in which all control variables are voltages, we now have $v_3(t)$ for $i_e(t)$, $v_5(t)$ for $i_q(t)$ but also $i_e(t)$ for $i_c(t)$. Because $v_3(t) = v_5(t) = v_{be}(t)$, we really need to determine only two control variables. In the frequency-domain, their first-order components are given by

$$V_{be,1}(\omega_0) = Z_{31}(\omega_0) \frac{I_S(\omega_0)}{2} \tag{5.123}$$

$$I_{e,1}(\omega_0) = G_e V_{be,1}(\omega_0) = G_e Z_{31}(\omega_0) \frac{I_S(\omega_0)}{2} \tag{5.124}$$

The fundamental component of the output voltage, $V_{2,1}(\omega)$, is

$$V_{2,1}(\omega_0) = Z_{21}(\omega_0)\frac{I_S(\omega_0)}{2} \tag{5.125}$$

Second-order nonlinear currents are then

$$I_{3,2}(\Delta\omega) = -2G_{e2}|Z_{31}(\omega_0)|^2\frac{|I_S|^2}{4} \tag{5.126}$$

$$I_{3,2}(\Sigma\omega) = -2G_{e2}Z_{31}(\omega_0)^2\frac{I_S^2}{4} \tag{5.127}$$

$$I_{3,2}(2\omega) = -G_{e2}Z_{31}(\omega_0)^2\frac{I_S^2}{4} \tag{5.128}$$

and

$$I_{4,2}(\Delta\omega) = -2\alpha_2 G_e^2|Z_{31}(\omega_0)|^2\frac{|I_S|^2}{4} \tag{5.129}$$

$$I_{4,2}(\Sigma\omega) = -2\alpha_2 G_e^2 Z_{31}(\omega_0)^2\frac{I_S^2}{4} \tag{5.130}$$

$$I_{4,2}(2\omega) = -\alpha_2 G_e^2 Z_{31}(\omega_0)^2\frac{I_S^2}{4} \tag{5.131}$$

and also

$$I_{5,2}(\Delta\omega) = -2j\Delta\omega C_{e2}|Z_{31}(\omega_0)|^2\frac{|I_S|^2}{4} \tag{5.132}$$

$$I_{5,2}(\Sigma\omega) = -2j\Sigma\omega C_{e2}Z_{31}(\omega_0)^2\frac{I_S^2}{4} \tag{5.133}$$

$$I_{5,2}(2\omega) = -2j\omega_0 C_{e2}Z_{31}(\omega_0)^2\frac{I_S^2}{4} \tag{5.134}$$

According to the narrowband excitation assumption, the second-order current component at $\Delta\omega$ generated by q_{be} becomes negligible in comparison to other second-order contributions. Thus, we will make $I_{5,2}(\Delta\omega) \approx 0$. The remaining nonlinear currents produce second-order control variables of

$$V_{be,2}(\Delta\omega) \approx Z_{33}(\Delta\omega)I_{3,2}(\Delta\omega) + Z_{34}(\Delta\omega)I_{4,2}(\Delta\omega) \tag{5.135}$$

$$V_{be,2}(\Sigma\omega) \approx Z_{33}(\Sigma\omega)I_{3,2}(\Sigma\omega) + Z_{34}(\Sigma\omega)I_{4,2}(\Sigma\omega) + Z_{35}(\Sigma\omega)I_{5,2}(\Sigma\omega) \tag{5.136}$$

$$V_{be,2}(2\omega) \approx Z_{33}(2\omega)I_{3,2}(2\omega) + Z_{34}(2\omega)I_{4,2}(2\omega) + Z_{35}(2\omega)I_{5,2}(2\omega) \tag{5.137}$$

and

$$\begin{aligned} I_{e,2}(\Delta\omega) &\approx G_e V_{be,2}(\Delta\omega) - I_{3,2}(\Delta\omega) \\ &= [G_e Z_{33}(\Delta\omega) - 1]I_{3,2}(\Delta\omega) + G_e Z_{34}(\Delta\omega)I_{4,2}(\Delta\omega) \end{aligned} \tag{5.138}$$

$$\begin{aligned} I_{e,2}(\Sigma\omega) &\approx G_e V_{be,2}(\Sigma\omega) - I_{3,2}(\Sigma\omega) \\ &= [G_e Z_{33}(\Sigma\omega) - 1]I_{3,2}(\Sigma\omega) + G_e Z_{34}(\Sigma\omega)I_{4,2}(\Sigma\omega) \\ &\quad + G_e Z_{35}(\Sigma\omega)I_{5,2}(\Sigma\omega) \end{aligned} \tag{5.139}$$

$$\begin{aligned} I_{e,2}(2\omega) &\approx G_e V_{be,2}(2\omega) - I_{3,2}(2\omega) \\ &= [G_e Z_{33}(2\omega) - 1]I_{3,2}(2\omega) + G_e Z_{34}(2\omega)I_{4,2}(2\omega) \\ &\quad + G_e Z_{35}(2\omega)I_{5,2}(2\omega) \end{aligned} \tag{5.140}$$

and an output second-order voltage component of

$$V_{2,2}(\Delta\omega) \approx Z_{23}(\Delta\omega)I_{3,2}(\Delta\omega) + Z_{24}(\Delta\omega)I_{4,2}(\Delta\omega) \tag{5.141}$$

$$V_{2,2}(\Sigma\omega) = Z_{23}(\Sigma\omega)I_{3,2}(\Sigma\omega) + Z_{24}(\Sigma\omega)I_{4,2}(\Sigma\omega) + Z_{25}(\Sigma\omega)I_{5,2}(\Sigma\omega) \tag{5.142}$$

$$V_{2,2}(2\omega) = Z_{23}(2\omega)I_{3,2}(2\omega) + Z_{24}(2\omega)I_{4,2}(2\omega) + Z_{25}(2\omega)I_{5,2}(2\omega) \tag{5.143}$$

Now, inband third-order distortion components at $2\omega_1 - \omega_2$ become

$$
\begin{aligned}
I_{3,3}(2\omega_1 - \omega_2) \approx & -2G_{e2}Z_{31}(\omega_0)^* \\
& \cdot [Z_{33}(2\omega)I_{3,2}(2\omega) + Z_{34}(2\omega)I_{4,2}(2\omega) + Z_{35}(2\omega)I_{5,2}(2\omega)]\frac{I_S^*}{2} \\
& - 2G_{e2}Z_{31}(\omega_0)[Z_{33}(\Delta\omega)I_{3,2}(\Delta\omega) + Z_{34}(\Delta\omega)I_{4,2}(\Delta\omega)]\frac{I_S}{2} \\
& - 3G_{e3}Z_{31}(\omega_0)|Z_{31}(\omega_0)|^2\frac{I_S|I_S|^2}{8} \qquad (5.144a)
\end{aligned}
$$

or

$$
\begin{aligned}
I_{3,3}(2\omega_1 - \omega_2) \approx & \; Z_{31}(\omega_0)|Z_{31}(\omega_0)|^2 \\
& \{2G_{e2}[G_{e2}(Z_{33}(2\omega) + 2Z_{33}(\Delta\omega)) + \alpha_2 G_e^2(Z_{34}(2\omega) \\
& + 2Z_{34}(\Delta\omega)) + 2j\omega_0 C_{e2}Z_{35}(2\omega)] - 3G_{e3}\}\frac{I_S|I_S|^2}{8} \qquad (5.144b)
\end{aligned}
$$

$$
\begin{aligned}
I_{4,3}(2\omega_1 - \omega_2) \approx & -2\alpha_2 G_e Z_{31}(\omega_0)^* \\
& \cdot \{[G_e Z_{33}(2\omega) - 1]I_{3,2}(2\omega) + G_e Z_{34}(2\omega)I_{4,2}(2\omega) \\
& + G_e Z_{35}(2\omega)I_{5,2}(2\omega)\}\frac{I_S^*}{2} \\
& - 2\alpha_2 G_e Z_{31}(\omega_0)\{[G_e Z_{33}(\Delta\omega) - 1]I_{3,2}(\Delta\omega) \\
& + G_e Z_{34}(\Delta\omega)I_{4,2}(\Delta\omega)\}\frac{I_S}{2} \\
& - 3\alpha_3 G_e^3 Z_{31}(\omega_0)|Z_{31}(\omega_0)|^2\frac{I_S|I_S|^2}{8} \qquad (5.145a)
\end{aligned}
$$

or

$$I_{4,3}(2\omega_1 - \omega_2) \approx Z_{31}(\omega_0)|Z_{31}(\omega_0)|^2$$
$$\{2\alpha_2 G_e[G_{e2}(G_e Z_{33}(2\omega) + 2G_e Z_{33}(\Delta\omega) - 3)$$
$$+ \alpha_2 G_e^2(G_e Z_{34}(2\omega) + 2G_e Z_{34}(\Delta\omega)) + 2j\omega_0 C_{e2} G_e Z_{35}(2\omega)]$$
$$- 3\alpha_3 G_e^3\} \frac{I_S|I_S|^2}{8} \quad (5.145b)$$

and

$$I_{5,3}(2\omega_1 - \omega_2) \approx -2j\omega_0 C_{e2} Z_{31}(\omega_0)^*$$
$$\cdot [Z_{33}(2\omega) I_{3,2}(2\omega) + Z_{34}(2\omega) I_{4,2}(2\omega) + Z_{35}(2\omega) I_{5,2}(2\omega)] \frac{I_S^*}{2}$$
$$- 2j\omega_0 C_{e2} Z_{31}(\omega_0)[Z_{33}(\Delta\omega) I_{3,2}(\Delta\omega) + Z_{34}(\Delta\omega) I_{4,2}(\Delta\omega)] \frac{I_S}{2}$$
$$- 3j\omega_0 C_{e3} Z_{31}(\omega_0)|Z_{31}(\omega_0)|^2 \frac{I_S|I_S|^2}{8} \quad (5.146a)$$

or even

$$I_{5,3}(2\omega_1 - \omega_2) \approx Z_{31}(\omega_0)|Z_{31}(\omega_0)|^2$$
$$\{2j\omega_0 C_{e2}[G_{e2}(Z_{33}(2\omega) + 2Z_{33}(\Delta\omega))$$
$$+ \alpha_2 G_e^2(Z_{34}(2\omega) + 2Z_{34}(\Delta\omega))$$
$$+ 2j\omega_0 C_{e2} Z_{35}(2\omega)] - 3j\omega_0 C_{e3}\} \frac{I_S|I_S|^2}{8} \quad (5.146b)$$

Finally, third-order output voltage distortion components can be calculated from these nonlinear currents as

$$V_{2,3}(2\omega_1 - \omega_2) \approx Z_{23}(\omega_0) I_{3,3}(2\omega_1 - \omega_2)$$
$$+ Z_{24}(\omega_0) I_{4,3}(2\omega_1 - \omega_2) \quad (5.147)$$
$$+ Z_{25}(\omega_0) I_{5,3}(2\omega_1 - \omega_2)$$

Following the calculations made in the precedent section for the FET amplifier, we have

$$P_L = \frac{1}{2} G_L(\omega_0)|Z_{21}(\omega_0)|^2 |I_S|^2 \tag{5.148}$$

for the fundamental output power per tone, and

$$P_{L_{2\Delta}} = 2G_L(\Delta\omega)|V_{2,2}(\Delta\omega)|^2 \tag{5.149}$$

$$P_{L_{2\Sigma}} = 2G_L(\Sigma\omega)|V_{2,2}(\Sigma\omega)|^2 \tag{5.150}$$

for the output powers of second-order distortion at the difference and sum frequencies, and also

$$P_{L_3} = 2G_L(\omega_0)|V_{2,3}(2\omega_1 - \omega_2)|^2 \tag{5.151}$$

for third-order distortion at $2\omega_1 - \omega_2$.

5.2.4.7 Second-Order Distortion Optimization in BJT-Based Small-Signal Amplifiers

Using the results derived above, we now calculate the corresponding output and input second-order intercept points. Following (5.95), we get

$$IP_{2\Delta} = \frac{1}{2} \frac{G_L(\omega_0)^2}{G_L(\Delta\omega)} \frac{|Z_{21}(\omega_0)|^4}{|Z_{31}(\omega_0)|^4} |G_{e2}Z_{23}(\Delta\omega) + \alpha_2 G_e^2 Z_{24}(\Delta\omega)|^{-2} \tag{5.152}$$

$$IP_{2\Sigma} = \frac{1}{2} \frac{G_L(\omega_0)^2}{G_L(\Sigma\omega)} \frac{|Z_{21}(\omega_0)|^4}{|Z_{31}(\omega_0)|^4} \\ |G_{e2}Z_{23}(\Sigma\omega) + \alpha_2 G_e^2 Z_{24}(\Sigma\omega) + j\Sigma\omega C_{e2} Z_{25}(\Sigma\omega)|^{-2} \tag{5.153}$$

or, with the power gain expression of (5.90),

$$IP_{2\Delta_i} = \frac{1}{2} \frac{G_{in}(\omega_0)G_L(\omega_0)}{G_L(\Delta\omega)} \frac{|Z_{11}(\omega_0)|^2 |Z_{21}(\omega_0)|^2}{|Z_{31}(\omega_0)|^4} \\ |G_{e2}Z_{23}(\Delta\omega) + \alpha_2 G_e^2 Z_{24}(\Delta\omega)|^{-2} \tag{5.154}$$

$$IP_{2\Sigma_i} = \frac{1}{2} \frac{G_{in}(\omega_0) G_L(\omega_0)}{G_L(\Sigma\omega)} \frac{|Z_{11}(\omega_0)|^2 |Z_{21}(\omega_0)|^2}{|Z_{31}(\omega_0)|^4} \tag{5.155}$$

$$|G_{e2} Z_{23}(\Sigma\omega) + \alpha_2 G_e^2 Z_{24}(\Sigma\omega) + j\Sigma\omega C_{e2} Z_{25}(\Sigma\omega)|^{-2}$$

Before we can proceed to the qualitative analysis of these intercept points, some simplifications have to be made to the Z-parameters' expressions.

Since both $|Y_L(\omega)|$ and $|Y_S(\omega)|$ are usually much greater than ωC_{bc} in the major part of the device's operating frequency range, the common Z-parameters denominator $\Delta(\omega)$ of (5.115) to (5.122) can be approximated by

$$\Delta(\omega) \approx -[(1 - \alpha_1) G_e + j\omega C_e + Y_S(1 + Z_e Y_{be})] Y_L \tag{5.156}$$

Because $Z_{11}(\omega)$ can be expressed as

$$Z_{11}(\omega) = -\frac{(Y_L + j\omega C_{bc})(1 + Z_e Y_{be})}{\Delta(\omega)} \approx -\frac{Y_L(1 + Z_e Y_{be})}{\Delta(\omega)} \tag{5.157}$$

and, for wideband systems where second-order distortion is important $Z_L(\omega_0) \approx Z_L(\Delta\omega) \approx Z_L(\Sigma\omega)$, $IP_{2\Sigma_i}$, for example, can be approximately expressed by

$$IP_{2\Sigma_i} \approx \frac{1}{2} G_{in}(\omega_0) \frac{|1 + Z_e(\omega_0) Y_{be}(\omega_0)|^2}{|Y_L(\omega_0)|^2} |\Delta(\Sigma\omega)|^2$$

$$\left| j\Sigma\omega(1 + Z_e Y_S) C_{e2} - \frac{G_{e2}}{G_e} [Y_S + j\Sigma\omega C_e (1 + Z_e Y_S)] \right.$$

$$\left. - \frac{\alpha_2}{\alpha_1} G_e [Y_{be} + Y_S(1 + Z_e Y_{be})] \right|^{-2} \tag{5.158}$$

Several interesting conclusions can now be drawn from this expression, which can also be extrapolated to $IP_{2\Delta_i}$.

The first one regards the distortion dependence on the output terminating impedance. Since we have assumed that $Y_L(\Sigma\omega) \approx Y_L(\omega_0)$ and $\Delta(\omega)$ is proportional to $Y_L(\omega)$, it is clear that second-order distortion will be a very weak function of $Y_L(\omega)$. In fact, (5.158) implies that only by the unconsidered and much smaller nonlinearities of $C_{bc}(v_{bc})$ and $i_c(v_{bc})$ could second-order distortion show variation with output termination. This is consistent with the fact that in a bipolar device the most important nonlinearities are associated to the device's input.

The second conclusion refers to the dependence on input termination. At the lower end of frequency band, input capacitance nonlinearity vanishes, and the term of $i_e(v_{be})$ is proportional to $Y_S(\Sigma\omega)$. Thus, and unless $(1-\alpha_1)G_e >> |Y_S(1+Z_e.G_e)|$, second-order distortion will be again a weak function of $Y_S(\omega)$. However, when Y_S is very low $(1-\alpha_1)G_e >> |Y_S(1+Z_e.G_e)|$, $IP_{2\Sigma_i}$ improves each time $Y_S(\omega)$ is decreased, up to the point where it becomes dominated by current gain nonlinearity, $\alpha(i_e)$.

The third implication of (5.158) deals with second-order distortion behavior with frequency. Since diffusion charge, $q_{be}(v_{be})$, is intimately related with $i_e(v_{be})$ nonlinearity by $C_e = G_e\tau_F$ and $C_{e2} = G_{e2}\tau_F$ (where τ_F is the BJT forward transit time), the nonlinear current contributions of these two dominant sources of distortion tend to cancel when $|\Sigma\omega C_e(1+Z_e.Y_S)| >> |Y_S|$, leaving only the remaining $-G_{e2}/G_eY_S$ term and $\alpha(i_e)$ contributions. This implies that second-order distortion of a bipolar device is likely to decrease with frequency, a phenomenon already predicted and experimentally reported by other workers [14, 15].

This rather remarkable behavior was also used to explain the surprisingly good linearity of modern HBT's although they are recognized to rely on the same strongly nonlinear exponential dependence of i_E on v_{BE} as BJTs [16]. In practice, however, gain optimization demands for an approximate input matching, which imposes $Y_S(\omega) = Y_{in}(\omega)^*$, where $Y_{in}(\omega)$ is given by

$$Y_{in}(\omega) = \frac{1}{Z_{11}(\omega)} - Y_S(\omega) \approx \frac{(1-\alpha_1)G_e + j\omega C_e}{(1+Z_eY_{be})} \tag{5.159}$$

Equation (5.159) shows that $Y_S(\omega)$ must track ωCe, and thus, exact cancellation will unfortunately not occur. Moreover, it is clear that even if the $|\omega C_e(1+Z_e.Y_S)| >> |Y_S|$ condition is met for the sum frequency, it cannot be guaranteed for the difference—at least when $\Delta\omega << \Sigma\omega$—and so such high-frequency cancellation effect will hardly be observed at the difference frequency.

Figure 5.15 summarizes the frequency variation of second-order distortion, arising from a transistor having linear or nonlinear base-emitter capacitance, for reference impedance terminations of $Z_S(\omega) = Z_L(\omega) = 50\Omega$. It includes a reference graph [Figure 5.15(a)] of fundamental output power per tone for both linear and nonlinear base-emitter capacitance, and reports second-order output intercept point at the sum and difference frequencies (to compensate for IP_{2i} changes induced by the input gain variation imposed by the $C_e/(1-\alpha)G_e$ input time-constant). By comparing the results of $IP_{2\Sigma}$ obtained for the device with linear and nonlinear capacitance, Figure 5.15(b) shows that $i_e(v_{be})$ is, in fact, progressively canceled by the nonlinear $q_{be}(v_{be})$ when frequency gets higher. As expected, Figure 5.15(c) shows that such a compensation effect does not take place for the difference frequency.

In terms of bias dependence, we should begin by noticing that G_e, G_{e2}, and C_{e2} are all approximately proportional to quiescent current. Therefore, because of $G_{in}(\omega_0)$ but also, with a smaller extent, $|1+Z_e(\omega)Y_{be}(\omega)|^2$ terms, $IP_{2\Sigma_i}$ grows

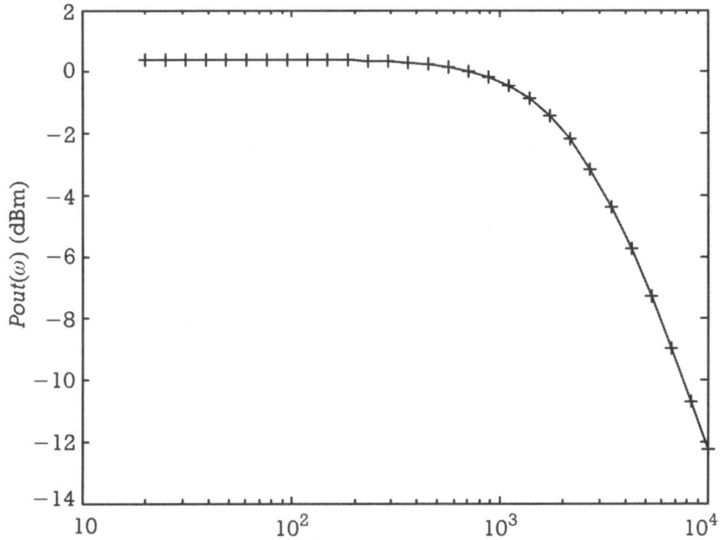

(a)

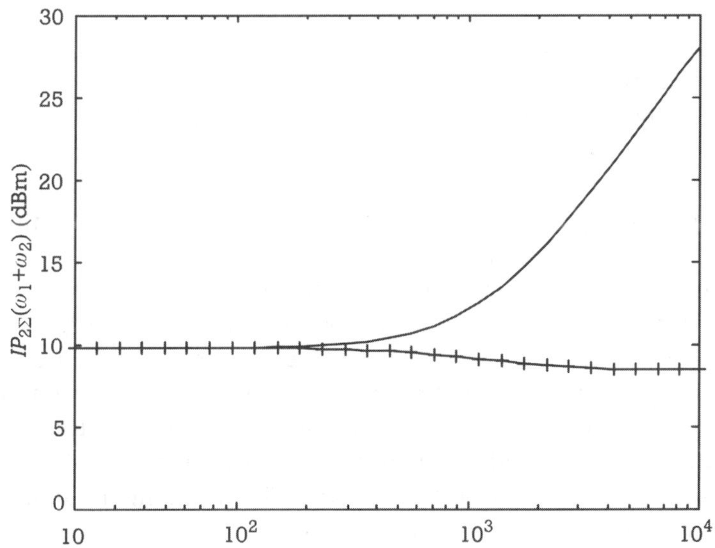

(b)

Figure 5.15 Simulated second-order output intercept point variation with frequency for a typical BJT device with linear (-+-) or nonlinear (–) base-emitter capacitance when $Z_S(\omega) = Z_L(\omega) = 50\Omega$. (a) Fundamental output power; (b) output second-order intercept point at the sum frequency ($2Freq + 10$ MHz) $IP_{2\Sigma}$; and (c) output second-order intercept point at the difference frequency (10 MHz), $IP_{2\Delta}$.

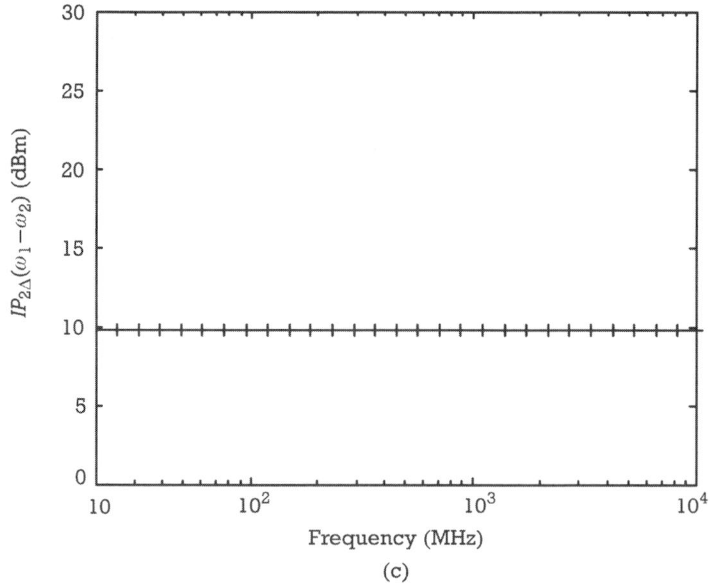

Figure 5.15 (continued).

with I_C bias, until high level injection effects become important, increasing $\alpha(i_e)$ nonlinearity.

Finally, let us make a remark on the importance of emitter impedance, Z_e, which has a current-series feedback effect on the BJT operation. Despite its reasonably low values, the dual of the Miller's theorem indicates that its effective value becomes multiplied by the device's current gain. In (5.158), this is modeled by the term $Z_e Y_{be}$, present in $|1 + Z_e(\omega_0) Y_{be}(\omega_0)|^2$ and in $\Delta(\Sigma\omega)$, which may be significantly greater than one, and thus contribute to reduce distortion. In HBTs, where the emitter low doping concentration produces emitter resistances on the order of 1 to 10Ω, this feedback effect may be a serious candidate for explaining these device's good linearity figures of merit [17]. Unfortunately, even though Z_e is sometimes used to get flat gain in broadband amplifiers, this form of feedback also degrades gain and stability margin. So, unless the loss of gain can be acceptable and stability is guaranteed by, for example, providing input and/or output resistive loading, these associated drawbacks reduce the use of emitter degeneration as a means to intentionally linearize the device.

5.2.4.8 Third-Order Distortion Optimization in BJT-Based Small-Signal Amplifiers

The increased complexity of third-order nonlinear current contributions, when compared to second-order ones, does not allow a general analysis of IP_3 similar to the one above performed for IP_2. Actually, third-order distortion is controlled by third-degree coefficients, but also by second-order mixing products arising from all nonlinearities. These may have magnitudes comparable to the ones associated with direct third-order mixing products, unless some special source terminations are considered. So, since it does not seem viable to present the whole third-order distortion picture, we will discuss two particular views that correspond to (1) very low frequencies, where $q_{be}(v_{be})$ effects can be neglected, and (2) high frequencies, where cancellation effects of $i_e(v_{be})$ and $q_{be}(v_{be})$ are evident. They will determine, therefore, the low and high frequency third-order distortion asymptotic characteristics.

Starting with very low frequencies (herein defined by Y_S, $(1-\alpha_1)G_e >> \omega C_e$ or $\omega < \omega_T/\beta$), it can be considered that second-order contributions of $i_e(v_{be})$ dominate over those of $q_{be}(v_{be})$ or $i_c(i_e)$. And so,

$$IP_3 \approx 2G_L(\omega_0)\frac{|Z_{21}(\omega_0)|^3}{|Z_{31}(\omega_0)|^3}\Big|Z_{23}(\omega_0)\{2G_{e2}^2[Z_{33}(2\omega)+2Z_{33}(\Delta\omega)]-3G_{e3}\}$$
$$+Z_{24}(\omega_0)\{2\alpha_2 G_e G_{e2}[G_e Z_{33}(2\omega)+2G_e Z_{33}(\Delta\omega)-3]-3\alpha_3 G_e^3\}\Big|^{-1} \quad (5.160)$$

and

$$IP_{3_i} \approx 2G_{in}(\omega_0)\frac{|Z_{11}(\omega_0)|^2|Z_{21}(\omega_0)|}{|Z_{31}(\omega_0)|^3}$$
$$\Big|Z_{23}(\omega_0)\{2G_{e2}^2[Z_{33}(2\omega)+2Z_{33}(\Delta\omega)]-3G_{e3}\} \quad (5.161)$$
$$+Z_{24}(\omega_0)\{2\alpha_2 G_e G_{e2}[G_e Z_{33}(2\omega)+2G_e Z_{33}(\Delta\omega)-3]-3\alpha_3 G_e^3\}\Big|^{-1}$$

In the very low frequency range, it can be assumed that $Y_S(\Delta\omega) \approx Y_S(\omega_0) \approx Y_S(2\omega) = Y_S$, that $Y_L(\Delta\omega) \approx Y_L(\omega_0) \approx Y_L(2\omega) = Y_L$, and that $\Delta(\Delta\omega) \approx \Delta(\omega_0) \approx \Delta(2\omega) \approx \Delta \approx -[(1-\alpha_1)G_e + Y_S(1+R_eG_e)]Y_L$, which allows (5.161) to be rewritten as

$$IP_{3_i} \approx \frac{2}{3}G_{in}(\omega_0)\frac{|1+R_eG_e|^2\alpha_1 G_e}{|Y_L|^3}$$
$$\Big|-\frac{\alpha_1 Y_S}{\Delta}\left[2G_{e2}^2\frac{R_eY_S+(1-\alpha_1)}{(1-\alpha_1)G_e+Y_S(1+R_eG_e)}-G_{e3}\right] \quad (5.162)$$
$$+\frac{G_e+Y_S(1+R_eG_e)}{[(1-\alpha_1)G_e+Y_S(1+R_eG_e)]Y_L}$$
$$\cdot\left[2\alpha_2 G_e G_{e2}\left(G_e\frac{R_eY_S+(1-\alpha_1)}{(1-\alpha_1)G_e+Y_S(1+R_eG_e)}-1\right)-\alpha_3 G_e^3\right]\Big|^{-1}$$

which shows that, according to what happened to IP_{2_i}, low-frequency IP_{3_i} will also be nearly independent on the output termination. Because $i_E(v_{BE})$ nonlinearity is dominant, and since an ideal junction exponential law would determine that $G_{e3} = 2/3(G_{e2}^2/G_e)$, it is interesting to investigate the conditions determining a null on the first term of (5.162). This is verified when $G_e < 1/(2R_e)$ or $I_C < \eta V_T/(2R_e)$, for an Y_S equal to

$$Y_S = \frac{2(1 - \alpha_1)G_e}{1 - 2R_e G_e} \tag{5.163}$$

Furthermore, since at low frequencies it can be assumed that both $Y_S(\omega)$ and $Y_L(\omega)$ are resistive, and the dominant first term passes through a zero (and then changes sign), it can be anticipated that there must be a Y_S not far from the one predicted by (5.163) that completely eliminates third-order distortion.

This is shown in Figure 5.16 where the low-frequency output third-order intercept point, IP_3, is plotted against Y_S.

This phenomenon, seen from the point of view of third-order distortion variation with quiescent current [18, 19], reflects the existence of a small-signal IMD sweet-spot. In reality, and despite the fact that the exponential $i_E(v_{BE})$ characteristic does not have any G_{e3} null, the device associated to its base-emitter junction driving

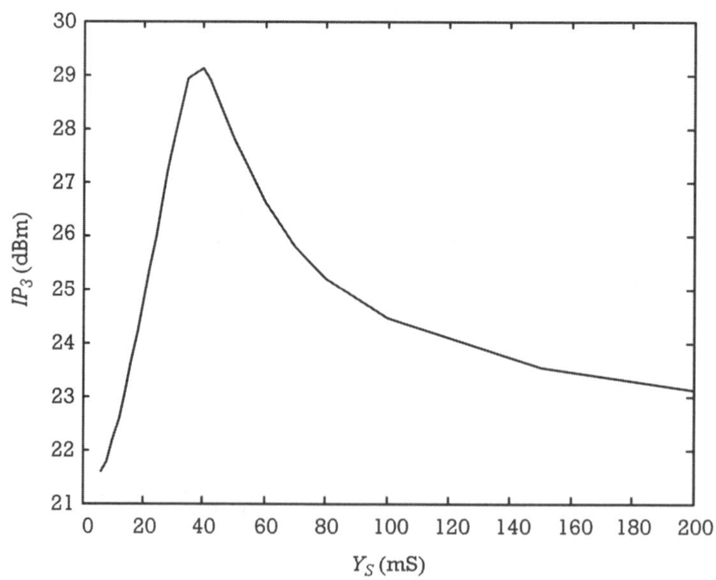

Figure 5.16 Low frequency output IP_3 variation with source admittance Y_S for a transistor of current gain $\beta = 72.5$, junction forward ideality factor $\eta = 1.24$, $R_e = 0.311\Omega$, and biased with $I_C \approx 20$ mA.

circuitry determines an $i_E(v_S)$ function that indeed presents such a null in its third-degree Taylor series coefficient as is depicted in Figure 5.17.

To understand this behavior let us derive $i_E(v_S)$ and its first three derivatives.

The analysis starts by assuming the device is biased with an ac decoupled dc voltage source V_{BB} of internal resistance R_B, which determines a quiescent emitter current I_E, base current $I_B = (1 - \alpha)I_E$ and base-emitter voltage V_{BE} that are approximately related by

$$I_E = I_S e^{\frac{V_{BE}}{\eta V_T}} = I_S e^{\frac{V_{BB} - R_B(1-\alpha)I_E - R_e I_E}{\eta V_T}} \tag{5.164}$$

Superimposed on this quiescent point, a small-signal source dc decoupled, of v_s voltage and R_S internal resistance, drives the device base-emitter junction without perturbing the bias point. Therefore, the composite driving signal will be $v_S = V_{BB} + v_s$ and produces an emitter current of

$$i_E = I_S e^{\frac{v_{BE}}{\eta V_T}} = I_S e^{\frac{v_S - R_B(1-\alpha)I_E - R_S(1-\alpha)(i_E - I_E) - R_e i_E}{\eta V_T}} \tag{5.165}$$

This is the sought expression of $i_E(v_S)$, although it is expressed in implicit form. To obtain the first-order coefficient of the Taylor series expansion, we make

$$\begin{aligned} G_{es} &\equiv \left.\frac{di_E(v_S)}{dv_S}\right|_{v_S = V_{BB}} \\ &= I_S e^{\frac{v_S - R_B(1-\alpha)I_E - R_S(1-\alpha)(i_E - I_E) - R_e i_E}{\eta V_T}} \\ &\quad \cdot \left.\frac{1 - [R_S(1-\alpha) + R_e]G_{es}}{\eta V_T}\right|_{v_S = V_{BB}} \end{aligned} \tag{5.166a}$$

or

$$G_{es} = I_E \frac{1 - [R_S(1-\alpha) + R_e]G_{es}}{\eta V_T} \tag{5.166b}$$

or even

$$G_{es} = \frac{I_E}{\eta V_T + [R_S(1-\alpha) + R_e]I_E} \tag{5.166c}$$

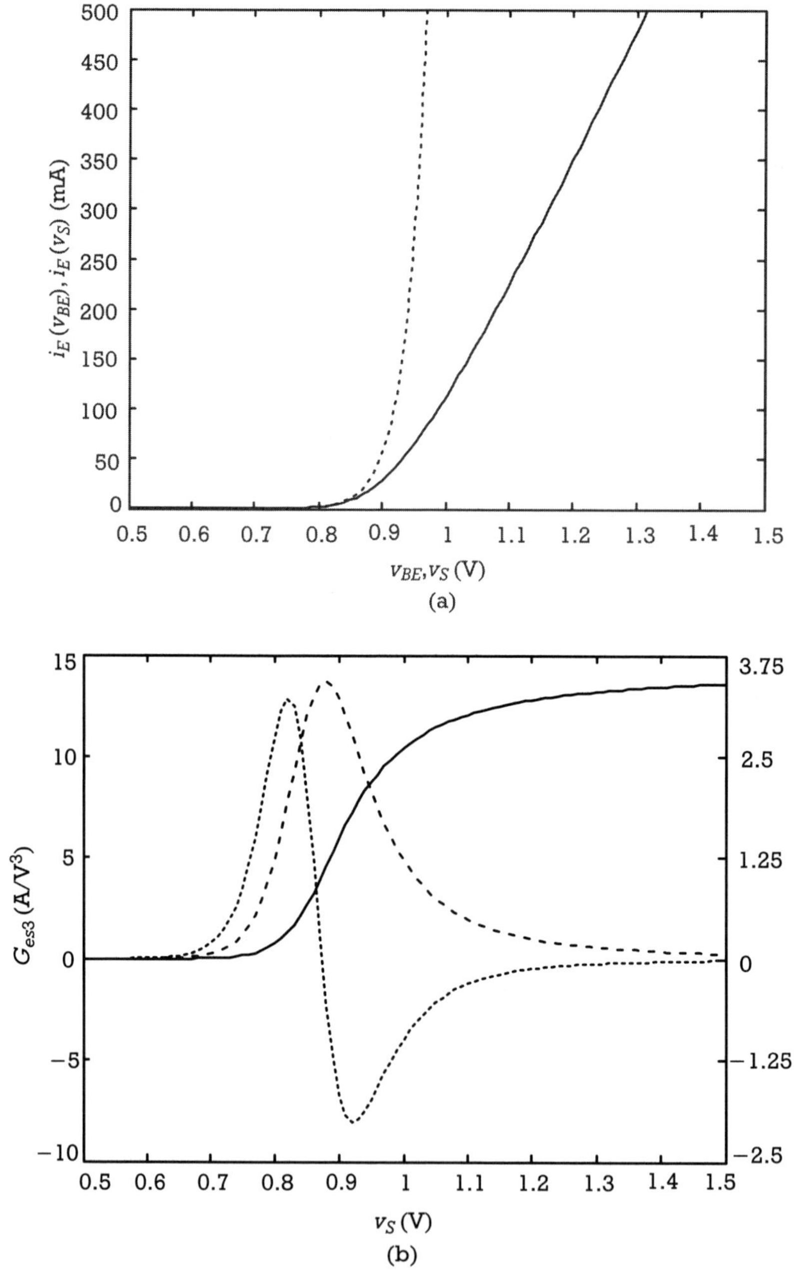

Figure 5.17 (a) Typical BJT emitter junction current as a function of intrinsic base-emitter voltage, $i_E(v_{BE})$ (--), and as a function of extrinsic driving voltage , $i_E(v_S)$ (-). Note the expected exponential characteristic of $i_E(v_{BE})$ and the linearized behavior of $i_E(v_S)$. (b) First three Taylor series expansion coefficients of $i_E(v_S)$, G_{es} (-), G_{es2} (--), and G_{es3} (..) as a function of v_S. Note the presence of a small-signal IMD sweet spot ($G_{es3} = 0$) near $v_S = 0.87$V.

Now, G_{es2} can be derived from $G_{es}(v_S)$ as

$$G_{es2} = \frac{1}{2}\frac{dG_{es}(v_S)}{dv_S}\bigg|_{v_S = V_{BB}} = \frac{1}{2}\frac{\eta V_T G_{es}}{\{\eta V_T + [R_S(1-\alpha) + R_e]I_E\}^2} \tag{5.167}$$

and G_{es3} from $G_{es2}(v_S)$, as

$$G_{es3} = \frac{1}{3}\frac{dG_{es2}(v_S)}{dv_S}\bigg|_{v_S = V_{BB}} \tag{5.168}$$

$$= \frac{1}{3}\eta V_T \frac{G_{es2}\{\eta V_T + [R_S(1-\alpha) + R_e]I_E\} - [R_S(1-\alpha) + R_e]G_{es}^2}{\{\eta V_T + [R_S(1-\alpha) + R_e]I_E\}^3}$$

The form of G_{es3} implies a small-signal IMD sweet-spot ($G_{es3} = 0$) when

$$I_E = \frac{1}{2}\frac{\eta V_T}{R_S(1-\alpha) + R_e} \tag{5.169}$$

or, since $G_e = I_E/(\eta V_T)$, when R_S is

$$R_S = \frac{1 - 2R_e G_e}{2(1-\alpha)G_e} \tag{5.170}$$

the result found in (5.163).

Some final notes on this analysis require that we take a closer look at the forms of G_{es} and I_E. For small I_E currents, where $[(1-\alpha_1)R_S + R_e]I_E << \eta V_T$, $v_{BE} \approx v_S$, I_E is an exponential function of v_S, $G_{es} \approx G_e$ and the device is as strongly nonlinear as expected when driven by a voltage source. However, when $[(1-\alpha_1)R_S + R_e]I_E >> \eta V_T$, v_{BE} is dominated by the voltage drop present in R_S and R_e, ($[R_S + (1+\beta)R_e]I_B$), i_B becomes dominated by this equivalent series resistance and the device behaves as if it had been linearized. In fact, this is probably the fundamental reason why a BJT or an HBT present such good linearity figures. Even though the output current of this type of device is a strongly nonlinear function of input voltage, it is almost ideally linear with input current, the way they are usually driven.

These results indicate that it would be possible to build a third-order, distortion-free, small-signal, low-frequency amplifier if the source admittance were carefully selected and maintained constant at the separation frequencies, the fundamental, and the second harmonics. Unfortunately, this ideal situation may not be so easily obtained in practice for three main reasons. First, it may be difficult to guarantee the required conditions in such a wide bandwidth, unless ω_1 and ω_2 are kept much below the transistor current gain limit. Second, the need for handling real signals of dense spectra determines a group of frequency separations so close to dc that $Y_S(\Delta\omega)$ cannot be given by (5.163), but is dictated by base bias circuitry. And, since the base-emitter junction is usually biased in a constant current mode, it may be anticipated that $Y_S(\text{dc})$ will certainly be much lower than the desired $Y_S(\Delta\omega)$. The last difficulty associated to that linearization process is related to the gain reduction determined by the input mismatch. Indeed, $Y_{in}(\omega)$ of (5.159) implies that input matching should be met for a significantly smaller source admittance of $Y_S \approx (1-\alpha)G_e/(1+R_eG_e)$. Nevertheless, it seems this subject should deserve attention, at least for low-frequency circuit designs where differential configurations can be made to circumvent many of these drawbacks [19].

In the high-frequency asymptote, that is, for frequencies such that $\omega > \omega_T/\beta$, but where C_{bc} effects (both linear and nonlinear) are still not evident, major second-order components at 2ω, arising from $i_E(v_{BE})$ and $q_{BE}(v_{BE})$ tend to cancel as already discussed. Furthermore, direct third-order components of these two nonlinearities, described by G_{e3} and C_{e3} coefficients, are proportional to

$$\begin{aligned} &-3Z_{31}(\omega_0)|Z_{31}(\omega_0)|^2[Z_{23}(\omega_0)G_{e3} + j\omega_0 Z_{25}(\omega_0)C_{e3}] \\ &\approx -3Z_{31}(\omega_0)|Z_{31}(\omega_0)|^2 \\ &\left[-\frac{\alpha_1[Y_S + j\omega_0 C_e(1+Z_eY_S)]}{\Delta(\omega_0)}G_{e3} + j\omega_0\frac{\alpha_1 G_e(1+Z_eY_S)}{\Delta(\omega_0)}C_{e3}\right] \end{aligned} \tag{5.171}$$

which also tend to cancel (when the previous condition of $\omega C_e(1+R_eY_S) >> Y_S$ is again verified), as $C_eG_{e3} = G_eC_{e3}$. So, as long as resistive inband distortion is controlled by G_{e3}, the capacitive nonlinearity will have a linearizing effect similar to the one studied for second-order distortion, and the remaining third-order distortion contributions will only be the ones due to the remaining terms on $-\alpha_1 Y_S/\Delta(\omega_0)$ and $i_C(i_E)$ nonlinearity. The high-frequency asymptote of IP_{3_i} will then be

$$\begin{aligned} IP_{3_i} \approx\ & 2G_{in}(\omega_0)\frac{|Z_{11}(\omega_0)|^2|Z_{21}(\omega_0)|}{|Z_{31}(\omega_0)|^3} \\ & \cdot\left|\frac{\alpha_1 Y_S(\omega_0)}{\Delta(\omega_0)}\left\{2G_{e2}^2\left(\frac{Y_L(2\omega)[Z_e(2\omega)Y_S(2\omega)+1-\alpha_1]}{\Delta(2\omega)}\right.\right.\right. \\ & \left.+2\frac{Y_L(\Delta\omega)[Z_e(\Delta\omega)Y_S(\Delta\omega)+1-\alpha_1]}{\Delta(\Delta\omega)}\right) \\ & + 4j\omega_0 G_{e2}C_{e2}\frac{Y_L(2\omega)[1+Z_e(2\omega)Y_S(2\omega)]}{\Delta(2\omega)} \\ & \left.-2G_{e2}\alpha_2 G_e^2\left(\frac{Y_L(2\omega)}{\Delta(2\omega)} + 2\frac{Y_L(\Delta\omega)}{\Delta(\Delta\omega)}\right) + 3G_{e3}\right\} \end{aligned}$$

$$+\frac{Y_{be}(\omega_0)+Y_S(\omega_0)[1+Z_e(\omega_0)Y_{be}(\omega_0)]}{\Delta(\omega_0)}$$

$$\cdot\left\{2\alpha_2 G_e\left[G_{e2}\left(G_e\frac{Y_L(2\omega)[Z_e(2\omega)Y_S(2\omega)+1-\alpha_1]}{\Delta(2\omega)}\right.\right.\right.$$

$$\left.+2G_e\frac{Y_L(\Delta\omega)[Z_e(\Delta\omega)Y_S(\Delta\omega)+1-\alpha_1]}{\Delta(\Delta\omega)}+3\right) \tag{5.172}$$

$$+2j\omega_0 C_{e2}G_e\frac{Y_L(2\omega)[1+Z_e(2\omega)Y_S(2\omega)]}{\Delta(2\omega)}$$

$$\left.\left.\left.-\alpha_2 G_e^3\left(\frac{Y_L(2\omega)}{\Delta(2\omega)}+2\frac{Y_L(\Delta\omega)}{\Delta(\Delta\omega)}\right)\right]+3\alpha_3 G_e^3\right\}\right|^{-1}$$

revealing independence on the output termination $Y_L(\omega)$. Obviously, this high-frequency cancellation effect suffers from the limitations already indicated for second-order distortion, and thus should be used carefully as an amplifier design tool. Nevertheless, Figure 5.18 and the works of Narayanan [14], Narayanan and Poon [15], and Maas, et al. [16] indeed confirm that a small-signal amplifier based on a bipolar device presents a certain form of high-frequency linearization characteristics.

However, at higher I_C bias currents, where second-degree coefficients tend to dominate over G_{e3}, the situation may change appreciably. Actually, there, third-order distortion arising, for example, from the difference frequency, opposes to the distortion generated by G_{e3}, and thus favors the distortion due to C_{e3}, which may produce an opposite effect as compared to the previous high-frequency cancellation. This is illustrated in Figure 5.19, in which now a BJT with nonlinear base-emitter capacitance produces more distortion than a similar transistor with a linear C_e.

In summary, and contrary to a FET-based small-signal amplifier, a circuit built over a bipolar transistor will have its major nonlinearities concentrated at the input port. The influence of source impedance will be much stronger then the load termination, and out-of-band impedances also have more significant impacts on inband distortion. A careful control of those terminations to even-order products may then enable a third-order IMD sweet-spot, which can be used as a promising linearizing tool.

Finally, the reactive nonlinearity associated with the base-emitter diffusion capacitance determines the amplifier's high-frequency behavior, reducing the level of even-order products and also, sometimes, of odd-order ones. Although these high-frequency cancellation effects can be associated with the referred small-signal IMD sweet-spot to provide useful startup design rules for highly linear circuits, the distortion behavior of a bipolar transistor is so involved that it may be wise

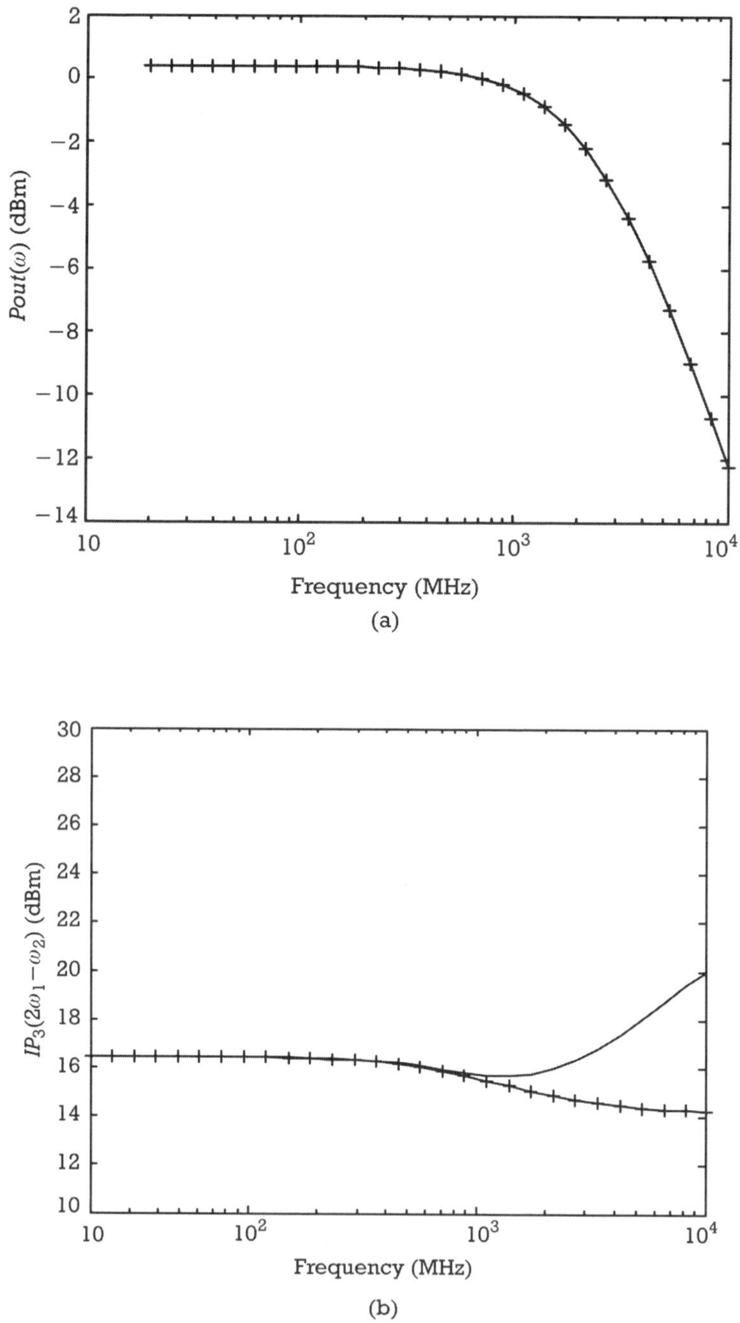

Figure 5.18 Output third-order intercept point variation with frequency for a device with linear (-+-) or nonlinear (-) base-emitter capacitance when $Z_S(\omega) = Z_L(\omega) = 50\Omega$. (a) Fundamental output power; and (b) output third-order intercept point at the inband intermodulation product $2\omega_1 - \omega_2$ (*Frequency* – 10 MHz), IP_3.

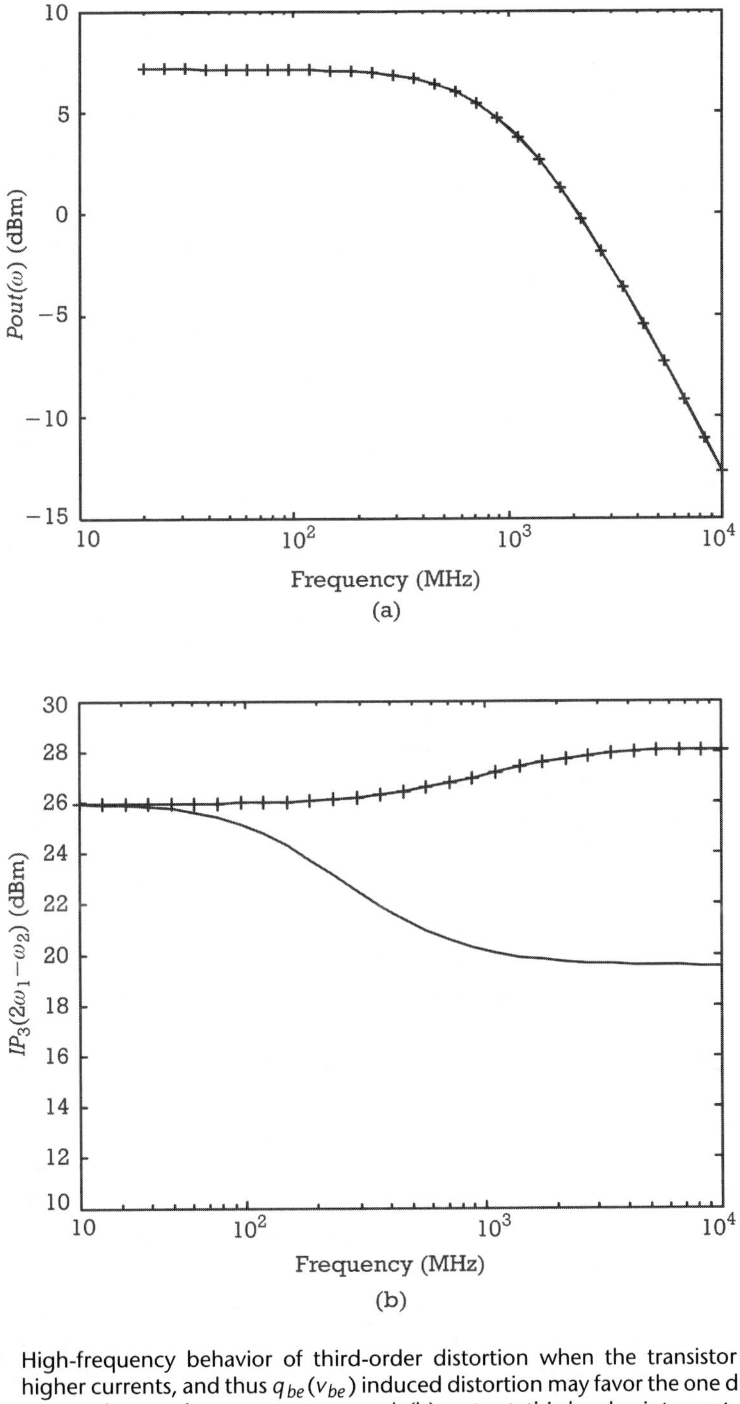

Figure 5.19 High-frequency behavior of third-order distortion when the transistor is biased at higher currents, and thus $q_{be}(v_{be})$ induced distortion may favor the one due to $i_e(v_{be})$. (a) Fundamental output power; and (b) output third-order intercept point at the inband intermodulation product $2\omega_1 - \omega_2$ (*Frequency* – 10 MHz), IP_3.

to use them carefully. In any case, those startup designs should always be validated and/or optimized by simulations of the circuit's full nonlinear model.

5.3 Linear Power Amplifier Design

5.3.1 Power Amplifier Concepts and Specifications

Contrary to the preceding small-signal amplifiers, in which gain, noise, and linearity were the primary performance goals, in a power amplifier (PA) everything is driven for absolute output power and efficiency. To realize this, we must keep in mind that activity is the ultimate characteristic of any amplifier. That is, an amplifier is a device meant to convert energy from an available source of power (the dc power supply in electric amplifiers) into signal energy. So, its merit should be measured in terms of signal added power, $P_a = P_{out} - P_{in}$, and not of gain, $G_P = P_{out}/P_{in}$. To understand this, two aspects of the power relations that take action in a PA should be clarified.

Although power gain and signal added power are closely related concepts, $G_P = 1 + P_a/P_{in}$ or $P_a = (G_P - 1)P_{in}$, they should be very well distinguished. As we will see later on, and even though it may seem counterintuitive at first sight, maximum absolute output power or added power optimization is by no means synonymous with linear maximized power transfer, matching, or small-signal gain. The reason for this is that while these three last concepts describe how infinitesimal small-signal power is transferred from the signal source to the PA load (concepts of ideal linear behavior), not taking into account the energy restrictions of the system, maximum output power is exactly determined by those large-signal (or nonlinear) effects (see Section 1.2). So, we will find that the conditions required for maximized PA output power capability or gain obtained under large-signal regimes (typically quiescent point and load impedance) do not necessarily maximize linear gain.

But, even if large-signal gain is considered, it may still not be as good PA figure of merit as absolute added power or output power. In fact, despite a gain of $G_P = 1{,}000$ (30 dB) generally causes a stronger impression as compared to a gain of only $G_P = 2$ (3 dB), we probably would change our minds if the former were obtained by adding a $P_a \approx 1$mW to a $P_{in} = 1$ μW, while the latter were achieved with a $P_a = 1$W and a $P_{in} = 1$W. And this would be even more convincing if we were told that both amplifiers relied on the same $P_{dc} = 2$W power supply.

Accordingly, as required output powers increase, gain looses importance as a valuable performance evaluation tool, being substituted by the actual absolute output power, added power, or, more commonly, *power added efficiency*. That very important PA figure of merit is defined as the ratio between signal added power and supply power being thus expressed by

$$PAE \equiv \frac{P_a}{P_{dc}} = \frac{P_{out} - P_{in}}{P_{dc}} \tag{5.173}$$

Another efficiency figure, named *dc conversion efficiency, collector efficiency,* or *drain efficiency* (whether the active device is a BJT or a FET), is defined as the ratio between output signal power and supply power and is thus expressed by

$$\eta \equiv \frac{P_{out}}{P_{dc}} \tag{5.174}$$

Although widely used in low-frequency or RF fields, η should be substituted by PAE whenever the amplifier's gain is so low that P_{in} represents a substantial part of P_{out}. For instance, our previous example of $P_{out} = 2\text{W}$ for a $P_{in} = 1\text{W}$ and a $P_{dc} = 2\text{W}$, which involves a moderate power added efficiency of $P_{AE} = 50\%$, would indicate the ideal value of $\eta = 100\%$, instead.

Energy conservation principle requires that the difference between signal input power plus dc source power, and signal output power must be transformed (or dissipated) into any other energy form. This can take place as either power delivered to the load at some other frequency components, or, more commonly, as heat. Therefore, efficiency optimization is a two-fold benefic process as it reduces dc power consumption and relaxes active devices' heat dissipation requirements.

5.3.2 Power Amplifier Design

Following what was already explained for small-signal amplifiers, power amplifier design process consists of determining the appropriate active device quiescent point and input and output terminating impedances.

So, we will begin our discussion by studying how the quiescent point and terminations affect PAE, P_{out}, and IMD, independently. This will provide us with a view of the possible trade-offs between these specification marks and, hopefully, design guidelines for obtaining optimized performance.

5.3.2.1 Maximum Power Added Efficiency

The most effective way of PAE optimization consists of selecting an appropriate PA operation class. For example, and referring to Figure 5.20, the highest PAE provided by a linear amplifier—like the ones studied in the preceding sections, whose active device is biased at Q, and signal excursion is maintained within the clipping free limits—is only 50%.

This PAE limit can be easily calculated if we notice that absorbed dc power, P_{dc} is

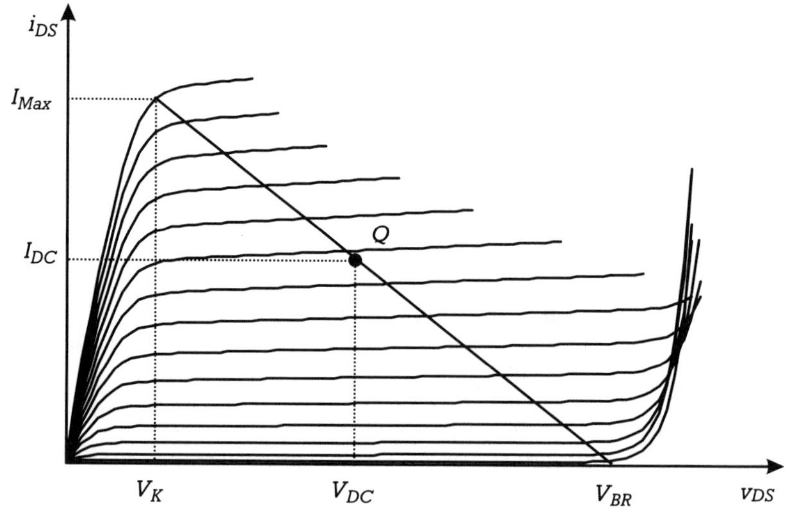

Figure 5.20 Bias point and load-line selection for class A linear power operation.

$$P_{dc} = V_{DC} I_{DC} \tag{5.175}$$

while output power cannot exceed the value imposed by output voltage and current excursion limits, which are

$$|V_{L_{Max}}| = \frac{V_{BR} - V_K}{2} = V_{DC} - V_K = V_{BR} - V_{DC} \tag{5.176}$$

and

$$|I_{L_{Max}}| = \frac{I_{Max}}{2} = I_{DC} \tag{5.177}$$

Therefore, assuming sinusoidal signal waveforms, we have

$$P_{out_{Max}} = \frac{1}{2}(V_{DC} - V_K) I_{DC} = \frac{1}{2} P_{dc} - \frac{1}{2} V_K I_{DC} \tag{5.178}$$

which leads to a maximum dc conversion efficiency of

$$\eta_{Max} = \frac{P_{out_{Max}}}{P_{dc}} = \frac{1}{2} - \frac{1}{2} \frac{V_K}{V_{DC}} \tag{5.179}$$

In an ideal circuit, in which $V_K = 0$, η_{Max} would be 0.5, which equals the anticipated 50% *PAE*, obtained when the power gain is very large.

To increase this efficiency value it is necessary to reduce dc power, which demands lower quiescent voltages and currents, and so, smaller signal excursion limits. But, since we want to keep output power, we have no alternative than to abandon our previous pure sinusoidal waveform conditions. That is, we are required to trade linearity for efficiency, moving from this form of linear amplification to other PA operation classes.

In order to review the traditional theory of PA operation class, we begin by assuming the ideal RF amplifier of Figure 5.21.

The PA is composed of a transistor, herein assumed as a simple piecewise linear transfer characteristic defined by

$$\begin{cases} i_O(t) = 0 & \text{if } v_I(t) < 0 \\ i_O(t) = G_m v_I(t) & \text{if } v_I(t) > 0 \end{cases} \tag{5.180}$$

and a sinusoidal waveform reconstruction filter, represented by the parallel LC tank. As a first-order approximation, it is assumed that the load quality factor of the filter is infinite, to guarantee that $v_L(t)$ and $i_L(t)$ are sinusoids, no matter the $i_O(t)$ waveform. The output RF choke, L_∞, behaves as an independent current source of I_{DC}, the average value of the output device current $i_O(t)$, while C_∞ is its dual, acting as an independent voltage source of V_{DC}.

If $v_I(t)$ is sinusoidal, $i_O(t)$ will be a clipped sinusoid as is represented in Figure 5.22.

For obvious reasons, $2\theta = \theta_2 - \theta_1$ is called the conduction angle, and it is used to define PA operation class. Therefore, the expressions we will derive for output

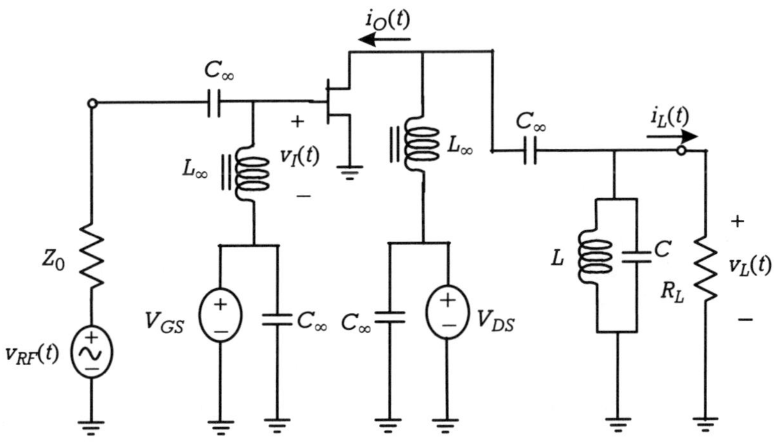

Figure 5.21 Simplified schematic diagram of an RF tuned power amplifier.

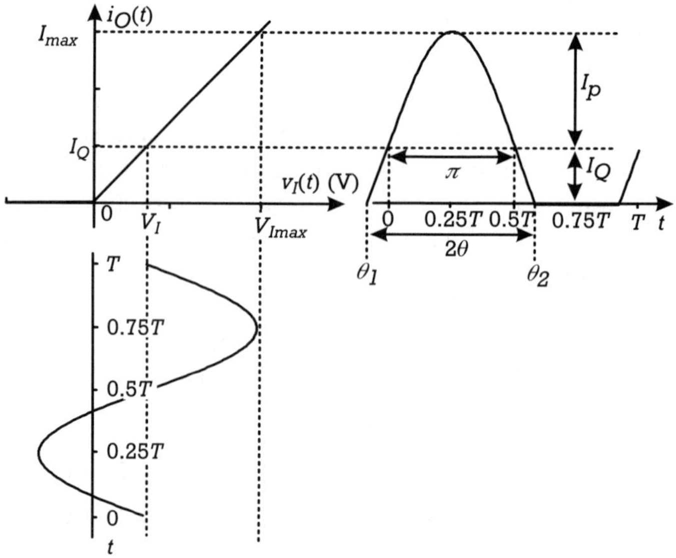

Figure 5.22 Linear piecewise amplification for PA conduction angle definition.

power, dc power, and efficiency will be expressed as functions of this conduction angle.

Assuming $v_I(t)$ is a biased sinusoid given by

$$v_I(t) = V_I + V_i \sin(\omega t) \tag{5.181}$$

the output device current will be

$$\begin{cases} i_O(t) = I_P \sin(\omega t) + I_Q & \text{if } \theta_1 \le \omega t \le \theta_2 \\ i_O(t) = 0 & \text{otherwise} \end{cases} \tag{5.182}$$

In this expression, $I_P = G_m V_i$ is the correspondent output peak amplitude, as if the $i_O(t)$ waveform had not been truncated. And $I_Q = G_m V_I$ is the generalized quiescent current, in the sense that it would be exactly the quiescent current if $i_O(t)$ dependence on $v_I(t)$ were linear in the whole range of input control voltages. In reality, the actual quiescent current will equal $I_Q = G_m V_I$ when V_I is positive, but null, and not negative as I_Q, when the device is biased below cut-off.

The conduction angle can be related to the average excitation's value and peak amplitude by

$$\begin{cases} 2\theta = 0 & \text{if } V_I < -V_i \\ 2\theta = 2\cos^{-1}\left(-\dfrac{V_I}{V_i}\right) & \text{if } -V_i \le V_I \le V_i \\ 2\theta = 2\pi & \text{if } V_I > V_i \end{cases} \tag{5.183}$$

Using these relations, and noting that the quiescent current can be expressed as $I_Q = -I_P \cos\,\theta$, maximum device current I_{Max} becomes

$$I_{Max} = \mathrm{I}_P(1 - \cos\,\theta) \tag{5.184}$$

To calculate the output power delivered to R_L and the absorbed power at dc, it is necessary to determine the amplitude of the fundamental and dc components of $i_O(t)$. Since $i_O(t)$ is periodic, it admits a Fourier series expansion such that

$$I_0 = \frac{I_P}{\pi}[\sin\,\theta - \theta\cos\,\theta] \tag{5.185}$$

and

$$I_1 = \frac{I_P}{2\pi}[2\theta - \sin\,2\theta] \tag{5.186}$$

are the desired current components at dc, $I_{DC} = I_0$, and the fundamental, $I_L = I_1$, respectively. So, then,

$$P_{dc} = V_{DC} I_{DC} = V_{DC}\frac{I_P}{\pi}[\sin\,\theta - \theta\cos\,\theta] \tag{5.187}$$

and, since the ideal maximum output voltage excursion is exactly V_{DC} (assuming $V_K = 0$) P_L will be expressed by

$$P_{L_{Max}} = \frac{1}{2}\frac{V_{DC}^2}{R_L} = \frac{1}{2}V_{DC} I_1 = \frac{1}{4\pi}V_{DC} I_P[2\theta - \sin\,2\theta] \tag{5.188}$$

Finally, the maximum efficiency for each conduction angle is achieved when the signal excitation is sufficient for driving the device to the maximum P_L:

$$\eta_{Max} = \frac{P_{L_{Max}}}{P_{dc}} = \frac{1}{4}\frac{2\theta - \sin\,2\theta}{\sin\,\theta - \theta\cos\,\theta} \tag{5.189}$$

This is the traditional expression for PA dc conversion efficiency. It indicates that the minimum value of η_{Max} is 50% when $2\theta = 2\pi$. It corresponds to the already known linear amplification mode, now called class A operation. Reducing the conduction angle increases η_{Max}. When $i_O(t)$ is exactly an half sinusoid, $2\theta = \pi$, $\eta_{Max} = \pi/4 = 78.5\%$, defining class B operation. Finally, if the conduction angle is less than π, $i_O(t)$ consists of short duration pulses, defining class C operation. dc conversion efficiency increases steadily as conduction angle goes down, and it tends to an asymptotic limit of $\eta_{Max} = 100\%$ when 2θ approximates zero.

Unfortunately, this ideal situation can never be even approximated in practice because (5.188) implies that $P_{L_{Max}}$ also vanishes when $2\theta \to 0$. Actually, since (5.184) can be rewritten as

$$I_P = \frac{I_{Max}}{1 - \cos\theta} \tag{5.190}$$

and I_{Max}, the allowed device's maximum current, is limited, η_{Max} only goes to 100% because an infinitesimal conduction angle produces simultaneously vanishing P_{dc} and $P_{L_{Max}}$ of

$$P_{dc} = \frac{V_{DC} I_{Max}}{\pi} \frac{\sin\theta - \theta\cos\theta}{1 - \cos\theta} \tag{5.191}$$

and

$$P_{L_{Max}} = \frac{1}{4\pi} V_{DC} I_{Max} \frac{2\theta - \sin 2\theta}{1 - \cos\theta} \tag{5.192}$$

Table 5.1 summarizes these results by showing $R_L = V_{DC}/I_1$, P_{dc}, $P_{L_{Max}}$, and η_{Max} for various conduction angles from class A to deep class C.

Because of the $P_{L_{Max}}$ collapse in deep class C, RF PAs are generally used somewhere between class A and class B, a broad operation zone denominated class AB, or just at the onset of class C.

Table 5.1 DC Power Consumption, Maximum Output Power, Collector Efficiency, and Load Resistance Versus PA Operation Class

2θ	R_L	P_{dc}	$P_{L_{Max}}$	η_{Max}
2π	$2\ V_{DC}/I_{MAX}$	$0.50\ V_{DC} I_{MAX}$	$0.25\ V_{DC} I_{MAX}$	50%
$3\pi/2$	$1.88\ V_{DC}/I_{MAX}$	$0.44\ V_{DC} I_{MAX}$	$0.26\ V_{DC} I_{MAX}$	60.1%
π	$2\ V_{DC}/I_{MAX}$	$0.32\ V_{DC} I_{MAX}$	$0.25\ V_{DC} I_{MAX}$	78.5%
$\pi/2$	$3.22\ V_{DC}/I_{MAX}$	$0.16\ V_{DC} I_{MAX}$	$0.15\ V_{DC} I_{MAX}$	94%
0	∞	0	0	100%

Actually, there are at least two more reasons that discourage use of deep class C.

The first one refers to power gain. Remember that signal drive must increase when the transistor is biased more below cut-off. That rise in excitation signal excursion really indicates more available source power for the same P_L power, and thus less gain. Hence, η_{Max} becomes more and more optimistic when compared to the real power added efficiency. Then, the PA designer rapidly concludes that if the active device is already short in gain (a situation particularly important in the higher bands of UHF, microwaves, and millimeter-waves), even the expected results of η_{Max} are illusive.

The second reason preventing the use of deep class C as a means to increase PA efficiency is due to transistors' breakdown effects. Deep class C requires quiescent points substantially bellow cut-off, along with high input voltage excursions. The combination of these usually produces so large reverse voltage peaks that the transistor is likely to enter breakdown or even to be destroyed.

Beyond I_{Max} and input signal excursion restrictions, there is another practical issue worth to discuss.

From the above, we have seen that PA operation mode is classified according to the conduction angle, 2θ, a concept intimately related to the piecewise linear model of (5.180) adopted for the active device. Unfortunately, actual transistors have smooth I/V characteristics, presenting no derivative discontinuity points as the one assumed at $v_I = 0$, which prevents the definition of 2θ, and thus the classification described above. In practice, what happens is that amplifiers biased near those idealized critical points are sometimes classified as class AB, class B, or even class C. Although this may not be very important in terms of P_{out} or PAE characteristics, it can be instrumental from the point of view of distortion. Therefore, in the following paragraphs we will generalize the above PA classification process.

Because we are particularly interested in the I/V characteristics presented in the vicinity of cut-off, we will concentrate on the subthreshold conduction of FET devices and in the exponential current-voltage relation of BJTs.

As we have seen in Chapter 4, subthreshold conduction of FETs can be modeled by a smooth turn-on function like

$$i_{DS} = K_I \ln\left[1 + e^{(v_{GS} - V_T)}\right] \tag{5.193}$$

in which K_I is a current scaling constant, V_T is the threshold voltage, and i_{DS} and v_{GS} are the output channel current and the input gate-source voltage, respectively. This function and its first three derivatives are plotted in Figure 5.23. Also shown in this figure is a piecewise linear approximation of i_{DS}. As we will see later on, the sign of the third-degree coefficient determines some important large-signal distortion characteristics. So, the critical point for distortion behavior is $v_{GS} = V_T$,

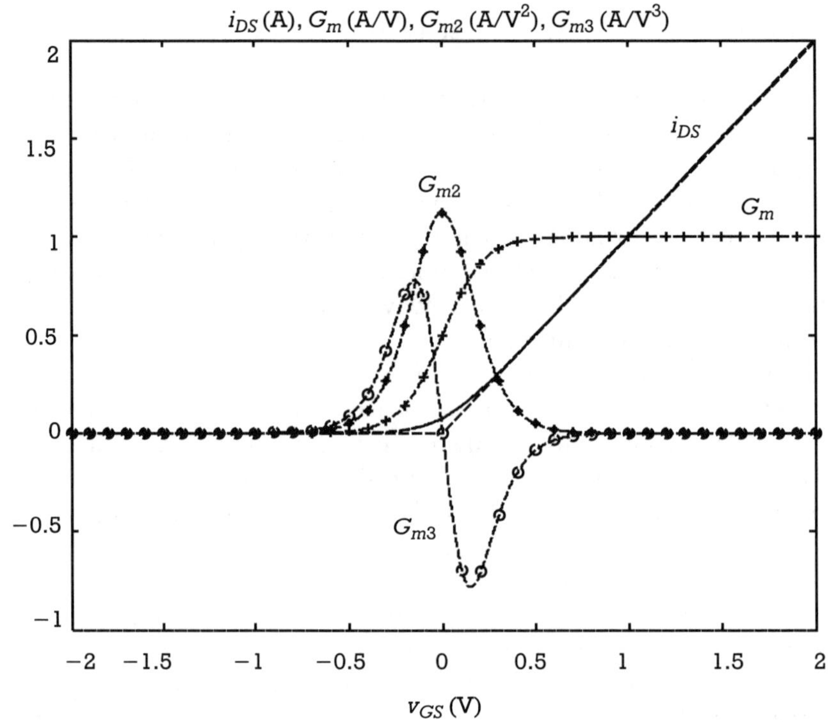

Figure 5.23 FET's channel current dependence on applied gate-source voltage (-) and its first three derivatives G_m (-+-), G_{m2}(-*-), and G_{m3}(-o-).

the third-order small-signal IMD sweet-spot, exactly the one corresponding to the piecewise linear approximation cut-off.

In practical FETs, i_{DS} conduction above threshold is not linear causing difficulties in identifying V_T. However, maximum G_{m2} or null G_{m3} are always precisely located points, and thus should be used for the definition of a generalized 2-θ conduction angle.

In the case of BJTs the situation is even more difficult because the $i_C(v_{BE})$ transfer current-voltage characteristic is an exponential in all v_{BE} domain, showing no critical points:

$$i_C = \alpha I_{ES}\left(e^{\frac{v_{BE}}{\eta V_T}} - 1\right) \tag{5.194}$$

where now V_T is the junction thermal voltage, η the ideality factor (do not confuse it with efficiency), and αI_{ES} is a current scaling factor. In real circuits, however, v_{BE} is never equal to the source voltage, v_S, because $i_B(v_{BE})$ is not negligible, and there always exists an equivalent series resistance in the BJT input mesh. Assuming the device is excited by a Thévenin equivalent source of internal resistance R_S, the

BJT base-spread resistance is R'_{bb} and emitter resistance is R_e, that equivalent series resistance would be

$$R = R_S + R'_{bb} + (1 + \beta)R_e \tag{5.195}$$

in which β is the BJT forward current gain $[\beta = i_C/i_B = \alpha/(1 - \alpha)]$. So, the actual relation between i_C and v_S will be (in implicit form)

$$i_C = \alpha I_{ES}\left(e^{\frac{v_s - Ri_C/\beta}{\eta V_T}} - 1\right) \tag{5.196}$$

which is plotted in Figure 5.24. Also shown in this figure are the first three derivatives of $i_C(v_S)$ and its piecewise linear approximation.

The resemblance of Figures 5.24 and 5.23 is evident, allowing for the conclusion that the critical point that should be used for the generalization of conduction angle is again the maximum of G_{m2} or the third-order small-signal IMD sweet-spot.

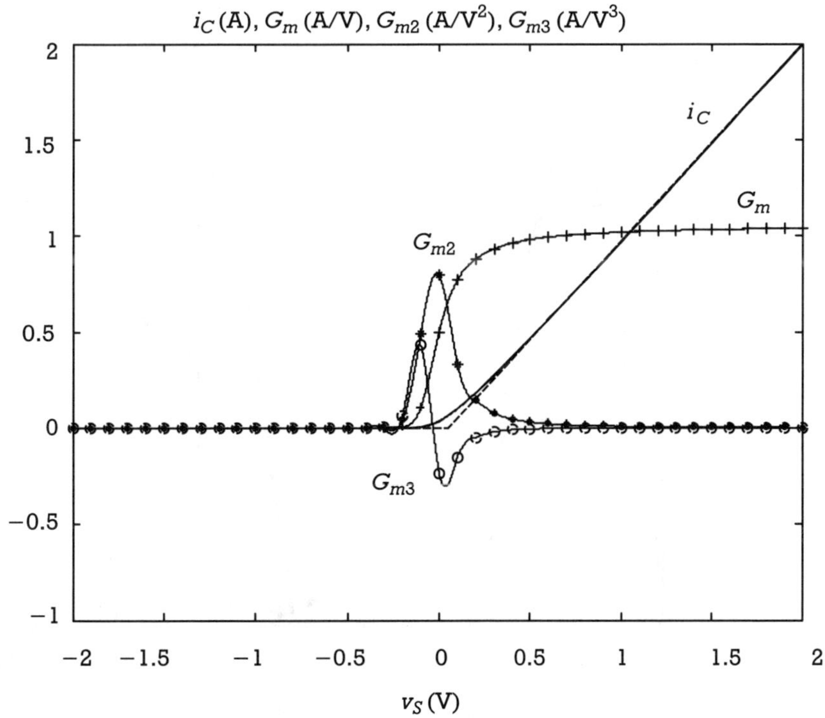

Figure 5.24 BJT's collector current dependence on applied v_S (-) and its first three derivatives G_m (-+-), G_{m2}(-*-), and G_{m3}(-o-).

Figure 5.25 shows typical current waveforms for the piecewise linear approximation of $i_{DS}(v_{GS})$ and $i_C(v_{BE})$, along with the ones obtained with the smooth models of (5.193) and (5.196).

The Fourier expansion coefficients numerically evaluated up to order three from these waveforms are depicted in Figure 5.26 for various generalized conduction angles. This allows the comparison of P_{dc} (Figure 5.27), $P_{L_{Max}}$ (Figure 5.28), and η_{Max} (Figure 5.29), obtained for various conduction angles and using the three considered models. As expected, there is no significant difference between the three models considered, which validates the common practice of using the theory developed under the piecewise linear approximation when the desired goals are only output power or efficiency.

To conclude the discussion on PA efficiency, we would like to address overdriven operation. First, we need to point out that we assumed the output signal excursion had a cut-off bottom limit, but showed no upper bound other than I_{Max}. In practical amplifiers, however, there is another boundary imposed by the triode zone in FETs, or by saturation in BJTs. For example, recalling Figure 5.20, we assumed v_{DS} was always greater than V_K. This permitted us to consider that the PA transistor could be represented by a simple transfer model, whose output

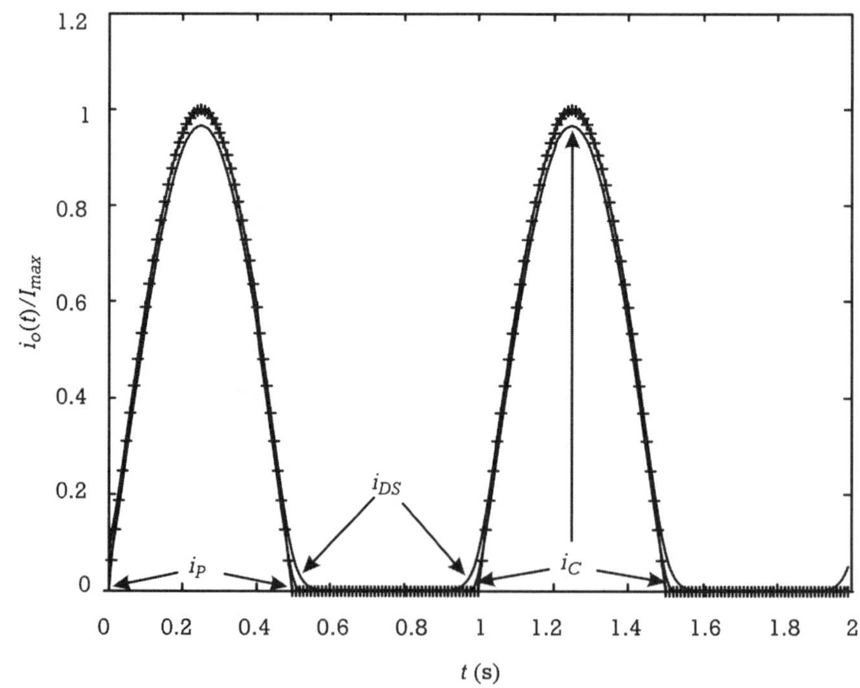

Figure 5.25 Typical normalized current waveforms, $i_O(t)/I_{Max}$, of a FET, $i_{DS}(t)$, a BJT, $i_C(t)$, and the piecewise linear model, $i_P(t)$.

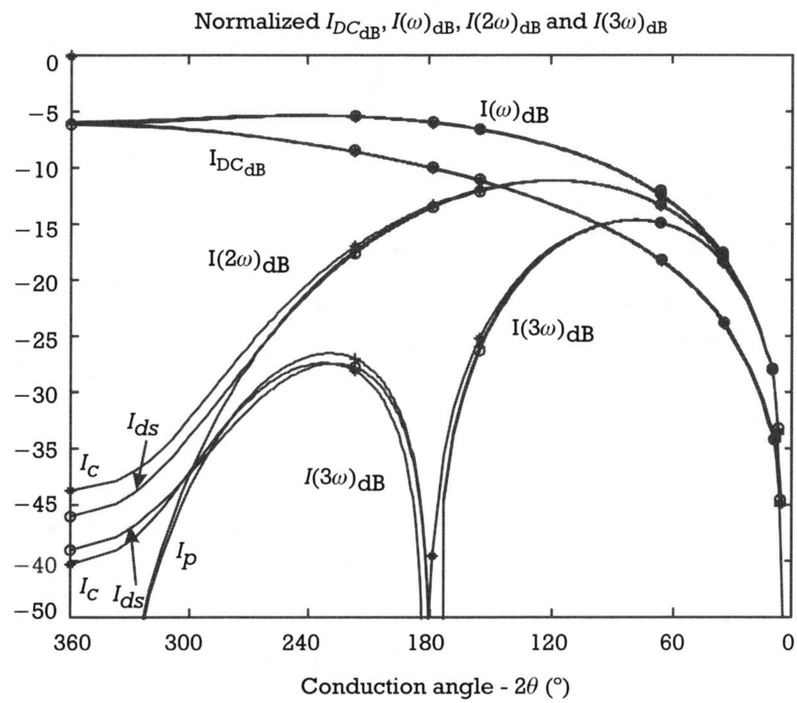

Figure 5.26 First four Fourier expansion coefficients of the normalized waveforms reported in Figure 5.25, $i_O(t)/I_{Max}$, versus generalized conduction angle.

current was only dependent on v_{GS}. But, if we now assumed the output condition determined by the dynamic load-line impedance $z_L(t)$,[4] and a typical hyperbolic tangent dependence of i_{DS} on v_{DS}, we would get an i_{DS} current that becomes strongly dependent on v_{DS}: $i_{DS}(v_{GS}, v_{DS})$ presents a sudden decrease when v_{DS} crosses V_K approaching zero, and in a way that is almost independent on v_{GS}.

Actually, as we have already seen, the similarity between the current-voltage characteristics of FETs and BJTs is so high, that their macroscopic behavior can be approximately described by a single set of empirical equations representing the device's output current i_O dependence on input and output voltages, v_I and v_O:

$$i_O(t) = K_I \ln\left[1 + e^{(K_{vi} v_I(t))}\right] \tanh\left[K_{vo} v_O(t)\right] \tag{5.197}$$

and

$$v_O(t) = V_{BR} - z_L(t) i_O(t) \tag{5.198}$$

4. Note that this dynamic load line impedance is a conceptual time-varying load resistance such that the output voltage, $v_O(t)$, can be treated as if it were given by $v_O(t) = V_{BR} - z_L(t) i_O(t)$.

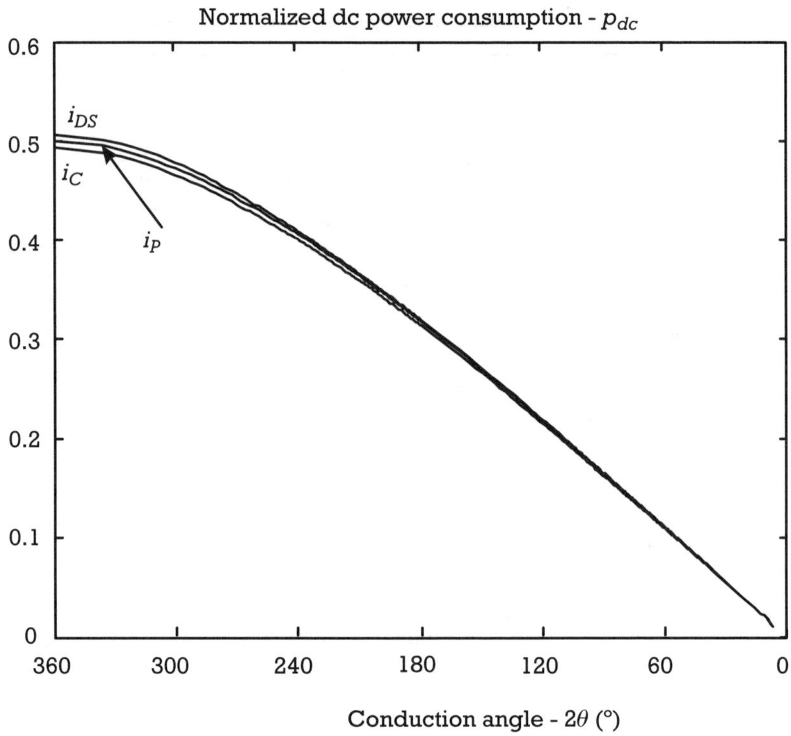

Figure 5.27 Normalized dc power consumption, $P_{dc}/(V_{DC}I_{Max})$, versus conduction angle for the three considered transfer models.

where V_{BR} is the maximum $v_O(t)$ voltage, usually selected close to the output breakdown voltage. Substituting (5.198) into (5.197), we obtain an expression for $i_O[v_I(t)]$

$$i_O[v_I(t)] = K_I \ln\left[1 + e^{(K_{vi}v_I(t))}\right] \tanh\left[K_{vo}(V_{BR} - z_L(t)i_O(t))\right] \quad (5.199)$$

Although (5.199) is nonlinear and in implicit form, it can be easily shown that it tends to its first factor, when $i_O(t)$ is very low, and governed by the hyperbolic tangent when $i_O(t)$ is high.

In fact, from very low v_I (near cut-off) to moderate v_I, i_O is itself low and v_O is far from knee voltage. The hyperbolic tangent argument is much greater than one, and $i_O[v_I]$ becomes only controlled by the input. This is the situation where the BJT is in its active region and the FET is in saturation.

On the other hand, when v_I is high enough, i_O is such that $V_{BR} - z_L(t)i_O(t)$ is low, the hyperbolic tangent starts to decrease, switching the control of $i_O[v_I]$ from the input to the output. The BJT enters saturation and the FET the triode

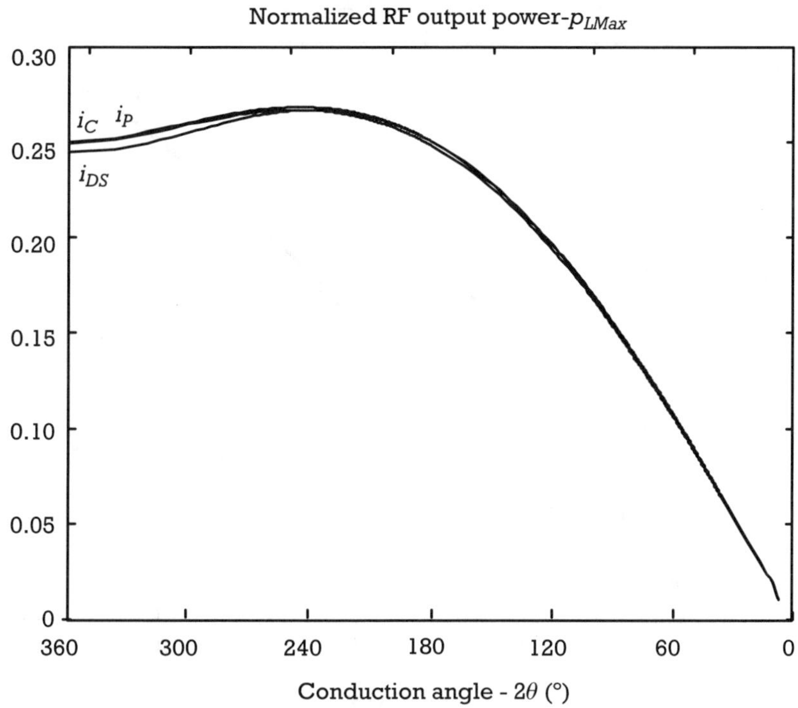

Figure 5.28 Normalized RF output power, $P_{LMax}/(V_{DC}I_{Max})$, versus conduction angle for the three considered transfer models.

region. In this high v_I and low v_O zone, the hyperbolic tangent tends to $K_{vo}[V_{BR} - z_L(t)i_O(t)]$ and the input-output characteristic function $i_O[v_I(t)]$ becomes

$$i_O[v_I(t)] = \frac{V_{BR}K_{vo}K_I \ln\left[1 + e^{(K_{vi}v_I(t))}\right]}{1 + K_{vo}K_I \ln\left[1 + e^{(K_{vi}v_I(t))}\right]z_L(t)} \tag{5.200}$$

which tends to an asymptotic constant limit of $V_{BR}/z_L(t)$.

Figure 5.30 is an example of such a saturated transfer characteristic, obtained by solving (5.199) with a Newton-Raphson iteration scheme.

With such a saturated $i_O(v_I)$ function, it is obvious that any of the previously studied PA classes will tend to a switching mode of operation if driving amplitude is sufficiently large. In this extreme case $i_O(t)$ becomes a square wave swinging from cut-off to $I_{Max} = (V_{BR} - V_K)/z_L(t)$, no matter the quiescent current. But, due to the bandpass filter present at the output, $v_O(t)$ still remains a sinusoid centered at $V_{DC} = (V_{BR} + V_K)/2$ and of amplitude $(V_{DC} - V_K) = (V_{BR} - V_K)/2$. The dc value, or bias point, of $i_O(t)$ is $I_{DC} = I_{Max}/2$ while its first-order Fourier component is

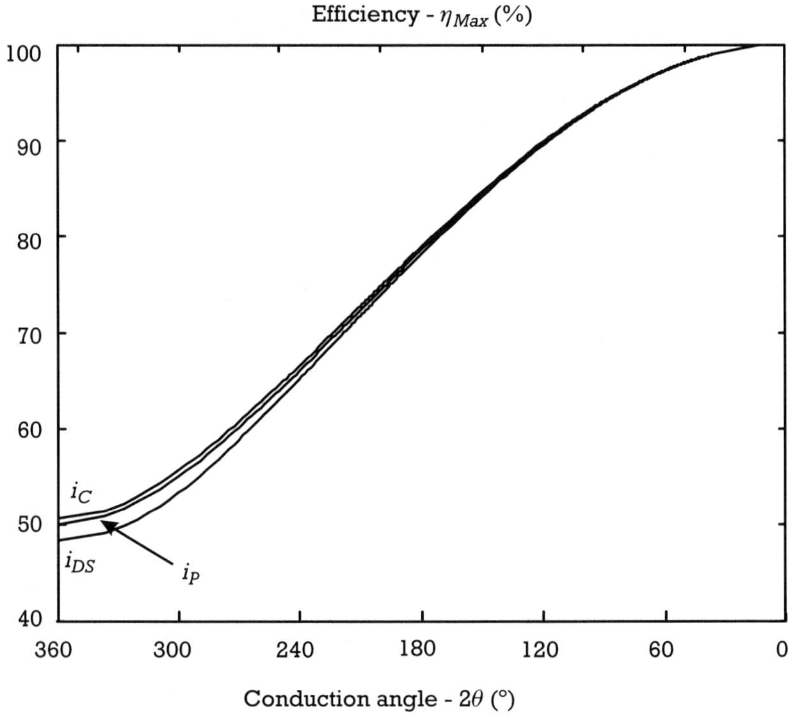

Figure 5.29 Maximum collector or drain efficiency, η_{Max}, versus conduction angle for the three considered transfer models.

$$I_1 = \frac{4}{\pi}\frac{V_{DC} - V_K}{R_L} = \frac{2}{\pi}\frac{V_{BR} - V_K}{R_L} \tag{5.201}$$

dc supply power will be

$$P_{dc} = V_{DC} I_{DC} = V_{DC}\frac{V_{BR} - V_K}{2R_L} = \frac{V_{BR}^2 - V_K^2}{4R_L} \tag{5.202}$$

and output fundamental power delivered to the load

$$P_L = \frac{1}{2} R_L I_1^2 = \frac{8}{\pi^2}\frac{(V_{DC} - V_K)^2}{R_L} = \frac{2}{\pi^2}\frac{(V_{BR} - V_K)^2}{R_L} \tag{5.203}$$

leading to an optimized efficiency of

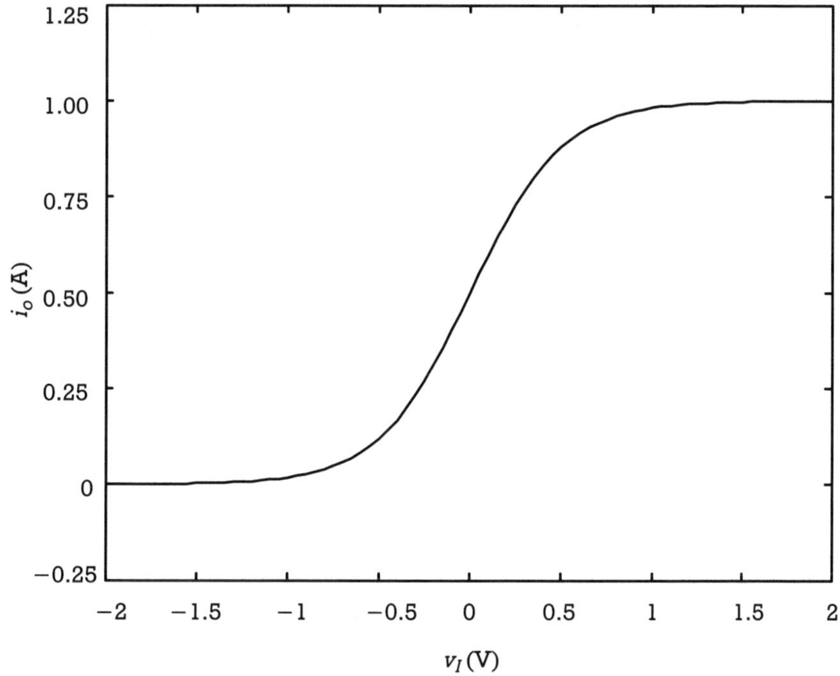

Figure 5.30 Illustrative example of a saturated input-output transfer characteristic.

$$\eta_{Max} = \frac{8}{\pi^2} \frac{V_{BR} - V_K}{V_{BR} + V_K} \tag{5.204}$$

or $\eta_{Max} \approx 81.1\%$ in the ideal case of $V_K = 0$.

Despite the increased efficiency, these modes of operation introduce further waveform distortion, and thus, their intermodulation characteristics will be substantially different from the ones shown by the correspondent unsaturated classes. So, to obviate any possible confusion, we will refer these operation modes by adding the term "overdriven" [3] to their original class names. For example, a class B amplifier in which the active device is showing half-wave cut-off clipping, but also knee-voltage (or saturation-voltage) clipping, will be said to operate in an overdriven class B mode.

The need for increased efficiencies, maintaining useful output power values, has led RF PA designers to approach true switching operation. There, transistor dissipation is reduced by allowing $v_O(t)$ voltage only at null current (cut-off) and non-null $i_O(t)$ current, I_{Max}, at approximately zero $v_O(t)$ voltage, V_K actually.

For that, the ideal resonant circuit of Figure 5.21 is replaced by complex reactive networks performing a precise harmonic loading control, and thus waveform shaping. Ideally, open circuits presented to odd-order harmonics and short circuits shown to even-order harmonics, create an almost squared device voltage waveform, which guarantees the desired high efficiency switching operation. In that case, the transistor's dissipated power falls to

$$P_{diss} = \frac{1}{2} V_K I_{Max} \tag{5.205}$$

which indicates an efficiency rise up to

$$\eta_{Max} = \frac{P_{dc} - P_{diss}}{P_{dc}} = 1 - \frac{V_K}{V_{DC}} = 1 - \frac{2V_K}{V_{BR} + V_K} \tag{5.206}$$

or $\eta_{Max} \approx 100\%$ if $V_K = 0$. This is the concept behind other saturation classes like class F. A review of these strongly nonlinear PA operation modes can be seen in [3, 20].

5.3.2.2 Output Power Capability

Having discussed PA efficiency optimization by quiescent point (or operation class) selection and waveform shaping, via harmonic loading control, we now move on to discuss maximized output power capability.

Figure 5.31 represents typical output I/V characteristics of a solid-state device.

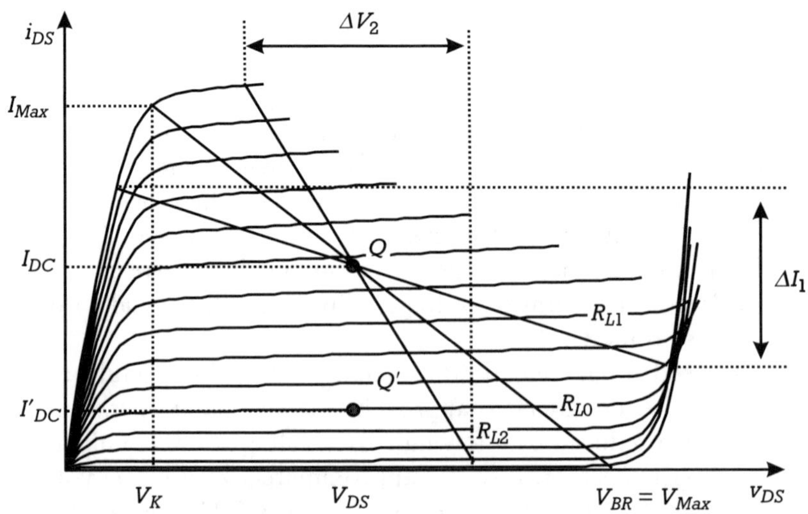

Figure 5.31 Example of FET's output I/V curves showing signal excursion dependence on intrinsic load impedance.

Returning to class A, since P_{out} results from the product of output voltage by output current excursion, it is obvious from Figure 5.31 that V_{DC} and I_{DC} must be selected as $V_{Max}/2$ and $I_{Max}/2$, respectively. If a minimum v_O voltage, V_K, is imposed by device constraints, and a minimum i_O current, I_{min}, had to be imposed because of linearity requirements, then the quiescent point Q should be selected as

$$Q = (V_{DC}, I_{DC}): V_{DC} = \frac{V_{BR} + V_K}{2}; I_{DC} = \frac{I_{Max} + I_{min}}{2} \qquad (5.207)$$

Any other quiescent point will be short in voltage, in current, or in both, leading to reduced output power capabilities.

These are general quiescent point conditions for maximum undistorted output power in class A operation. However, if linearity can be exchanged for maximized efficiency, then alternative points of lower quiescent current, but similar V_{DC} voltage, could be tried. An example of these is Q', a quiescent point intended for class AB operation.

Appropriate quiescent point selection is a necessary, but not sufficient, condition for maximum P_{out}. In fact, only the load-line determined by R_{L_0} uses the available voltage and current excursions, $\Delta V = V_{BR} - V_K$ and $\Delta I = I_{Max}$. An higher load resistance, R_{L_1}, will allow a smaller ΔI, ΔI_1, while a lower load resistance, R_{L_2}, will inevitably lead to a smaller ΔV, ΔV_2.

Following the derivation presented in previous section, maximum P_{out} will be given by

$$P_{L_{Max}} = \frac{1}{8}\Delta V \Delta I = \frac{1}{8} G_{L_0} \Delta V^2 = \frac{1}{8} R_{L_0} \Delta I^2 \qquad (5.208)$$

while R_{L_1} and R_{L_2} will only provide

$$P_{L_1} = \frac{1}{8}\Delta V \Delta I_1 = \frac{1}{8} G_{L_1} \Delta V^2 \qquad (5.209)$$

and

$$P_{L_2} = \frac{1}{8}\Delta V_2 \Delta I = \frac{1}{8} R_{L_2} \Delta I^2 \qquad (5.210)$$

If we now imagine that P_{L_1} and P_{L_2} are equal to a specified output power P_L (necessarily lower than $P_{L_{Max}}$), then G_{L_1} is the lowest load conductance that meets the P_L spec, and R_{L_2} is the lowest load resistance to accomplish the same goal. So, getting a P_{out} greater or equal than P_L demands for load terminations whose conductance is $G_L > G_{L_1}$ and whose resistance is $R_L > R_{L_2}$. In a Smith chart of reference impedance $Z_0 = 1/Y_0$, the first condition defines the interior of a constant conductance circle: $g_{L_1} = G_{L_1}/Y_0$, while the second limits the interior of a constant resistance circle: $r_{L_2} = R_{L_2}/Z_0$, as exemplified in Figure 5.32.

This procedure, known as the *Cripps method* or *load-line theory* [3, 21], permits the construction of a series of (g_{L_1}, r_{L_2}) plots for a correspondent array of desired output powers, which can be used as good estimates of constant P_{out} load-pull contours.

Actually, load-pull contours such as the one of Figure 5.32 define load impedances, Z_{L_i}, as seen from the intrinsic device. These are the impedances applied directly across the voltage-dependent current source representing $i_O(v_I, v_O)$ [i.e., $i_{DS}(v_{GS}, v_{DS})$ in the common-source FET nonlinear equivalent circuit model, or $i_C(v_{BE}, v_{CE})$ in a common-emitter BJT model]. Finding the correspondent extrinsic

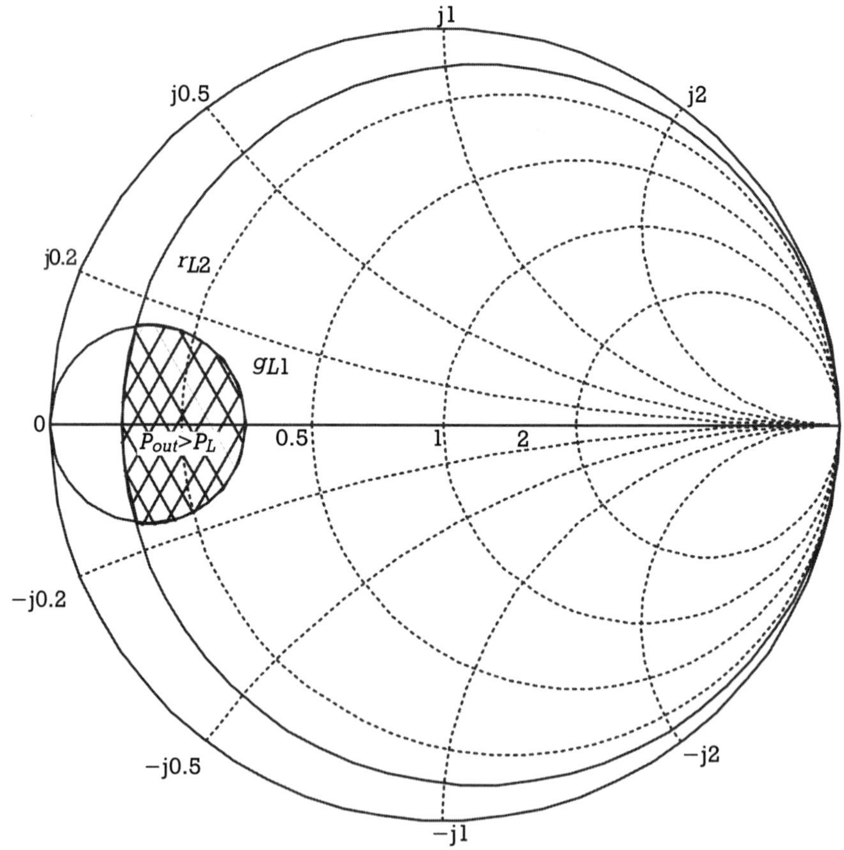

Figure 5.32 Smith chart zone of load impedances leading to an output power greater or equal than P_L.

load impedances (that must be presented to the external terminals of the device), Z_L, consists of solving an inverse matching problem. Indeed, and contrary to what we are used to do, we now need to determine the impedance Z_L with which we must terminate the drain network, so that the intrinsic impedance will be Z_{L_i}. To exemplify, let us consider the simplified FET drain parasitics network shown in Figure 5.33.

Since Z_{L_i} can be calculated from Z_L as

$$Z_{L_i} = \left\{\left[\left(Z_L^{-1} + j\omega C_{pd}\right)^{-1} + j\omega L_d\right]^{-1} + j\omega C_{ds}\right\}^{-1} \tag{5.211}$$

then, the desired Z_L must be given by

$$Z_L = \left\{\left[\left(Z_{L_i}^{-1} - j\omega C_{ds}\right)^{-1} - j\omega L_d\right]^{-1} - j\omega C_{pd}\right\}^{-1} \tag{5.212}$$

which, in the Smith chart, corresponds to the mapping depicted in Figure 5.34.

If a different PA operation class, other than class A, is adopted, optimum R_L can be estimated from the load voltage, $v_L(t)$, and current $i_L(t)$, in terms of the conduction angle. For that, we should remember that the maximum output power, $P_{L_{max}}$, (5.192), is delivered to a load resistance in which the voltage has an amplitude of $V_{L_{max}} = V_{DC}$ and the current $I_{L_{Max}}$ is

$$I_{L_{Max}} = I_{1_{Max}} = \frac{I_{Max}}{2\pi} \frac{2\theta - \sin 2\theta}{1 - \cos \theta} \tag{5.213}$$

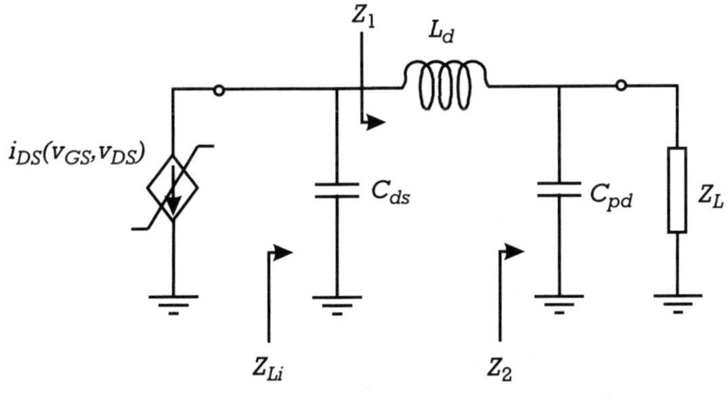

Figure 5.33 Illustrative example of drain parasitics network for load-pull estimation.

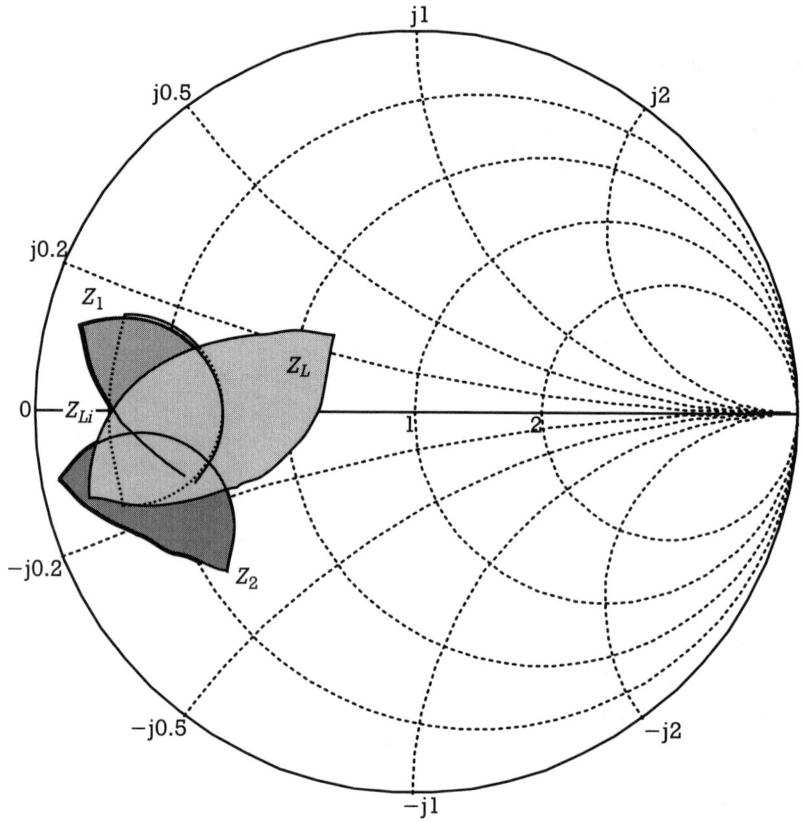

Figure 5.34 Successive load-pull plot transformations imposed by the drain parasitics network depicted in Figure 5.33.

Therefore, R_L must be given by

$$R_L = \frac{V_{L_{Max}}}{I_{L_{Max}}} = 2\pi \frac{V_{DC}}{I_{Max}} \frac{1 - \cos\theta}{2\theta - \sin 2\theta} \tag{5.214}$$

Those values are depicted in Figure 5.35.

One interesting thing to note on these results is the predicted infinite value of R_L when $2\theta \rightarrow 0$. In that situation (i.e., when the device is biased deeper in class C), $I_Q = -I_P \cos\theta$ gets increasingly lower negative values (higher and higher in modulus), while V_{DC} remains constant. Therefore, the load-line slope rises steadily, which could, at first glance, contradict the infinite limit of R_L. The solution to this apparent quandary stands on the fact that only at class A does the load-line slope equal $-1/R_L$, because, only in that situation do $i_O(t)$ and $i_L(t)$, and $v_O(t)$ and $v_L(t)$ share the same sinusoidal format. In any other operation class $v_O(t)$,

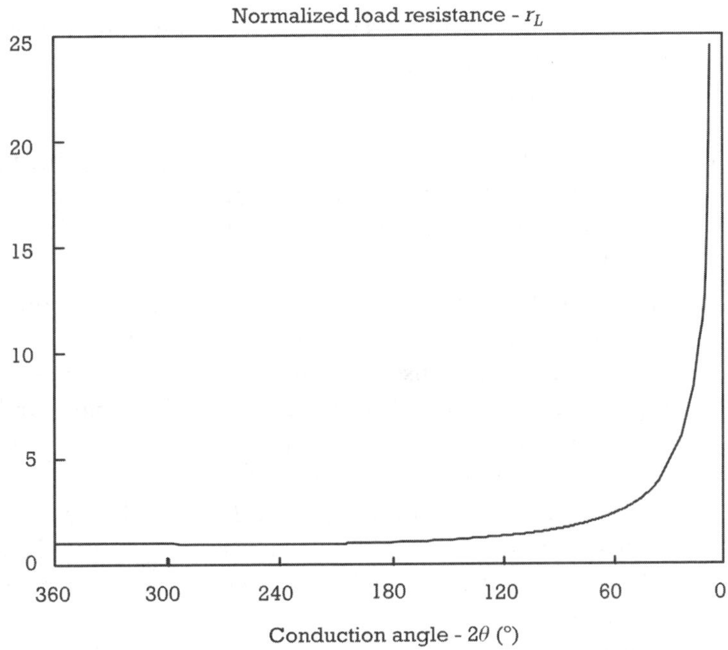

Figure 5.35 Normalized load resistance, $R_L/(2V_{DC}/I_{Max})$, for optimized output power versus conduction angle.

$v_L(t)$, and $i_L(t)$ keep their sinusoidal waveform, while $i_O(t)$ becomes a truncated sinusoid, acquiring a rich harmonic content.

To close this discussion on PA load impedance selection for maximized output power capability, it is interesting to note that R_L (and thus Z_{L_0}) was only determined from maximized signal excursion criteria, ignoring any output matching conditions. In fact, typical intrinsic R_L for class A, AB, or B operation, tend to be close to $R_L = V_{Max}/I_{Max}$, a value much lower than the one required for output matching or optimized gain ($R_{L_g} \approx R_{ds} = 1/G_{ds}$). Thus, PAs are amplifiers of low gain and poor output matching. Compared to high gain small-signal amplifiers, they present lower output powers when driving level is very small, but maintain their gain up to a much higher output power. Following a P_{in}/P_{out} characteristic, PAs behave as long distance runners in the sense that they begin presenting lower P_{out} for very small P_{in}, but then overcome the performance of small-signal amplifiers when the output of these becomes to get more and more compressed.

Fortunately, as long as the device can be approximately considered unilateral, P_{out} shows no appreciable dependence on source termination. Therefore, Γ_S is usually selected as Γ_{in}^*, to compensate for the PA low gain, relaxing driving level requirements for rated output power. A lower driving level is also beneficial for meeting power added efficiency specs, and for lowering driver stages' P_{out} requirements.

To summarize the conclusions presented on power amplifier design, let us write them in the form of a procedure:

1. Considering the desired PA specifications, choose between a design optimized for linearity or efficiency. In case of the former, adopt a class A or class AB design, and reserve class C or mostly overdriven operation for the second.
2. Select an appropriate active device and its quiescent point. This small-signal operation point, which must be located comfortably well inside the device's safe operating area (usually imposed by operating temperature and thus dc power dissipation), determines conduction angle (and consequently PA operation class) and maximum output power capability. In a PA architecture like the one depicted in Figure 5.21, V_{DC} should equal the midpoint between knee voltage and breakdown, while I_Q should be chosen according to the adopted operation class. If I_{Max} is the maximum allowed device current and 2θ the conduction angle, quiescent current can be given by $I_Q = I_{Max} \cos\theta/(\cos\theta - 1)$.
3. Determine optimum intrinsic R_L, from (5.214), for maximized output current and voltage swing, and therefore output power capability. Obtain the extrinsic Z_L required to produce the calculated R_L by deembedding the device's output parasitics. If maximized output power is not necessarily a goal, draw a set of output power load-pull contours (using approximate load-line theory, nonlinear simulation, or laboratory load-pull measurements) in the Γ_L Smith chart, along with the load-pull contours for some other competing specification (e.g., output match or linear gain).
4. Design an output matching circuit capable of transforming the PA terminal impedance (typically 50Ω) into the calculated Z_L. In case of high efficiency PA designs, specially using overdriven operation classes, this output matching circuit must also guarantee certain load impedances at the harmonics. This provides the required current and voltage waveform shaping and also helps in filtering out those high harmonic contents from the PA output terminal.
5. After knowing Z_L, determine Γ_{in} correspondent to this Γ_L, and choose $\Gamma_S = \Gamma_{in}^*$ (unless input stability considerations do not allow that) to optimize power gain. In case of quasilinear amplifiers like class A designs, this Γ_{in} can be easily derived from S-parameter relations. Otherwise, it must be measured (a source-pull contour is also sometimes used for that) or simulated under nominal driving power. Finally, design the required input matching network that should transform the PA source terminal impedance into the required Γ_S.

5.3.3 Nonlinear Distortion in Power Amplifiers

Depending on the mode of operation, and in what nonlinear distortion is concerned, power amplifiers can be divided into two major groups.

In one of these groups stands the class A and some class AB power amplifiers, where their quasilinear operation allows a study based on a weakly nonlinear model. Volterra series can then be used, enabling the extraction of general qualitative and quantitative information.

The other group includes all class B and class C, along with any other overdriven or switching mode power amplifiers. The strongly nonlinear operation that characterizes those amplifier circuits obviates the use of Volterra series (see Section 3.2.5). So, previously measured behavioral models, or numerical simulation techniques, have to be used instead, and closed-form expressions for IMD prediction are no longer possible. General qualitative conclusions become much more difficult, if not impossible, and nothing more is left than to rely on intuition for extending particular IMD behavior observed in a certain circuit to similar ones.

The following sections' objective is to discuss nonlinear distortion presented by these amplifier circuits. Their distortion characteristics are first analyzed, and then some design rules for prescribed distortion performance, output power, or power added efficiency are finally presented.

5.3.3.1 Quasilinear Power Amplifiers

High dynamic range small-signal amplifiers and class A linear power amplifiers are essentially the same circuits, except that they are optimized for different specifications. Therefore, the analysis developed under Section 5.2.4 is directly applicable to unsaturated class A and some class AB power amplifiers.

Starting with quiescent point selection, we have seen that, if the active device presents a high-gain small-signal IMD sweet-spot (like in some GaAs MESFETs), then it should be biased there. The only disadvantages referred to that bias option were possible reduced signal excursion and poor noise figure. If the latter one is irrelevant now, the former is basic for class A power amplifiers. Indeed, an IMD sweet-spot associated with very high quiescent currents can dramatically reduce maximum output power capability and efficiency. For a class AB power amplifier this subject is completely out of the question as that power amplifier operation class has already imposed the quiescent point at a much lower quiescent current. If good efficiency and third-order IMD performance should be simultaneously obtained, then a bias point close to the first third-order IMD sweet-spot could be tried. In that case, we would end up designing a class B amplifier, and large-signal simulations should be used to verify IMD levels in the whole desired range of output power.

Turning our attention to load impedance selection, we again face an important trade-off between output power and power added efficiency on one side, and third-order linearity on the other. If the compensation of third-order IMD produced on contributions arising from the FET's transconductance and cross coefficients' nonlinearities were beneficial for gain optimization, now it moves against the output power and power added efficiency enhancement we seek. As output power maximization usually requires low intrinsic load resistances, power amplifier voltage gains are substantially lower than in small-signal amplifiers. So, in those Γ_L zones, nonlinear distortion tends to be dominated by transconductance nonlinearity and IMD performance becomes even more insensitive to load impedance terminations.

As concluded in Sections 5.2.4.7 and 5.2.4.8, distortion in BJT amplifiers is almost independent of Γ_L.

Finally, source termination is still a free parameter that can be used for input match and thus transducer power gain optimization.

5.3.3.2 Large-Signal Power Amplifiers

Designing a nonclass A power amplifier for a prescribed IMD performance, or even simply predicting its IMD, is an incomparably more difficult task than doing the same thing for a quasilinear amplifier. This is a consequence of the fact that such a PA is a strongly nonlinear device. Volterra series is helplessly inaccurate, and we loose our powerful analytical model that previously led us to all the qualitative design rules discussed in preceding sections.

If we want to keep our goal of giving broad insights onto PA distortion, and thus refraining from the temptation of presenting a group of particular results extracted from specific circuits, we have no other way than to rely on some very simple, but general, theoretical results. Although those results should, by no means, be used as a substitute to the traditional design method based on computer aided IMD predictions or laboratory tests, they might serve as a good starting point or, at least, provide qualitative boundaries to the complex large-signal PA IMD behavior. These are the goals of the approach followed in this section.

Envisaging Large-Signal Power Amplifier Distortion

For developing a simplified theoretical model of PA large-signal distortion, let us start by some crude mathematical statements that say that if our output current versus input voltage nonlinear characteristics, $i_O[v_I(t)]$, is excited by a sinusoid

$$v_I(t) = V_i \cos(\omega t) \tag{5.215}$$

it will produce an inband response (that naturally includes the linear response and the distortion) of

$$i_O(t) = I_1(V_i) \cos[\omega t + \phi_1(V_i)] \tag{5.216}$$

Alternatively, if it is excited by an equal-amplitude two-tone signal,

$$v_I(t) = V_i \cos(\omega_1 t) + V_i \cos(\omega_2 t) \tag{5.217}$$

it will generate a fundamental, or signal output, of

$$i_{OS}(t) = I_{s1}(V_i)\cos[\omega_1 t + \phi_{s1}(V_i)] + I_{s2}(V_i)\cos[\omega_2 t + \phi_{s2}(V_i)] \tag{5.218}$$

and a series of distortion sidebands represented by

$$\begin{aligned} i_{OD}(t) = {} & I_{d21}(V_i)\cos[(2\omega_1 - \omega_2)t + \phi_{d21}(V_i)] \\ & + I_{d12}(V_i)\cos[(2\omega_2 - \omega_1)t + \phi_{d12}(V_i)] \\ & + I_{d32}(V_i)\cos[(3\omega_1 - 2\omega_2)t + \phi_{d32}(V_i)] \\ & + I_{d23}(V_i)\cos[(3\omega_2 - 2\omega_1)t + \phi_{d23}(V_i)] + \ldots \end{aligned} \tag{5.219}$$

Since $[I_1(V_i), \phi_1(V_i)]$ is the first-order component of the $i_O[V_i \cos(\omega t)]$ Fourier series decomposition, it can be obtained from the characteristic function as

$$I_1(V_i)e^{j\phi_1(V_i)} = \frac{2}{T}\int_{-T/2}^{T/2} i_O[V_i \cos(\omega t)]e^{-j\omega t}\, dt \tag{5.220}$$

while $[I_s(V_i), \phi_s(V_i)]$ and $[I_{d21}(V_i), \phi_{d21}(V_i)]$, $[I_{d32}(V_i), \phi_{d32}(V_i)]$, etc. are the correspondent coefficients of a bidimensional Fourier series:

$$I_{s1}(V_i)e^{j\phi_{s1}(V_i)} = \frac{2}{T_1 T_2}\int_{-T_1/2}^{T_1/2}\int_{-T_2/2}^{T_2/2} i_O[V_i \cos(\omega_1 \tau_1) + V_i \cos(\omega_2 \tau_2)] \, e^{-j\omega_1 \tau_1}\, d\tau_1\, d\tau_2 \tag{5.221a}$$

$$I_{s2}(V_i)e^{j\phi_{s2}(V_i)} = \frac{2}{T_1 T_2}\int_{-T_1/2}^{T_1/2}\int_{-T_2/2}^{T_2/2} i_O[V_i \cos(\omega_1 \tau_1) + V_i \cos(\omega_2 \tau_2)] \, e^{-j\omega_2 \tau_2}\, d\tau_1\, d\tau_2 \tag{5.221b}$$

$$I_{d_{n_1 n_2}}(V_i)e^{j\phi_{d_{n_1 n_2}}(V_i)} = \frac{2}{T_1 T_2}\int_{-T_1/2}^{T_1/2}\int_{-T_2/2}^{T_2/2} i_O[V_i \cos(\omega_1 \tau_1) + V_i \cos(\omega_2 \tau_2)] \, e^{-jn_1\omega_1 \tau_1}\, e^{-jn_2\omega_2 \tau_2}\, d\tau_1\, d\tau_2 \tag{5.221c}$$

Unfortunately, these expressions are not of much use for the large majority of practical PAs. If only by chance these integrals could be analytically evaluated in case $i_O[v_I(t)]$ were known, the truth is that even this characteristic function cannot be expressed in analytical form. So, any efforts of extracting qualitative information of PA IMD demands for some simplifying assumptions.

The first simplification we will make is to consider that (except for its bandpass behavior), $i_O[v_I(t)]$ is memoryless. Regarding the discussion of Section 3.1.1, this implies that the amplifier resonant circuits must be perfectly tuned, the out-of-band terminations are purely resistive, and the active device has, itself, no reactive elements.

The second simplification determines that $i_O[v_I(t)]$ must be smooth on the domain where it is controlled by the input, so that it presents no discontinuities in the function or in its derivatives, also allowing a Volterra series with a small number of terms (in fact, a power series) description of the PA small-signal behavior.

Under these assumptions, it can be shown that the two-tone response is symmetrical and purely real [22]: $I_{s1}(V_i)e^{j\phi_{s1}(V_i)} = I_{s2}(V_i)e^{j\phi_{s2}(V_i)} = S(V_i)$, $I_{d21}(V_i)e^{j\phi_{d21}(V_i)} = I_{d12}(V_i)e^{j\phi_{d12}(V_i)} = D(V_i)$ and that it can be obtained from the sinusoidal response $I_1(V_i)$ (actually the PA AM/AM characteristic) by [23]

$$S(V_i) = \frac{1}{\pi}\int_0^{\pi} I_1[2V_i \cos\theta] \cos\theta \, d\theta \tag{5.222}$$

and

$$D(V_i) = \frac{1}{\pi}\int_0^{\pi} I_1[2V_i \cos\theta] \cos(3\theta) \, d\theta \tag{5.223}$$

Beyond the immediate interest of these expressions, in the sense that they allow the extraction of two-tone behavior from the output, under sinusoidal excitation (for example, measured or easily simulated AM/AM characteristics), they permitted the derivation of the following relation between PA's output signals, $S(V_i)$, and distortion, $D(V_i)$ [23]:

$$D(V_i) = S(V_i) - \frac{4}{V_i^3}\int_0^{V_i} v^2 S(v) \, dv \tag{5.224}$$

which is of paramount importance in predicting two-tone IMD behavior produced in a general memoryless nonlinearity.

First of all, even the possibility of deriving it gives us information. It states that the distortion $D(V_i)$ can be directly obtained from the knowledge of only the fundamentals, $S(V_i)$, which is another way of saying that the manner the output signal components deviate from linearity shares the natural origin of the studied sideband distortion. Furthermore, (5.224) also affirms that if the system can present distortion-free operation it must obey

$$\int_0^{V_i} v^2 S(v)\, dv = \frac{V_i^3}{4} S(V_i) \tag{5.225}$$

or

$$V_i^2 S(V_i) = \frac{V_i^3}{4} \frac{dS(V_i)}{dV_i} + \frac{3}{4} V_i^2 S(V_i) \tag{5.226}$$

or even

$$\frac{d}{dV_i}\left[\frac{S(V_i)}{V_i}\right] = 0 \tag{5.227}$$

which reaffirms, at large-signal operation, the conclusion that linearizing gain or reducing nonlinear distortion are, in fact, two different aspects of the same reality.[5]

This condition can actually be verified in two different situations. Either the system is always ideally linear, and so $S(V_i)$ is proportional to V_i, or it may present some driving amplitudes V_{i0} in which it behaves as if it were locally linear.

As we already know from Chapter 1, due to energy limitations of the PA dc power supply, the first hypothesis can never be fully achieved, although, in the small-signal region, it corresponds to the linear case where the characteristic function presents a vanishing third-order Taylor series expansion coefficient. It may then be viewed as if the PA were biased in a small-signal IMD sweet-spot.

The second situation, more relevant under large-signal operation, where the first terms of the Taylor expansion are no longer capable of representing $i_O[v_I(t)]$ with enough accuracy, is verified for a certain excitation level V_{i0} where the PA large-signal gain versus P_{in} characteristic presents a minimum, or, as is more commonly observed, a maximum, and the $P_{IMD}(P_{in})$ curve shows a sudden minimum.

5. This should be of no surprise if we remember that sideband distortion and signal-correlated distortion share the same natural origin, and that this signal-correlated distortion must be responsible for the required output power saturation.

These critical points, usually observed in the PA onset of saturation, provide high signal-to-distortion ratios at high output power and PAE, being therefore very attractive as a linear PA design tool. To distinguish them from the small-signal IMD sweet-spots, obtained at certain bias points and independent of small-signal level, they were called in the literature "*large-signal IMD sweet-spots*" [24].

Studying these large-signal IMD sweet-spots to gather qualitative information on their origin, which, hopefully, may lead to their control, is the objective of the following discussion.

According to what we already know, a large-signal IMD sweet spot, V_{i0}, must correspond to a point of driving level in which PA AM/AM characteristic must be preceded by gain expansion and followed by gain compression or vice-versa. Although it may not be easy to find such a point for a general PA characteristic function, $i_O[v_I(t)]$, one thing is certain: $S(V_i)$ must saturate when the dc supply becomes short in supplied power [i.e., when $i_O(t)$ tends to a square wave]. There, $S(V_i)$ tends to a constant, S_s, and $D(V_i)$ becomes, from (5.224),

$$D(V_i) = S_S - \frac{4}{V_i^3}\int_0^{V_i} v^2 S_S \, dv = -\frac{1}{3}S_s \tag{5.228}$$

indicating that P_{IMD} tends to a saturated asymptotic value that is about 9.5 dB below the saturated output signal power, and in opposite phase to the signal. But, we have seen in Section 5.3.2.1 that PA saturation could be directly related to saturated $I_O[v_I(t)]$ characteristic, and that to the switching of $i_O(t)$ control from the PA circuit's input mesh, to its output mesh.

So, we should expect a large-signal IMD sweet-spot taking place for a V_{i0} which approximately imposes $V_o = V_{DC} - V_K$ (i.e., for which the output swing leaves saturated region of FETs to enter in their triode region, or leaves the active region of BJTs to enter saturation), whenever the device is biased in a bias point of positive third-order $i_O[v_I(t)]$ Taylor series expansion coefficients. Actually, if we admit the continuity of the PA's intermodulation response, and we know that it starts, at small-signal levels, by reinforcing the fundamentals and ends in opposite phase to them, at very large-signal regimes, it must pass through an IMD null somewhere in between.

On the contrary, if the device is biased in a point of negative third-order Taylor series coefficient, a sudden increase in IMD is to be expected at the onset of output power saturation.

To test those hypotheses, let us take again the general PA characteristic function of Section 5.3.2.1 (see Figure 5.30). Expanding such a $i_O[v_I(t)]$ in a Taylor series up to, let us say, fifth order, we get

$$i_O[v_I(t)] = I_0 + G_1 v_i(t) + G_2 v_i(t)^2 + G_3 v_i(t)^3 + G_4 v_i(t)^4 + G_5 v_i(t)^5 \tag{5.229}$$

in which the $v_i(t) = v_I(t) - V_I$ is the signal deviation from the bias point V_I. In (5.229), the various coefficients are defined by

$$G_n = \frac{1}{n!} \frac{\partial^n i_O[v_I]}{\partial v_I^n} \bigg|_{v_I = V_I} \tag{5.230}$$

and the odd-order ones are plotted in Figure 5.36 along with $i_O(v_I)$. Since for very low v_I, $i_O(v_I)$ is only controlled by the input, its coefficients are similar to the ones shown by an isolated FET [small-signal Taylor series expansion of $i_{DS}(v_{GS})$: G_m, G_{m2}, and G_{m3}], or voltage driven BJT [small-signal Taylor series expansion of $i_E(v_S)$: G_{es}, G_{es2}, and G_{es3}]. This is why G_3 is positive before cut-off (class C mode), it is null near cut-off (class B mode), and then becomes negative (class AB and class A modes). The following negative and positive large lobes do not correspond to any physically reasonable bias point, but simply to the strong nonlinear effect of the output boundary appearing at very high $i_O(v_I)$ currents.

Figure 5.37(a–e) presents the output power [corresponding to $S(V_i) - P_L(V_i)$], intermodulation distortion power [corresponding to $D(V_i) - P_{IMD}(V_i)$], power

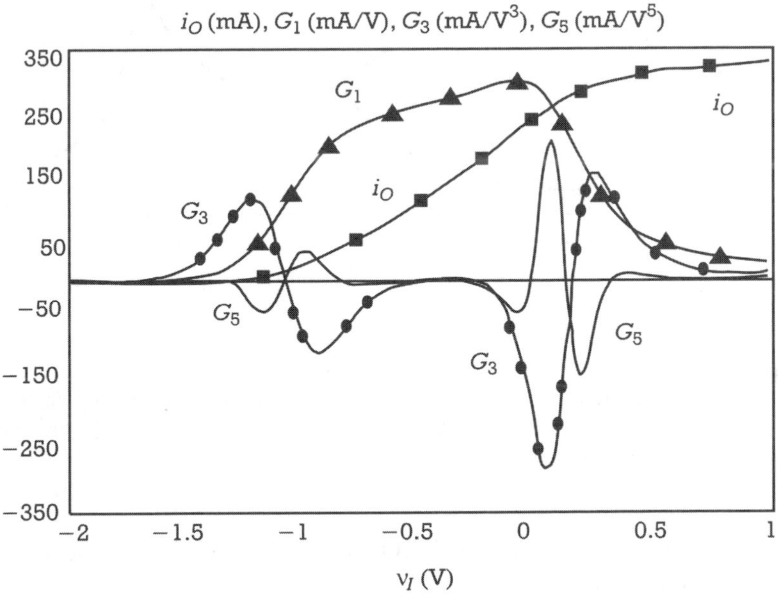

Figure 5.36 Illustrative characteristic function of the nonlinearity used in the circuit under study, i_O, and odd-order Taylor series expansion coefficients, G_1, G_3, and G_5, versus input bias.

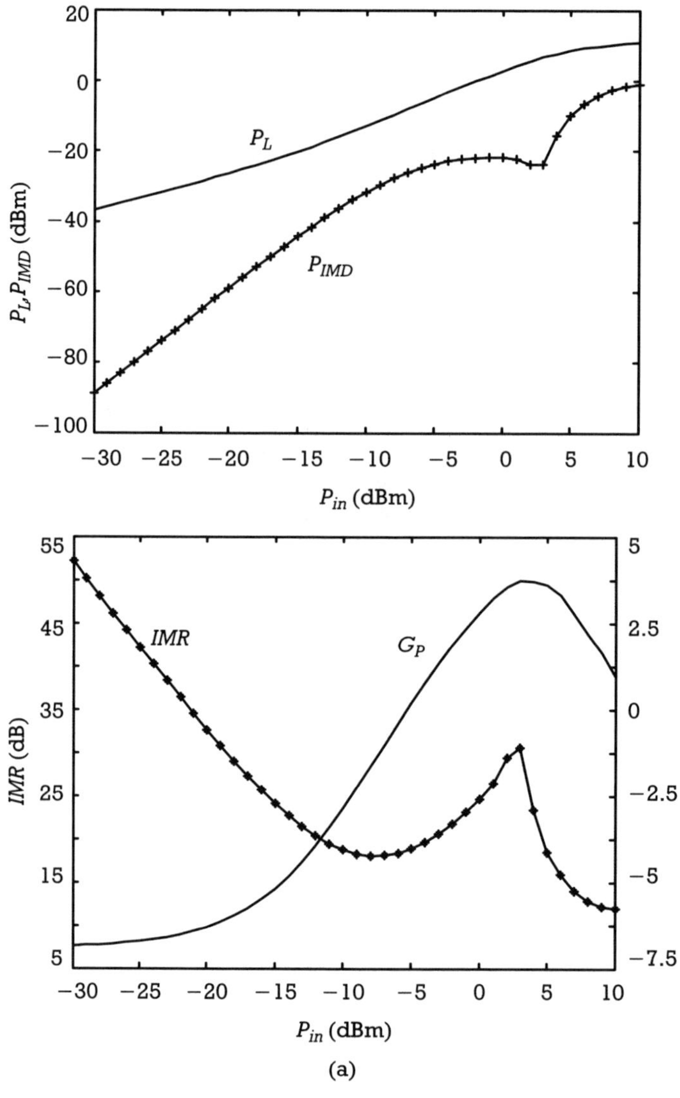

Figure 5.37 Fundamental output power, P_L; IMD power, P_{IMD}; power gain, G_P; and intermodulation ratio, *IMR*, for various quiescent points determining (a) class C, (b) class AB, (c) class A, (d) class B in a small-signal IMD sweet-spot, and (e) class AB, but very close to class B.

gain $[G_p(V_i)]$, and signal-to-IMD ratio $[IMR(V_i)]$, of a power amplifier biased for several quiescent points.

In Figure 5.37(a) the device was biased for class C with $V_I = -1.20$V, where G_3 is clearly positive. $S(V_i)$ and $G_p(V_i)$ show the expected small-signal gain expansion, typical of unsaturated class C PAs, which is then followed by a maximum of $G_p(V_i)$ and the corresponding IMD null or high *IMR*. This large-signal IMD

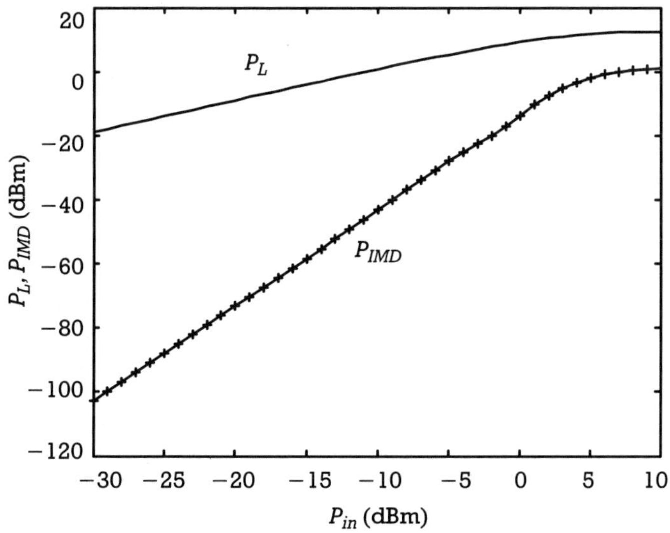

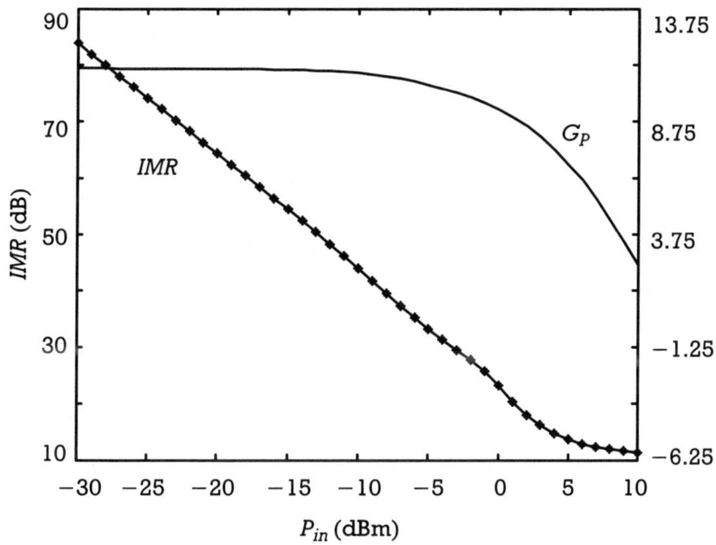

(b)

Figure 5.37 (continued).

sweet-spot appears at a driving amplitude of approximately $V_i = 1.19\text{V}$ ($P_{in} = 2.5$ dBm), which was found to be coherent with the expected input voltage swing necessary to drive the device to its knee voltage. As expected, after the sweet-spot, the *IMR* tends to the 9.5-dB asymptotic value.

In Figure 5.37(b) the device was biased at $V_I = -0.68\text{V}$, clearly in class AB operation mode. The steadily compressing behavior of $S(V_i)$ is typical of that PA operation mode. Because G_3 is already negative, input small-signal distortion adds

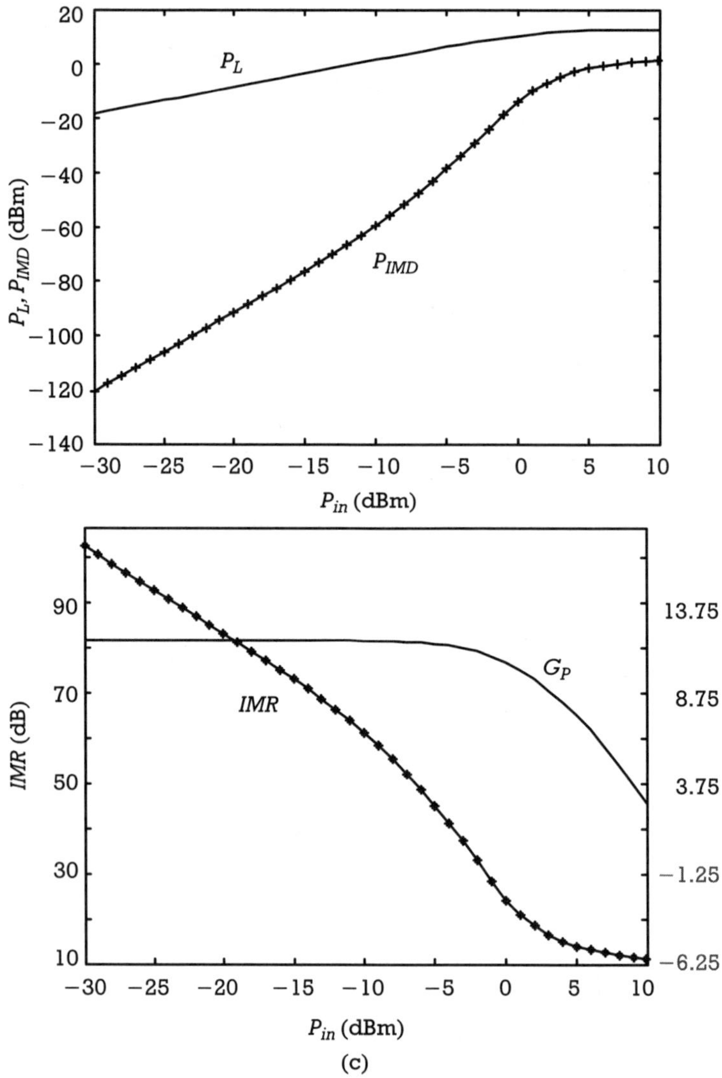

Figure 5.37 (continued).

in phase to large-signal output distortion causing the sudden increase in $D(V_i)$ when the device enters saturation. After that, *IMR* tends again to its asymptotic value of 9.5 dB.

Figure 5.37(c) corresponds to a class A PA. The constant gain associated to very large *IMR* is typical of this "linear" mode of operation. But, even a class A PA becomes strongly nonlinear when overdriven, as is shown by the obvious increase of $D(V_i)$ seen for V_i over 0.84V ($P_{in} = -0.5$ dBm). This is a situation similar to the class AB PA as they both share a quiescent point where G_3 is negative.

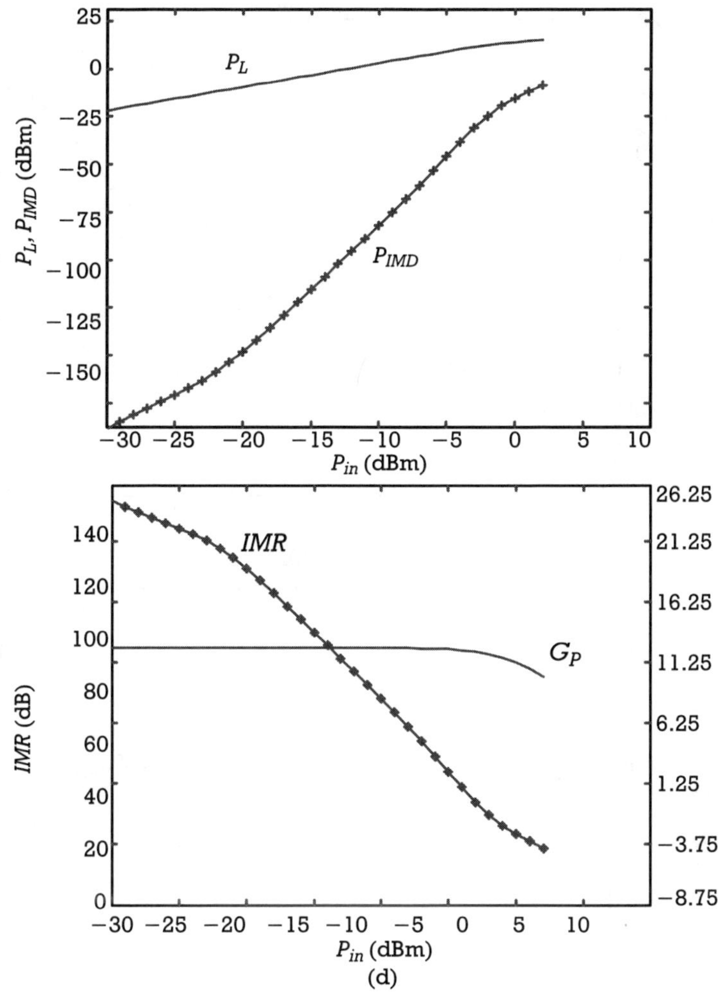

Figure 5.37 (continued).

In Figure 5.37(d) the device was biased exactly at $V_I = -1.04$V [i.e., the small-signal IMD sweet-spot ($G_3 = 0$)], determining, therefore, ideal class B operation. As expected, small-signal $G_p(V_i)$ is almost constant, $D(V_i)$ is extremely low and shows a rise of about 5 dB/dB, a clear indication of uncovered fifth-order distortion. Although, in the present case, there is no large-signal IMD sweet-spot, in a real PA it is very hard to tell what will actually happen. Indeed, near the small-signal IMD sweet-spot the ratio and relative signs of third and higher order distortion components vary so much that it becomes very difficult to predict the magnitude and phase of IMD before reaching the onset of saturation. And this may generate $P_{IMD}(P_{in})$ patterns even more complex than the ones already discussed.

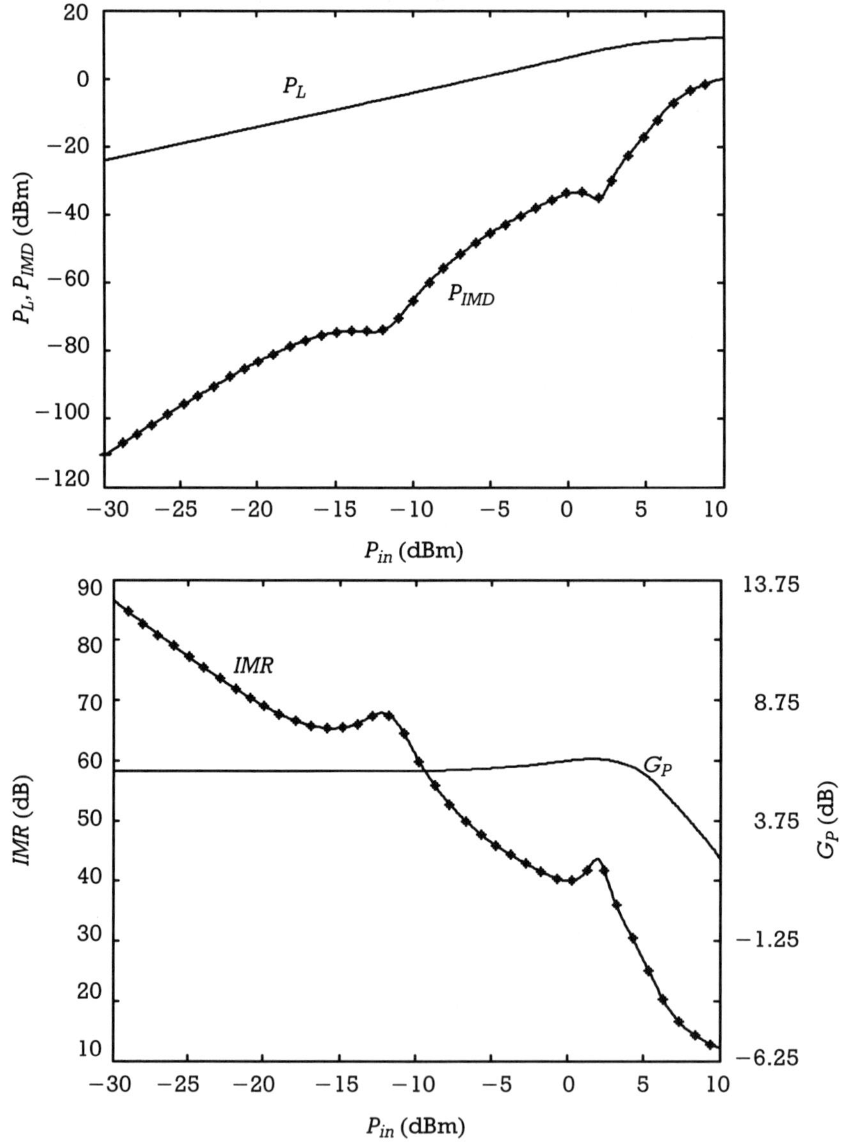

(e)

Figure 5.37 (continued).

That is what we want to illustrate with the last case shown in Figure 5.37(e), where a small, but already negative G_3 (corresponding to class AB) is first canceled by the higher order contributions, thus creating a new IMD null quite far from saturation. But, since after this excitation level IMD becomes controlled by those higher order components, it also faces an inversion of phase [accompanied by a change from gain compression to gain expansion in the $S(V_i)$ pattern], thus still enabling the previously studied large-signal IMD sweet-spot at the onset of saturation. Although such a situation may sound, at this time, not much more than a theoretical curiosity, hardly (if ever) observed in MESFET based PAs, it was already experimentally reported in PAs using Si MOSFET and LDMOS devices [25], where it is quite common. It may therefore be used to build very linear and highly efficient MOSFET PAs.

In summary, we have seen that the interaction between small- and large-signal distortion can, indeed, largely determine the shape of $D(V_i)$, inducing regions of sudden IMR degradation, but also of IMR improvement. In both cases, these critical V_{i0} driving levels are determined by the onset of saturation, which is naturally dependent on the voltage gain, $A_v = -G_1 R_L$, and the small-signal IMD level (i.e., G_3). Since G_1 and G_3 are strongly dependent on the quiescent point, V_I, it should cause no surprise that this V_I can be used to control V_{i0}.

Another way to control the sweet-spot position, would be by changing the value of R_L and therefore the gain of the PA. That is exactly what can be concluded from the observation of Figure 5.38, which illustrates the impact of R_L on large-signal IMD sweet-spot position, and IMD level. This figure shows that higher R_L values, producing more horizontal load lines, and so crossing a smaller number of v_I I/V curves before reaching knee voltage (indicating a reduced v_{GS} voltage swing), indeed generate sweet-spots at lower V_{i0} driving levels.

After this introduction to large-signal IMD behavior, it is now important to discuss the limitations imposed by the simplifying model assumptions. We will begin by relaxing the first assumption: absence of memory.

Memory can have its origin in the device or terminating impedances, and it can have a short or long duration, wherever it is compared to the carrier or the envelope period, respectively.

Following the Volterra series analysis undertaken on small-signal amplifiers (Section 5.2.4), the active device can present reactive elements that, combined with the PA input and output matching networks, create short memory effects. But, it can also present low-frequency dispersion or self-heating time constants that, along with the PA bias circuitry, are comparable with the period of the difference frequency of the two-tone excitation. Finally, there are also equivalent circuit elements, both in the device and the matching networks, that may present reactive impedances to the even-order harmonics (mainly the second-harmonic), which are then reflected onto inband IMD by even-order mixing.

To evaluate the impact of terminating impedances on large-signal IMD performance, it is better to consider a real power amplifier microwave circuit, like the one referred to as the DUT in the laboratory setup of Figure 5.39. Being a MESFET-based amplifier, it is supposed that the device's nonlinearities are concentrated at the output circuit [mainly the $i_{DS}(v_{GS}, v_{DS})$ channel current], and therefore it is

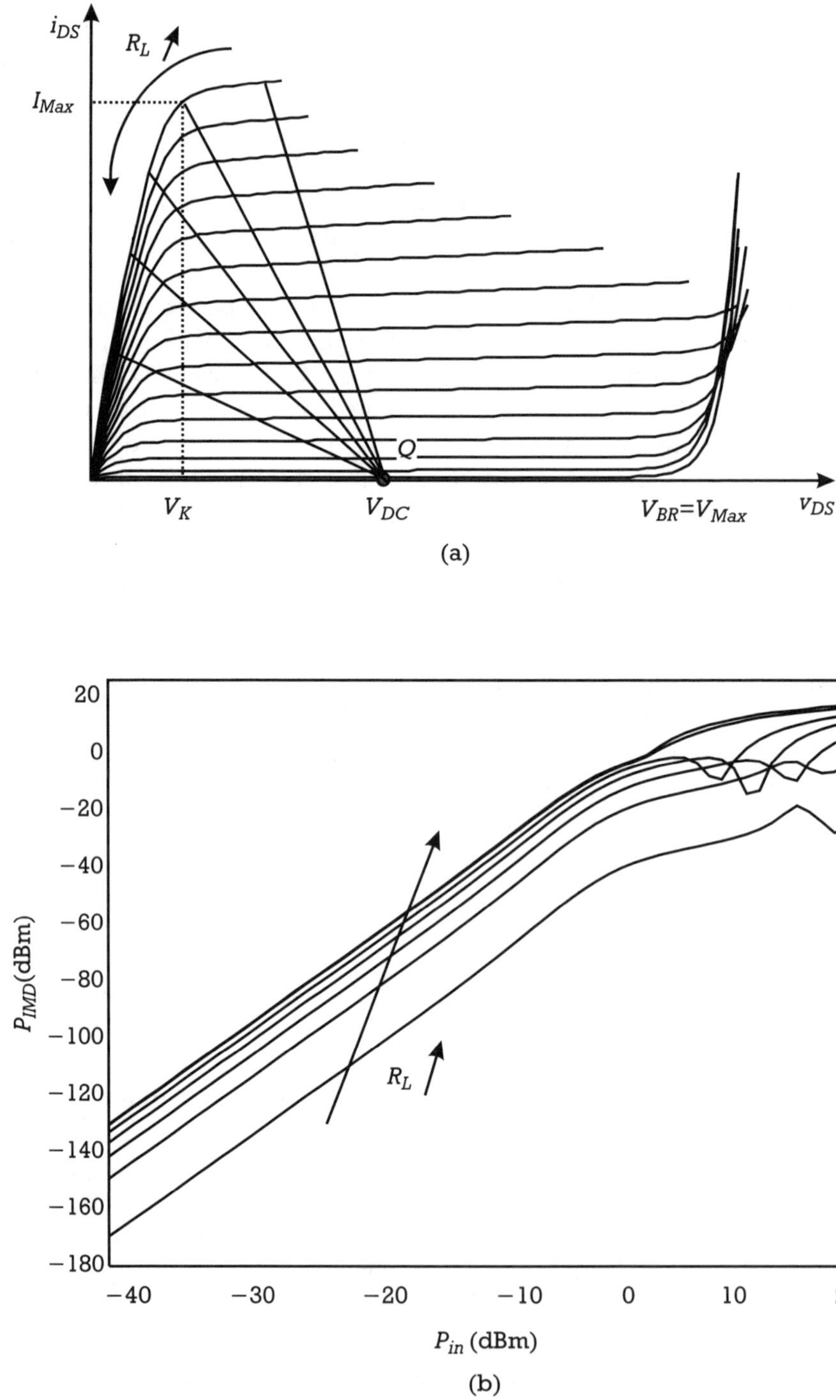

Figure 5.38 Impact of resistive PA load resistance, R_L, on large-signal IMD behavior.

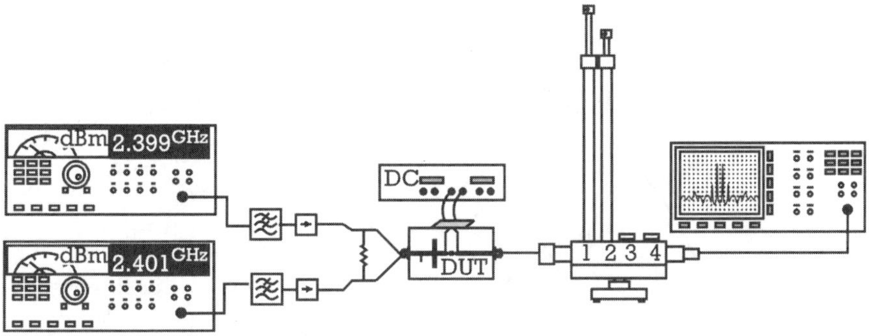

Figure 5.39 Typical laboratory setup used for evaluating the impact of load impedance on large-signal IMD performance.

expected that only the load impedance will have a noticeable effect. That is the reason why we concentrated the following analysis in the amplifier's output mesh, and we have replaced it by a convenient one-port network.

Beginning by the termination presented to the fundamental, Figure 5.40 illustrates the result of a simulated load-pull of $S(V_i)$ and $IMR(V_i)$ at the onset of saturation, and in a quiescent condition where this PA is expected to present a large-signal IMD sweet-spot. Since, as we have just studied, load impedance controls large-signal sweet-spot position, these load-pull results were obtained by appropriately adjusting input power, so that the $IMR(V_i)$ values were all measured at the sweet-spot.

The first conclusion that can be drawn from Figure 5.40 is that the best $IMR(V_i)$ load impedance values define an approximate straight line, which is close to a Smith chart diameter. In fact, it is nothing but the line of real impedances slightly rotated by the MESFET output parasitics, such that the intrinsic load impedance presented at the equivalent circuit drain-source current source terminals is purely real. Actually, that is why the load of highest output power also falls over that line. So, optimized IMD results correspond to deeper sweet-spot valleys, which are obtained under ideal memoryless conditions.

When intrinsic load impedance shows notable reactive components, providing the circuit with memory effects at the fundamental, small-signal input and large-signal output generated distortion can no longer be in exact opposite phase, and sweet-spot cancellation ceases to be perfect. The large-signal IMD sweet-spot valley is less pronounced, and measured $IMR(V_i)$ is lower. This is the justification for the $IMR(V_i)$ decrease observed in Figure 5.40, when load impedance is located farther and farther from the referred optimum line.

Advancing now to study the effect of out-of-band load impedance terminations, we considered three groups of impedances at baseband, second-harmonic, and third and higher harmonics. Because both RF power dissipation and waveform

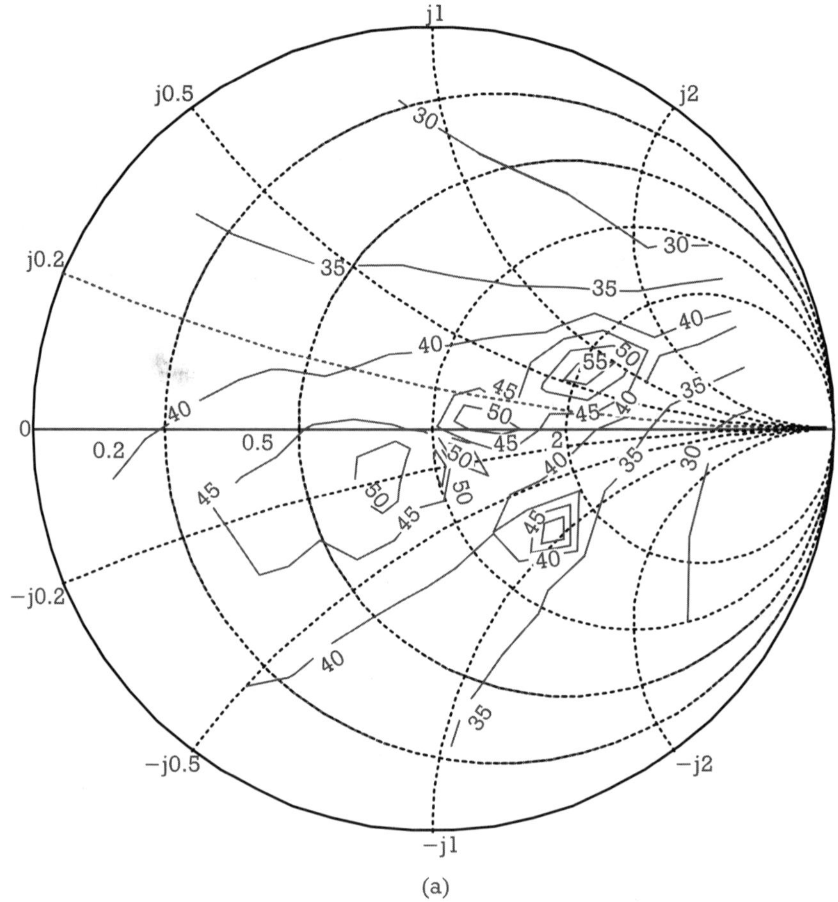

Figure 5.40 Load-pull contours of the output terminating impedance at the fundamental signals: (a) *IMR* simulated at a large-signal IMD sweet-spot, and (b) associated P_{out}.

shaping for optimized efficiency criteria demand for reactive out-of-band impedances, only purely imaginary impedances of $\Gamma_L = 1$, $\Gamma_L = j$, $\Gamma_L = -1$, and $\Gamma_L = -j$ were considered. When one of the out-of-band impedances was varied, the others were kept fixed at either an open or short-circuit, as indicated in Table 5.2.

Results shown in Figure 5.41(a, b) clearly indicate that the load impedance presented to the beat-frequency $(\omega_2 - \omega_1)$ has a major impact, not only on the deepness of the IMD valley, but even in its position and existence, as is reflected in the compression or expansion characteristics evidenced by the output power at the fundamental. Moreover, Figure 5.41(c, d) show that second-harmonic termination can only have any noticeable effect if baseband impedance is set to a short-circuit. If the difference frequency termination is left open, third-order distortion

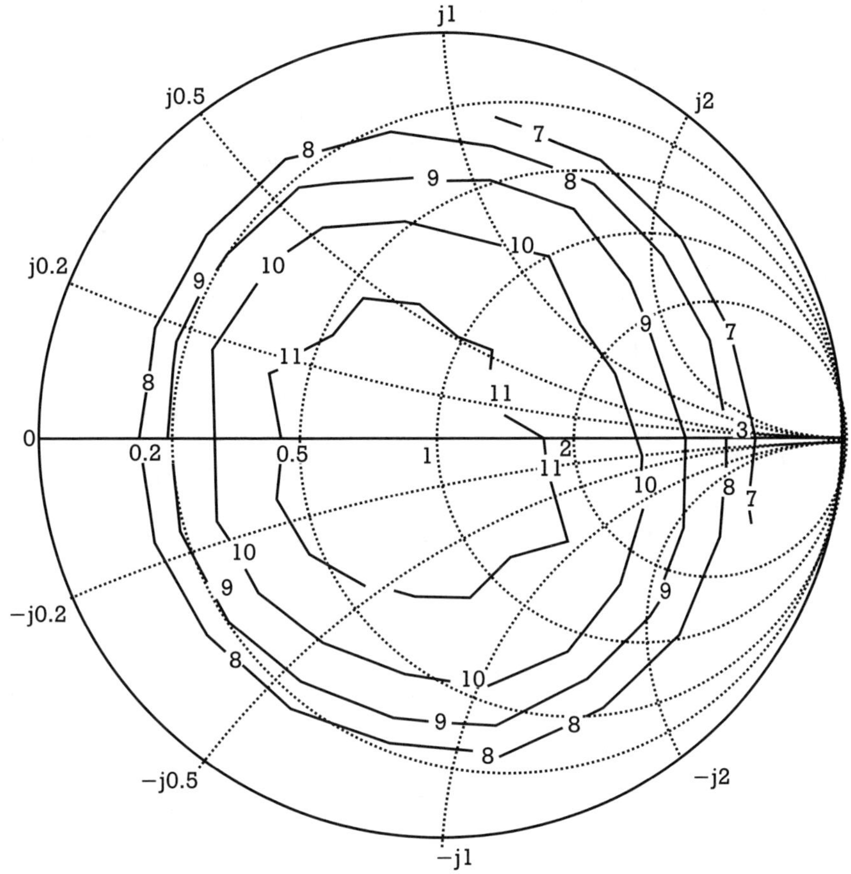

(b)

Figure 5.40 (continued).

Table 5.2 Different Out-of-Band Load Termination Settings Used for Getting the Results of Figure 5.41

Load Termination Settings	Γ_L *@ Baseband* $(\omega_2 - \omega_1)$	Γ_L *@ Second-Harmonic* (2ω)	Γ_L *@ Third and Higher Harmonics* $(>2\omega)$
(a)	Variable	$1\angle 180^\circ$	$1\angle 180^\circ$
(b)	Variable	$1\angle 0^\circ$	$1\angle 180^\circ$
(c)	$1\angle 180^\circ$	Variable	$1\angle 180^\circ$
(d)	$1\angle 0^\circ$	Variable	$1\angle 180^\circ$
(e)	$1\angle 180^\circ$	$1\angle 180^\circ$	Variable
(f)	$1\angle 0^\circ$	$1\angle 180^\circ$	Variable

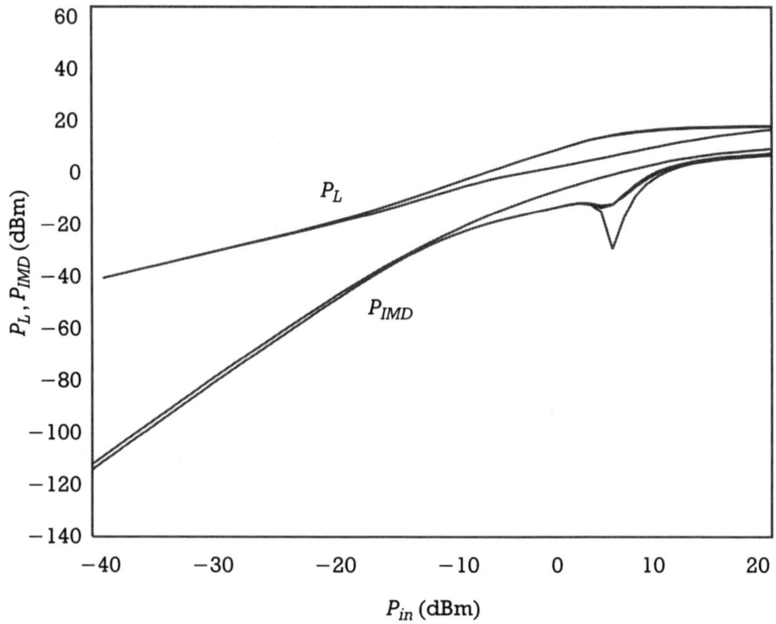

(a)

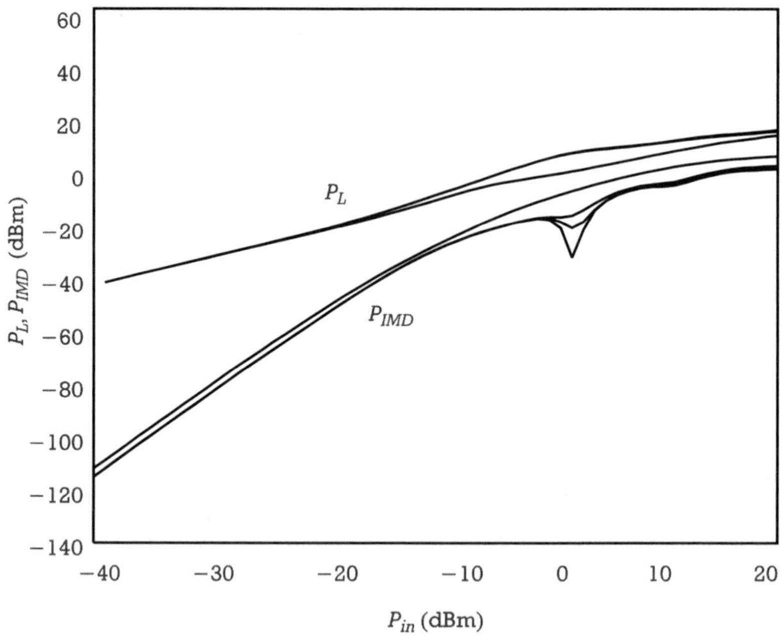

(b)

Figure 5.41 Fundamental output power, P_L, and IMD power, P_{IMD}, for each matching network load-pulling: (a, b) baseband; (c, d) second-harmonic; and (e, f) third-, fourth-, fifth-, and upper harmonics, according to Table 5.2.

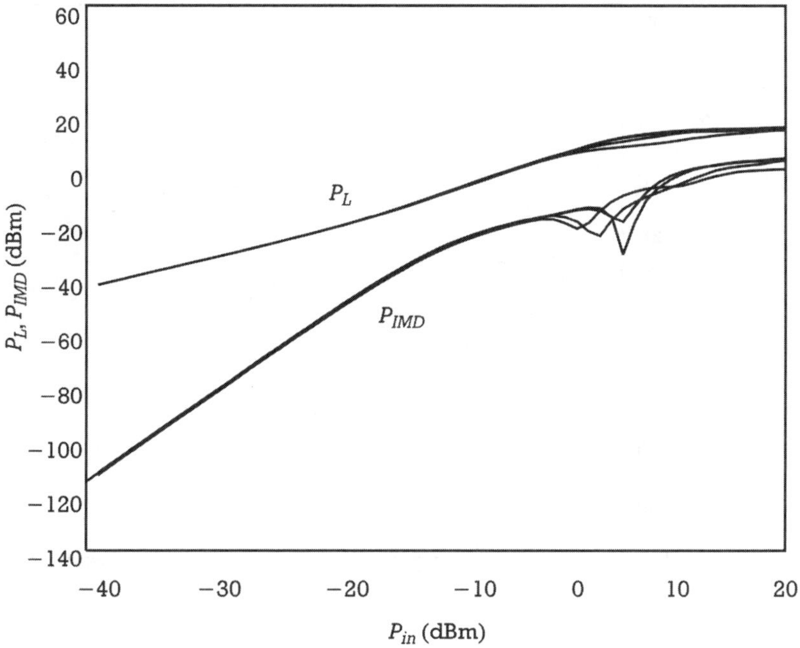

(c)

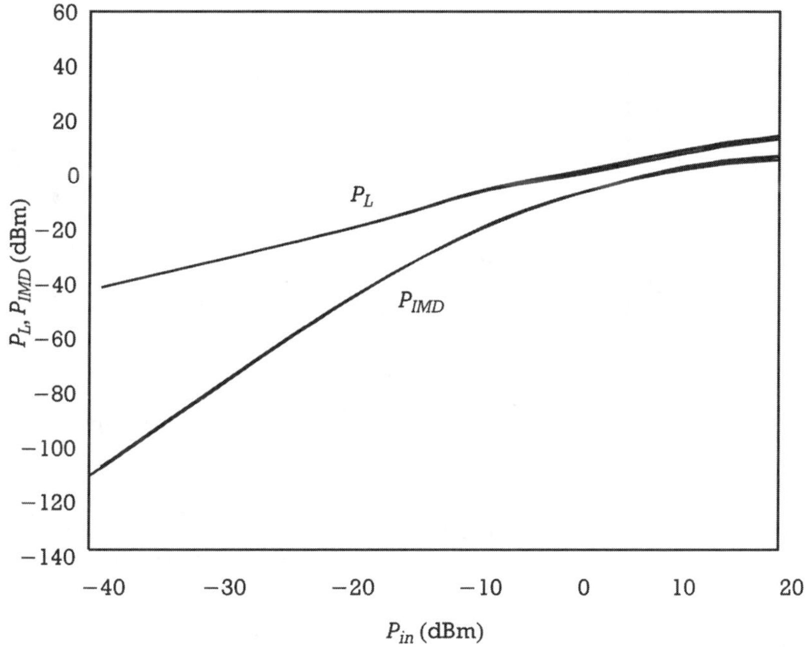

(d)

Figure 5.41 (continued).

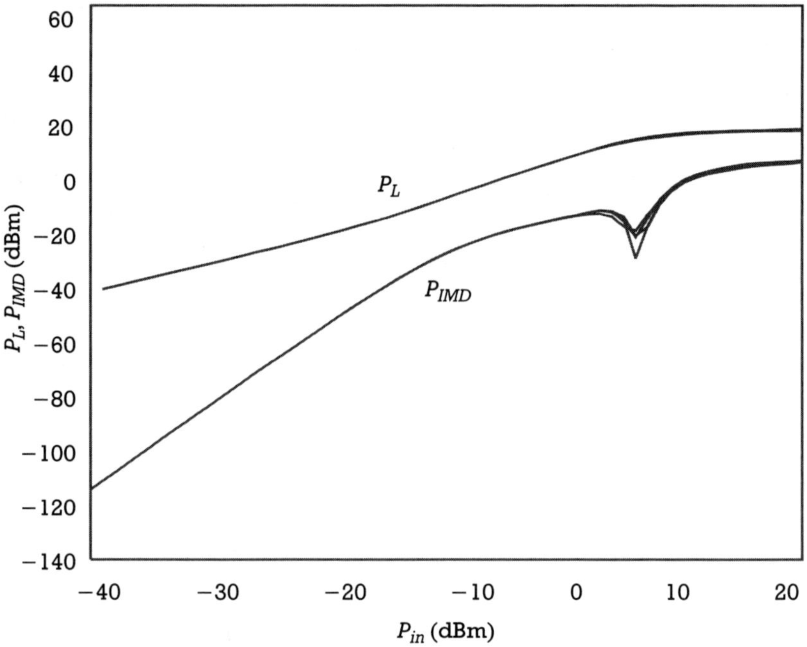

(e)

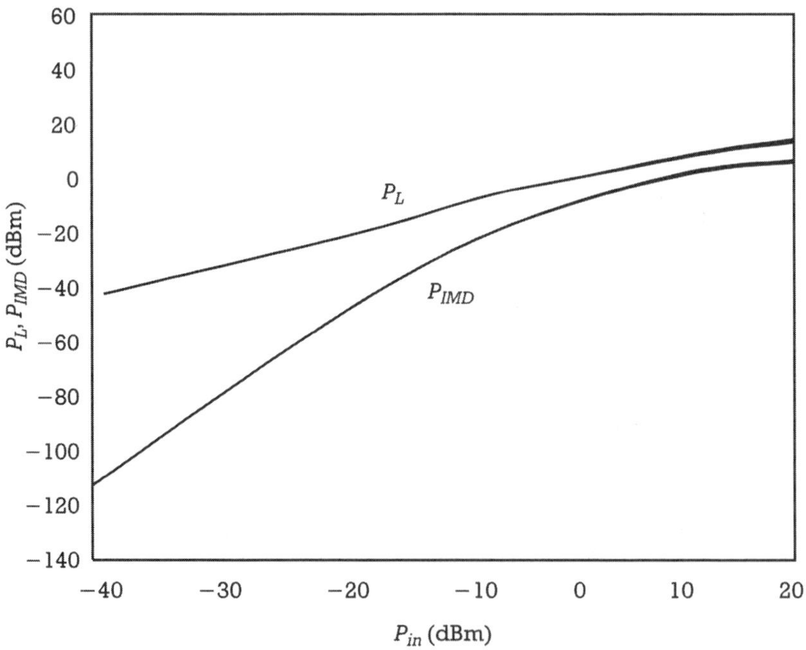

(f)

Figure 5.41 (continued).

created by remixing high baseband output voltage with the fundamentals dominates any possible third-order distortion created by remixing second-harmonics with the fundamentals.

This strong impact of baseband termination on inband IMD characteristics is not particular to the present amplifier under study but is, indeed, quite common to a broad range of topologies and even transistor types. In fact, it has been extensively used as a practical tool for amplifier IMD control and efficiency enhancement, by injecting a processed version of the envelope at the input or the output. In the first case, we have a group of PA linearizer circuits operating by envelope feedback [3], while, in the second, various forms of envelope injection known as envelope restoration or envelope bias modulation [1] were proposed.

Although in the present case baseband termination was only varied when the second-harmonic was kept fixed at a short or open-circuit, and vice-versa, it has often been observed, and theoretically explained, that reactive baseband terminations can also interact with reactive second-harmonic impedances to produce asymmetry in the IMD upper and lower sidebands [10]. Very briefly, this could be attributed to the fact that $Z_L(\omega_2 - \omega_1)$ must be complex conjugate to $Z_L(\omega_1 - \omega_2)$, while $Z_L(2\omega_1) \approx Z_L(2\omega_2)$ and $Z_L(\omega_1) \approx Z_L(\omega_2)$. Therefore, if all the out-of-band terminations have strong imaginary parts, they will determine approximately complex conjugate third-order IMD components, proportional to $Z_L(\omega_2 - \omega_1)Z_L(\omega_2)$ and $Z_L(\omega_1 - \omega_2)Z_L(\omega_1)$, adding in amplitude and phase to components proportional to $Z_L(2\omega_2)Z_L(-\omega_1)$ and $Z_L(2\omega_1)Z_L(-\omega_2)$, which present similar imaginary parts. So, while one of these pairs of IMD components produce cumulative imaginary parts in its corresponding IMD sideband, the other pair generates compensating imaginary parts, thus creating the observed IMD power asymmetries.

Finally, load impedance terminations at the third or higher harmonics have an almost negligible impact on inband IMD, as seen from Figure 5.41(e, f). This result should cause no surprise, as these terminations can only control fifth or higher order inband products, which are expected to have a magnitude much smaller than the dominating third-order ones.

Before concluding this discussion on PA large-signal distortion, it is convenient to make a brief remark about multitone or randomly modulated stimuli, since they are the actual type of signals handled by real PA circuits. Comparing them to the two-tone excitation considered up to now, there are two important differences. The first one is their time-varying amplitude, or envelope, and the second refers to their frequency content.

Contrary to a two-tone, whose envelope is a deterministic sinusoid, a real bandpass telecommunications excitation has an envelope depending on both the particular statistics of the modulating signal, and the modulation technique in use. So, there may be excitations, as those of the GSM mobile system [26], in which the amplitude envelope is almost constant, or others, like the third generation

UMTS system [27], where the amplitude envelope varies significantly. So, although the conclusions drawn for small-signal IMD sweet-spots under two-tone excitation can be directly extrapolated to these complex modulated signals, the conclusions associated to the large-signal IMD sweet-spots cannot. While, in the first case, the linearization effect was a consequence of the elimination of the source of distortion (nullifying Volterra series third-order nonlinear transfer function, responsible for small-signal inband distortion), in the second, we only manage to cancel contributions of various orders, and at a particular voltage swing. In a varying amplitude envelope signal, however, such a strict voltage swing is indefinable, but statistically reached depending on the excitation power. Actually, when average input excitation is increased from small-signal towards saturation, the probability with which the envelope takes that critical value starts by being very small, reaches a maximum for a certain stimulus level, and dropping again after that. So, we can no longer have these sharp IMD minima (nor the also observed abrupt rises in distortion) on P_{IMD} versus P_{in} curves, but simply diffuse zones of distortion output where its power rises more slowly (or rapidly) then the typical small-signal 3 dB/dB slope as illustrated Figure 5.42.

Where the frequency content is concerned, it is obvious that a two-tone excitation has only one baseband component at $\omega_2 - \omega_1$, whereas a bandpass noise signal of Bw bandwidth determines a continuous baseband spectrum from dc to Bw. Therefore, we can no longer talk about a single complex number representing the baseband terminating impedance, but must take track of all effects that might result from the whole possible combinations of even-order mixing products falling in this continuous baseband. Therefore, if that terminating impedance varies significantly, it is likely that the average result will again contribute to smooth the former $P_{IMD}(P_{in})$ abrupt changes.

5.4 Linear Mixer Design

Although the title of this section specifically refers to mixers, and we will adopt this example for the explanation, its material is applicable to a wider range of circuits, such as modulators and demodulators, waveform samplers, RF switches, and even controlled attenuators. In fact, every one of these devices can be viewed as a time-varying mildly nonlinear circuit in which the flow of one signal is controlled by the time variation of another one. For instance, in the case of mixers or samplers, the controlling signal is the local oscillator (LO) or sampling clock, while the stimulus is the RF signal, and the output is the intermediate-frequency (IF) or sampled waveform. In RF switches or variable attenuators, the LO role is played by the control signal, although it varies in a much slower way than usual mixer LOs. And, since most of the modulators and demodulators can be implemented with RF mixers, or analog multipliers, they can be also analyzed with the following

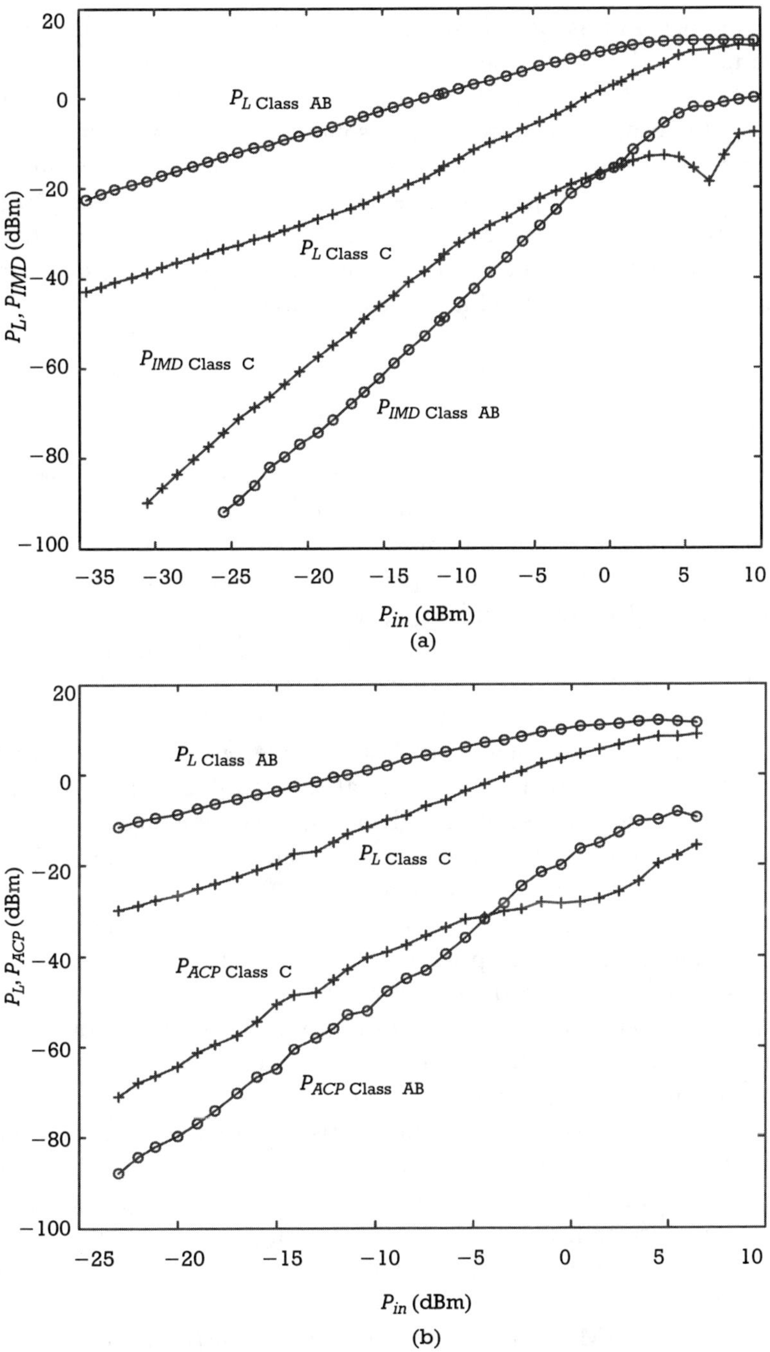

Figure 5.42 (a) Fundamental output power, P_L, and IMD power, P_{IMD}, for a two-tone excitation; and (b) adjacent-channel power, P_{ACP}, for a noise signal, driving a PA biased at class AB and class C.

techniques. In this latter case, however, it may not be so clear which signal should we take as the LO or the RF. If one of the inputs has so stronger amplitude that it is reasonable to admit it drives the circuit into a distinct nonlinear regime, while the other constitutes only a small perturbation to that operating point, the attribution of roles is obvious. Otherwise, the quasilinear techniques of the following sections are not applicable, demanding for a true multitone large-signal analysis.

Furthermore, in all the addressed circuits, it is assumed that the two inputs are uncorrelated, in the sense that their frequencies are not harmonically related, and so there is no definable relation between their phases.

5.4.1 General Mixer Design Concepts

One of the applications where mixer nonlinear distortion deserves special attention is downconversion of heterodyne telecommunications receivers. There, the mixer is usually preceded by a low-noise preamplifier and coarse channel-selection filter, and drives a high-gain narrowband IF amplifier. Contrary to these blocks, which may be very linear, the mixer is, necessarily, a nonlinear device, because only a nonlinearity can operate the desired frequency translation. So, it is quite likely that it will be the mixer that determines the receiver's overall distortion performance.

In order to prevent an intolerable mixer distortion level, it must be driven by an RF signal of sufficiently small amplitude, which indicates that the gain of the preamplifier must be kept at reasonable low values. This, in turn, implies that the preamplifier gain may not be high enough to desensitize the overall receiver's noise factor from the mixer's noise contribution. And so, mixer noise factor is another characteristic to take into account.

Due to the downconversion operated onto the receiver's excitation, signal conditioning at IF is, in principle, much easier than at RF. Consequently, it can be expected that good IF amplifier gain and noise factor performance are not too difficult to get, which then implies that mixer gain may not be so important as its noise or distortion. And, in reality, many good mixer designs do not present power gain from RF to IF, but loss, instead.

Other important mixer characteristics are port-to-port isolation and image rejection. In the former, we want to guarantee, for example, that no LO signal appears at the RF or IF ports in order to prevent desensitization of the preamplifier or IF amplifier, or even its radiation through the receiving antenna. In the latter, we ought to prevent noise present at the image frequency, IM, from being converted onto the desired IF. This is particularly important in multichannel systems where the undesired IM channel (one which is the mirrored image of the RF across the LO symmetry axis) may have an amplitude even higher than the sought channel. As a matter of fact, the main role of the input coarse channel-selection filter is IM rejection. Because this IM frequency is located only two times the IF apart from the RF input, available IM rejection filter cut-off slopes may dictate the IF frequency value, and so the number of different IF chains in the receiver's architecture.

There are many distinct mixer topologies, which differ from the selected nonlinearity or the number of equal coupled nonlinearities. According to the selected nonlinearity, there are diode mixers, normally using high-speed Schottky diodes, or transistor mixers, either based on FETs or BJTs. Referring to the number of those nonlinear devices needed, the mixer can be unbalanced (only one device), singly balanced (at least two equal devices), or doubly balanced (at least four equal devices). However, every one of these follows a common design procedure, which will be exemplified for one simple, but also of practical relevance, unbalanced topology: the active FET mixer. After that, nonlinear distortion mechanisms in general diode mixers will also be briefly addressed.

5.4.2 Illustrative Active FET Mixer Design

The FET active mixer topology under study is depicted in Figure 5.43. It includes LO, RF, and IF matching networks, which also serve as appropriate terminations to the other signals, and gate-source and drain-source bias circuits. In the present

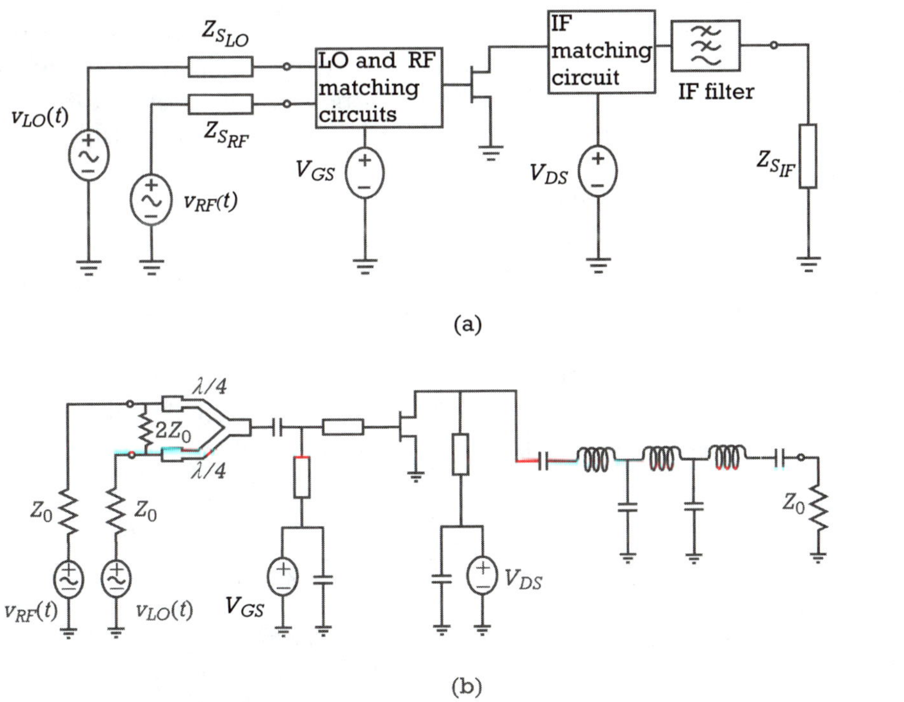

Figure 5.43 (a) General FET active mixer topology. (b) Schematic diagram of the S-band MESFET active gate mixer used for the practical examples of this section.

example, it is assumed that the mixer will be operated as a downconverter where $\omega_{IF} = \omega_{RF} - \omega_{LO}$ and $\omega_{RF} > \omega_{LO}$.

The IF matching circuit must also filter out any other mixer output spurious signals, like LO, RF, or IF harmonics, and their possible beat products. In a downconverter, the IF matching circuit should, therefore, include a lowpass or bandpass filter. Looking into the RF matching network, it should be clear that it has to help the necessary LO-RF isolation, and to provide the eventual IM rejection.

In this type of FET mixer, both the LO excitation and the RF input are applied to the gate terminal, while the IF output is collected from the drain. Since the RF is a small $v_{gs}(t)$ excitation, and the LO is a strong $v_{GS}(t)$ modulating signal, the following $i_{DS}(t)$ expression clearly indicates that a maximized conversion efficiency is obtained when the LO induced variation of $g_m(t)$ is higher:

$$\begin{aligned} i_{DS}(t) &= I_{DS}(t) + g_m(t)v_{gs}(t) + g_{ds}(t)v_{ds}(t) + \ldots \\ &\approx I_{DS}[v_{GS_{LO}}(t)] + g_m[v_{GS_{LO}}(t)]\,V_{gs_{RF}}\cos(\omega_{RF}t) \\ &\quad + g_{ds}[v_{GS_{LO}}(t)]\,V_{ds_{RF}}\cos(\omega_{RF}t) + \ldots \end{aligned} \tag{5.231}$$

where it was assumed that no LO signal appears at the drain.

The form of the output I/V curves of a FET indicates that this maximum $g_m(t)$ variation is obtained when the device is biased in its saturation region ($V_{DS} > V_K$). Moreover, to prevent the generation of nonlinear distortion in the strongly nonlinear triode zone, the device should be kept always inside that region of high v_{DS} voltages. That can be guaranteed by using a very low load resistance to the LO signal, typically a short-circuit.

Biased in the saturation region, the mixer behaves as a time-varying transconductance—$g_m(t)$—and thus may be active. That is, unlike diode or FET resistive mixers, which always show conversion loss, a FET active gate mixer can provide conversion gain.

Turning our attention to the gate port, it is also clear that the desired maximized variation of $g_m(t)$ with the LO requires that $v_{GS}(t)$ should be biased in a point, V_{GS}, where $\partial G_m/\partial v_{GS}\big|_{v_{GS}=V_{GS}}$ is larger. Since

$$\left.\frac{\partial G_m}{\partial v_{GS}}\right|_{\substack{v_{GS}=V_{GS}\\ v_{DS}=V_{DS}}} = \left.\frac{\partial^2 i_{DS}}{\partial v_{GS}^2}\right|_{\substack{v_{GS}=V_{GS}\\ v_{DS}=V_{DS}}} = 2!\,G_{m2}(V_{GS}, V_{DS}) \tag{5.232}$$

we conclude that minimum conversion loss should be reached when biasing the FET for its $G_{m2}(V_{GS})$ maximum. Although G_{m2} is a small-signal expansion parameter, while the LO excitation will certainly present a large swing in comparison to the dc quiescent point, this result is consistent with the empirical knowledge that maximum conversion efficiency is obtained if the FET is biased near turn-on:

$V_{GS} \approx V_T$. In terms of the desired RF termination at the drain, a short-circuit should be, again, and in principle, the best choice. Actually, any possible RF component present at $v_{DS}(t)$ could also generate distortion in the FET's $i_{DS}(v_{DS})$ nonlinearities.

After this introductory discussion, let us now carry on a quantitative analysis of the circuit. Since the mixer is supposed to handle an RF signal of much lower amplitude than the bias or LO driving, it can be analyzed as a time-varying mildly nonlinear system. Therefore, the adopted analysis method will follow the time-varying Volterra theory presented in Section 3.2.3.

Figure 5.44 represents an equivalent circuit model of our FET active mixer, in which the input and output matching circuits were lumped with the gate and drain bias circuits into $Y_1(\omega)$ and $Y_2(\omega)$ embedding admittances.

5.4.2.1 Large-Signal Mixer Analysis Under Local Oscillator Excitation

According to Section 3.2.3, the mixer analysis procedure starts by the determination of the time-varying quiescent point. For that, the RF small perturbation is ignored (i.e., I_{RF} is assumed zero), and a harmonic balance analysis of the circuit is performed.

According to the piecewise HB analysis presented in Section 3.3.5, our network involves two port sources, $F = 2$, three nonlinear elements, $M = 3$, and three controlling port voltages, $L = 3$. But, since all these controlling ports are shared by the nonlinear elements, and the output port coincides with the dc drain excitation, the total number of required network ports is five, as depicted in Figure 5.44. Following the same procedure of Section 3.3.5, we now augment the linear network with the current sources, $i_{LO}(t)$, I_{GS}, and I_{DS}, to end up in a system of $3(2K+1)$ equations per $3(2K+1)$ unknowns. It relates $\mathbf{I}_{NL}[\mathbf{V}(\omega)] = [I_3(\omega) I_4(\omega) I_5(\omega)]^{\mathrm{T}}$ to $\mathbf{V}(\omega) = [V_3(\omega) V_4(\omega) V_5(\omega)]$, where $I_3(\omega) = j\omega Q_{gs}[V_3(\omega)]$, $I_4(\omega) = I_{ds}[V_3(\omega), V_4(\omega)]$, and $I_5(\omega) = j\omega Q_{gd}[V_5(\omega)]$, since $V_3(\omega) = V_{gs}(\omega)$, $V_4(\omega) = V_{ds}(\omega)$, and $V_5(\omega) = V_{gd}(\omega)$.

When the HB solution is reached, we get a frequency-domain description of the control voltages, from which all network branch currents or node voltages can be determined. In particular, the current entering the extrinsic gate terminal, $I_{in}(\omega)$, and the voltage at that node, $V_{in}(\omega)$, define the input impedance at all LO harmonics, $k\omega_p$, $k = -K, \ldots, 0, \ldots, K$:

$$Z_{in}(k\omega_p) \equiv \frac{V_{in}(k\omega_p)}{I_{in}(k\omega_p)} \tag{5.233}$$

LO efficiency demands for a conjugate match at the fundamental, and reactive source impedances at all other harmonics, typically short-circuits. So, the gate matching network should be designed to transform the LO source impedance (generally a 50-Ω resistance) into $Z_S(\omega_p) = Z_{in}(\omega_p)^*$ and $Z_S(k\omega_p) \approx 0$ for $k > 1$.

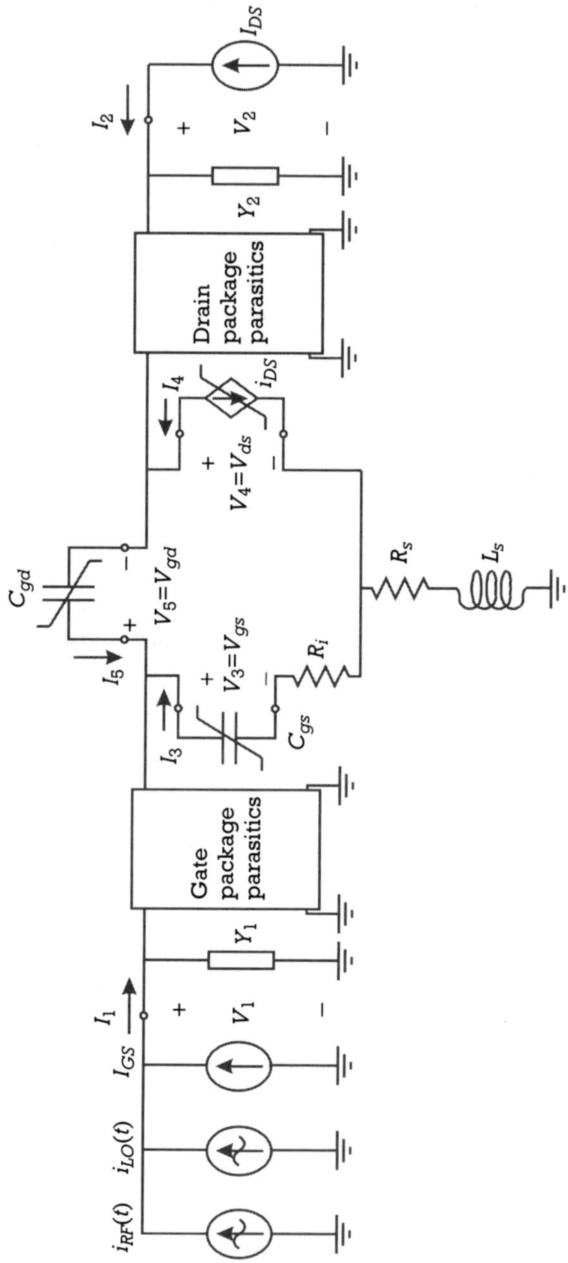

Figure 5.44 Active gate FET mixer equivalent circuit used for nonlinear analysis.

In FET mixers, and as long as LO pumping does not lead to gate-channel junction conduction or breakdown, $Z_{in}(\omega)$ remains approximately unchanged. So, finding optimum matching conditions is probably not too difficult.

On the contrary, when the nonlinearity is a diode or bipolar transistor, $Z_{in}(\omega)$ changes significantly with LO drive, and so obtaining a conjugate match may require some iterations. That is clear with an unbiased p-n or Schottky junction that shows an almost pure reactance (due to the junction depletion charge) when LO is too small to induce forward conduction. This inhibits efficient power transfer to the device, and the pumping regime remains practically unchanged. However, when some power begins to be delivered to the junction, its impedance starts to show some resistive component and an adjustment on LO source impedance is required. Then, additional power can be delivered, the input impedance gets progressively more resistive, and a new adjustment is needed. Those steps should then be repeated until a matching condition, under the desired LO drive, is reached.

5.4.2.2 Mixer Small-Signal Linear Analysis

The second step in our mixer design procedure consists of determining the best bias voltages, V_{GS} and V_{DS}, LO driving amplitude, and terminating impedances for conversion efficiency maximization.[6]

Optimizing conversion performance requires the inclusion of RF excitation. For that, the control voltages obtained from HB simulations are converted to the time-domain (a result readily available from mixed-mode HB, time-step integration, or shooting-Newton simulators), and then used to obtain the time-varying linear description of the mixer circuit. So, following the procedure described in Section 3.2.3, $v_{GS}(t)$, $v_{DS}(t)$, and $v_{GD}(t)$ are substituted into the first-degree Taylor series expansions of the FET:

$$i_{gs}(t) = \frac{d}{dt}[q_{gs1}(t)v_{gs}(t)] \tag{5.234}$$

$$= \sum_{k_1=-K}^{K} \sum_{k_2=-K}^{K} \sum_{b=-B}^{B} Q_{gs1_{k_1}} V_{gs_{b,k_2}} j[(k_1+k_2)\omega_p + \omega_b] e^{j[(k_1+k_2)\omega_p+\omega_b)]t}$$

$$i_{ds}(t) = g_m(t)v_{gs}(t) + g_{ds}(t)v_{ds}(t)$$

$$= \sum_{k_1=-K}^{K} \sum_{k_2=-K}^{K} \sum_{b=-B}^{B} G_{m_{k_1}} V_{gs_{b,k_2}} e^{j[(k_1+k_2)\omega_p+\omega_b)]t} \tag{5.235}$$

$$+ \sum_{k_1=-K}^{K} \sum_{k_2=-K}^{K} \sum_{b=-B}^{B} G_{ds_{k_1}} V_{ds_{b,k_2}} e^{j[(k_1+k_2)\omega_p+\omega_b)]t}$$

6. In accordance to our goal of dynamic range optimization, a low mixer noise factor should also be addressed. However, the analysis of noise in mixers is quite involved, and falls outside the scope of this text. The interested reader is kindly suggested to consult [26].

and

$$i_{gd}(t) = \frac{\mathrm{d}}{\mathrm{d}t}[q_{gd1}(t)v_{gd}(t)] \tag{5.236}$$

$$= \sum_{k_1=-K}^{K} \sum_{k_2=-K}^{K} \sum_{b=-B}^{B} Q_{gd1_{k_1}} V_{gd_{b,k_2}} j[(k_1 + k_2)\omega_p + \omega_b] e^{j[(k_1+k_2)\omega_p+\omega_b)]t}$$

which provide the linear responses to the multitone small-signal RF perturbation:

$$i_{RF}(t) = \sum_{q=-Q}^{Q} I_{rf_q} e^{j\omega_q t} \tag{5.237}$$

As stated in Section 3.2.3, (5.234) to (5.236) can be expressed in conversion matrix format, to read as

$$\mathbf{I}_{gs} = j\mathbf{\Omega}\ .x\ \mathbf{Q}_{gs}\mathbf{V}_{gs} \tag{5.238}$$

$$\mathbf{I}_{ds} = \mathbf{G}_m\mathbf{V}_{gs} + \mathbf{G}_{ds}\mathbf{V}_{ds} \tag{5.239}$$

and

$$\mathbf{I}_{gd} = j\mathbf{\Omega}\ .x\ \mathbf{Q}_{gd}\mathbf{V}_{gd} \tag{5.240}$$

Still following Section 3.2.3, a conversion matrix formulation can also be used to describe all other linear elements of the FET equivalent circuit. So, although quite laborious to be treated by hand, a conversion matrix formulation of the circuit of Figure 5.44 would be straightforward, leading to an [Y] or [S] parameter description, in which all matrix elements are, themselves, conversion matrices.

That would be the standard direction to follow. And, nowadays, a few commercial HB simulators, which have already incorporated a specific mixer analysis mode, can be used to facilitate this task.

Nevertheless, we will keep our hand analysis adopting three simplifying assumptions.

The first one states that the FET is approximately unilateral to the IF. This is an acceptable supposition, at least for reasonably low frequencies, and whenever the input LO and RF matching circuits are capable of presenting the desired short-circuit to the IF signal. The effect of that belief is the absence of any IF signal component at $v_{GS}(t)$, and thus $v_{gs_{IF}}(t) \approx 0$. If that assumption can be extrapolated to the rest of mixing products (a realistic guess in low-frequency designs), then $v_{gs}(t)$ will only have non-null components at ω_{RF} and ω_{IM}.

The second assumption admits that the output IF matching circuit presents an approximate short-circuit to the RF signal and all high-frequency mixing products, so that we will have $V_{ds}(k\omega_p + \omega_{IF}) \approx 0$ for $k \neq 0$.

The third assumption consists of neglecting $c_{gs}(t)$ and $c_{gd}(t)$ mixing effects (i.e., admitting linear C_{gs} and C_{gd}).

These assumptions lead to an ideal FET active mixer, in which the only time-varying elements are $g_m(t)$ and $g_{ds}(t)$.

In terms of its mixing properties, the mixer becomes the time-varying four-pole of Figure 5.45. In this equivalent circuit, all drain parasitics plus $Y_2(\omega)$ and C_{ds} were lumped into a single $Y_d(\omega)$, and the gate parasitics plus C_{gs} and $Y_1(\omega)$ were lumped into $Y_g(\omega)$. The Miller equivalents of source parasitics and C_{gd} were also included in these two terminating admittances.

This time-varying four-pole can now be described by the following admittance conversion matrix system:

$$\begin{bmatrix} I_{gs_{-K}} \\ \vdots \\ I_{gs_{-1}} \\ I_{gs_0} \\ I_{gs_1} \\ \vdots \\ I_{gs_K} \end{bmatrix} = \begin{bmatrix} 0 \\ \vdots \\ 0 \\ 0 \\ 0 \\ \vdots \\ 0 \end{bmatrix} \tag{5.241a}$$

and

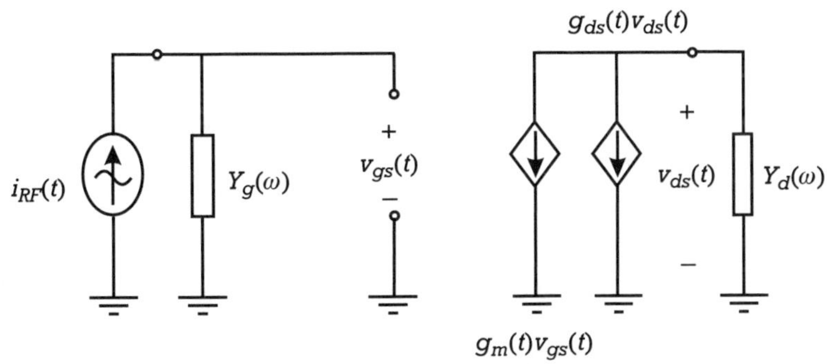

Figure 5.45 Simplified linear time-variant equivalent circuit of the FET active mixer of Figure 5.44.

$$\begin{bmatrix} I_{ds_{-K}} \\ \vdots \\ I_{ds_{-1}} \\ I_{ds_0} \\ I_{ds_1} \\ \vdots \\ I_{ds_K} \end{bmatrix} = \begin{bmatrix} G_{m_0} & G_{m_{-1}} & \cdots & G_{m_{-K}} & \cdots & \cdots & G_{m_{-2K}} \\ \vdots & \vdots & \vdots & \vdots & \vdots & \vdots & \vdots \\ \cdots & G_{m_1} & G_{m_0} & G_{m_{-1}} & \cdots & G_{m_{-K}} & \cdots \\ G_{m_K} & \cdots & G_{m_1} & G_{m_0} & G_{m_{-1}} & \cdots & G_{m_{-K}} \\ \cdots & G_{m_K} & \cdots & G_{m_1} & G_{m_0} & G_{m_{-1}} & \cdots \\ \vdots & \vdots & \vdots & \vdots & \vdots & \vdots & \vdots \\ G_{m_{2K}} & \cdots & \cdots & G_{m_K} & \cdots & G_{m_1} & G_{m_0} \end{bmatrix} \begin{bmatrix} V_{gs_{-K}} \\ \vdots \\ V_{gs_{-1}} \\ V_{gs_0} \\ V_{gs_1} \\ \vdots \\ V_{gs_K} \end{bmatrix}$$

$$+ \begin{bmatrix} G_{ds_0} & G_{ds_{-1}} & \cdots & G_{ds_{-K}} & \cdots & \cdots & G_{ds_{-2K}} \\ \vdots & \vdots & \vdots & \vdots & \vdots & \vdots & \vdots \\ \cdots & G_{ds_1} & G_{ds_0} & G_{ds_{-1}} & \cdots & G_{ds_{-K}} & \cdots \\ G_{ds_K} & \cdots & G_{ds_1} & G_{ds_0} & G_{ds_{-1}} & \cdots & G_{ds_{-K}} \\ \cdots & G_{ds_K} & \cdots & G_{ds_1} & G_{ds_0} & G_{ds_{-1}} & \cdots \\ \vdots & \vdots & \vdots & \vdots & \vdots & \vdots & \vdots \\ G_{ds_{2K}} & \cdots & \cdots & G_{ds_K} & \cdots & G_{ds_1} & G_{ds_0} \end{bmatrix} \begin{bmatrix} V_{ds_{-K}} \\ \vdots \\ V_{ds_{-1}} \\ V_{ds_0} \\ V_{ds_1} \\ \vdots \\ V_{ds_K} \end{bmatrix}$$

$$\Leftrightarrow \mathbf{I}_{ds} = \mathbf{G}_m \mathbf{V}_{gs} + \mathbf{G}_{ds} \mathbf{V}_{ds} \tag{5.241b}$$

where a Fourier expansion from $-2K$ to $2K$ was adopted in the $\mathbf{G}_m$ and $\mathbf{G}_{ds}$ conversion matrices, since it is more common in mixer literature and because of its increased accuracy. Simplicity of notation also determined that in (5.241) and in following expressions I_{ds_k}, I_{gs_k}, V_{gs_k}, or V_{ds_k} represent the whole set of mixing products arising from any of the IF ω_b (or RF ω_q) tones with the kth harmonic component of the local oscillator. According to the conventions adopted in Section 3.2.3, the organization of frequency mixing components in all $\mathbf{I}_{ds}$, $\mathbf{V}_{gs}$, and $\mathbf{V}_{ds}$ arrays are such that, for example, $\mathbf{I}_{ds}$ in (5.241) reads as

$$\begin{array}{c} \\ -K \\ \vdots \\ -1 \\ 0 \\ +1 \\ \vdots \\ +K \end{array} \begin{array}{c} \begin{array}{ccccccc} -B & & \cdots & -1 & +1 & \cdots & B \end{array} \\ \begin{bmatrix} I_{ds}[-(K-1)\omega_{LO} - \omega_{RF_B}] & \cdots & & & \cdots & I_{ds}[-(K+1)\omega_{LO} + \omega_{RF_B}] \\ \vdots & & \vdots & \vdots & & \vdots \\ I_{ds}(-\omega_{RF_B}) & \cdots & I_{ds}(-\omega_{RF_1}) & I_{ds}(-\omega_{IM_1}) & \cdots & I_{ds}(-\omega_{IM_B}) \\ I_{ds}(-\omega_{IF_B}) & \cdots & I_{ds}(-\omega_{IF_1}) & I_{ds}(\omega_{IF_1}) & \cdots & I_{ds}(\omega_{IF_B}) \\ I_{ds}(\omega_{IM_B}) & \cdots & I_{ds}(\omega_{IM_1}) & I_{ds}(\omega_{RF_B}) & \cdots & I_{ds}(\omega_{RF_B}) \\ \vdots & & \vdots & \vdots & & \vdots \\ I_{ds}[(K+1)\omega_{LO} - \omega_{RF_B}] & \cdots & & & \cdots & I_{ds}[(K-1)\omega_{LO} + \omega_{RF_B}] \end{bmatrix} \end{array} \tag{5.242}$$

So, looking at the general output ω_{IF} components we obtain

$$
\begin{aligned}
I_{ds}(\omega_{IF}) = {} & G_{m_K} V_{gs}[-(K+1)\omega_{LO} + \omega_{RF}] + \ldots \\
& + G_{m_1} V_{gs}(-\omega_{IM}) + G_{m_0} V_{gs}(\omega_{IF}) + G_{m_{-1}} V_{gs}(\omega_{RF}) + \ldots \\
& + G_{m_{-K}} V_{gs}[(K-1)\omega_{LO} + \omega_{RF}] \\
& + G_{ds_K} V_{ds}[-(K+1)\omega_{LO} + \omega_{RF}] + \ldots \\
& + G_{ds_1} V_{ds}(-\omega_{IM}) + G_{ds_0} V_{ds}(\omega_{IF}) + G_{ds_{-1}} V_{ds}(\omega_{RF}) + \ldots \\
& + G_{ds_{-K}} V_{ds}[(K-1)\omega_{LO} + \omega_{RF}]
\end{aligned} \tag{5.243}
$$

But, since the assumed unilateral characteristics of our particular mixer lead to a $v_{gs}(t)$ that contains no components either than ω_{RF} and ω_{IM}, and because $v_{ds}(t)$ is nearly short-circuited for all mixing products except the IF, $I_{ds}(\omega_{IF})$ can be simplified to

$$
I_{ds}(\omega_{IF}) \approx G_{m_1} V_{gs}(-\omega_{IM}) + G_{m_{-1}} V_{gs}(\omega_{RF}) + G_{ds_0}(\omega_{IF}) V_{ds}(\omega_{IF}) \tag{5.244}
$$

This expression indicates that conversion gain optimization requires a (V_{GS}, V_{DS}) bias point and a LO drive that maximize first-order Fourier component of our time-varying transconductance $g_m(t)$. That is, conversion efficiency is optimized whenever $g_m(t)$ presents an amplitude as high as possible, and with odd symmetry. Relating this to the device's $G_m(v_{GS})$ variation, we can conclude that maximum conversion gain must be associated to bias points of rapidly varying $G_m(v_{GS})$, or high $G_{m2}(v_{GS})$. Although this conclusion was determined by the approximate unilateral characteristics of our idealized mixer, it has a wider application because, in general, the amplitude of mixing components tends to diminish with increasing order.

Figure 5.46 shows simulated conversion gain variation versus gate-source bias, V_{GS}, of the MESFET active mixer of Figure 5.43, when drain bias and local oscillator drive are kept constant. Figure 5.47 is a plot of the $G_{m2}(V_{GS})$ of the MESFET device used. The similar behavior of $G_{m2}(V_{GS})$ and conversion loss validates our qualitative statement on that direction.

To get deep insight into this frequency conversion process, Figure 5.48 represents $G_m(V_{GS})$, $G_{m_1}(V_{GS})$, and also three $g_m[v_{GS_{LO}}(t)]$ waveforms obtained at $V_{GS} = -1.3$V, $V_{GS} = -1$V, and $V_{GS} = -0.6$V. Observing those curves, and realizing that $v_{GS_{LO}}(t)$ is sinusoidal, it is now clear that maximum conversion efficiency should be obtained near $V_{GS} = -1$V, since it is the point where transconductance is an almost ideal odd function of v_{GS}, therefore maximizing its odd-order Fourier components and, in particular, G_{m_1}.

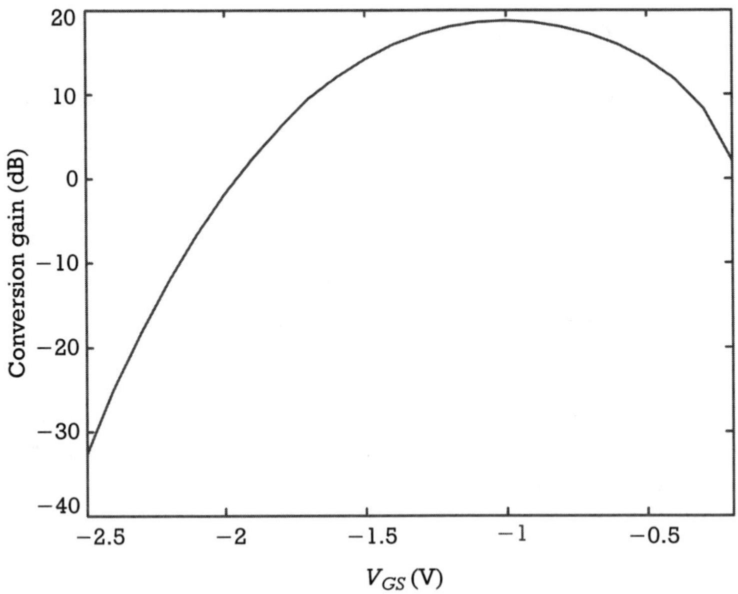

Figure 5.46 Typical MESFET active mixer conversion gain versus V_{GS} gate bias for constant V_{DS} bias and LO drive level.

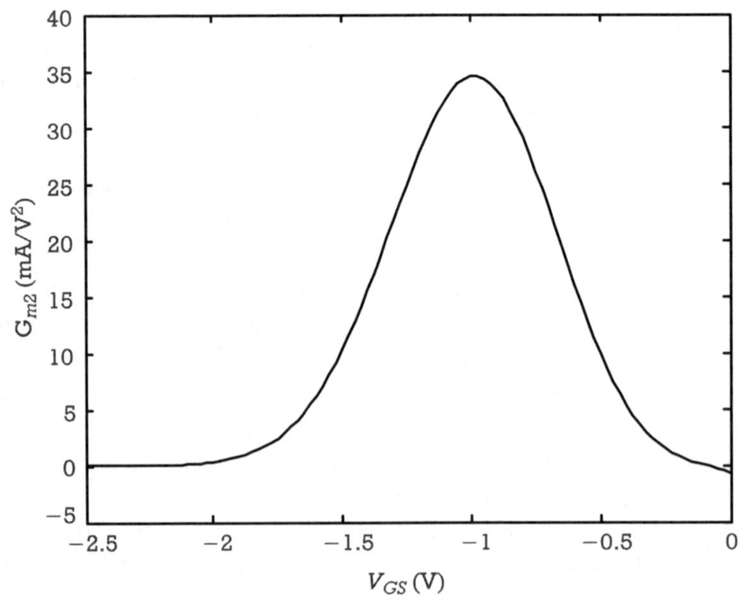

Figure 5.47 $G_{m2}(V_{GS})$ of the device used in the mixer of Figure 5.46, showing that optimized conversion performance is attained near G_{m2} maximum.

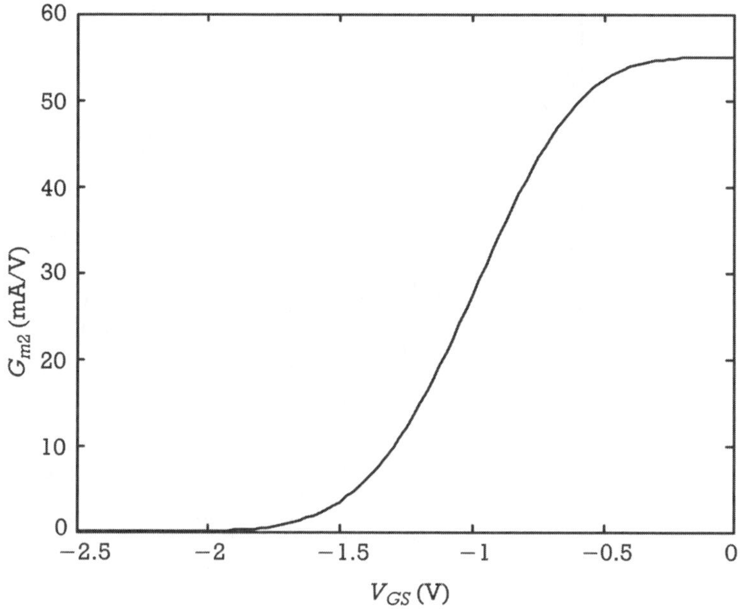

(a)

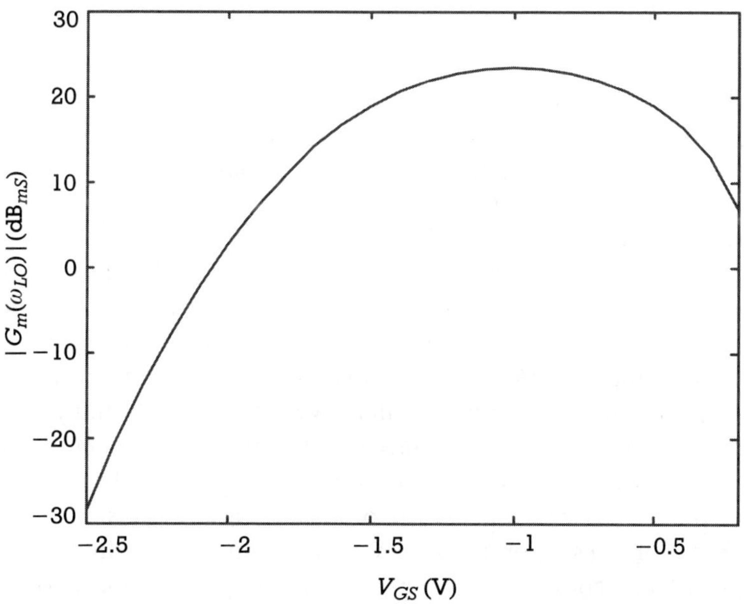

(b)

Figure 5.48 (a) $G_m(V_{GS})$ of the MESFET used in our active gate mixer implementation. (b) Variation of first-order Fourier component of $g_m(t)$, G_{m_1}, with V_{GS} bias. (c) Examples of time-varying $g_m(t)$ waveforms obtained for $V_{GS} = -1.3$V (..), $V_{GS} = -1$V (-), and $V_{GS} = -0.6$V (--).

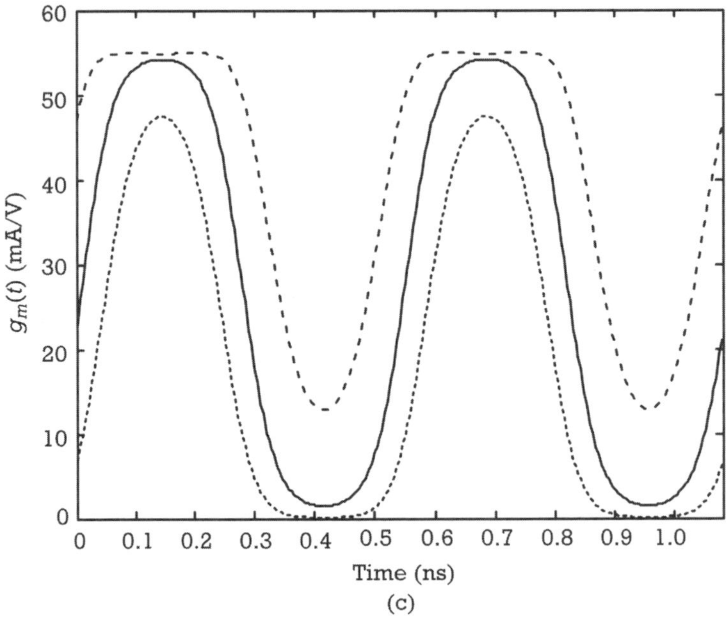

Figure 5.48 (continued).

Figure 5.49 refers to conversion gain variation versus LO drive level, when $V_{DS}=3$V and V_{GS} has the best value suggested by Figure 5.46. It shows a continuous increase in conversion gain until a point where it tends to saturate. This result is coincident with the knowledge gathered from previous figures which showed that, biased for maximum G_{m2}, G_{m_1} increases steadily with LO power until a point where $g_m(t)$ becomes clamped at zero below the threshold voltage (current cut-off), and at its approximately constant maximum value (corresponding to the v_{GS} zone where i_{DS} varies in an nearly linear manner with v_{GS}).

Returning to (5.241), it also indicates that, although sharing the same physical ports at the gate and drain terminals, we may conceive many different mixer ports corresponding to the various mixing products, $\omega_{b,k} = k\omega_p + \omega_b$: one input port for the RF signal at ω_{RF}, one output for the IF signal falling at ω_{IF}, and many other ports corresponding to the other mixing products.

Each of these ports sees its own terminating admittances, $Y_g(\omega_{b,k})$ and $Y_d(\omega_{b,k})$, and contributes to determine mixer performance. For example, terminating admittances at RF and IF reflect input and output power transfer, and thus have a direct impact on mixer conversion gain. But, power loss at any spurious frequency also leads to degradation of conversion characteristics, demanding a careful load admittance selection. Typically, these spurious contributions are minimized by choosing $Y_g(\omega_{b,k})$ and $Y_d(\omega_{b,k})$ as short-circuits, although there may

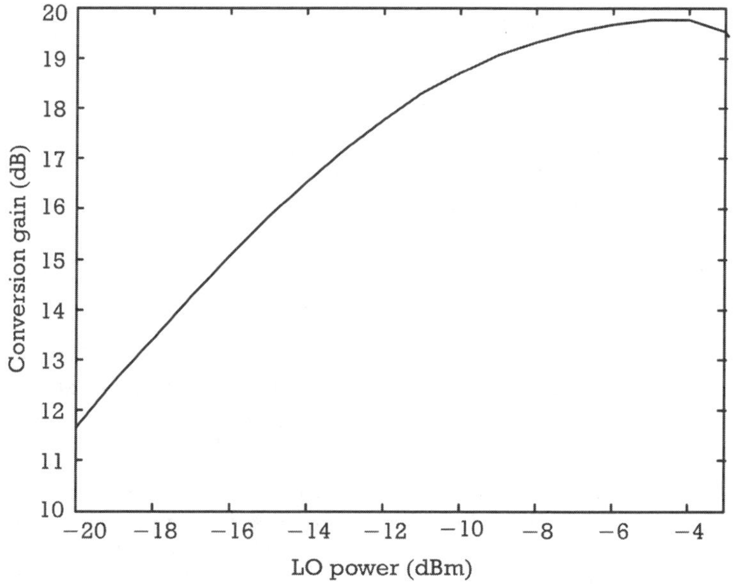

Figure 5.49 MESFET active mixer conversion gain versus LO drive level. $V_{GS} = -1$V and $V_{DS} = 3$V.

be other terminations that alter conversion gain, noise, or even intermodulation performance [28, 29].

From the large set of spurious mixing products, the terminating admittances at the image frequency, ω_{IM}, are the ones having a greater impact in the generality of mixers, thus deserving a special attention. When the IF frequency is very low, compared to the RF, the IM frequency is not significantly different from the RF. Therefore, its termination will not be dramatically distinct from the one at ω_{RF}. However, when ω_{IF} is considerably high, the mixer will probably incorporate an image rejection filter that will independently impose $Y_g(\omega_{IM})$. In that case, this filter should be designed, in combination with the RF and LO matching networks, to present the desired $Y_g(\omega_{IM})$ [as $V_{ds}(\omega_{IM})$ can be easily short-circuited]. This process of mixer performance optimization, known as *image enhancement,* should be used with care as it may lead to reduced conversion loss but at the expense of worse noise or intermodulation performance [29].

Since we have eliminated both feedback and input nonlinearity from our FET active mixer circuit, only $Y_d(\omega_{IM})$ would have any impact. But, since drain was already short-circuited at ω_{RF} and ω_{IM}, we may expect that our simplified design be insensitive even to ω_{IM} terminations.

In general mixer design, and after the terminations at the spurious frequencies have been selected, it is necessary to design RF and IF matching networks to provide the required input and output matching. Contrary to LO matching, which involved a nonlinear regime where optimum driving impedance was dependent on pumping level, RF and IF matching is a linear process, similar to the one found in small-signal amplifiers. Therefore, it is interesting to use that solid two-port linear theory in mixer design. For obtaining such a two-port description, the $(2K + 1) \times (2K + 1)$ conversion matrix system of (5.241) is reduced to a 2×2 admittance matrix of a network that was augmented with all but ω_{RF} and ω_{IF} terminating admittances, as described in any linear mixer design text [28, 30].

5.4.2.3 Mixer Small-Signal Distortion Analysis

Distortion analysis in mixers is significantly more complex than in amplifiers, due to the time-varying nature of the circuit. Therefore, it is also in an earlier stage of development. Published data on this subject usually addresses particular circuits, most of the time resulting from numerical simulations or laboratory measurements. Qualitative interpretations are rare, and so results cannot be easily extrapolated to other mixer topologies and even less to different mixer devices. To complement that, in the following paragraphs, we will restrict the analysis to very simple situations, just to bring a first insight into the controlling mechanisms of mixer distortion.

Following the procedure of Section 3.2.3, nonlinear distortion analysis is performed in a similar way as linear mixer analysis. It begins by extending the device's mildly nonlinear voltage-dependent charge and current Taylor series expansions from degree one up to the desired degree, typically three. So, now, (5.234) to (5.236) should read as

$$i_{gs}(t) = \frac{\mathrm{d}}{\mathrm{d}t}[q_{gs1}(t)v_{gs}(t) + q_{gs2}(t)v_{gs}(t)^2 + q_{gs3}(t)v_{gs}(t)^3] \tag{5.245}$$

$$\begin{aligned} i_{ds}(t) = {} & g_m(t)v_{gs}(t) + g_{ds}(t)v_{ds}(t) \\ & + g_{m2}(t)v_{gs}(t)^2 + g_{md}(t)v_{gs}(t)v_{ds}(t) + g_{d2}(t)v_{ds}(t)^2 \\ & + g_{m3}(t)v_{gs}(t)^3 + g_{m2d}(t)v_{gs}(t)^2 v_{ds}(t) + g_{md2}(t)v_{gs}(t)v_{ds}(t)^2 \\ & + g_{d3}(t)v_{ds}(t)^3 \end{aligned} \tag{5.246}$$

$$i_{gd}(t) = \frac{\mathrm{d}}{\mathrm{d}t}[q_{gd1}(t)v_{gd}(t) + q_{gd2}(t)v_{gd}(t)^2 + q_{gd3}(t)v_{gd}(t)^3] \tag{5.247}$$

Knowing the time-domain waveforms of $v_{GS}(t)$ and $v_{DS}(t)$, determined by the large LO excitation, and the laws with which those Taylor series coefficients vary with the control voltages, it is straightforward to obtain the time-varying coefficients' description and their corresponding conversion matrices.

A full time-varying Volterra series analysis of the circuit would be similar to the time-invariant study already presented for the small-signal amplifiers of Section 5.2.4.2, with the exception that the quiescent point is no longer a fixed (V_{GS}, V_{DS}) pair, but the time-varying $[v_{GS}(t), v_{DS}(t)]$ imposed by the dc bias supplies plus the LO pumping. Each element of the circuit must then be substituted by its conversion matrix, and the linear circuit [Y], [Z], or [S] matrix becomes a matrix of conversion matrices. This shows that even the most simple mixer topologies are too involved to be analyzed by hand, unless they are based in very few nonlinearities and their equivalent circuits include one single node (plus the reference) or mesh.

To enable a qualitative small-signal distortion study, we reiterate the approximations above assumed for the linear mixer analysis, which allowed the treatment of the full four-pole FET as the unilateral one of Figure 5.45. Furthermore, assuming the FET is always kept in saturation, we may expect that $G_{ds}(v_{GS})$ behaves as a linear conductance, which can be lumped into $Y_d(\omega)$, and that the $i_{ds}(v_{gs})$ nonlinearity will dominate over the $q_{gs}(v_{gs})$ and $q_{gd}(v_{gd})$ charge nonlinearities, the i_{ds} current nonlinearities of the output or even the ones represented by the cross-coefficients. In this case, the only time-varying nonlinear element is $i_{DS}[v_{GS}(t)]$, and (5.246) can be reduced to

$$i_{ds}(t) = g_m(t)v_{gs}(t) + g_{m2}(t)v_{gs}(t)^2 + g_{m3}(t)v_{gs}(t)^3 \tag{5.248}$$

where $v_{gs}(t)$ is assumed to be determined by a two-tone excitation at $\omega_{RF_1}(\omega_{IF_1})$ and $\omega_{RF_2}(\omega_{IF_2})$.

The analysis follows the method of nonlinear currents and starts by calculating the first-order control voltage vector, $\mathbf{V}_{gs1}$, and the output voltage vector, $\mathbf{V}_{ds1}$.

Since the linear analysis of the circuit of Figure 5.45 leads to

$$\mathbf{V}_{gs1} = \mathbf{Z}_g .x\ \mathbf{I}_s \tag{5.249a}$$

or

$$\mathbf{V}_{gs1} = \begin{bmatrix} V_{gs1_{-2,-K}} & \cdots & V_{gs1_{2,-K}} \\ \vdots & & \vdots \\ V_{gs1_{-2,0}} & \cdots & V_{gs1_{2,0}} \\ \vdots & & \vdots \\ V_{gs1_{-2,K}} & \cdots & V_{gs1_{2,K}} \end{bmatrix}$$

$$= \begin{bmatrix} Z_g[-(K-1)\omega_{LO} - \omega_{RF_2}] & \cdots & Z_g[-(K+1)\omega_{LO} + \omega_{RF_2}] \\ \vdots & & \vdots \\ Z_g(-\omega_{IF_2}) & \cdots & Z_g(\omega_{IF_2}) \\ \vdots & & \vdots \\ Z_g[(K+1)\omega_{LO} - \omega_{RF_2}] & \cdots & Z_g[(K-1)\omega_{LO} + \omega_{RF_2}] \end{bmatrix}$$

$$.x \begin{bmatrix} 0 & \cdots & 0 \\ \vdots & & \vdots \\ I_{s_{-2}} & \cdots & I_{s_2} \\ \vdots & & \vdots \\ 0 & \cdots & 0 \end{bmatrix} \tag{5.249b}$$

$$\mathbf{I}_{ds1} = \mathbf{G}_m \mathbf{V}_{gs1} \tag{5.250a}$$

or

$$\mathbf{I}_{ds1} = \begin{bmatrix} I_{ds1_{-2,-K}} & \cdots & I_{ds1_{2,-K}} \\ \vdots & & \vdots \\ I_{ds1_{-2,0}} & \cdots & I_{ds1_{2,0}} \\ \vdots & & \vdots \\ I_{ds1_{-2,K}} & \cdots & I_{ds1_{2,K}} \end{bmatrix} = \begin{bmatrix} G_{m_0} & \cdots & G_{m_{-K}} & \cdots & G_{m_{-2K}} \\ \vdots & & & & \vdots \\ G_{m_K} & \cdots & G_{m_0} & \cdots & G_{m_{-K}} \\ \vdots & & & & \vdots \\ G_{m_{2K}} & \cdots & G_{m_K} & \cdots & G_{m_0} \end{bmatrix} \mathbf{V}_{gs1} \tag{5.250b}$$

the first-order control voltage, or output voltage, is

$$\mathbf{V}_{ds1} = -\mathbf{Z}_d \; .x \; \mathbf{I}_{ds1} \tag{5.251a}$$

or

$$\mathbf{V}_{ds1} = \begin{bmatrix} V_{ds1_{-2,-K}} & \cdots & V_{ds1_{2,-K}} \\ \vdots & & \vdots \\ V_{ds1_{-2,0}} & \cdots & V_{ds1_{2,0}} \\ \vdots & & \vdots \\ V_{ds1_{-2,K}} & \cdots & V_{ds1_{2,K}} \end{bmatrix} \tag{5.251b}$$

$$= - \begin{bmatrix} Z_d[-(K-1)\omega_{LO} - \omega_{RF_2}] & \cdots & Z_d[-(K+1)\omega_{LO} + \omega_{RF_2}] \\ \vdots & & \vdots \\ Z_d(-\omega_{IF_2}) & \cdots & Z_d(\omega_{IF_2}) \\ \vdots & & \vdots \\ Z_d[(K+1)\omega_{LO} - \omega_{RF_2}] & \cdots & Z_d[(K-1)\omega_{LO} + \omega_{RF_2}] \end{bmatrix} .x \; \mathbf{I}_{ds1}$$

in which $Z_g(\omega) \equiv 1/Y_g(\omega)$ and $Z_d(\omega) \equiv 1/Y_d(\omega)$.

When squared, first-order control voltage at the gate, $\mathbf{V}_{gs_1}$, will produce second-order intermodulation products, $V^{(2)}_{gs1_{c,k}}$, corresponding to all possible combinations of two different $\omega_{b,k}$ frequencies. These $V_{gs1_{c,k}}^{(2)}$ will span from $\omega_{-4,k}$ to $\omega_{4,k}$, and result in second-order nonlinear current components in $g_{m2}(t)$, which will be given by

$$\mathbf{I}_{ds2} = \begin{bmatrix} I_{ds2_{-4,-K}} & \cdots & I_{ds2_{4,-K}} \\ \vdots & & \vdots \\ I_{ds2_{-4,0}} & \cdots & I_{ds2_{4,0}} \\ \vdots & & \vdots \\ I_{ds2_{-4,K}} & \cdots & I_{ds2_{4,K}} \end{bmatrix}$$

$$= \begin{bmatrix} G_{m2_0} & \cdots & G_{m2_{-K}} & \cdots & G_{m2_{-2K}} \\ \vdots & & & & \vdots \\ G_{m2_K} & \cdots & G_{m2_0} & \cdots & G_{m2_{-K}} \\ \vdots & & & & \vdots \\ G_{m2_{2K}} & \cdots & G_{m2_K} & \cdots & G_{m2_0} \end{bmatrix} \begin{bmatrix} V_{gs1_{-4,-K}}^{(2)} & \cdots & V_{gs1_{4,-K}}^{(2)} \\ \vdots & & \vdots \\ V_{gs1_{-4,0}}^{(2)} & \cdots & V_{gs1_{4,0}}^{(2)} \\ \vdots & & \vdots \\ V_{gs1_{-4,K}}^{(2)} & \cdots & V_{gs1_{4,K}}^{(2)} \end{bmatrix} \tag{5.252}$$

As long as the application of Volterra series nonlinear currents method is concerned, the second-order equivalent circuit of Figure 5.45 has only one excitation source, $\mathbf{I}_{ds2}$. So, second-order output intermodulation products come directly as

$$\mathbf{V}_{ds2} = \begin{bmatrix} V_{ds2_{-4,-K}} & \cdots & V_{ds2_{4,-K}} \\ \vdots & & \vdots \\ V_{ds2_{-4,0}} & \cdots & V_{ds2_{4,0}} \\ \vdots & & \vdots \\ V_{ds2_{-4,K}} & \cdots & V_{ds2_{4,K}} \end{bmatrix} = -\mathbf{Z}_d \,.x\, \mathbf{I}_{ds2} \tag{5.253}$$

Because, in general, entries of highest amplitude are usually the ones at ω_{RF} and ω_{IF}, second-order intermodulation components at the difference frequency, $\omega_{IF_{2\Delta}} = \omega_{IF_1} - \omega_{IF_2} = \omega_{RF_1} - \omega_{RF_2}$, will be dominated by mixing products between ω_{RF} and ω_{IF}:

$$\begin{aligned} I_{ds2}\left(\omega_{IF_{2\Delta}}\right) \approx\ & G_{m2_1} V_{gs1}^{(2)}\left(\omega_{IF_1} - \omega_{RF_2}\right) + G_{m2_0} V_{gs1}^{(2)}\left(\omega_{IF_1} - \omega_{IF_2}\right) \\ & + G_{m2_0} V_{gs1}^{(2)}\left(\omega_{RF_1} - \omega_{RF_2}\right) + G_{m2_{-1}} V_{gs1}^{(2)}\left(\omega_{RF_1} - \omega_{IF_2}\right) \end{aligned} \tag{5.254}$$

while the ones at the sum frequency $\omega_{IF_{2\Sigma}} = \omega_{IF_1} + \omega_{IF_2}$ will be dominated by

$$I_{ds2}\left(\omega_{IF_{2\Sigma}}\right) \approx G_{m2_0} V_{gs1}^{(2)}\left(\omega_{IF_1} + \omega_{IF_2}\right) + G_{m2_{-1}} V_{gs1}^{(2)}\left(\omega_{RF_1} + \omega_{IF_2}\right) \quad (5.255)$$
$$+ G_{m2_{-1}} V_{gs1}^{(2)}\left(\omega_{IF_1} + \omega_{RF_2}\right) + G_{m2_{-2}} V_{gs1}^{(2)}\left(\omega_{RF_1} + \omega_{RF_2}\right)$$

However, in the particular case of our mixer, it is expected that $V_{gs1}(\omega_{RF})$ dominates over any other gate voltage component, and thus (5.254) and (5.255) simply become

$$I_{ds2}\left(\omega_{IF_{2\Delta}}\right) \approx G_{m2_0} V_{gs1}^{(2)}\left(\omega_{RF_1} - \omega_{RF_2}\right) \quad (5.256)$$

and

$$I_{ds2}\left(\omega_{IF_{2\Sigma}}\right) \approx G_{m2_{-2}} V_{gs1}^{(2)}\left(\omega_{RF_1} + \omega_{RF_2}\right) \quad (5.257)$$

Equations (5.254) and (5.256) show that second-order distortion in a general mixer is controlled by various Fourier components of the time-varying second-degree nonlinear coefficient. So, minimizing this form of out-of-band distortion requires that one determines the dominant terms of (5.254) and (5.255), and then adopt a time-varying quiescent point that reduces the correspondent Fourier component. Equations (5.256) and (5.257) indicate that in our simplified mixer model optimum second-order intermodulation distortion at the difference frequency is obtained when $G_{m2}(dc)$ is minimum, while an optimum at the sum frequency (or any of the second harmonics of the output, $2\omega_{IF_1}$, $2\omega_{IF_2}$) would require a minimum of $G_{m2}(2\omega_{LO})$.

Figures 5.50 and 5.51 illustrate the dependence of second-order distortion at $\omega_{IF_{2\Delta}}$ and $\omega_{IF_{2\Sigma}}$ on V_{GS} bias point. For comparison purposes, these figures also represent the magnitude of the $G_{m2}(dc)$ and $G_{m2}(2\omega_{LO})$ Fourier coefficients.

Note that, contrary to what we had in small-signal amplifiers, in which G_{m2} controlled second-order distortion of both $\omega_{2\Delta}$ and $\omega_{2\Sigma}$ outputs, in a mixer these are determined by distinct $G_{m2}(dc)$ and $G_{m2}(2\omega_{LO})$ Fourier coefficients.

Figure 5.52 shows HB simulation results of second-order distortion at $\omega_{IF_{2\Delta}}$ and $\omega_{IF_{2\Sigma}}$ of our active mixer prototype, versus LO power level, when the device is biased at the point of maximum conversion gain, $V_{GS} = -1$V. The comparison of these results with the respective $G_{m2}(dc)$ and $G_{m2}(2\omega_{LO})$ shows that these Fourier coefficients can, indeed, predict the level of second-order distortion. In practice, this should be expected up to the level where $v_{DS_{LO}}(t)$ excursion enters the FET's triode zone, creating, this way, other mixing effects in the device's output [$g_{ds}(t)$, $g_{md}(t)$, and $g_{d2}(t)$].

According to what we found in Section 3.2.3, third-order intermodulation arises directly from the cube of first-order components, $V_{gs_1}(k\omega_{LO} + \omega_{IF})$,

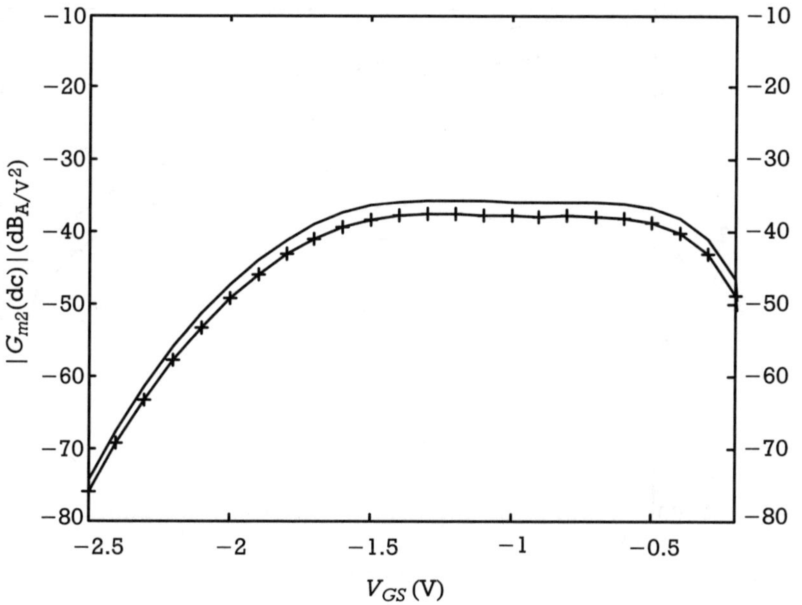

Figure 5.50 Dependence of second-order distortion at $\omega_{IF_{2\Delta}}$ on V_{GS} bias point. Note the strong correlation between the amplitude of this distortion component (-+-) and $G_{m2}(dc)$ (-).

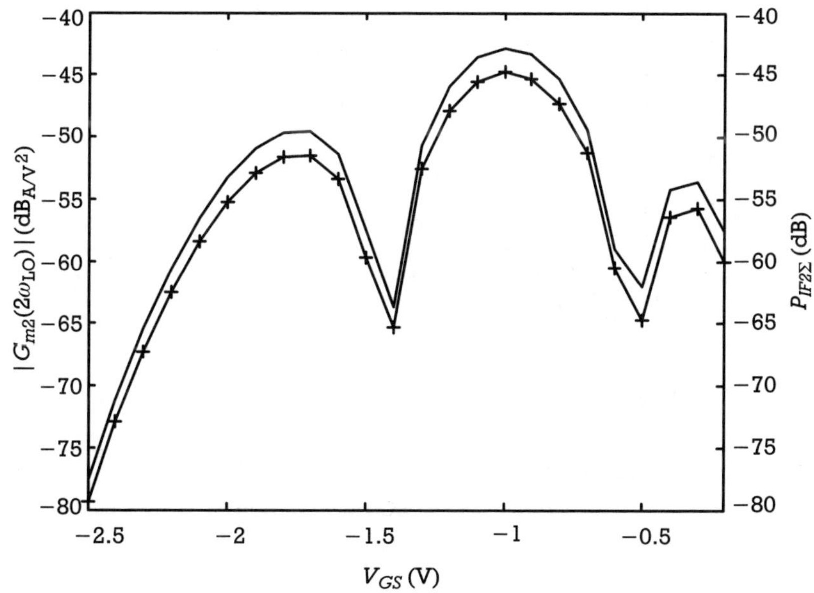

Figure 5.51 Dependence of second-order distortion at $\omega_{IF_{2\Sigma}}$ on V_{GS} bias point. Note the strong correlation between the amplitude of this distortion component (-+-) and $G_{m2}(2\omega_{LO})$ (-).

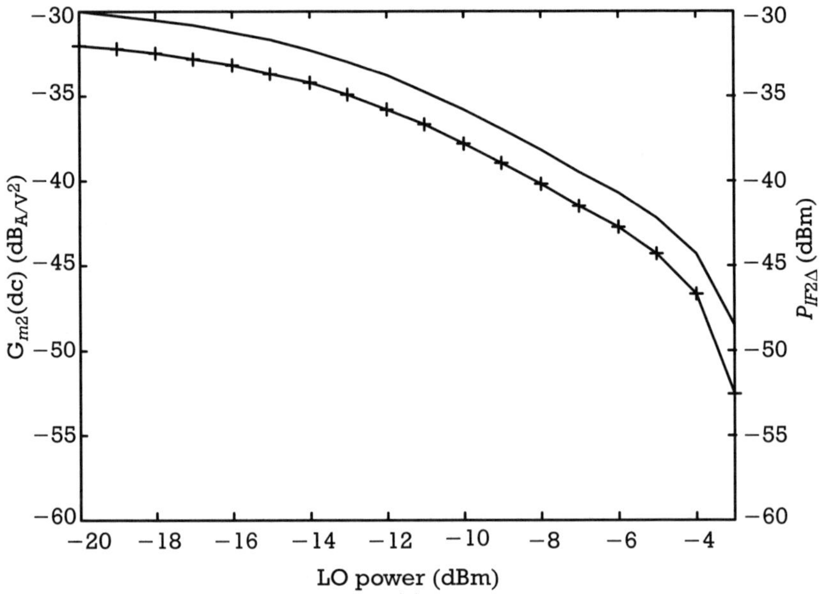

(a)

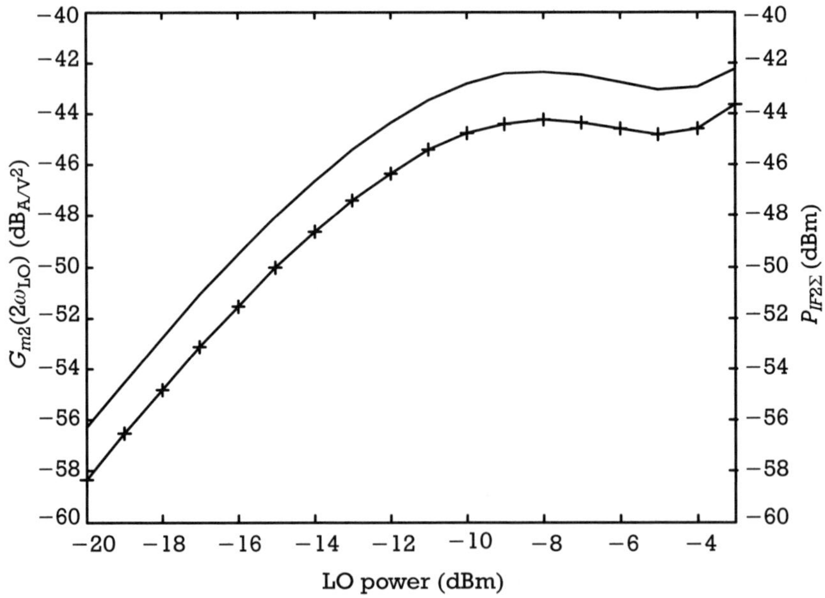

(b)

Figure 5.52 Dependence of second-order distortion at (a) $\omega_{IF_{2\Delta}}$ (-+-), and (b) $\omega_{IF_{2\Sigma}}$ (-+-) on LO available power level, when the device is biased at the point of maximum conversion efficiency, $V_{GS} = -1$V. For comparison purposes, G_{m2}(dc) (-) and $G_{m2}(2\omega_{LO})$ (-) were also plotted in (a) and (b), respectively.

$V_{gs1_{d,k}}^{(3)}$, and from mixing products between these fundamentals and second-order ones, $V_{gs12_{d,k}}^{(3)}$:

$$\mathbf{I}_{ds3} = \begin{bmatrix} I_{ds3_{-8,-K}} & \cdots & I_{ds3_{8,-K}} \\ \vdots & & \vdots \\ I_{ds3_{-8,0}} & \cdots & I_{ds3_{8,0}} \\ \vdots & & \vdots \\ I_{ds3_{-8,K}} & \cdots & I_{ds3_{8,K}} \end{bmatrix}$$

$$= 2 \begin{bmatrix} G_{m2_0} & \cdots & G_{m2_{-K}} & \cdots & G_{m2_{-2K}} \\ \vdots & & & & \vdots \\ G_{m2_K} & \cdots & G_{m2_0} & \cdots & G_{m2_{-K}} \\ \vdots & & & & \vdots \\ G_{m2_{2K}} & \cdots & G_{m2_K} & \cdots & G_{m2_0} \end{bmatrix} \begin{bmatrix} V_{gs12_{-8,-K}}^{(3)} & \cdots & V_{gs12_{8,-K}}^{(3)} \\ \vdots & & \vdots \\ V_{gs12_{-8,0}}^{(3)} & \cdots & V_{gs12_{8,0}}^{(3)} \\ \vdots & & \vdots \\ V_{gs12_{-8,K}}^{(3)} & \cdots & V_{gs12_{8,K}}^{(3)} \end{bmatrix}$$

$$+ \begin{bmatrix} G_{m3_0} & \cdots & G_{m3_{-K}} & \cdots & G_{m3_{-2K}} \\ \vdots & & & & \vdots \\ G_{m3_K} & \cdots & G_{m3_0} & \cdots & G_{m3_{-K}} \\ \vdots & & & & \vdots \\ G_{m3_{2K}} & \cdots & G_{m3_K} & \cdots & G_{m3_0} \end{bmatrix} \begin{bmatrix} V_{gs1_{-8,-K}}^{(3)} & \cdots & V_{gs1_{8,-K}}^{(3)} \\ \vdots & & \vdots \\ V_{gs1_{-8,0}}^{(3)} & \cdots & V_{gs1_{8,0}}^{(3)} \\ \vdots & & \vdots \\ V_{gs1_{-8,K}}^{(3)} & \cdots & V_{gs1_{8,K}}^{(3)} \end{bmatrix} \tag{5.258}$$

Therefore, third-order output intermodulation products, will be given by

$$\mathbf{V}_{ds3} = \begin{bmatrix} V_{ds3_{-8,-K}} & \cdots & V_{ds3_{8,-K}} \\ \vdots & & \vdots \\ V_{ds3_{-8,0}} & \cdots & V_{ds3_{8,0}} \\ \vdots & & \vdots \\ V_{ds3_{-8,K}} & \cdots & V_{ds3_{8,K}} \end{bmatrix} = -\mathbf{Z}_d \,.x\, \mathbf{I}_{ds3} \tag{5.259}$$

Repeating the approximate analysis done for second-order distortion, we conclude that, in general time-varying circuits, third-order intermodulation components at $\omega_{IF_3} = 2\omega_{IF_1} - \omega_{IF_2}$ will be dominated by the following mixing products between ω_{RF} and ω_{IF}:

$$
\begin{aligned}
I_{ds3}\left(\omega_{IF_3}\right) \approx\ & 2G_{m2_0} V_{gs12}^{(3)}\left(2\omega_{IF_1} - \omega_{IF_2}\right) + 2G_{m2_{-1}} V_{gs12}^{(3)}\left(2\omega_{RF_1} - \omega_{RF_2}\right) \\
& + 2G_{m2_{-1}} V_{gs12}^{(3)}\left(\omega_{RF_1} + \omega_{IF_1} - \omega_{IF_2}\right) + 2G_{m2_{-2}} V_{gs12}^{(3)}\left(2\omega_{RF_1} - \omega_{IF_2}\right) \\
& + G_{m3_1} V_{gs1}^{(3)*}\left(2\omega_{IF_1} - \omega_{RF_2}\right) + G_{m3_0} V_{gs1}^{(3)}\left(\omega_{RF_1} + \omega_{IF_1} - \omega_{RF_2}\right) \\
& + G_{m3_0} V_{gs1}^{(3)}\left(2\omega_{IF_1} - \omega_{IF_2}\right) + G_{m3_{-1}} V_{gs1}^{(3)}\left(2\omega_{RF_1} - \omega_{RF_2}\right) \\
& + G_{m3_{-1}} V_{gs1}^{(3)}\left(\omega_{RF_1} + \omega_{IF_1} - \omega_{IF_2}\right) + G_{m3_{-2}} V_{gs1}^{(3)}\left(2\omega_{RF_1} - \omega_{IF_2}\right)
\end{aligned}
\tag{5.260}
$$

According to (5.260), inband third-order distortion is again controlled by various Fourier coefficients of $g_{m2}(t)$ and $g_{m3}(t)$. However, noticing that because the terms affected by $g_{m2}(t)$ arise from mixing weak second-order distortion components with the excitation, while the ones affected by $g_{m3}(t)$ operate directly on the RF input, it may be expected that these latter contributions will determine mixer inband distortion. Furthermore, if only terms including ω_{RF} components are considered, as is the case of our FET active mixer, then we can conclude that third-order distortion at $\omega_{IF_3} = 2\omega_{RF_1} - \omega_{RF_2}$ is governed by $G_{m3}(\omega_{LO})$, becoming

$$
I_{ds3}\left(\omega_{IF_3}\right) \approx G_{m3_{-1}} V_{gs1}^{(3)}\left(2\omega_{RF_1} - \omega_{RF_2}\right) \tag{5.261}
$$

This is actually the outcome of Figure 5.53, where third-order distortion and $G_{m3}(\omega_{LO})$ magnitudes are plotted against V_{GS} bias.

If third-order harmonic distortion at $3\omega_{IF}$ were compared to third-order distortion at $2\omega_{IF_1} - \omega_{IF_2}$, then we would again conclude that, while the latter is determined by $G_{m3}(\omega_{LO})$, the former is controlled by the different Fourier coefficient of $G_{m3}(3\omega_{LO})$, as is depicted in Figure 5.54.

Comparing those results with third-order distortion observed in FET-based small-signal amplifiers, one surprising conclusion arises. If one of those amplifiers were biased in a point of maximum G_m variation with v_{GS}, and thus maximum G_{m2}, we would expect a very good third-order distortion, as a maximum G_{m2} necessarily corresponds to a null G_{m3} (i.e., to a small-signal IMD sweet-spot). Figure 5.55(a) indeed shows that $G_{m3}|_{V_{GS}=-1V} = 0$. But, the curious situation is that the analysis of our mixer does not lead to a minimum of third-order distortion, but to a maximum instead! The reason for this fact is apparent in Figure 5.55(b) which shows three $g_{m3}[v_{GS_{LO}}(t)]$ waveforms obtained when the mixer is biased before the null of G_{m3}, exactly in the null, and after it. Actually, what is happening is that, due to the almost ideal odd symmetry of G_{m3} in the null, $g_{m3}(t)$ waveform presents strong odd-order Fourier components, therefore determining a maximum of FET gate mixer third-order inband or harmonic distortion.

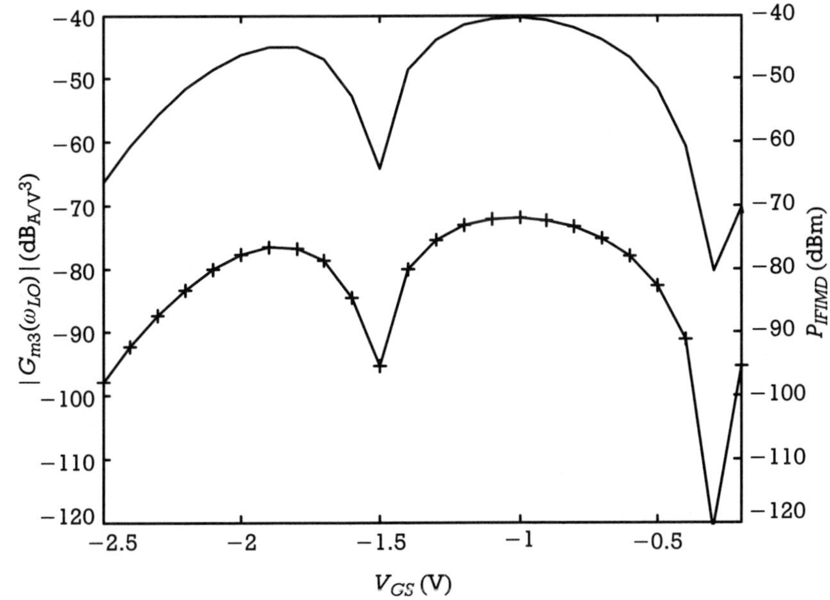

Figure 5.53 Dependence of third-order distortion at ω_{IF} on V_{GS} bias point. Note the strong correlation between the amplitude of this distortion component (-+-) and $G_{m3}(\omega_{LO})$ (-).

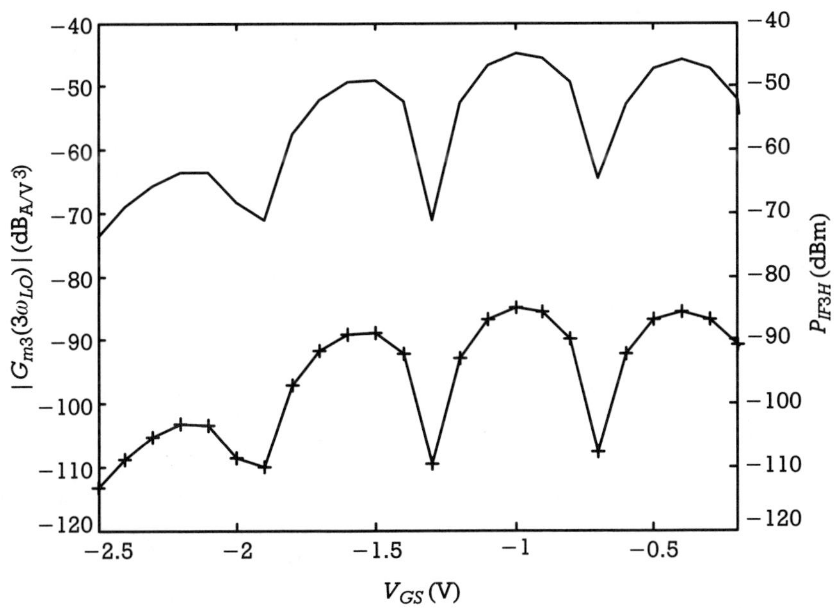

Figure 5.54 Dependence of third-order harmonic distortion at $3\omega_{IF}$ on V_{GS} bias point. Note that it is clearly governed by the Fourier coefficient of $G_{m3}(3\omega_{LO})$.

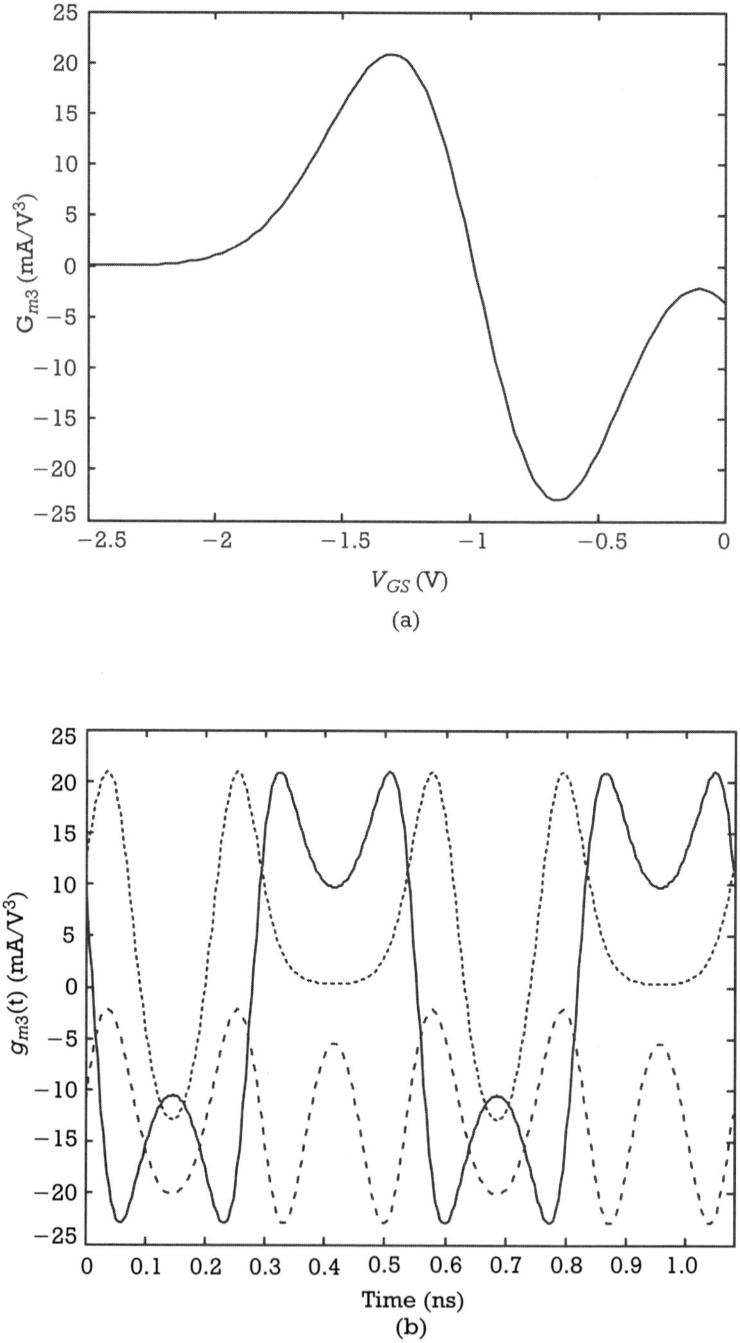

Figure 5.55 (a) Variation of the third-degree Taylor series coefficient with V_{GS} bias for the MESFET used in our active mixer. (b) Three $g_{m3}[v_{GS_{LO}}(t)]$ waveforms obtained for $V_{GS} = -1.5$V (..), $V_{GS} = -1$V (-), and $V_{GS} = -0.3$V (--).

For completeness, Figure 5.56 shows HB simulations of third-order inband and harmonic distortion of our MESFET active mixer, versus LO power level, when the device is biased at the point of maximum conversion gain, $V_{GS} = -1\text{V}$. According to what we have said about second-order distortion results, the range of validity of the analysis made for third-order distortion is again limited by the contribution of mixing components generated in the FET's output [$g_{ds}(t)$, $g_{md}(t)$, $g_{d2}(t)$, $g_{m2d}(t)$, $g_{md2}(t)$, and $g_{d3}(t)$].

5.4.2.4 Mixer Large-Signal Distortion Analysis

As discussed in Section 3.2.5, the main limitation of any Volterra series model (either it is time-invariant or time-varying) is the restricted range of allowed excitation drive levels. For normal RF circuits this is located near the 1-dB compression point. Although frequency converters operated into saturation are rare, this might not be the case for mixers used as modulators or PLL phase detectors, for example. Handling those situations requires the use of a multitone large-signal analysis procedure like the ones referred to in Section 3.3.4.

Considering the various simulation methods described in Chapter 3, there may be some different opinions on what type of technique offers the best compromise between flexibility, simulation time, computer memory storage requirements, and accuracy. It seems, however, that for mixer large-signal distortion analysis, the decision must be taken from the mixed-mode harmonic-Newton algorithm, either it is based on the multidimensional Fourier transform (MDFT-HB) or on some appropriate artificial frequency mapping (AFM-HB). Because the computational workload of MDFT-HB implementations increases dramatically with the number of input tones, it is probable that mapping methods will play the major role for this job.

To study the distortion behavior under saturation, the available power of a two-tone excitation at $\omega_{RF_1} = 1{,}795$ MHz and $\omega_{RF_2} = 1{,}805$ MHz was swept from -30 dBm up to 0 dBm, while the power of local oscillator signal ($\omega_{LO} = 1{,}850$ MHz) was kept constant at -5 dBm.

For this simulation seven harmonics were considered for each of the three signals. The handled spectrum resulted from the diamond-diamond truncation scheme of Section 3.3.4.3, where $|k_1 \omega_{LO}| + |k_2 \omega_{RF_1}| + |k_3 \omega_{RF_2}| < 9$. Although it may appear unreasonable to use the same number of harmonics for both the LO and RF excitations, one must realize that near saturation there is no significant difference between the LO and of the RF signals. In fact, at those drive levels, the specific role of the LO (as a time-varying quiescent point) and the RF (small-signal perturbation of the time-varying quiescent point) is lost. For example, in the present case of a FET active gate mixer this situation is reached when the RF drive is so high that $v_{GS_{RF}}(t)$ voltage swing is similar to the one imposed by the local oscillator.

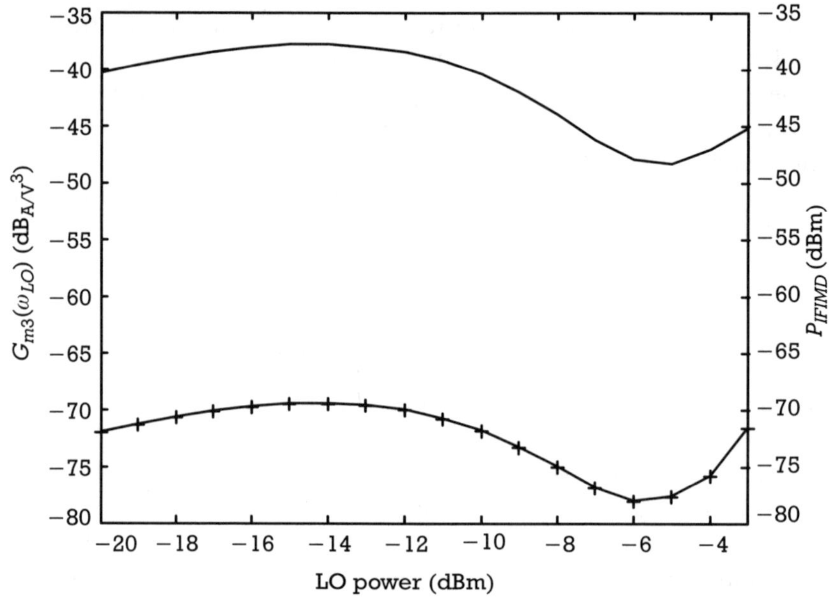

(a)

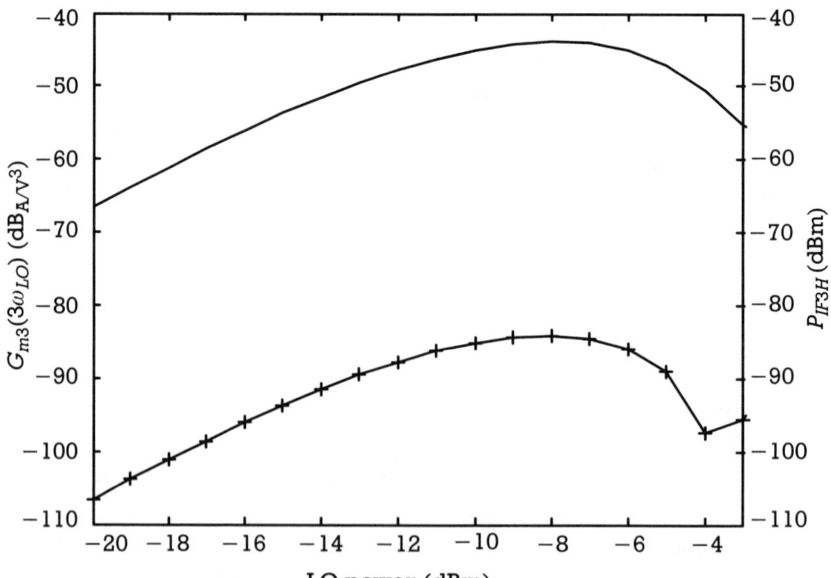

(b)

Figure 5.56 Dependence of third-order distortion at (a) $2\omega_{RF_1} - \omega_{RF_2}$ (-+-), and (b) $3\omega_{RF_1}$ (-+-) on LO available power level, when the device is biased at the point of maximum conversion efficiency, $V_{GS} = -1$V. For comparison purposes, $G_{m3}(\omega_{LO})$ (-) and $G_{m3}(3\omega_{LO})$ (-) were also plotted in (a) and (b), respectively.

Figure 5.57 presents simulation results of output IF power at ω_{IF_1}, even-order distortion at the difference frequency $\omega_{IF_1} - \omega_{IF_2}$, and also output third-order distortion power at $2\omega_{IF_1} - \omega_{IF_2}$. Similarly to what we had already observed for the power amplifier, the distortion begins presenting a 2 and 3 dB/dB behavior at small-signal levels and then tends to saturation. The extrapolated output second-order and third-order intercept points are $IP_{2\Delta} = 23.9$ dBm and $IP_3 = 23.5$ dBm, respectively, while the 1-dB compression point is close to $P_{1dB} = 3.8$ dBm.

As expected, the mixer output departs from its quasilinear behavior for early input power levels and shows strong compression when RF available power becomes comparable with LO oscillator power (–5 dBm).

5.4.3 Intermodulation Distortion in Diode Mixers

Although transistor devices are preferred in monolithic integrated circuit designs, Schottky diodes are still one of the most popular technologies in microwave and wireless hybrid mixers. And, even though their design methodology, or nonlinear distortion analysis, closely follows the one already presented for the FET mixer, its practical importance demands at least a glimpse in this text. Therefore, the following analysis is a compromise between a long and detailed study, and an absolute lack of reference to that pillar of mixer technology.

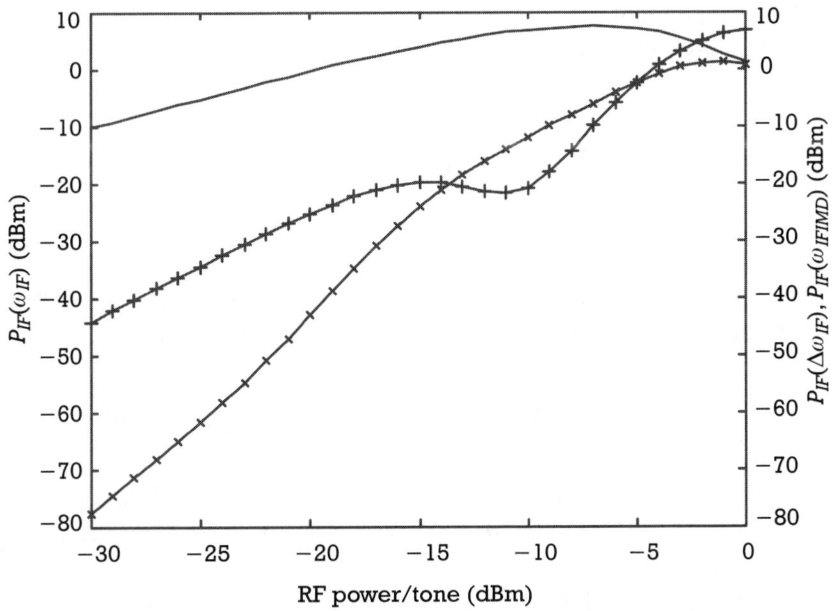

Figure 5.57 HB simulation results of output IF power at ω_{IF_1} (-), even-order distortion at the difference frequency $\omega_{IF_1} - \omega_{IF_2}$ (-+-), and output third-order distortion power at $2\omega_{IF_1} - \omega_{IF_2}$ (-x-). xx axis is the available input RF power level per tone.

Except at the highest end of millimeter-wave band, diode mixers are rarely made of a single device. Actually, they are usually associated in singly balanced pairs, or doubly balanced quads. Nevertheless, these composite configurations can always be divided in two symmetrical circuits of singly balanced arrangements or single devices, and finally, analyzed as individual diodes. The characteristics obtained with that isolated device can then be extrapolated to the final balanced circuit via the conclusions derived in Section 5.5.

The adopted topology for the present study is the opposite-phase singly balanced diode mixer of Figure 5.58(a). It is composed of a 180-degree transformer hybrid, in which the LO is driving in phase both nonlinear devices, while the RF signal is applied in opposite phase. At the IF port, the diode output currents are simply added in phase and finally filtered by the lowpass IF matching circuit. The grounded quarter-wavelength stubs, present between the diodes and the transformer, provide the necessary dc current path and are approximately short-circuits to the IF frequency. On the other hand, it is also supposed that the open-circuited stub and the grounded capacitor, at the input of the IF filter, provide a short-circuit to both the RF, LO signals, and all harmonics.

For the purpose of nonlinear analysis, the single diode circuits of Figure 5.58(b, c) can be represented as one of the alternative Norton or Thévenin equivalents depicted in Figure 5.59.

Assuming the diode is an ideal Schottky junction composed of an exponential voltage-dependent current source, plus a nonlinear junction capacitance modeled by a voltage-dependent depletion charge, we have

$$i_D(t) = \left[I_S e^{\frac{v_D(t)}{\eta V_T}} - 1\right] + \frac{\mathrm{d}}{\mathrm{d}t}\left[Q_{j0}\left(1 - \frac{v_D(t)}{V_{bi}}\right)^{1-\gamma}\right] = i_{NL}(t) + \frac{\mathrm{d}q_{NL}(t)}{\mathrm{d}t} \tag{5.262}$$

Therefore, the circuit response to the large-signal LO excitation corresponds to the solution of the following harmonic balance equation:

$$Y_s(\omega)\mathbf{V}_d(\omega) + \mathbf{I}_{nl}(\omega) + j\boldsymbol{\Omega}\mathbf{Q}_{nl}(\omega) - \mathbf{I}_s(\omega) = 0 \tag{5.263}$$

in which $\mathbf{V}_d(\omega)$, $\mathbf{I}_{nl}(\omega)$, and $\mathbf{Q}_{nl}(\omega)$ are the discrete Fourier transform vectors of the $v_D(t)$, $i_{NL}(t)$, and $q_{NL}(t)$, respectively, and $\mathbf{I}_S(\omega)$ is the LO sinusoidal excitation.

After the $\mathbf{V}_d(\omega) \leftrightarrow v_D(t)$ is obtained for the $i_S(t)$ LO stimulus, the diode current and charge can be obtained from (5.262). Then, the diode current and charge constitutive relations can be expanded in two Taylor series around this time-varying quiescent point:

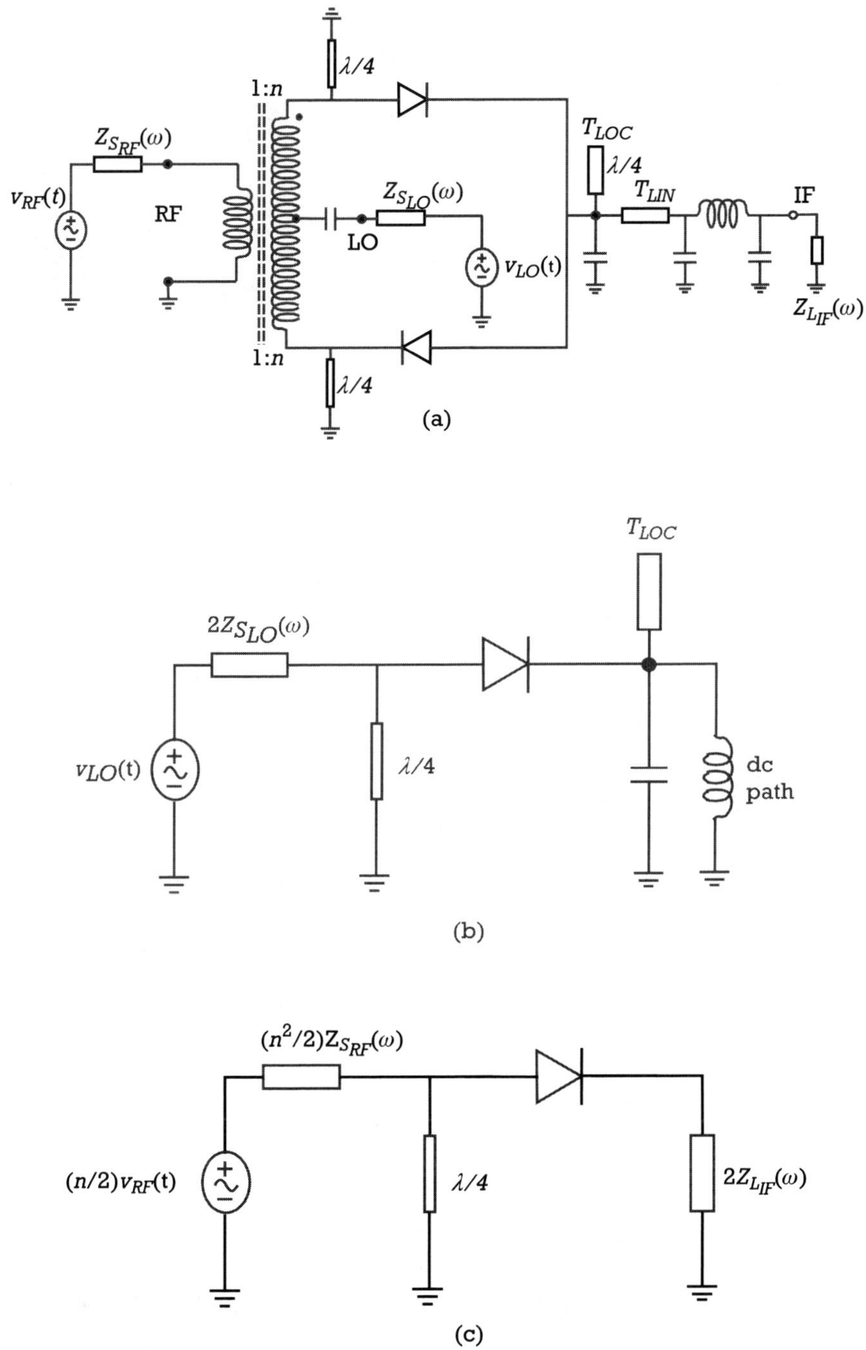

Figure 5.58 (a) Singly balanced diode mixer topology under study. (b) Single diode equivalent mixer circuit for large-signal time-varying quiescent point calculation. (c) Single diode equivalent mixer circuit for small-signal RF excitation.

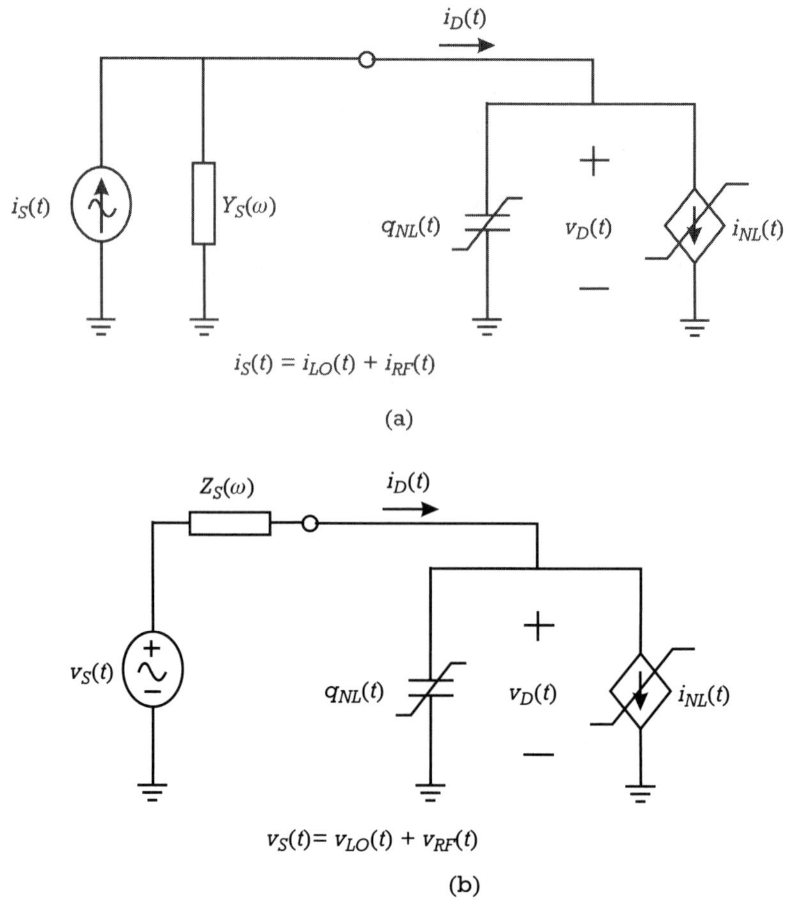

Figure 5.59 (a) Norton and (b) Thévenin equivalent circuits of the single diode mixer used in the nonlinear analysis.

$$i_{nl}(t) = g_{d1}(t)v_d(t) + g_{d2}(t)v_d(t)^2 + g_{d3}(t)v_d(t)^3 \tag{5.264}$$

and

$$q_{nl}(t) = c_{d1}(t)v_d(t) + c_{d2}(t)v_d(t)^2 + c_{d3}(t)v_d(t)^3 \tag{5.265}$$

If now the $g_{d,n}(t)$ and $c_{d,n}(t)$ are expanded in discrete Fourier series, the time-domain products $g_{d,n}(t)v_{d_{RF}}(t)^n$ and $c_{d,n}(t)v_{d_{RF}}(t)^n$ of (5.264) and (5.265) can be represented in conversion matrix form as in the usual small-signal time-varying system analysis. So, if the RF signal is a two-tone excitation, the first-order conversion matrix equation of the circuit becomes

$$\mathbf{Y}_s .x\ \mathbf{V}_{d1} + \mathbf{G}_{d1}\mathbf{V}_{d1} + j\boldsymbol{\Omega}\ .x\ \mathbf{C}_{d1}\mathbf{V}_{d1} = \mathbf{I}_s \tag{5.266}$$

where the conversion matrices and current and voltage vectors have the form already described for the MESFET gate mixer.

The application of this first-order small-signal diode voltage, $V_{d1}(\omega)$, into (5.264) and (5.265) generates second-order nonlinear current and charge components of $i_{nl2}(t) = g_{d2}(t)v_{d1}(t)^2$ and $q_{nl2}(t) = c_{d2}(t)v_{d1}(t)^2$, which will generate second-order diode voltages determined by

$$\mathbf{Y}_s .x\ \mathbf{V}_{d2} + \mathbf{G}_{d1}\mathbf{V}_{d2} + j\boldsymbol{\Omega}\ .x\ \mathbf{C}_{d1}\mathbf{V}_{d2} = -\mathbf{I}_{nl}^{(2)} - j\boldsymbol{\Omega}\ .x\ \mathbf{Q}_{nl}^{(2)} \tag{5.267}$$

Solving (5.267) for $V_{d2}(\omega)$ enables the calculation of third-order nonlinear currents, and then third-order diode voltages, as in the usual time-varying Volterra series analysis.

Although conceptually simple, this analysis is particularly involved in the present case. In fact, since the diode current and charge exponential nonlinearities operate over a control voltage that contains components at virtually all mixing products, there is no easy way of identifying the dominant contributors for the resulting diode current or voltage. Note that, contrary to what we have assumed for the MESFET active gate mixer, we no longer have a transfer nonlinearity of isolated input and output ports, but a two-terminal nonlinearity in which a common port is shared between the input and output. The full conversion matrix description is thus required, and a qualitative hand analysis is impossible.

So, in order to provide the reader with a basic understanding of the origins of nonlinear distortion in a diode mixer, we will continue the analysis in two levels. We will adopt a simplified purely resistive model to study the nonlinear distortion generation process, which will be afterwards complemented by three-tone harmonic balance simulation results obtained from the full diode mixer model.

5.4.3.1 Nonlinear Distortion Analysis of a Simplified Diode Mixer Model

For this simplified diode mixer analysis we assume a purely resistive model, in which $q_{NL}(t) = 0$ and $Z_s(\omega) = 1/Y_s(\omega) = R_S$ is constant in frequency. Assuming that the reverse current is negligible, the approximate mesh equation that describes the circuit of Figure 5.59(b) is thus

$$v_S(t) \approx R_S i_D(t) + \eta V_T \ln\left[\frac{i_D(t)}{I_S}\right] \tag{5.268}$$

or

$$i_D(t) \approx I_S e^{\frac{v_S(t) - R_S i_D(t)}{\eta V_T}} \tag{5.269}$$

which shows that, similarly to what we have already done for the BJT-based small-signal amplifier in Section 5.2.4, we can conceive the circuits of Figure 5.59(a, b) as being based on a new nonlinearity composed of the diode and the resistor R_S. In this way, (5.269) is the implicit-form current/voltage relationship describing that new nonlinearity, $i_D(t) = f[v_S(t)]$. Since, now, the mixer nonlinearity is driven by an ideal voltage source, its input port is automatically decoupled from the output. In the small-signal analysis, the new input voltage has only non-null components at $\pm\omega_{RF}$ (which are a priori known) and the conversion matrix formulation becomes straightforward one-dimensional products.

The large-signal analysis under LO excitation is now a simple calculation of $i_D(t) = f[v_S(t)]$ when $v_S(t) = V_S + V_{S_{LO}} \cos(\omega_{LO} t)$. As seen in Section 5.2.4, $f[v_S(t)]$ is exponential near conduction threshold, some $V_\gamma = 0.6\text{V}$ to 0.7V, and tends to become linearly dependent on $v_S(t)$, actually $i_D(t) = 1/R_S[v_S(t) - V_\gamma]$, when forward conduction is limited by series resistance R_S. So, biased at $V_S = 0\text{V}$ and excited with an LO amplitude large enough to drive the diode into strong conduction, the $i_D(t)$ current waveform will be almost zero for $v_S(t) < V_\gamma$, an exponential of the sinusoidal $v_S(t)$ in the vicinity of conduction threshold, and then sinusoidal for large and positive $v_S(t)$. Biased at $V_S < 0$ the diode remains cut-off at a major portion of the $v_S(t)$ period and conversion efficiency drops. Finally, biased at $V_S > V_\gamma$, the diode is driven to strong conduction in a significant part of the sinusoidal LO. $i_D(t)$ is almost linearly dependent on $v_S(t)$, the circuit lacks in mixing nonlinearity and conversion efficiency drops again.

This behavior of bell shaped conversion efficiency is, again, a consequence of the form of $G_{d2} = (1/2)\partial i_D/\partial v_S$ versus bias, as was shown in Figure 5.17(b).

Following the approximate analysis previously undertaken for the MESFET active mixer, Figures 5.60 through 5.62 show conversion loss, second-order distortion at $\omega_{IF_1} - \omega_{IF_2}$ and $\omega_{IF_1} + \omega_{IF_2}$, and third-order distortion at $2\omega_{IF_1} - \omega_{IF_2}$ and $3\omega_{IF_1}$, obtained from our simplified analysis and the full three-tone harmonic balance simulations of the circuits of Figure 5.58(b, c).[7]

Those curves show a rather complex nonlinear distortion behavior versus bias, presenting some peaks, but also very deep nulls. And, although in real circuits a higher LO drive level plus reactive terminating impedances and depletion junction capacitance may soften these strong distortion power variations, there may be still some space for bias optimization. Unfortunately, since minimum conversion loss must also be sought, this bias margin becomes restricted to no more than a few hundred millivolts; in our case, about $0.3\text{V} < V_S < 0.45\text{V}$. And, in that region, it seems inband distortion passes through a valley, while out-of-band distortion presents a reasonably wide plateau.

7. Although these results can be extrapolated for any single diode mixer, the reader should keep in mind that, in singly or doubly balanced configurations, second-order distortion will be strongly reduced by the inherent cancellation process provided by those balanced mixer arrangements.

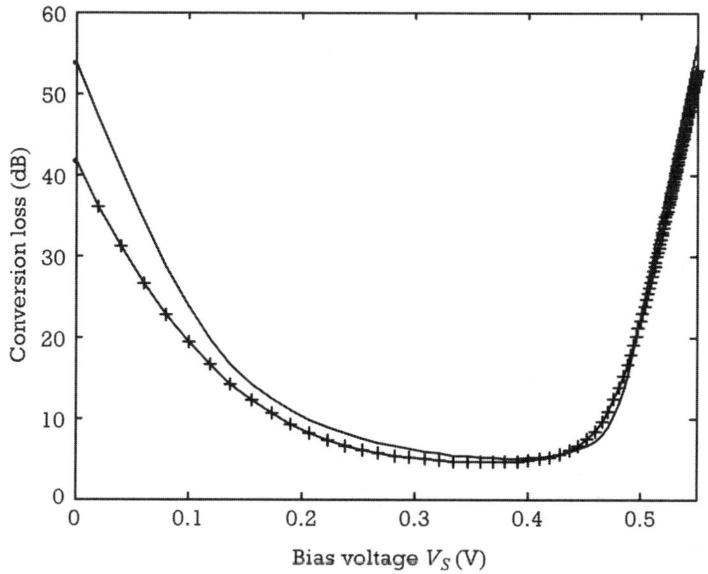

Figure 5.60 Diode mixer conversion loss versus V_S bias, for constant LO drive level, obtained from the simplified resistive model (-) and harmonic balance simulations of the full nonlinear model (-+-).

Where conversion efficiency and distortion versus LO drive are concerned, a similar qualitative interpretation can be proposed. We suppose that bias is left at $V_S = 0$V (a typical situation in most practical diode mixers, but probably far from being the best as suggested in Figure 5.60 where a value near $V_S = 0.3$V to 0.4V would be better) but LO drive is increased from a small amplitude excitation towards a large magnitude significantly higher than V_γ.

Starting at very low LO signal level, where $V_s(\omega_{LO}) << V_\gamma$ the mixing process can only be produced in the depletion capacitance. Conversion efficiency will be very low, as will be the level of all other mixing products. When LO level is increased up to V_γ, $i_D(t) = f[v_S(t)]$ has an exponential characteristic and conversion efficiency presents a gentle increase. Distortion products also rise as a consequence of the increased G_{d2} and G_{d3}, and so of $G_{d2}(dc)$, $G_{d2}(2\omega_{LO})$, $G_{d3}(\omega_{LO})$, and $G_{d3}(3\omega_{LO})$. However, when the diode becomes strongly driven, its current waveform approximates a series of sinusoidal pulses, the device is progressively acting as a linear (although time-varying) switch, conversion loss tends to a constant and nonlinear distortion vanishes.

This is what can be observed in Figures 5.63 through 5.65, where the conversion loss and distortion behaviors of the simplified resistive circuit and the ones of the full nonlinear model are plotted versus LO drive level.

Conversion loss stabilizes for LO drive levels higher than about −2 dBm (+1 dBm in the singly balanced configuration), a region where there is still significant

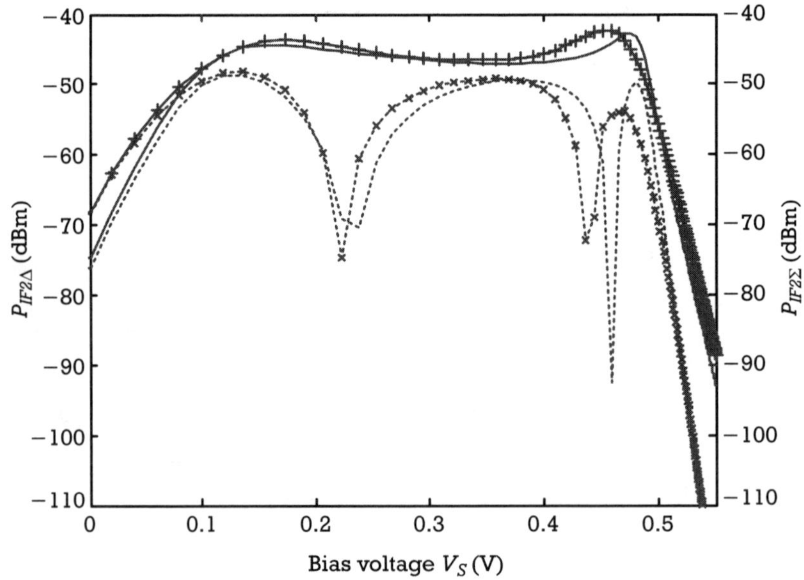

Figure 5.61 Diode mixer second-order nonlinear distortion at the difference, $\omega_{IF_1} - \omega_{IF_2}$, and the sum, $\omega_{IF_1} + \omega_{IF_2}$, frequencies, versus V_S bias, for constant LO drive level, obtained from the simplified resistive model and harmonic balance simulations of the full nonlinear model. $\omega_{IF_1} - \omega_{IF_2}$ results of the simplified resistive model (-) and $\omega_{IF_1} - \omega_{IF_2}$ results of the full model (-+-). $\omega_{IF_1} + \omega_{IF_2}$ results of the simplified resistive model (..) and $\omega_{IF_1} + \omega_{IF_2}$ results of the full model (..x..).

variation of second and third-order out-of-band distortion levels, whereas inband distortion power is already in its monotonic descending zone. So, in a practical mixer this may be investigated as a clue for possible linearity optimization.

In summary, and similarly to what was done for the MESFET active mixer, a qualitative interpretation of the shapes of conversion loss, second and third-order distortion versus bias and LO level is possible, taken the discussed $i_D(t)$ time-varying quiescent point and the G_{d1}, G_{d2}, and G_{d3} obtained from a successive differentiation of (5.269). That helps discerning the distortion origins of a mixing diode, and thus can be used as a guide in the process of diode mixer linearity optimization. Actually, note that even though these conclusions were obtained from a very simplified analysis, they are in very good agreement with the observations made in a real X-band diode mixer [29].

5.5 Nonlinear Distortion in Balanced Circuits

To conclude this chapter on nonlinear distortion in typical microwave and wireless circuits, it is convenient to study balanced combinations of similar nonlinearities. We will see that such arrangements offer some attractive properties on this respect.

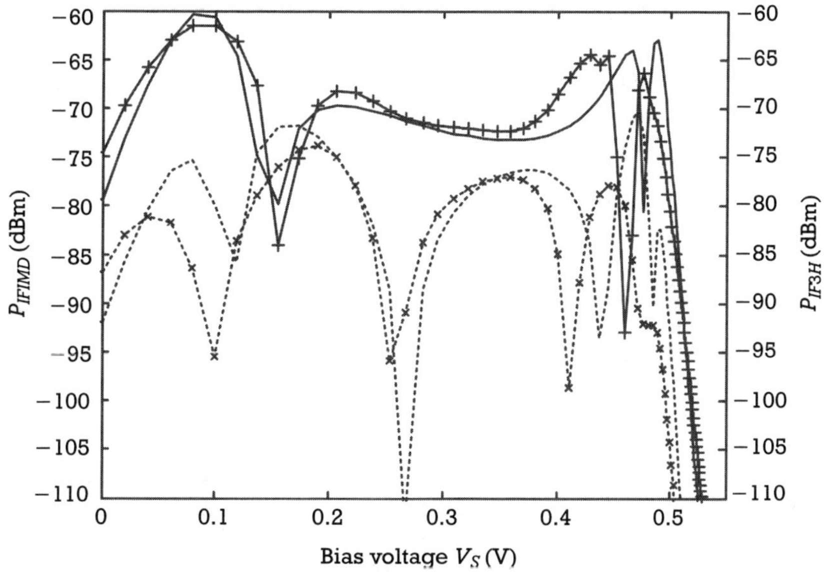

Figure 5.62 Diode mixer third-order nonlinear distortion at $2\omega_{IF_1} - \omega_{IF_2}$, and $3\omega_{IF_1}$ versus V_S bias, for constant LO drive level, obtained from the simplified resistive model and harmonic balance simulations of the full nonlinear model. $2\omega_{IF_1} - \omega_{IF_2}$ results of the simplified resistive model (-) and $2\omega_{IF_1} - \omega_{IF_2}$ results of the full model (-+-). $3\omega_{IF_1}$ results of the simplified resistive model (..) and $3\omega_{IF_1}$ results of the full model (..x..).

Balanced arrangements are common in both amplifiers (mainly power amplifiers) and mixers. In the former, they provide increased output power capabilities (as compared to single devices), port matching, and distortion, while in mixers they can also offer improved port-to-port isolation and spurious mixing products' rejection [30].

Multiple amplifier topologies usually rely on the combination of two devices, or in connections of various device pairs, either they are balanced or simple in-phase parallel structures. On the contrary, typical balanced mixer arrangements usually show only two (singly balanced topology) or four devices (doubly balanced topology).

Where the type of power splitter and power combiner are concerned, these connections can be constituted by a simple in-phase parallel addition, an opposite-phase arrangement, or a quadrature connection.

Although the following analysis is directed to the nonlinear distortion properties, the interested reader can obtain further information on the advantages of microwave balanced circuits in, for example, [30].

5.5.1 Distortion in Multiple-Device Amplifier Circuits

The common topology adopted for the composite amplifier analysis is depicted in Figure 5.66. It is composed of two identical amplifiers, A and B, connected by a

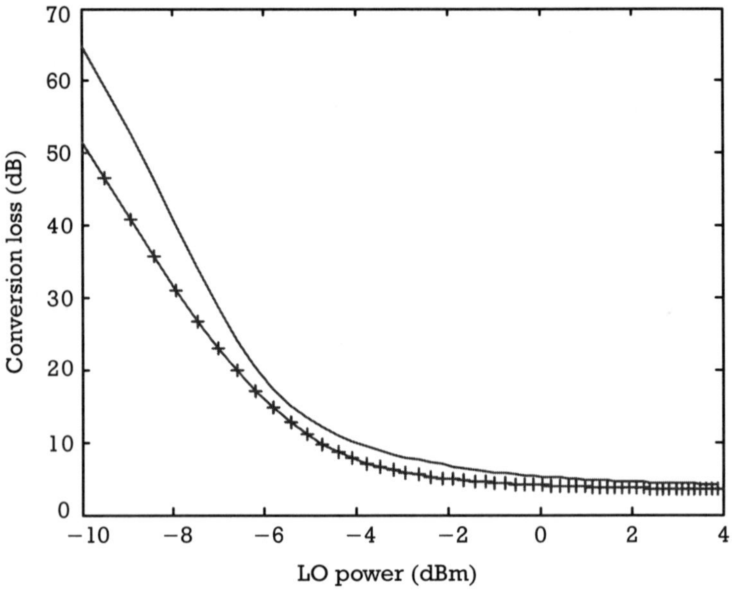

Figure 5.63 Diode mixer conversion loss versus LO drive level, for constant $V_S = 0V$ bias, obtained from the simplified resistive model (-) and harmonic balance simulations of the full nonlinear model (-+-).

power splitter, P_S, and a power combiner, P_C. For the sake of analysis' simplicity (but without loss of generality), we suppose that the amplifiers are treated as memoryless transconductance nonlinearities, in which the input signal voltages are $v_{i_A}(t)$ and $v_{i_B}(t)$, while the correspondent output currents are $i_{o_A}(t)$ and $i_{o_B}(t)$, respectively. So, assuming a power series model, the input-output relationship of those devices can be represented by

$$i_{o_{A,B}}(t) = G_{m1} v_{i_{A,B}}(t) + G_{m2} v_{i_{A,B}}(t)^2 + G_{m3} v_{i_{A,B}}(t)^3 + \ldots = \sum_{n=1}^{\infty} G_{mn} v_{i_{A,B}}(t)^n \tag{5.270}$$

5.5.1.1 In-Phase Amplifier Arrangement

If both the power splitter and combiner are equal 3 dB division and in-phase devices like, for example, the Wilkinson divider/combiner, then $v_{i_A}(t) = v_{i_B}(t)$ and so $i_{o_A}(t) = i_{o_B}(t)$. Since these currents are merely added in-phase, the fundamental or nth-order distortion power is doubled, as compared to a single isolated device. The even or odd-order distortion properties of this pair are thus 3-dB scaled replicas of the ones presented by amplifiers A and B.

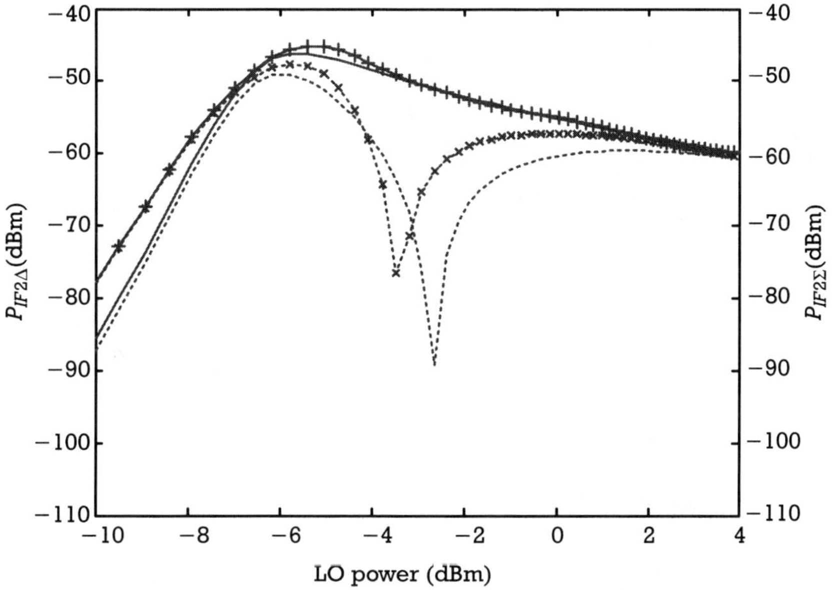

Figure 5.64 Diode mixer second-order nonlinear distortion at the difference, $\omega_{IF_1} - \omega_{IF_2}$, and the sum, $\omega_{IF_1} + \omega_{IF_2}$, frequencies, versus LO drive level, for constant $V_S = 0$V bias, obtained from the simplified resistive model and harmonic balance simulations of the full nonlinear model. $\omega_{IF_1} - \omega_{IF_2}$ results of the simplified resistive model (-), and $\omega_{IF_1} - \omega_{IF_2}$ results of the full model (-+-). $\omega_{IF_1} + \omega_{IF_2}$ results of the simplified resistive model (..), and $\omega_{IF_1} + \omega_{IF_2}$ results of the full model (..x..).

5.5.1.2 Opposite-Phase Amplifier Arrangement

The opposite-phase amplifier arrangement is a very important case, representing many circuit topologies found in practice, like the usual class B or class AB push-pull amplifier and the emitter-coupled (or source-coupled) differential pair. In low-frequency integrated circuit (IC) designs, the 180-degree power splitter can be obtained from an emitter-follower (source-follower) and common-emitter (common-source) stage (taken from the emitter and collector of a single transistor), while in RF circuits it is usually made from a broadband transformer hybrid or even a ring (also know as the *rat-race*) hybrid [30].

In this case, $v_{i_A}(t) = -v_{i_B}(t)$, and so,

$$i_{o_A}(t) = \sum_n G_{mn} v_{i_A}(t)^n \tag{5.271}$$

$$i_{o_B}(t) = \sum_n G_{mn} v_{i_B}(t)^n = \sum_n (-1)^n G_{mn} v_{i_A}(t)^n \tag{5.272}$$

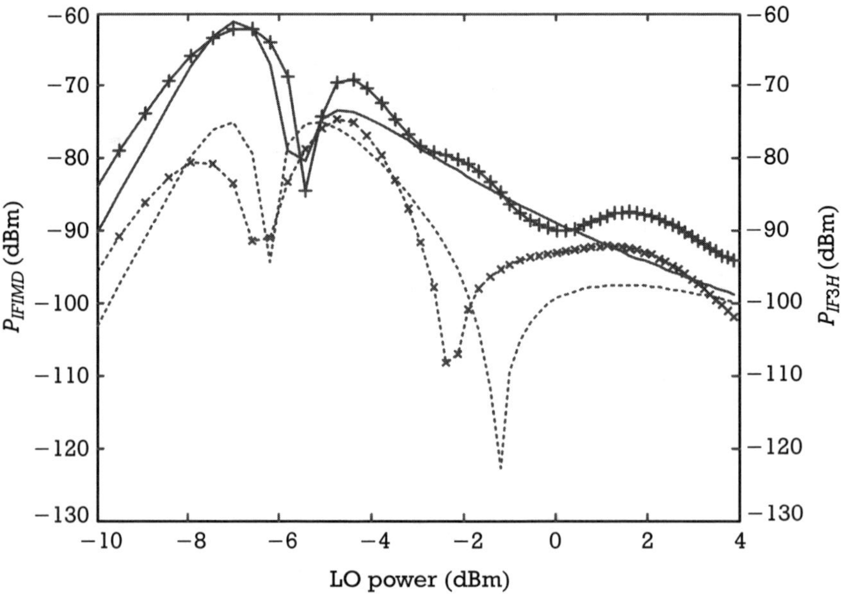

Figure 5.65 Diode mixer third-order nonlinear distortion at $2\omega_{IF_1} - \omega_{IF_2}$, and $3\omega_{IF_1}$ versus LO drive level, for constant $V_S = 0$V bias, obtained from the simplified resistive model and harmonic balance simulations of the full nonlinear model. $2\omega_{IF_1} - \omega_{IF_2}$ results of the simplified resistive model (-), and $2\omega_{IF_1} - \omega_{IF_2}$ results of the full model (-+-). $3\omega_{IF_1}$ results of the simplified resistive model (..), and $3\omega_{IF_1}$ results of the full model (..x..).

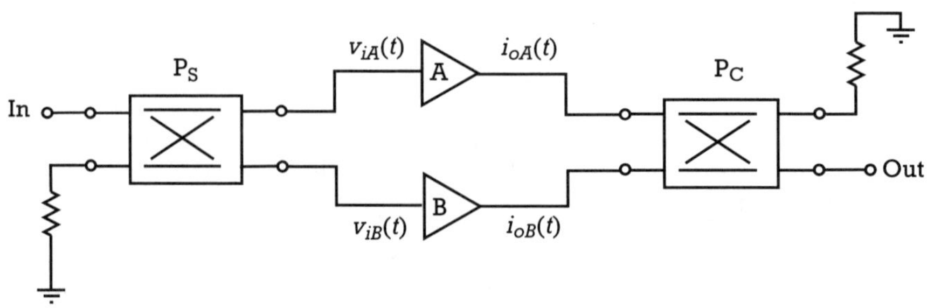

Figure 5.66 General connection of an amplifier pair.

which shows that, if those currents are subtracted in a similar 180-degree hybrid, the fundamental and odd-order distortion will double, but even-order distortion cancels exactly.

Obviously, this analysis requires that the ideal subtraction expected from the 180-degree hybrid is guaranteed for the fundamental, but also for any of the distortion products. And, if this is reasonably true for the broadband active IC designs, it may fail for some of the higher order harmonics or very low-frequency components in the transformer hybrid, and is virtually false for any microwave hybrid made of transmission lines. In fact, remember that, in this latter case, a 180-degree phase shift at the fundamentals, ω_1 and ω_2, is transformed into 360 degrees at the second harmonics or sum frequency, $\omega_1 + \omega_2$, or into an almost zero phase shift at the beat product $\omega_1 - \omega_2$.

Alternatively, if the devices are symmetrical (as is the typical IC case of push-pull output complementary pairs based on CMOS transistors or PNP-NPN BJT pair), then,

$$i_{o_A}(t) = \sum_n G_{mn} v_{i_A}(t)^n \tag{5.273}$$

$$i_{o_B}(t) = \sum_n -G_{mn} v_{i_B}(t)^n = \sum_n (-1)^{n+1} G_{mn} v_{i_A}(t)^n \tag{5.274}$$

proving that the same effect can be obtained without the necessity of another 180-degree hybrid at the output.

In effect, it is this even-order nonlinear distortion cancellation that allows signal reconstruction from the two half-wave sinusoidal currents produced in the complementary pair alternate operation.

5.5.1.3 Quadrature-Phase Amplifier Arrangement

The quadrature-phase amplifier uses branch-line, coupled-line, or Lange-coupler 90-degree hybrids [30]. Since these are distributed element devices, this amplifier connection is almost restricted to microwave and millimeter-wave bands. A quadrature hybrid presents a phase shift of 90 degrees between its two outputs, and thus an equal amplitude two-tone excitation for the amplifiers can be represented by

$$v_{i_A}(t) = V_i \cos(\omega_1 t) + V_i \cos(\omega_2 t) \tag{5.275}$$

$$v_{i_B}(t) = V_i \cos\left(\omega_1 t + \frac{\pi}{2}\right) + V_i \cos\left(\omega_2 t + \frac{\pi}{2}\right) \tag{5.276}$$

Therefore, their current outputs become

$$i_{o_A}(t) = \sum_n G_{mn} V_i^n [\cos(\omega_1 t) + \cos(\omega_2 t)]^n \tag{5.277}$$

$$i_{o_B}(t) = \sum_n G_{mn} V_i^n \left[\cos\left(\omega_1 t + \frac{\pi}{2}\right) + \cos\left(\omega_2 t + \frac{\pi}{2}\right)\right]^n \tag{5.278}$$

showing that the fundamental output must be built from a 90-degree combination of $i_{O_A}(t)$ and $i_{O_B}(t)$.

Now, looking into the second-order distortion at, for example, the difference frequency, $\omega_1 - \omega_2$, and second-harmonic $2\omega_1$ (also illustrative of the behavior at the sum frequency $\omega_1 + \omega_2$), we see that

$$i_{O_A}(t) = \frac{1}{2} G_{m2} V_i^2 \{2 \cos[(\omega_1 - \omega_2)t] + \cos(2\omega_1 t)\} \tag{5.279}$$

$$i_{O_B}(t) = \frac{1}{2} G_{m2} V_i^2 \{2 \cos[(\omega_1 - \omega_2)t] + \cos(2\omega_1 t + \pi)\} \tag{5.280}$$

Since we have already added a 90-degree phase shift to $i_{O_A}(t)$, for recovering fundamental signal, the addition of the quadrature version of $i_{O_A}(t)$ and $i_{O_B}(t)$ will not cancel these second-order products. Because the fundamental currents add in phase, total output will be the double of each amplifier current. And, as second-order distortion current components add in quadrature, their sum will only be $1/\sqrt{2}$ of their amplitude addition. Therefore, and although this 90-degree amplifier arrangement cannot provide total second-order distortion cancellation, it still offers a 3-dB increase in signal-to-distortion power ratio.

For the third-order distortion at, for example, $2\omega_1 - \omega_2$ (also illustrative of any product of the form $\omega_i + \omega_j - \omega_k$) and $3\omega_1$ (or any product given by $\omega_i + \omega_j + \omega_k$), we have

$$i_{O_A}(t) = \frac{1}{4} G_{m3} V_i^3 \{3 \cos[(2\omega_1 - \omega_2)t] + \cos(3\omega_1 t)\} \tag{5.281}$$

$$i_{O_B}(t) = \frac{1}{4} G_{m3} V_i^3 \left\{3 \cos\left[(2\omega_1 - \omega_2)t + \frac{\pi}{2}\right] + \cos\left(3\omega_1 t + \frac{3\pi}{2}\right)\right\} \tag{5.282}$$

This shows that none of the third-order inband distortion products are cancelled, but $2\omega_1 + \omega_2$, $2\omega_2 + \omega_1$, $3\omega_1$, and $3\omega_2$ are rejected.

Again, note that these results were obtained under the assumption that both input and output quadrature hybrids show the ideal 90-degree phase shift at all mixing products. And, this is usually not the case, at least for the common distributed element implementations. So, any practical balanced amplifier analysis should be first preceded by a study of the broadband properties of the hybrid in use and then adapt the present conclusions to that particular circuit.

5.5.2 Distortion in Multiple-Device Mixer Circuits

The topology adopted for our balanced mixer circuits closely follows the one already presented for the composite amplifier. However, and as seen in Figure 5.67,

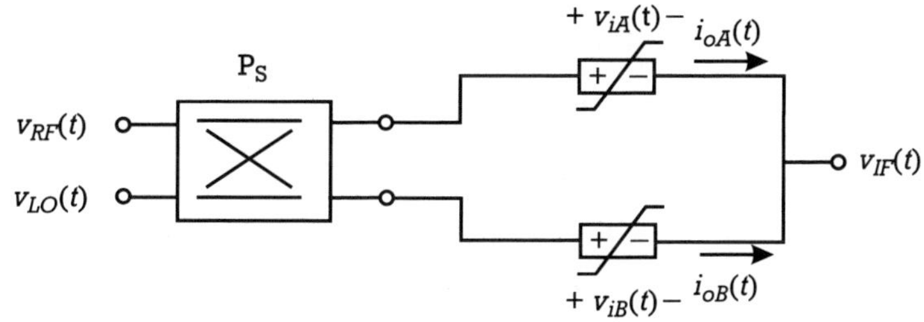

Figure 5.67 General singly balanced mixer topology used in distortion analysis.

now both power splitter hybrid inputs are used, and a simple current addition is provided at the IF output. Actually, since the inputs of 180-degree hybrids and 90-degree hybrids are mutually isolated, they constitute ideal ports for the local oscillator and the RF signal. The absence of another hybrid at the output is justified by what is usually encountered in practical singly balanced mixers.

5.5.2.1 Opposite-Phase Singly Balanced Mixers

When a 180-degree hybrid is used, one of the input signals is applied in opposite-phase to the nonlinearities, while the other is applied in-phase. Therefore, two different cases must be considered. In the first one, we assume the LO is applied in phase and the RF in opposite-phase, and in the other the LO and RF inputs are interchanged.

If the large-signal LO is applied in phase, the time-varying coefficients of (5.270), $g_{mn}(t)$, are equal and described by

$$g_{mn}(t) = \sum_{k=-K}^{K} G_{mn_k} e^{jk\omega_{LO} t} \tag{5.283}$$

while the opposite-phase RF small-signal components will be

$$v_{A_{RF}}(t) = V_{RF} \cos\left(\omega_{RF_1} t\right) + V_{RF} \cos\left(\omega_{RF_2} t\right) \tag{5.284}$$

and

$$v_{B_{RF}}(t) = V_{RF} \cos\left(\omega_{RF_1} t + \pi\right) + V_{RF} \cos\left(\omega_{RF_2} t + \pi\right) \tag{5.285}$$

So, the output currents at the fundamental IF component, for example $\omega_{RF_1} - \omega_{LO}$, are given by

$$i_{o_{A_{IF}}}(t) = V_{RF}\left|G_{m1_{-1}}\right|\cos\left[\left(\omega_{RF_1} - \omega_{LO}\right)t\right] \tag{5.286}$$

$$i_{o_{B_{IF}}}(t) = V_{RF}\left|G_{m1_{-1}}\right|\cos\left[\left(\omega_{RF_1} - \omega_{LO}\right)t + \pi\right] \tag{5.287}$$

A simple addition of $i_{o_{A_{IF}}}(t)$ and $i_{o_{B_{IF}}}(t)$ thus requires that one of the nonlinearities (e.g., B) is mounted with reversed polarity, so that $i_{o_{B_{IF}}}(t)$ is reversed. Alternatively, two identical nonlinearities can be used, but $i_{o_{A_{IF}}}(t)$ and $i_{o_{B_{IF}}}(t)$ must be subtracted in another 180-degree hybrid. In such cases, where, for example, $i_{o_B}(t)$ is reversed, second-order distortion at $\omega_{IF_1} - \omega_{IF_2}$ and $2\omega_{IF_1}$ becomes

$$i_{o_{A_2}}(t) = G_{m2_0}V_{RF}^2\cos\left[\left(\omega_{RF_1} - \omega_{RF_2}\right)t\right] + \frac{1}{2}\left|G_{m2_{-2}}\right|V_{RF}^2\cos\left[\left(2\omega_{RF_1} - 2\omega_{LO}\right)t\right] \tag{5.288}$$

$$i_{o_{B_2}}(t) = -G_{m2_0}V_{RF}^2\cos\left[\left(\omega_{RF_1} - \omega_{RF_2}\right)t\right] - \frac{1}{2}\left|G_{m2_{-2}}\right|V_{RF}^2\cos\left[\left(2\omega_{RF_1} - 2\omega_{LO}\right)t + \pi\right] \tag{5.289}$$

being therefore canceled when added.

The output third-order distortion currents at $2\omega_{IF_1} - \omega_{IF_2}$ and $3\omega_{IF_1}$ will be

$$i_{o_{A_3}}(t) = \frac{3}{4}\left|G_{m3_{-1}}\right|V_{RF}^3\cos\left[\left(2\omega_{RF_1} - \omega_{RF_2} - \omega_{LO}\right)t\right] + \frac{1}{4}\left|G_{m3_{-3}}\right|V_{RF}^3\cos\left[\left(3\omega_{RF_1} - 3\omega_{LO}\right)t\right] \tag{5.290}$$

$$i_{o_{B_3}}(t) = -\frac{3}{4}\left|G_{m3_{-1}}\right|V_{RF}^3\cos\left[\left(2\omega_{RF_1} - \omega_{RF_2} - \omega_{LO}\right)t + \pi\right] - \frac{1}{4}\left|G_{m3_{-3}}\right|V_{RF}^3\cos\left[\left(3\omega_{RF_1} - 3\omega_{LO}\right)t + 3\pi\right] \tag{5.291}$$

proving that they are treated in the same way as the fundamental IF components.

If now the nonlinearities were driven in opposite-phase by the LO and in-phase by the RF, we would have a similar situation.

In summary, the 180-degree singly balanced mixer requires that the two individual nonlinearities are mounted in reverse polarity—unless their output currents are subtracted by another 180-degree hybrid—provides cancellation of second-order products, but no rejection of third-order ones.

5.5.2.2 Quadrature-Phase Singly Balanced Mixers

In the presence of a 90-degree hybrid, one nonlinearity (lets say, A) is driven by in-phase RF and quadrature-phase LO signals, while the other is driven by quadrature-phase RF and in-phase LO. So, their output fundamental IF currents at, for example, ω_{IF_1}, will be

$$i_{o_{A_{IF}}}(t) = V_{RF}\left|G_{m1_{-1}}\right| \cos\left[\left(\omega_{RF_1} - \omega_{LO}\right)t - \frac{\pi}{2}\right] \tag{5.292}$$

$$i_{o_{B_{IF}}}(t) = V_{RF}\left|G_{m1_{-1}}\right| \cos\left[\left(\omega_{RF_1} - \omega_{LO}\right)t + \frac{\pi}{2}\right] \tag{5.293}$$

which are 180 degrees out-of-phase, therefore requiring the reverse polarity of one of the nonlinear devices.

In such a circuit, second-order currents at $\omega_{IF_1} - \omega_{IF_2}$ and $2\omega_{IF_1}$ are given by

$$\begin{aligned} i_{o_{A_2}}(t) &= G_{m2_0} V_{RF}^2 \cos\left[\left(\omega_{RF_1} - \omega_{RF_2}\right)t\right] \\ &+ \frac{1}{2}\left|G_{m2_{-2}}\right| V_{RF}^2 \cos\left[\left(2\omega_{RF_1} - 2\omega_{LO}\right)t - \pi\right] \end{aligned} \tag{5.294}$$

$$\begin{aligned} i_{o_{B_2}}(t) &= -G_{m2_0} V_{RF}^2 \cos\left[\left(\omega_{RF_1} - \omega_{RF_2}\right)t\right] \\ &- \frac{1}{2}\left|G_{m2_{-2}}\right| V_{RF}^2 \cos\left[\left(2\omega_{RF_1} - 2\omega_{LO}\right)t + \pi\right] \end{aligned} \tag{5.295}$$

and are thus ideally rejected.

Third-order IF currents at $2\omega_{IF_1} - \omega_{IF_2}$ and $3\omega_{IF_1}$ will be

$$\begin{aligned} i_{o_{A_3}}(t) &= \frac{3}{4}\left|G_{m3_{-1}}\right| V_{RF}^3 \cos\left[\left(2\omega_{RF_1} - \omega_{RF_2} - \omega_{LO}\right)t - \frac{\pi}{2}\right] \\ &+ \frac{1}{4}\left|G_{m3_{-3}}\right| V_{RF}^3 \cos\left[\left(3\omega_{RF_1} - 3\omega_{LO}\right)t - \frac{3\pi}{2}\right] \end{aligned} \tag{5.296}$$

$$\begin{aligned} i_{o_{B_3}}(t) &= -\frac{3}{4}\left|G_{m3_{-1}}\right| V_{RF}^3 \cos\left[\left(2\omega_{RF_1} - \omega_{RF_2} - \omega_{LO}\right)t + \frac{\pi}{2}\right] \\ &- \frac{1}{4}\left|G_{m3_{-3}}\right| V_{RF}^3 \cos\left[\left(3\omega_{RF_1} - 3\omega_{LO}\right)t + \frac{3\pi}{2}\right] \end{aligned} \tag{5.297}$$

being therefore both added in a similar way as the IF fundamentals.

In conclusion, the distortion properties of a quadrature-phase singly balanced mixer are similar to the ones already studied for the 180-degree arrangement, in the sense that, again, IF second-order distortion components are rejected and third-order ones are not.

5.5.2.3 Opposite-Phase Doubly Balanced Mixers

The preceding sections have shown that symmetric connections of two similar devices, driven with the appropriate phased signals, acquired some special distortion properties. And even though we have treated these devices as being simple nonlinearities, there is no essential reason to keep such a restriction. In fact, when amplifier arrangements were studied, there was nothing that prevented the use of one of our amplifier pairs as the device of another symmetric arrangement.

This hierarchical way of building balanced configurations is particularly useful in mixers and modulators, resulting in some of its major implementations as the SSB, BPSK, or QPSK modulators, the image-rejection mixer, and the most common doubly balanced mixer. Even though there are many possible implementations of this doubly balanced topology, it is mostly seen as the Schottky diode ring, the diode star [28], or the four-quadrant analog multiplier (a three differential pair arrangement know as the Gilbert cell, when implemented in BJT technology) [31].

As an example, Figures 5.68 through 5.70 represent a schematic diagram of the diode ring mixer, and two equivalent circuits valid when the LO is applied at port P_1 or P_2, RF at port P_2 or P_1, and IF is collected from P_3; and when the LO is applied at port P_3 and the RF and IF are the two other ports.

When the controlling LO signal is applied to, say, port P_1, the diode pairs (D_1, D_2) and (D_3, D_4) will be driven in an opposite way. So, the diode ring mixer can be represented as an in-phase combination of the outputs of two opposite-phase RF driven singly balanced configurations (of symmetric nonlinearites) as shown in Figure 5.69.

When the controlling LO signal is applied at port P_3, it is now the diode pairs (D_1, D_3) and (D_2, D_4) that are being driven in opposite-phase. So, if now the RF is applied to, say, port P_1, the ring can be viewed as an opposite-phase combination of the outputs of two in-phase RF driven singly balanced configurations as shown in Figure 5.70. Note that, according to what we have already seen in Section 5.5.2.1, because each of the singly balanced configurations uses similar (not symmetric) nonlinearities, their currents must be subtracted in another 180-degree hybrid transformer [conversely to what we saw in Figure 5.69(a) where these currents were simply added in a parallel connection].

In conclusion, because these doubly balanced mixers can be viewed as opposite-phase arrangements of two opposite-phase singly balanced configurations, they offer further rejection of second-order distortion products, but no cancellation of third-order ones.

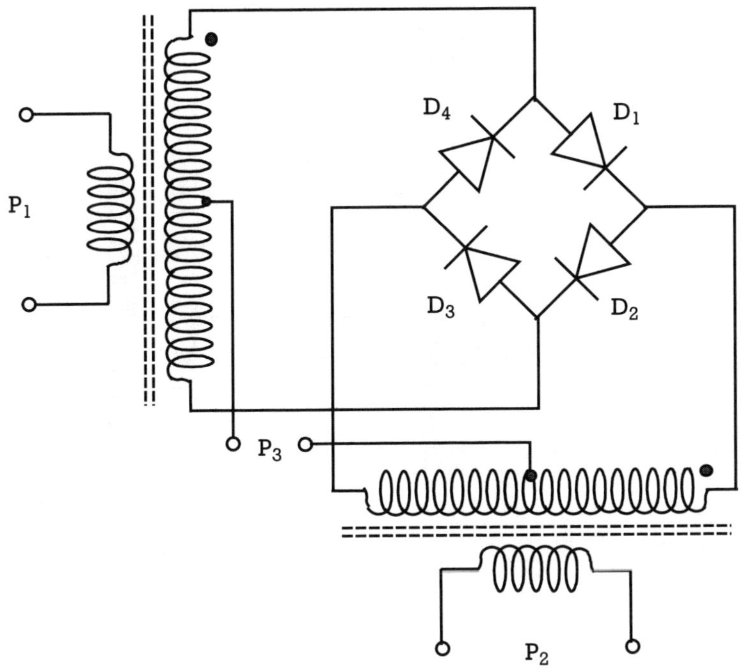

(a)

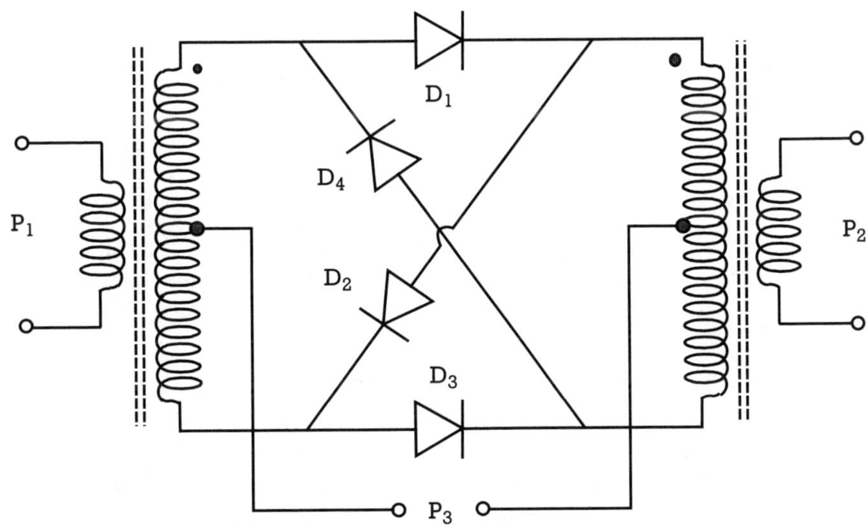

(b)

Figure 5.68 Two equivalent schematic diagrams of the common doubly balanced mixer implemented with a diode ring.

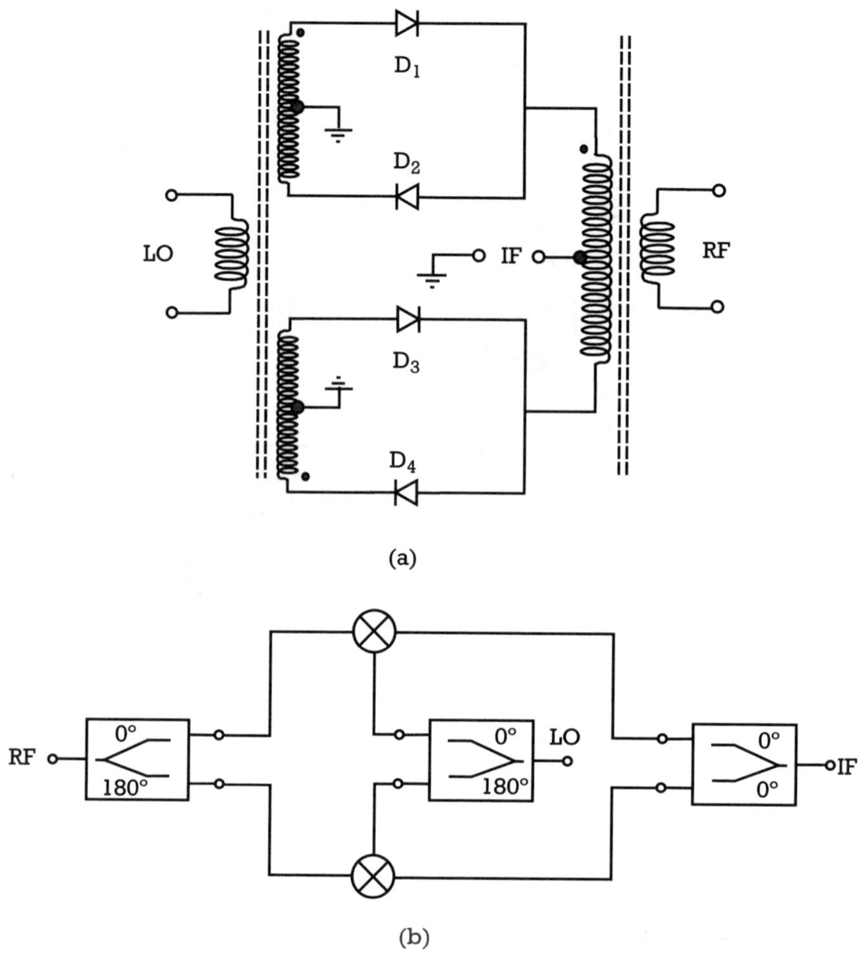

Figure 5.69 (a) Equivalent circuit of the diode ring mixer when the LO and RF signals are applied at ports P_1 and P_2, and the IF is collected from P_3. (b) Block diagram representation showing that this diode ring mixer driven as in (a) can be viewed as a symmetric combination of two opposite-phase singly balanced mixers.

However, since the LO is always driving two nonlinearities at a time (instead of only one, as in singly balanced configurations), the RF voltage must be divided between these two equal nonlinear devices. Therefore, the doubly balanced mixer is expected to provide better intermodulation performance at the expense of requiring more LO power, than its singly balanced counterparts.

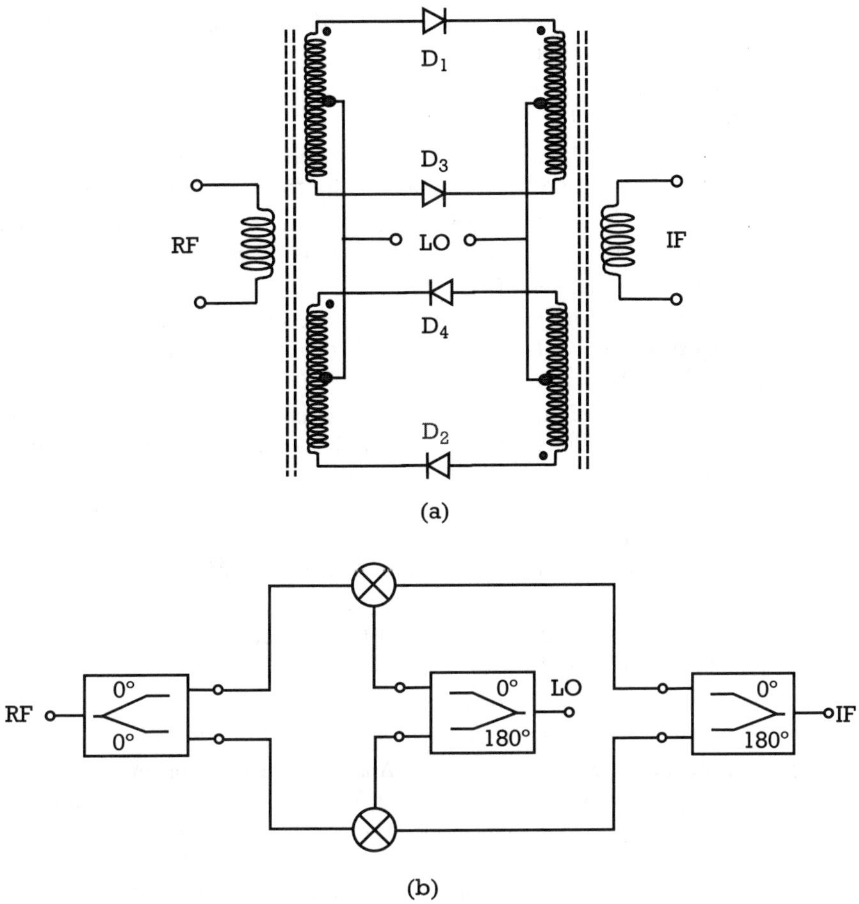

Figure 5.70 (a) Equivalent circuit of the diode ring mixer when the LO is applied to port P_3, and the RF and IF are applied to, or connected from, the other two ports. (b) Block diagram showing that this diode ring mixer driven as in (a) can be viewed as a symmetric combination of two opposite-phase singly balanced mixers.

References

[1] Kenington, P. B., *High-Linearity RF Amplifier Design,* Norwood, MA: Artech House, 2000.

[2] Pothecary, N., *Feedforward Linear Power Amplifiers,* Norwood, MA: Artech House, 1999.

[3] Cripps, S., *RF Power Amplifiers for Wireless Communications,* Norwood, MA: Artech House, 1999.

[4] Smith, J., *Modern Communications Circuits,* London: McGraw Hill, 1986.

[5] Maas, S. A., "Third-Order Intermodulation Distortion in Cascaded Stages," *IEEE Microwave and Guided Wave Letters,* Vol. 5, No. 6, June 1995, pp. 189–191.

[6] Gonzalez, G., *Microwave Transistor Amplifiers—Analysis and Design,* Second Edition, London: Prentice-Hall, 1997.

[7] Abrie, P. L., *Design of RF and Microwave Amplifiers and Oscillators,* Norwood, MA: Artech House, 1999.

[8] Engberg, J., "Simultaneous Input Power Match and Noise Optimization Using Feedback," *Proc. 4th European Microwave Conf.*, Montreaux, September 1974, pp. 385–389.

[9] Garcia, J. A., et al., "Characterizing the Gate-to-Source Nonlinear Capacitor Role on GaAs FET IMD Performance," *IEEE Transactions on Microwave Theory and Tech.*, Vol. 46, No. 12, 1998, pp. 2344–2355.

[10] Carvalho, N. B., and J. C. Pedro, "A Comprehensive Explanation of Distortion Sideband Asymmetries," *IEEE Transactions on Microwave Theory and Tech.*, Vol. 50, No. 9, 2002, pp. 2090–2101.

[11] Ho, C. Y., and D. Burges, "Practical Design of 2-4 GHz Low Intermodulation Distortion GaAs FET Amplifiers with Flat Gain Response and Low Noise Figure," *Microwave Journal,* Vol. 26, No. 2, 1983, pp. 91–104.

[12] Pedro, J. C., and J. Perez, "Accurate Simulation of GaAs MESFET's Intermodulation Distortion Using a New Drain-Source Current Model," *IEEE Transactions on Microwave Theory and Techniques,* Vol. 42, No. 1, 1994, pp. 25–33.

[13] Crosmun, A. M., and S. A. Maas, "Minimization of Intermodulation Distortion in GaAs MESFET Small-Signal Amplifiers," *IEEE Transactions on Microwave Theory and Techniques,* Vol. 37, No. 9, 1989, pp. 1411–1417.

[14] Narayanan, S., "Transistor Distortion Analysis Using Volterra Series Representation," *Bell System Technical Journal,* Vol. 46, No. 5, 1967, p. 991.

[15] Narayanan, S., and H. C. Poon, "Analysis of Distortion in Bipolar Transistors Using Integral Charge Control Model and Volterra Series," *IEEE Transactions on Circuit Theory,* Vol. 20, No. 4, 1973, pp. 341–351.

[16] Maas, S. A., B. L. Nelson, and D. L. Tait, "Intermodulation in Heterojunction Bipolar Transistors," *IEEE Transactions on Microwave Theory and Techniques,* Vol. 40, No. 3, 1992, pp. 442–448.

[17] Asbeck, P., "HBT Linearity and Basic Linearization Approaches," *Workshop on Advances in Amplifier Linearization, 1998 MTT-S International Microwave Symposium Dig.*, Baltimore, June 1998.

[18] Reynolds, J., "Nonlinear Distortions and Their Cancellation in Transistors," *IEEE Transactions on Electron Devices,* Vol. 12, No. 11, 1995, pp. 595–599.

[19] Van Der Heijden, M. P., H. C. De Graaff, and C. N. De Vreede, "A Novel Frequency-Independent Third-Order Intermodulation Distortion Cancellation Technique for BJT Amplifiers," *IEEE Journal of Solid-State Circuits,* Vol. 37, No. 9, 2002, pp. 1176–1183.

[20] Krauss, H., C. Bostian, and F. Raab, *Solid State Radio Engineering,* New York: John Wiley & Sons, 1980.

[21] Cripps, S., "A Method for the Prediction of Load-Pull Power Contours in GaAs MESFETs," *Proc. 1983 MTT-S International Microwave Symposium Dig.*, Boston, June 1983, pp. 221–223.

[22] Blachman, N., "Detectors, Bandpass Nonlinearities, and Their Optimization: Inversion of the Chebyshev Transform," *IEEE Transactions on Information Theory,* Vol. 17, No. 4, 1971, pp. 398–404.

[23] Ballesteros, E., F. Pérez, and J. Perez, "Analysis and Design of Microwave Linearized Amplifiers Using Active Feedback," *IEEE Transactions on Microwave Theory and Techniques,* Vol. 36, No. 3, 1988, pp. 499–504.

[24] Carvalho, N. B., and J. C. Pedro, "Large and Small Signal IMD Behavior of Microwave Power Amplifiers," *IEEE Transactions on Microwave Theory and Techniques,* Vol. 47, No. 12, 1999, pp. 2364–2374.

[25] Fager, C., et al., "Prediction of IMD in LDMOS Transistor Amplifiers Using a New Large-Signal Model," *IEEE Transactions on Microwave Theory and Techniques,* Vol. 50, No. 12, 2002, pp. 2834–2842.

[26] Rappaport, T. S., *Wireless Communications—Principles and Practice,* London: Prentice Hall, 1996.

[27] Muratore, F., *UMTS: Mobile Communications for the Future,* New York: John Willey & Sons, 2001.

[28] Maas, S. A., *Microwave Mixers,* Norwood, MA: Artech House, 1986.

[29] Maas, S. A., "Two-Tone Intermodulation in Diode Mixers," *IEEE Transactions on Microwave Theory and Techniques,* Vol. 35, No. 3, 1987, pp. 307–314.

[30] Maas, S. A., *Nonlinear Microwave Circuits,* Norwood, MA: Artech House, 1988.

[31] Gilbert, B., "The MICROMIXER, A Highly Linear Variant of the Gilbert Mixer Using a Bisymmetric Class-AB Input Stage," *IEEE Journal on Solid State Circuits,* Vol. 32, No. 9, 1997, pp. 1412–1423.

List of Acronyms

ac	alternate current (usually the signal component)
ACP	adjacent-channel power
ACPR	adjacent-channel power ratio
AFM	artificial frequency mapping
AGC	automatic gain control
AM	amplitude modulation
APFT	almost periodic Fourier transform
ATT	attenuator
BiCMOS	bipolar and complementary MOSFET technology
BJT	bipolar junction transistor
Bw	bandwidth
CAD	computer-aided design
CCPR	cochannel power ratio
CMOS	complementary MOSFET
CW	continuous wave (a nonmodulated sinusoid or carrier)
dc	direct current (usually the bias)
DFT	discrete Fourier transform
DR	dynamic range
DUT	device under test
ETHB	envelope transient harmonic balance
FDTD	finite differences in time-domain
FET	field effect transistor
FFT	fast Fourier transform
HB	harmonic balance
HBT	heterojunction bipolar transistor
HEMT	high electron mobility transistor
HFET	heterojunction field effect transistor

IC	integrated circuit
IDFT	inverse discrete Fourier fransform
IF	intermediate frequency (usually taken as the output of a mixer)
IFFT	inverse fast Fourier transform
IM	image frequency
IMD	intermodulation distortion (usually the sideband distortion)
IMR	intermodulation distortion ratio
JFET	junction field effect transistor
LDMOS	laterally diffused MOSFET
LNA	low-noise amplifier
LO	local oscillator (usually the pumping or control signal of a mixer)
MDFT	multidimensional discrete Fourier transform
MODFET	modulation doped field effect transistor
MESFET	metal-semiconductor field effect transistor
MMIC	monolithic microwave integrated circuit
MOSFET	metal-oxide field effect transistor
MPDE	multirate partial differential equation
NF	noise factor or noise figure
NLTF	nonlinear transfer function
NPR	noise power ratio
ODE	ordinary differential equation
PA	power amplifier
PAE	power added efficiency
PDC	dc supply power
PDE	partial differential equation
PIM	passive intermodulation
PM	phase modulation
PSD	power spectral density
RBw	resolution bandwidth
RF	radio frequency (sometimes taken as the input signal of a mixer)
SB	spectral balance (sometimes also called frequency-domain harmonic balance)
SCDR	power spectral density distortion ratio
SINAD	signal-to-noise and distortion ratio
SNR	signal-to-noise ratio
SPICE	simulation program with integrated circuit emphasis

THD	total harmonic distortion
TWTA	traveling wave tube amplifier

Notation Conventions

In the mathematical formulations throughout this book, the notation conventions used for circuit variables (either voltages or currents) are as described below, unless otherwise stated. So, for instance, a certain output node voltage can be referred to as:

V_O	Quiescent point
$v_o(t)$	Signal component, assumed as a deviation from the fixed quiescent point, V_O
$v_O(t)$	Composite signal (signal and quiescent point): $v_O(t) = v_o(t) + V_O$
V_o	Frequency-domain component of thc composite signal
$\mathbf{V}_o$	Vector representation of the complete frequency-domain composite signal

About the Authors

José Carlos Pedro was born in Espinho, Portugal, on March 7, 1962. He received an electronics and telecommunications engineering degree from the University of Aveiro, Portugal, in 1985, and his Ph.D. in electrical engineering from the same university in 1993. In 2002, he received the Degree of Agregação, the highest Portuguese academic level.

From 1985 to 1993, Professor Pedro was an assistant lecturer at the University of Aveiro, where he became an assistant professor in 1993. Currently, he is an associate professor at the same university and a senior research scientist at the Telecommunications Institute-Aveiro, where he heads the Microwave Electronics Group.

As a professor, he has been responsible for several electrical engineering degree courses and undergraduate students' final projects. He supervised or cosupervised seven Ph.D. and M.Sc. students and was appointed the director of the electronics and telecommunications engineering degree.

Professor Pedro's main scientific research interests include the development of computer-aided design tools for nonlinear electronic circuits design, telecommunication systems identification, active device modeling, and the analysis and design of various nonlinear microwave and optoelectronics circuits, particularly the design of highly linear multicarrier power amplifiers and mixers. As a result of the various R&D projects in which he has been involved, he has authored or coauthored more than 60 papers in international journals and symposia.

As a result of these academic and research activities, Professor Pedro has been invited to give several invited talks at international symposia, to present tutorials in similar academic research groups, and to participate in the jury of various national and international M.Sc. and Ph.D. examinations. He is also a reviewer for various prestigious scientific international journals and conferences.

Professor Pedro received the Marconi Young Scientist Award in 1993 and the 2000 Institution of Electrical Engineers (IEE), U.K., Measurement Prize. Beyond that, two of his Ph.D. students have received the first and third prizes of the IEEE International Microwave Symposium Student Paper Contest.

Professor Pedro is also a senior member of the IEEE.

Nuno Borges Carvalho was born in 1972 in Luanda, Portugal. From 1990 to 1995, he studied telecommunications and electrical engineering at Aveiro University, Portugal, where he concentrated his study on RF and microwave circuits and systems. After finishing his studies at Aveiro University in 1995, he received his engineer licenciatura degree and was awarded the 1995 Aveiro University Prize and the Portuguese Engineering Association Prize for the best student at Aveiro University in 1995.

After his studies, Professor Carvalho remained at Aveiro University in the Telecommunications Institute, investigating nonlinear techniques for circuit CAD and power amplifier linearity optimization. After completing his dissertation, "Nonlinear Distortion Study of Multi-Carrier Large Signal RF Amplifiers," in 2000, he received his Ph.D.

In 1998, Professor Carvalho received third place at the Student Paper Competition presented at the IEEE International Microwave Symposium for his scientific work on the paper "Simulation of Multi-Tone IMD Distortion and Spectral Regrowth Using Spectral Balance." In 2000, he was one of the recipients of the 2000 IEE Measurement Prize, and he was also nominated as an assistant professor at the Aveiro University, a position that he still holds today.

As a professor, he has been responsible for several electrical engineering degree courses and undergraduate students' final projects. He also supervised or cosupervised Ph.D. and M.Sc. students.

He has published 50 scientific papers in such publications as *IEEE Transactions on Microwave Theory and Techniques, IEEE Transactions on Instrumentation and Measurement, IEEE Transactions on Circuits and Systems, IEEE Microwave Symposium,* and the *European Microwave Conference.* He holds one national patent on nonlinear distortion measurement, and is responsible for the RF Laboratory at the Telecommunications Institute-Aveiro.

Professor Carvalho is a member of the editorial board of the *IEEE Transactions on Microwave Theory and Techniques* and has served as a technical reviewer for several scientific journals as well as for several conferences. He is also a member of the IEEE and of the Engineer Portuguese Association.

Index

J

K

L

Q

Recent Titles in the Artech House Microwave Library

Advanced Techniques in RF Power Amplifier Design, Steve C. Cripps

Automated Smith Chart, Version 4.0: Software and User's Manual, Leonard M. Schwab

Behavioral Modeling of Nonlinear RF and Microwave Devices, Thomas R. Turlington

Computer-Aided Analysis of Nonlinear Microwave Circuits, Paulo J. C. Rodrigues

Design of FET Frequency Multipliers and Harmonic Oscillators, Edmar Camargo

Design of Linear RF Outphasing Power Amplifiers, Xuejun Zhang, Lawrence E. Larson, and Peter M. Asbeck

Design of RF and Microwave Amplifiers and Oscillators, Pieter L. D. Abrie

Distortion in RF Power Amplifiers, Joel Vuolevi and Timo Rahkonen

EMPLAN: Electromagnetic Analysis of Printed Structures in Planarly Layered Media, Software and User's Manual, Noyan Kinayman and M. I. Aksun

Feedforward Linear Power Amplifiers, Nick Pothecary

Generalized Filter Design by Computer Optimization, Djuradj Budimir

High-Linearity RF Amplifier Design, Peter B. Kenington

Intermodulation Distortion in Microwave and Wireless Circuits, José Carlos Pedro and Nuno Borges Carvalho

Lumped Elements for RF and Microwave Circuits, Inder Bahl

Microwave Circuit Modeling Using Electromagnetic Field Simulation, Daniel G. Swanson, Jr. and Wolfgang J. R. Hoefer

Microwave Component Mechanics, Harri Eskelinen and Pekka Eskelinen

Microwave Engineers' Handbook, Two Volumes, Theodore Saad, editor

Microwave Filters, Impedance-Matching Networks, and Coupling Structures, George L. Matthaei, Leo Young, and E.M.T. Jones

Microwave Materials and Fabrication Techniques, Third Edition, Thomas S. Laverghetta

Microwave Mixers, Second Edition, Stephen A. Maas

Microwave Radio Transmission Design Guide, Trevor Manning

Microwaves and Wireless Simplified, Thomas S. Laverghetta

Neural Networks for RF and Microwave Design, Q. J. Zhang and K. C. Gupta

Nonlinear Microwave and RF Circuits, Second Edition, Stephen A. Maas

QMATCH: Lumped-Element Impedance Matching, Software and User's Guide, Pieter L. D. Abrie

Practical RF Circuit Design for Modern Wireless Systems, Volume I: Passive Circuits and Systems, Les Besser and Rowan Gilmore

Practical RF Circuit Design for Modern Wireless Systems, Volume II: Active Circuits and Systems, Rowan Gilmore and Les Besser

Radio Frequency Integrated Circuit Design, John Rogers and Calvin Plett

RF Design Guide: Systems, Circuits, and Equations, Peter Vizmuller

RF Measurements of Die and Packages, Scott A. Wartenberg

The RF and Microwave Circuit Design Handbook, Stephen A. Maas

RF and Microwave Coupled-Line Circuits, Rajesh Mongia, Inder Bahl, and Prakash Bhartia

RF and Microwave Oscillator Design, Michal Odyniec, editor

RF Power Amplifiers for Wireless Communications, Steve C. Cripps

RF Systems, Components, and Circuits Handbook, Ferril Losee

Stability Analysis of Nonlinear Microwave Circuits, Almudena Suárez and Raymond Quéré

TRAVIS 2.0: Transmission Line Visualization Software and User's Guide, Version 2.0, Robert G. Kaires and Barton T. Hickman

Understanding Microwave Heating Cavities, Tse V. Chow Ting Chan and Howard C. Reader

For further information on these and other Artech House titles, including previously considered out-of-print books now available through our In-Print-Forever® (IPF®) program, contact:

Artech House Publishers
685 Canton Street
Norwood, MA 02062
Phone: 781-769-9750
Fax: 781-769-6334
e-mail: artech@artechhouse.com

Artech House Books
46 Gillingham Street
London SW1V 1AH UK
Phone: +44 (0)20 7596-8750
Fax: +44 (0)20 7630 0166
e-mail: artech-uk@artechhouse.com

Find us on the World Wide Web at:
www.artechhouse.com